Extending Mathematics

Volume 2

An 'A' level course in pure mathematics

T. J. Heard
D. R. Martin

1478
1978

Oxford University Press

Oxford University Press,
Walton Street, Oxford OX2 6DP

Oxford London Glasgow New York
Toronto Melbourne Wellington Cape Town
Ibadan Nairobi Dar Es Salaam Lusaka Addis Ababa
Kuala Lumpur Singapore Jakarta Hong Kong Tokyo
Delhi Bombay Calcutta Madras Karachi

ISBN 0 19 914049 9
© Oxford University Press 1978

Printed in Great Britain by
Billing and Sons Ltd., Guildford and London

Preface

This book continues the work started in *Extending Mathematics Volume 1* by T. J. Heard (Oxford University Press 1974). The complete course gives thorough coverage of the pure mathematics of the current single subject G.C.E. A and S level syllabuses, though we have gone beyond these when this seemed desirable mathematically. Consequently we also cover the double subject syllabuses, except for a few topics.

As well as paying attention to mathematics as a coherent discipline we have tried to include as many applications as possible, since for many readers the main attraction of mathematics is probably its usefulness. The difficulty is that many applications require specialist background knowledge which cannot be given in the space available. As a compromise we have included some questions which assume a modest knowledge of mechanics or of probability and statistics; these are indicated by (M) or (P) respectively. We have also taken into account the availability of calculators, which leads to an emphasis on approximate methods.

Each section of text includes questions to reinforce understanding (many of these are suitable for oral work in class), and is followed by a set of graded exercises. The order of topics is one which we have found suitable in practice, though of course there are many other possibilities. The large miscellaneous exercise at the end reviews the whole course in sets of questions.

We are grateful for permission to use questions from past G.C.E. A level and university scholarship examinations, which are indicated by the following code:

The Oxford and Cambridge Schools Examination Board [O & C]
 Mathematics in Education and Industry Project [MEI]
 School Mathematics Project [SMP]
The University of London [L]
The Oxford Local Examinations Board [O]
The University of Cambridge Local Examinations Syndicate [C]
The Joint Matriculation Board [JMB]
The Northern Ireland Examination Board [NI]
Oxford Entrance Scholarship [OS]
Cambridge Entrance Scholarship [CS]
International Mathematical Olympiad [IMO]

Finally we are pleased to record our thanks to Mr Hugh Neill for suggesting many improvements to the original manuscript, to Mr Roger Webb of the Oxford University Press for his patience and encouragement, and to our families, colleagues and pupils for their help and remarkable forbearance during the development of the course.

City of London School   T. J. Heard
1977   D. R. Martin

Contents

| Contents | iv |
| List of symbols | vii |

1 Proof
| 1.1 Mathematical proof | 1 |
| 1.2 Mathematical induction | 7 |

2 Approximate methods I
2.1 Second derivatives	15
2.2 Locating the roots of an equation	23
2.3 Linear interpolation	28
2.4 Newton-Raphson approximations	32
2.5 Maclaurin and Taylor approximations	37
2.6 Simpson's rule	46

3 Vectors I
3.1 Scalars and vectors: basic ideas	54
3.2 Vectors in 2 and 3 dimensions	61
3.3 Lines in three dimensions	68
3.4 Ratio Theorem	76

4 Calculus I
4.1 Integration by parts	82
4.2 Inverse functions	86
4.3 Inverse trigonometric functions	89
4.4 Integration using inverse trigonometric functions	92
4.5 Differentiation of implicit functions	97
4.6 Differentiation with a parameter	100
4.7 Hyperbolic functions	104

Complex numbers I

5.1	The development of the number system	111
5.2	Working with complex numbers	113
5.3	Representing complex numbers geometrically	117
5.4	The triangle inequalities	124
5.5	The polar form of complex numbers	126
5.6	Multiplication in the Argand diagram	130
5.7	De Moivre's Theorem	134

Rational functions

6.1	Polynomials	140
6.2	Properties of the roots of polynomial equations	147
6.3	Rational functions and their graphs	151
6.4	Inequalities	159
6.5	Partial fractions	163
6.6	Uses of partial fractions	168

Differential equations

7.1	Introduction	173
7.2	Graphical method	174
7.3	First-order differential equations	177
7.4	Simple harmonic motion	182
7.5	Linear equations	186
7.6	Step-by-step methods	192

Vectors II

8.1	Scalar product and projections	199
8.2	Planes	206
8.3	Further lines and planes	212
8.4	Differentiation of a vector	218
8.5	Vector product	224

Matrices

9.1	Linear transformations	230
9.2	Transpose	239
9.3	Determinants	242
9.4	The inverse of a 3×3 matrix	248
9.5	Linear equations: the singular case	253
9.6	Systematic elimination	259
9.7	Eigenvectors and eigenvalues	266

10 Coordinate geometry
10.1	Polar coordinates	276
10.2	The parabola	282
10.3	The ellipse	292
10.4	The hyperbola	304
10.5	The family of conics	315

11 Approximate methods II
11.1	Errors	321
11.2	Iteration	328
11.3	Convergence of series	337
11.4	Power series	345

12 Calculus II
12.1	Properties of definite integrals	351
12.2	Mean values	358
12.3	Finding the area of a sector	363
12.4	Some important limits	367
12.5	Improper integrals	370
12.6	Reduction formulae	375

13 Complex number II
13.1	Roots of complex numbers	380
13.2	Solving polynomial equations	386
13.3	Functions of a complex variable	391
13.4	The complex number field	396
13.5	Groups	402

Miscellaneous exercise	412
Answers	434
Index	474

List of symbols

$\equiv$	is identically equal to	$\geqslant$	is greater than or equal to
$\approx$	is approximately equal to	$\leqslant$	is less than or equal to
$>$	is greater than	$\Leftarrow$	is implied by
$<$	is less than	$\{...\}$	set
$\Rightarrow$	implies	$:$	such that
$\Leftrightarrow$	is equivalent to	$\cap$	intersection
$\in$	belongs to	$\subseteq$	is a subset of
$\cup$	union	f^{-1}	inverse of function f
$\emptyset$	empty set	$\perp$	is perpendicular to
$\mapsto$	maps to	Δ	determinant
$\rightarrow$	tends to	Σ	sum
δ	increment	$[f(x)]_a^b$	$f(b) - f(a)$
$!$	factorial	j	$\sqrt{-1}$
$\int$	integral	$\|z\|$	modulus of complex number z
$\sqrt{}$	non-negative square root	A_n	cofactor of a_n
z^*	complex conjugate of z	D.P.	decimal places
$=$	is equal to	S.F.	significant figures
$\neq$	is not equal to	$\mathbf{a.b}$	scalar product

$\square$	end of solution	$|\mathbf{a}|$	magnitude of vector $\mathbf{a}$
$\mathbf{M}^T$	transpose of matrix $\mathbf{M}$	$\mathbf{a} \times \mathbf{b}$	vector product
$|\mathbf{M}|$	determinant of matrix $\mathbf{M}$	P'	negation of statement
$\mathbf{PQ}$	vector represented by directed line segment $\overrightarrow{PQ}$		
$\hat{\mathbf{a}}$	unit vector in the same direction as $\mathbf{a}$		
$x = a\,(b)\,c$	values of x from a to c by steps of b		
$^{n}C_{r}$	number of combinations of r objects chosen from n		
$\mathbb{R}$	set of all real numbers		
$\mathbb{R}^2$	set of all ordered pairs of real numbers		
$\mathbb{R}^3$	set of all ordered triples of real numbers		
$\mathbb{C}$	set of all complex numbers		

1 Proof

1.1 Mathematical proof

One characteristic which distinguishes mathematics from other disciplines is the use of proof. A scientific theory may be supported by thousands of observations, but it can never be regarded as proved, since there is always the possibility that some contrary evidence will turn up. For example, Newton's theory of gravitation was thought to be thoroughly satisfactory for over two hundred years, but was eventually superseded by Einstein's theory of relativity, which may itself give way to a better theory in due course. In contrast, we can be absolutely sure that every positive integer can be expressed as the sum of at most four perfect squares, because Lagrange proved this in 1770.

What is a proof? Basically it is a sequence of logical steps or *deductions* which lead from a set of known statements or theorems to the new statement which is being proved. The first person to argue in this way is traditionally supposed to have been the Greek philosopher Thales (*c*.600 B.C.); the properties which were proved at this stage were simple and 'obvious' (e.g. the base angles of an isosceles triangle are equal), so the original purpose of a proof was more to show how the property was related to other properties than to convince doubters of its truth. The idea of proof was one of the great contributions of the Greeks to mathematics. For more than two thousand years the geometry and arithmetic text called *The Elements*, written by Euclid of Alexandria around 300 B.C., was held to be a perfect example of how a body of mathematics should be organized into a deductive logical system. The following classic proof comes from *The Elements*.

Example 1: Prove that there are infinitely many prime numbers.

Solution: Let p_n be the nth prime number (so that $p_1 = 2, p_2 = 3, p_3 = 5$, etc.) and let $X = p_1 p_2 p_3 \ldots p_n + 1$.
X is either prime or composite. If X is prime, it is a prime larger than p_n. If X is composite, then it must be exactly divisible by a prime. This prime divisor cannot be $p_1, p_2, \ldots,$ or p_n, since each of these leaves remainder 1 when it divides X, and so there must be another prime, larger than p_n. In both cases there is a prime larger than p_n, so there cannot be a last (i.e. largest) prime number. Therefore there are infinitely many prime numbers. ☐

It is important to realize that systematic organization comes *after* the results are known. Mathematical discovery does not usually come by logical deduction; new truths are more often found by experiment with particular examples, by analogy, by intuition or by inspiration. Large areas of mathematics, such as

1

algebra and calculus, grew and were used for many years before they were put on a sound logical basis.

There are degrees of strictness or rigour in proof. A proof which was once acceptable may well be criticized by the next generation for its hidden assumptions or looseness of reasoning. In a similar way, some of the arguments used at school level, particularly in parts of calculus, have to be recast more precisely in a more advanced course. The difference usually lies in what can be taken for granted at the start of the proof. The basic assumptions from which a theory is developed are called *axioms*; in §13.5 we shall look at some deductions that can be made from one particular set of simple axioms. For the moment however, we shall concentrate on the commonest patterns of reasoning which can be used whatever the starting points.

Most arguments advance by steps of the form 'if the statement P is true then it follows that the statement Q is true'; this is abbreviated to 'if P then Q' or 'P implies Q', or in symbols '$P \Rightarrow Q$'. Notice especially that the soundness of this step does not depend on whether P is in fact true. For example the argument

$$\text{one orange costs £5} \Rightarrow \text{four oranges cost £20}$$

is valid whatever the actual price of oranges.

An alternative way of writing $P \Rightarrow Q$ is $Q \Leftarrow P$ ('Q is implied by P').

Qu. 1.   Insert the correct symbol $\Rightarrow$ or $\Leftarrow$ between these statements.

    (i) $x = 1 \qquad x^2 - 3x + 2 = 0$;

    (ii) $x^2 > 4 \qquad x < -2$;

    (iii) $\tan \theta = -1 \qquad \theta = \dfrac{7\pi}{4}$;

    (iv) P is the midpoint of $AB \qquad PA = PB$.

The statement $Q \Rightarrow P$ is the *converse* of $P \Rightarrow Q$. If $P \Rightarrow Q$ is a valid statement, its converse may or may not be valid too.

Qu. 2.   For each of the following state whether or not the converse is true.

    (i) $x = 5 \Rightarrow x^2 = 25$;

    (ii) n is divisible by 3 $\Rightarrow$ the sum of the digits of n is divisible by 3;

    (iii) $\cos A < 0 \Rightarrow A$ is the largest angle of triangle ABC.

One common error of reasoning is to establish the converse instead of what is required, as in this 'proof' of the identity $\sec^2\theta + \operatorname{cosec}^2\theta \equiv \sec^2\theta \operatorname{cosec}^2\theta$.

$$\text{'}\sec^2\theta + \operatorname{cosec}^2\theta \equiv \sec^2\theta \operatorname{cosec}^2\theta \Rightarrow \frac{\sec^2\theta + \operatorname{cosec}^2\theta}{\sec^2\theta \operatorname{cosec}^2\theta} \equiv 1$$

$$\Rightarrow \frac{1}{\operatorname{cosec}^2\theta} + \frac{1}{\sec^2\theta} \equiv 1$$

$$\Rightarrow \sin^2\theta + \cos^2\theta \equiv 1, \text{ which is true.'}$$

A valid proof can be constructed easily from this by reversing the argument, starting with the known identity $\sin^2\theta + \cos^2\theta \equiv 1$ and deducing that $\sec^2\theta + \text{cosec}^2\theta \equiv \sec^2\theta\,\text{cosec}^2\theta$. The danger of the '... which is true' form of argument is shown by the following:

$$\text{'}\quad 1 = 2$$
$$\Rightarrow 2 = 1$$

Adding these gives $3 = 3$, which is true. Hence $1 = 2$.'

When $P \Rightarrow Q$ and $Q \Rightarrow P$ we say that P is *equivalent* to Q, and write $P \Leftrightarrow Q$. The difference between implication and equivalence is important in the following example.

Example 2: Solve the equation $2\sqrt{x} - \sqrt{(x+5)} = 1$.

Solution: $2\sqrt{x} - \sqrt{(x+5)} = 1 \quad\Leftrightarrow\quad 2\sqrt{x} - 1 = \sqrt{(x+5)}$

$$\Rightarrow 4x - 4\sqrt{x} + 1 = x + 5 \quad \text{(squaring)}$$
$$\Leftrightarrow 3x - 4 = 4\sqrt{x}$$
$$\Rightarrow 9x^2 - 24x + 16 = 16x \quad \text{(squaring again)}$$
$$\Leftrightarrow 9x^2 - 40x + 16 = 0$$
$$\Leftrightarrow (x-4)(9x-4) = 0$$
$$\Leftrightarrow x = 4 \quad \text{or} \quad \tfrac{4}{9}.$$

Thus $2\sqrt{x} - \sqrt{(x+5)} = 1 \Rightarrow x = 4$ or $\tfrac{4}{9}$, but the implication is not necessarily reversible. We must therefore test these two possible solutions by substitution:

$x = 4 \Rightarrow 2\sqrt{x} - \sqrt{(x+5)} = 2 \times 2 - 3 = 1$, so $x = 4$ is a solution

$x = \tfrac{4}{9} \Rightarrow 2\sqrt{x} - \sqrt{(x+5)} = 2 \times \tfrac{2}{3} - \tfrac{7}{3} = -1$, so $x = \tfrac{4}{9}$ is not a solution. $\square$

The *negation* of a statement P is the statement P' which is false when P is true and true when P is false. For example, the negation of '$x = 3$' is '$x \neq 3$', and the negation of 'C lies on AB produced' is 'C does not lie on AB produced'. By negating the negation we return to the original statement: $(P')'$ is the same as P. This simple observation provides a powerful method of proof, called *proof by contradiction*. In order to prove the truth of P, we show that P' is false.

Example 3: Prove that every prime number greater than 3 is of the form $6n \pm 1$, where n is a positive integer.

Solution (i): Let P be the statement 'every prime number greater than 3 is of the form $6n \pm 1$'.

Then P' is the statement 'it is false that every prime number greater than 3 is of the form $6n \pm 1$' or equivalently 'there is at least one prime number greater than 3 not of the form $6n \pm 1$'.

Therefore $P' \Rightarrow$ there is a prime number, a say, such that $a > 3$ and $a \neq 6n \pm 1$.

$\Rightarrow a = 6n$ or $a = 6n \pm 2$ or $a = 6n + 3$

$\Rightarrow a$ is divisible by 6 or by 2 or by 3

$\Rightarrow a$ is not prime, which contradicts the definition of a.

Therefore P' is false, and therefore P is true.   □

Qu. 3.   Why is the statement 'every prime number is of the form $6n \pm 1$' false?

The converse of Example 3 states that 'every number of the form $6n \pm 1$ is a prime greater than 3'. Taking $n = 1, 2, 3, 4$, gives the following values for $6n \pm 1$: $5, 7, 11, 13, 17, 19, 23, 25$. Since 25 is not prime, this converse is false, and we have shown this by producing the single *counter-example* $n = 4$. A counter-example is a particular case which disproves a general proposition.

Qu. 4.   Find a counter-example to disprove each of the following.
(i) $x^2 \geqslant x$;   (ii) $a + b > 2\sqrt{(ab)}, a, b$ positive;
(iii) $2n^2 + 11$ is prime.

If $P \Rightarrow Q$ and Q is false, then P must also be false (since otherwise it would imply that Q is true). In other words, if $P \Rightarrow Q$ then $Q' \Rightarrow P'$. Replacing P by Q' and Q by P' in this, we have also that if $Q' \Rightarrow P'$, then $(P')' \Rightarrow (Q')'$, i.e. $P \Rightarrow Q$. Therefore $(P \Rightarrow Q) \Leftrightarrow (Q' \Rightarrow P')$. The statement $Q' \Rightarrow P'$ is called the *contrapositive* of $P \Rightarrow Q$. For example '$x = 1 \Rightarrow x^2 - 3x + 2 = 0$' is equivalent to its contrapositive '$x^2 - 3x + 2 \neq 0 \Rightarrow x \neq 1$'.

Qu. 5.   Write down the contrapositives of the other statements of Qu. 1.

We can consider Example 3 again in terms of the contrapositive.

Solution (ii) of Example 3:   Let P be the statement 'x is a prime number greater than 3' and let Q be the statement 'x is of the form $6n \pm 1$'. Then

$Q' \Rightarrow x$ is not of the form $6n \pm 1$

$\Rightarrow x = 6n$ or $x = 6n \pm 2$ or $x = 6n \pm 3$

$\Rightarrow x$ is divisible by 6 or by 2 or by 3

$\Rightarrow x$ is not a prime number greater than 3

$\Rightarrow P'$.

So $Q' \Rightarrow P'$ and therefore $P \Rightarrow Q$.   □

When constructing a proof, be careful to look out for any exceptional cases or special values which have to be dealt with separately. Two such errors which commonly occur are

(i) multiplying an inequality through by a common factor which may be negative;

(ii) dividing by an expression which may be zero.

See if you can spot the fallacy in Example 4.

Example 4: Prove that $2 = -1$.

Solution:

$$x = 2 \Rightarrow \qquad 3x = x + 4$$
$$\Rightarrow \qquad 9x^2 = x^2 + 8x + 16$$
$$\Rightarrow 9x^2 - 18x = x^2 - 10x + 16$$
$$\Rightarrow 9x\,(x-2) = (x-8)(x-2)$$
$$\Rightarrow \qquad 9x = x - 8$$
$$\Rightarrow \qquad 8x = -8$$
$$\Rightarrow \qquad x = -1.$$

Therefore $\qquad 2 = -1.$ ☐

Exercise 1A

1. Prove that $\theta = \dfrac{8n + 1}{4}\pi \Rightarrow \cos\theta = \sin\theta$, where n is an integer, and disprove the converse.

2. With the notation of the figure,

   P is on the circular arc ACB

   $\Rightarrow$ angle $APB = \dfrac{\pi}{3}$.

   Is the converse true?

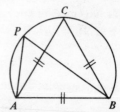

3. State carefully and prove the negation of the statement '$n^2 - n + 41$ is prime for all positive integers n'.

4. Given that 'Q is true if P is true' is the same as '$P \Rightarrow Q$' and also that 'Q is true only if P is true' is the same as '$Q \Rightarrow P$', insert the appropriate words (*if*, *only if*, or *if and only if*) between the following statements.

   (i) x is even x is a multiple of 4;

   (ii) $y = \sin\dfrac{12\pi}{7}$ $0 \leqslant y + 1 \leqslant 1$;

   (iii) n is divisible by 7 n is divisible by 49;

   (iv) n is divisible by 7 n^2 is divisible by 7.

5. If $P \Rightarrow Q$, we say that P is a *sufficient* condition for Q; if $P \Leftarrow Q$, we say that P is a *necessary* condition for Q. For each of the following, say

whether the first statement is a necessary or sufficient condition for the second (or both or neither).

   (i) John plays the piano, John is a concert pianist.

  (ii) There are more than 8 people in this room, there are at least 9 people in this room.

 (iii) $ABCD$ is a rhombus, the diagonals of $ABCD$ bisect each other.

 (iv) $ABCD$ is a parallelogram, the diagonals of $ABCD$ are perpendicular.

6. The sides of triangle ABC are a, b, c. Find in terms of a, b, c, a necessary and sufficient condition for angle A to be obtuse.

7. Solve the equation $\sqrt{(3x + 1)} = 9 - x$.

8. Solve the equation $\sqrt{(3x)} - \sqrt{(x + 1)} = 1$.

9. Prove that $3 - \sqrt{(2x + 3)} = 4 + \sqrt{(x + 1)} \Rightarrow x = -1$ or 3. What are the solutions of $3 - \sqrt{(2x + 3)} = 4 + \sqrt{(x + 1)}$?

10. Prove that $\dfrac{a}{b} = \dfrac{c}{d} \Leftrightarrow \dfrac{a + 2c}{b + 2d} = \dfrac{a - c}{b - d}$ in general, and state any restrictions which must be placed on the values of a, b, c, d.

11. If $\mathbf{A} = \begin{pmatrix} a & b \\ c & d \end{pmatrix}$ then the trace of $\mathbf{A}$ is the sum of the elements along the leading diagonal, and we write $\operatorname{tr}(\mathbf{A}) = a + d$. If $\mathbf{B}$ is also a 2×2 matrix, for each of the following either prove the statement or disprove it by giving a counter-example.

   (i) $\operatorname{tr}(\mathbf{A} + \mathbf{B}) = \operatorname{tr}(\mathbf{A}) + \operatorname{tr}(\mathbf{B})$;   (ii) $\operatorname{tr}(\mathbf{AB}) = \operatorname{tr}(\mathbf{A}) \operatorname{tr}(\mathbf{B})$;

 (iii) $\operatorname{tr}(\mathbf{AB}) = \operatorname{tr}(\mathbf{BA})$.

12. Explain the fallacy in the following.

$$\begin{aligned}
\text{`}\tan^2 x = \sec^2 x - 1 \quad &\Rightarrow & \tan x &= (\sec^2 x - 1)^{\frac{1}{2}} \\
&\Rightarrow & 2 + \tan x &= 2 + (\sec^2 x - 1)^{\frac{1}{2}} \\
&\Rightarrow & (2 + \tan x)^2 &= \{2 + (\sec^2 x - 1)^{\frac{1}{2}}\}^2
\end{aligned}$$

Putting $x = \dfrac{3\pi}{4}$ gives $\qquad\qquad 1 = 9$.'

13. Prove that, if n is a positive integer greater than one, $n^5 - n$ is a multiple of 5. Hence, or otherwise, prove that all square numbers are of the form $5r$ or $5r \pm 1$, where r is an integer.   [JMB]

14. In this question we write $a \equiv b$ if the integers a and b have the same remainder when divided by the integer m.

   (i) Prove that $a \equiv b \Leftrightarrow a = b + qm$ for some integer q.

  (ii) Prove that $a_1 \equiv b_1$ and $a_2 \equiv b_2 \Rightarrow a_1 a_2 \equiv b_1 b_2$.

 (iii) Give a counter-example to show that it is not true that $ka \equiv kb$ and $k \neq 0 \Rightarrow a \equiv b$.

15. Prove that $f(n) \equiv an^2 + bn + c$ if and only if $f(n) - f(n - 1) \equiv 2an - a + b$, where a, b, c are constants and n takes positive integer values.

To what extent does this remain true if n is replaced by x, any positive real number?

16. State the Remainder Theorem for polynomials. If a and n are integers greater than 1 prove that
    (i) if $a \neq 2$ then $a^n - 1$ is not a prime number;
    (ii) if n is not a prime number, then nor is $2^n - 1$.
    What follows from these statements if $a^n - 1$ is prime?   [MEI]

17. Integers of the form $4n + 1$ are called type A; those of the form $4n - 1$ are called type B.

    (i)   Prove that every odd integer is either of type A or of type B.
    (ii) Prove that the product of two type A integers is itself of type A. Complete the table for types of products.
    (iii) Prove that any integer of type B must have an odd number of prime factors of type B.

X	A	B
A	A	
B		

18. (i) Use Question 17 and an adaptation of Euclid's proof (Example 1) to prove that there are infinitely many primes of type B. [Hint: let $Y = 4q_1 q_2 \ldots q_n - 1$, where $q_1, q_2, \ldots, q_n$ are primes of type B.]
    (ii) Explain why a similar method fails to prove that there are infinitely many primes of type A. Does it follow that there are only a finite number of such primes?

1.2 Mathematical induction

The ability to spot a pattern and then use intelligent guesswork is of great importance in mathematics. As a simple example, consider the following sequence.

$$1 - \frac{1}{2^2} = \frac{3}{4}$$

$$\left(1 - \frac{1}{2^2}\right)\left(1 - \frac{1}{3^2}\right) = \frac{3}{4} \times \frac{8}{9} = \frac{2}{3}$$

$$\left(1 - \frac{1}{2^2}\right)\left(1 - \frac{1}{3^2}\right)\left(1 - \frac{1}{4^2}\right) = \frac{2}{3} \times \frac{15}{16} = \frac{5}{8}$$

$$\left(1 - \frac{1}{2^2}\right)\left(1 - \frac{1}{3^2}\right)\left(1 - \frac{1}{4^2}\right)\left(1 - \frac{1}{5^2}\right) = \frac{5}{8} \times \frac{24}{25} = \frac{3}{5}.$$

Qu. 1.   Work out the next two steps in this sequence.

If we let $f(n)$ stand for the product of $n - 1$ factors

$$\left(1 - \frac{1}{2^2}\right)\left(1 - \frac{1}{3^2}\right)\left(1 - \frac{1}{4^2}\right) \cdots \left(1 - \frac{1}{n^2}\right)$$

then we have the following results

n	$f(n)$
2	$\frac{3}{4}$
3	$\frac{2}{3} = \frac{4}{6}$
4	$\frac{5}{8}$
5	$\frac{3}{5} = \frac{6}{10}$

Writing all the fractions with even denominators immediately draws attention to an obvious pattern: $f(n) = \dfrac{n+1}{2n}$ for $n = 2, 3, 4, 5$.

Qu. 2. Check from Qu. 1. that $f(n) = \dfrac{n+1}{2n}$ for $n = 6, 7$ also.

This strikingly simple formula for $f(n)$ naturally leads us to guess that

$$\left(1 - \frac{1}{2^2}\right)\left(1 - \frac{1}{3^2}\right)\left(1 - \frac{1}{4^2}\right) \cdots \left(1 - \frac{1}{n^2}\right) = \frac{n+1}{2n} \qquad \text{(A)}$$

for all integers $n \geqslant 2$.

We have reached this conjecture by examining a number of particular cases, and generalizing from what we find. This process is known as *induction*; it contrasts with the method of deriving results by arguing logically from known general results, which is called *deduction*.

Of course there is a problem with induction: our guess may be wrong. A well known example of a wrong guess is provided by the Fermat numbers $2^{2^n} + 1$ $(n = 0, 1, 2, \ldots)$. Fermat noticed that the first five of these, $3, 5, 17, 257, 65\,537$, are prime numbers. He tried to find a divisor of the next number $2^{2^5} + 1 = 4\,294\,967\,297$ by a variety of ingenious methods, but failed. So in 1640 he made the conjecture that all the Fermat numbers are prime. This attracted considerable attention for the next hundred years, until in 1739 Euler found that 641 is a factor of $2^{2^5} + 1$. Despite extensive investigation, no further prime Fermat numbers have been found, so now the opposite hypothesis, that there are no prime Fermat numbers other than the first five, seems to be more favoured.

Consider again the conjecture (A). We have verified that (A) is true for $n = 2, 3, \ldots, 7$, but how can we show that (A) is true for *all* integers $n \geqslant 2$?

If statement (A) is true when $n = k$, then

$$\left(1 - \frac{1}{2^2}\right)\left(1 - \frac{1}{3^2}\right) \cdots \left(1 - \frac{1}{k^2}\right) = \frac{k+1}{2k}.$$

Multiplying both sides by the next factor $\left(1 - \dfrac{1}{(k+1)^2}\right)$ we have

$$\left(1 - \frac{1}{2^2}\right)\left(1 - \frac{1}{3^2}\right) \cdots \left(1 - \frac{1}{k^2}\right)\left(1 - \frac{1}{(k+1)^2}\right) = \frac{k+1}{2k}\left(1 - \frac{1}{(k+1)^2}\right).$$

If (A) is also true when $n = k + 1$, this right-hand side should equal $\dfrac{(k+1)+1}{2(k+1)}$, i.e. $\dfrac{k+2}{2(k+1)}$. We next show that the right-hand side does reduce to this:

$$\frac{k+1}{2k}\left(1 - \frac{1}{(k+1)^2}\right) = \frac{k+1}{2k}\left(\frac{(k+1)^2 - 1}{(k+1)^2}\right)$$

$$= \frac{k+1}{2k} \cdot \frac{k^2 + 2k}{(k+1)^2}$$

$$= \frac{k+2}{2(k+1)},$$

so that (A) is also true when $n = k + 1$.

Since (A) is true when $n = 7$ it follows (taking $k = 7$) that (A) is true when $n = 8$; hence (taking $k = 8$) (A) is true when $n = 9$, and so on. By this means we can show that (A) is true for any integer n however large. Thus (A) is true for all $n \geqslant 2$.

The method of proof we have used here involves two stages.

(1)   We first show that the statement to be proved, in this case (A), is true for some value of n.

(2)   Next we show that if (A) is true for any particular value of n, $n = k$ say, then it is true for the next value, $n = k + 1$.

Using (2) we can then establish the truth of (A) for each successive value of n, starting from the value in (1). We can compare this form of proof to the process of climbing a ladder: if we can (1) reach a starting place somewhere on the ladder and (2) get from one rung to the next, then we can climb as far as we like up the ladder. The next example uses the same form of proof.

Example 1: Prove that for all positive integers n

$$1^2 + 2^2 + 3^2 + \ldots + n^2 = \tfrac{1}{6}n(n+1)(2n+1) \tag{B}$$

Solution: (1) The statement (B) is true when $n = 1$, since $1^2 = \tfrac{1}{6} . 1 . 2 . 3$.
(2) If (B) is true when $n = k$,

i.e. if $1^2 + 2^2 + \ldots + k^2 = \tfrac{1}{6}k(k+1)(2k+1)$

then $1^2 + 2^2 + \ldots + k^2 + (k+1)^2 = \tfrac{1}{6}k(k+1)(2k+1) + (k+1)^2$

$$= (k+1)\,[\tfrac{1}{6}k(2k+1) + k + 1]$$

$$= \tfrac{1}{6}(k+1)\,[k(2k+1) + 6k + 6]$$

$$= \tfrac{1}{6}(k+1)(2k^2 + 7k + 6)$$

$$= \tfrac{1}{6}(k+1)(k+2)(2k+3),$$

so that (B) is true when $n = k + 1$.

From (1) (B) is true when $n = 1$.
From (2), taking $k = 1$, (B) is true when $n = 2$.

From (2), taking $k = 2$, (B) is true when $n = 3$, and so on.
Therefore (B) is true for all positive integers n.   □

This method of proof is called *proof by induction* (or mathematical induction, or complete induction). The word induction is used here because many of the results which can be discovered by the sort of inductive investigation given at the beginning of this section can be proved easily by this method, though proof by induction can be used in many other cases too, as Examples 2, 3, 4 show.

The method of proof by induction can be summarized as follows.

The statement (S) involving the positive integer n is true for all $n \geqslant n_0$ provided that
    (1) (S) is true when $n = n_0$
and (2) if (S) is true when $n = k$ then (S) is true when $n = k + 1$.

Both parts of the proof are essential.

Qu. 3.   Prove that the statement

$$1 + 2 + 3 + \ldots + n = n^2 \qquad (C)$$

is true when $n = 1$, but that

$$(C) \text{ true for } n = k \not\Rightarrow (C) \text{ true for } n = k + 1.$$

Qu. 4.   Prove that if the statement

$$1 + 2 + 3 + \ldots + n = \tfrac{1}{2}(n + \tfrac{1}{2})^2$$

is true for $n = k$ then it is true for $n = k + 1$. Explain why the statement is not true for any value of n.

Qu. 5.   What is the correct formula for the sum $1 + 2 + 3 + \ldots + n$? Prove this by induction.

Example 2: Prove that if $x > -1$ and $x \neq 0$, then for integers $n \geqslant 2$

$$(1 + x)^n > 1 + nx \qquad (S)$$

Solution: (1) When $n = 2$, $(1 + x)^n = 1 + 2x + x^2$
$$> 1 + 2x \quad \text{since } x^2 > 0.$$
            Thus (S) is true when $n = 2$.
     (2) [We want to show that $(1 + x)^k > 1 + kx$
$$\Rightarrow (1 + x)^{k+1} > 1 + (k + 1)x.]$$
           $(1 + x)^k > 1 + kx$

$$\Rightarrow (1 + x)^{k+1} > (1 + kx)(1 + x),$$
           multiply both sides by $1 + x$, which is positive since $x > -1$

$$(1 + kx)(1 + x) = 1 + (k + 1)x + kx^2$$
$$> 1 + (k + 1)x \quad \text{since } kx^2 > 0$$

Hence from (1) and (2) by induction $(1 + x)^n > 1 + nx$ for all $n \geqslant 2$.   □

Example 3: Show that $4^n + 6n - 1$ is divisible by 9 for all $n \geqslant 1$.

Solution: (1) When $n = 1$, $4^n + 6n - 1 = 4 + 6 - 1 = 9$, which is divisible by 9.

(2) [We want to show that $4^k + 6k - 1$ is divisible by 9 $\Rightarrow 4^{k+1} + 6(k + 1) - 1$ is divisible by 9.]

$$4^{k+1} + 6(k + 1) - 1 = 4.4^k + 6k + 5$$
$$= 4(4^k + 6k - 1) - 18k + 9.$$

Therefore $4^k + 6k - 1$ is divisible by 9

$$\Rightarrow \qquad 4^k + 6k - 1 = 9m, \text{ where } m \text{ is some integer}$$
$$\Rightarrow 4^{k+1} + 6(k + 1) - 1 = 4.9m - 18k + 9$$
$$= 9(4m - 2k + 1)$$
$$\Rightarrow 4^{k+1} + 6(k + 1) - 1 \text{ is divisible by 9.}$$

Hence from (1) and (2) by induction, $4^n + 6n - 1$ is divisible by 9 for all $n \geqslant 1$.   □

Example 4: The sequence of numbers $u_1, u_2, u_3, \ldots$ is defined by
$$u_1 = 6, u_2 = 20, u_{n+2} = 6u_{n+1} - 8u_n \ (n \geqslant 1).$$
Prove that $u_n = 2^n(1 + 2^n)$.

Solution: We modify the method of induction in this case.

Let (S) be the statement $u_n = 2^n(1 + 2^n)$.

(1) When $n = 1$, $2^n(1 + 2^n) = 2(1 + 2) = 6 = u_1$,
when $n = 2$, $2^n(1 + 2^n) = 4(1 + 4) = 20 = u_2$,
so (S) is true when $n = 1, 2$.

(2) If (S) is true when $n = k$ *and* when $n = k + 1$ then
$$u_{k+2} = 6.2^{k+1}(1 + 2^{k+1}) - 8.2^k(1 + 2^k)$$
$$= 2^{k+2}[3(1 + 2^{k+1}) - 2(1 + 2^k)]$$
$$= 2^{k+2}(3 + 3.2^{k+1} - 2 - 2^{k+1})$$
$$= 2^{k+2}(1 + 2.2^{k+1})$$
$$= 2^{k+2}(1 + 2^{k+2}),$$

so that (S) is true when $n = k + 2$.

Thus if (S) is true for two consecutive values of n, it is true also for the next value of n. But (S) is true for the consecutive values $n = 1, 2$. Hence by induction (S) is true for all $n \geqslant 1$.   □

Exercise 1B

1—7. By considering the first few values of n, guess a general formula for u_n and prove it by mathematical induction.

1. $u_n = 1 + 3 + 5 + \ldots + (2n - 1)$. [This was the first example of proof by induction ever published, by Francesco Maurolico in 1575.]

2. $u_n = \dfrac{1}{1.2} + \dfrac{1}{2.3} + \dfrac{1}{3.4} + \ldots + \dfrac{1}{n(n + 1)}$.

3. $u_n = \dfrac{1}{3} + \dfrac{1}{15} + \dfrac{1}{35} + \ldots + \dfrac{1}{4n^2 - 1}$.

4. $u_n = 1.1! + 2.2! + 3.3! + \ldots + n.n!$.

5. $u_n = \left(1 - \dfrac{4}{1}\right)\left(1 - \dfrac{4}{9}\right)\left(1 - \dfrac{4}{25}\right) \ldots \left(1 - \dfrac{4}{(2n - 1)^2}\right)$.

6. $u_{n+1} = \dfrac{u_n}{u_n + 1}, u_1 = 1$.

7. $u_{n+1} = 2u_n + 1, u_1 = 1$.

8. Prove by induction that $1^3 + 2^3 + 3^3 + \ldots + n^3 = \frac{1}{4}n^2(n + 1)^2$.

9. Prove by induction that $\displaystyle\sum_{r=0}^{n-1} x^r = \dfrac{1 - x^n}{1 - x}$ $(x \neq 1)$.

10. By differentiating in Question 9 find $\displaystyle\sum_{r=0}^{n-1} rx^{r-1}$. Give a proof by induction of the same result.

11. Use the pattern
$$1^2 = 1$$
$$1^2 - 2^2 = -(1 + 2)$$
$$1^2 - 2^2 + 3^2 = (1 + 2 + 3)$$
$$1^2 - 2^2 + 3^2 - 4^2 = -(1 + 2 + 3 + 4)$$
to find a formula for $1^2 - 2^2 + 3^2 - 4^2 + \ldots + (-1)^{n-1} n^2$.
Prove your formula by induction.

12. Prove that $\cos \theta \cos 2\theta \cos 4\theta \ldots \cos 2^n\theta = \dfrac{\sin 2^{n+1}\theta}{2^{n+1} \sin \theta}$
($\theta \neq m\pi$ for integer m).

13. Use induction to prove the Binomial Theorem, $(a + b)^n = \displaystyle\sum_{r=0}^{n} {}^nC_r a^{n-r} b^r$,
where ${}^nC_r = \dfrac{n!}{r!\,(n - r)!}$. [Remember that ${}^{n+1}C_{r+1} = {}^nC_r + {}^nC_{r+1}$.]

14. By writing $x^{n+1} = x^n.x$ and using the rule for differentiating a product, prove by induction that $y = x^n \Rightarrow \dfrac{dy}{dx} = nx^{n-1}$ for all integers $n \geqslant 1$.

15. Prove that $\displaystyle\sum_{r=1}^{n} r \sin r\theta = \dfrac{(n + 1) \sin n\theta - n \sin (n + 1) \theta}{4 \sin^2 \dfrac{\theta}{2}}$.

16. Find the smallest integer n_0 for which $3^n < n!$. Prove that $3^n < n!$ for all $n \geqslant n_0$.

17. Find the set of positive integers n for which $2^n \geqslant n^2$.

18. If the numbers $a_1, a_2, \ldots, a_n$ all lie between 0 and 1, prove that
$(1 - a_1)(1 - a_2) \ldots (1 - a_n) \geqslant 1 - a_1 - a_2 - \ldots - a_n$.

19. Examine the divisors of $11^n + 3.4^{n-1}$ for $n = 1, 2, 3, 4$. State a general result based on this evidence, and prove it by induction.

20. Prove that $11^{n+2} + 12^{2n+1}$ is divisible by 133 for $n \geqslant 0$.

21. Prove that (i) $3.7^{2n} + 1$ is divisible by 4;
              (ii) $7^{2n} + 16n - 1$ is divisible by 64.

22. If $u_{n+2} = 5u_{n+1} - 6u_n$ with $u_1 = 1, u_2 = 5$, prove that $u_n = 3^n - 2^n$.

23. If $u_{n+2} = 10u_{n+1} - 25u_n$ with $u_1 = 0, u_2 = 1$, prove that $u_n = (n-1)5^{n-2}$.

24. If $u_{n+2} = 6u_{n+1} - 4u_n$ with $u_1 = 6$ and $u_2 = 28$, prove that
$u_n = (3 + \sqrt{5})^n + (3 - \sqrt{5})^n$.
Deduce that the least integer greater than $(3 + \sqrt{5})^n$ is divisible by 2^n.

25. (i) Into how many sections is a straight line divided by n distinct points marked on it?
   (ii) Let s_n be the number of regions into which a plane is divided by n distinct lines, of which no two are parallel and no three are concurrent. Prove by induction that $s_n = \frac{1}{2}n(n+1) + 1$.
   (iii) Let t_n be the number of regions into which space is divided by n distinct planes, of which no two are parallel, no three meet in a line, and no four meet in a point. Explain why $t_{n+1} = t_n + s_n$. Find a formula for t_n, and prove it by induction.

26. (P) A member of a committee has to vote repeatedly on a particular issue. He never abstains, but he may change his mind between one vote and the next. The probability that he votes 'yes' on the nth occasion is p_n, and the probability that he changes his mind between one vote and the next is p.
   (i) Explain why $p_{n+1} = p_n(1-p) + (1-p_n)p$
   and deduce that $p_{n+1} = p + (1-2p)p_n$.
   (ii) Prove by induction that $p_n = \frac{1}{2} + (p_1 - \frac{1}{2})(1 - 2p)^{n-1}$.
   (iii) Describe the committee member's behaviour (a) when $p = 0$, (b) when $p = 1$. Check that this is correctly predicted by the formula for p_n in these particular cases.
   (iv) Show that if $0 < p < 1$, then p_n tends to a limit as $n \to \infty$, and find this limit. What does this suggest about the tactics that a minority on the committee favouring the issue should pursue?

27. Comment on the following 'proof' of the theorem:
'In every examination all candidates get the same number of marks' (S).
Proof by induction on the number of candidates n.
   (1) When $n = 1$ there is only one candidate, so (S) is true.
   (2) Suppose (S) is true when $n = k$. Consider an examination with $k + 1$ candidates $A, B, C, D, \ldots$. The k candidates $B, C, D, \ldots$ have the same mark, since (S) is true when $n = k$. Therefore B and C have the

same mark. The k candidates $A, C, D, \ldots$ have the same mark, so A and C have the same mark. Therefore $A, B, C, D, \ldots$ all have the same mark, and (S) is true when $n = k + 1$.

Hence by induction (S) is true for all $n \geqslant 1$.

28. In the great temple of Benares beneath the dome which marks the centre of the world rests a brass plate in which are fixed three diamond needles, each a cubit high and as thick as the body of a bee. On one of these needles, at the creation, God placed sixty-four discs of pure gold, the largest disc resting on the brass plate, and the others getting smaller and smaller up to the top one. This is the tower of Bramah. Day and night unceasingly priests transfer the discs from one diamond needle to another according to the fixed immutable laws of Bramah, which require that a priest on duty must not move more than one disc at a time and that he must place this disc on a needle so that there is no smaller disc below it. When sixty-four discs shall have been thus transferred from the needle on which at the creation God placed them to one of the other needles, tower, temple and Brahmins alike will crumble into the dust and with a thunder clap the world will vanish. (*La Nature*, Paris, 1884.) How many separate transfers of single discs are required to effect the transfer of the tower of discs? Hint: assume there are n discs and by experimentation and induction prove a formula for the number of transfers required.   [OS]

2 Approximate methods I

2.1 Second derivatives

If we know the displacement s metres after t seconds of a body moving along a straight line, we can find its velocity v m s^{-1} by differentiating the function $t \mapsto s$, giving $v = \dfrac{ds}{dt}$. The acceleration a m s^{-2} is then the derivative of v with respect to t, $a = \dfrac{dv}{dt}$. So we obtain the acceleration by differentiating the displacement function: $t \mapsto s$ twice; we say that a is the *second derivative* of s with respect to t. In Leibniz's notation, this second derivative is $\dfrac{d}{dt}\left(\dfrac{ds}{dt}\right)$, which is usually condensed to $\dfrac{d^2s}{dt^2}$. The derivative of $\dfrac{d^2s}{dt^2}$ with respect to t is the third derivative, written $\dfrac{d^3s}{dt^3}$; this is the rate of change of acceleration, which is of interest in, for example, research on the medical effects of space flight.

Qu. 1. Sketch a displacement-time graph showing positive velocity with negative acceleration.

Qu. 2. Sketch a velocity-time graph showing negative velocity with negative acceleration and positive rate of change of acceleration.

Qu. 3. In January 1976 a 'quality' national newspaper carried the headline 'Rate of decline of economy is accelerating'. What, if anything, do you think this means?

In a similar way the first, second, third, ..., nth derivatives of y with respect to x are denoted by $\dfrac{dy}{dx}, \dfrac{d^2y}{dx^2}, \dfrac{d^3y}{dx^3}, \ldots, \dfrac{d^ny}{dx^n}$.

For example, if $y = x^3 + \dfrac{1}{x}$ then $\dfrac{dy}{dx} = 3x^2 - \dfrac{1}{x^2}$

$$\dfrac{d^2y}{dx^2} = 6x + \dfrac{2}{x^3}$$

$$\dfrac{d^3y}{dx^3} = 6 - \dfrac{6}{x^4}.$$

Alternatively we may use the Lagrange dash notation, in which the successive derivatives of $f(x)$ are denoted by $f'(x), f''(x), f'''(x), \ldots, f^{(n)}(x)$. For the rest of this section we shall concentrate on the second derivative, which gives useful information about the graph of $y = f(x)$.

The second derivative $\frac{d^2y}{dx^2}$ is the rate

of change of the gradient $\frac{dy}{dx}$, so that

if $\frac{d^2y}{dx^2}$ is positive for a particular value

$x = a$, then $\frac{dy}{dx}$ increases as x increases

through a. Possible shapes of the
curve in the neighbourhood of $x = a$

are shown in Fig. 1 (where $\frac{dy}{dx} > 0$

at $x = a$) and Fig. 2 (where $\frac{dy}{dx} < 0$

at $x = a$). In both cases the curve is
said to be *concave upwards* in the
neighbourhood of $x = a$. Similarly if

$\frac{d^2y}{dx^2} < 0$ when $x = a$ then $\frac{dy}{dx}$ is

decreasing as x increases through a,
and the curve is *concave downwards*
in the neighbourhood of $x = a$, as in
Fig. 3 and Fig. 4.

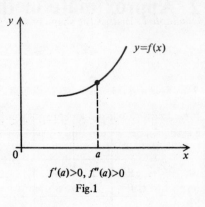

$f'(a) > 0,\ f''(a) > 0$
Fig.1

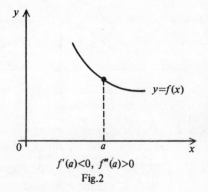

$f'(a) < 0,\ f''(a) > 0$
Fig.2

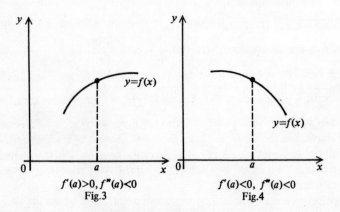

$f'(a) > 0, f''(a) < 0$ $f'(a) < 0, f''(a) < 0$
Fig.3   Fig.4

All four types of shape occur in the following example.

Example 1: Discuss the shape of the curve $y = x^3 - 3x^2 + 6$.

Solution: $y = x^3 - 3x^2 + 6 \Rightarrow \dfrac{dy}{dx} = 3x^2 - 6x = 3x(x-2)$

$$\Rightarrow \dfrac{d^2y}{dx^2} = 6x - 6$$

	$x<0$	$x=0$	$0<x<1$	$x=1$	$1<x<2$	$x=2$	$x>2$
y		6		4		2	
$\dfrac{dy}{dx}$	$+$	0	$-$	$-$	$-$	0	$+$
$\dfrac{d^2y}{dx^2}$	$-$	$-$	$-$	0	$+$	$+$	$+$

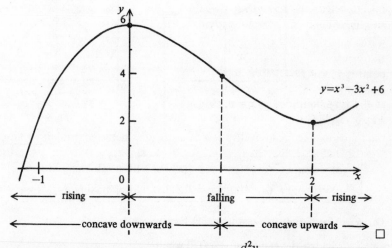

As x increases through the value 1 in Example 1, $\dfrac{d^2y}{dx^2}$ changes from negative through zero at $x = 1$ to positive, and the concavity changes from downwards to upwards. A point such as this, at which $\dfrac{d^2y}{dx^2} = 0$ and changes sign, is called a *point of inflexion*. At a point of inflexion there is a change in the concavity of the curve, and the tangent crosses the curve there. The gradient of the curve at a point of inflexion may be positive, zero, or negative, and the second derivative may be increasing or decreasing, so there are altogether six possible types of point of inflexion.

Qu. 4. Illustrate these six cases by sketch graphs.

Qu. 5.   Show that if $y = x^4$, then $\dfrac{d^2y}{dx^2} = 0$ when $x = 0$. Sketch the curve.

Does $x = 0$ give a point of inflexion? (This emphasizes that $\dfrac{d^2y}{dx^2}$

must change sign at a point of inflexion.)

The second derivative can also be useful in deciding whether a stationary point
(at which $\dfrac{dy}{dx} = 0$) is a local maximum or a local minimum. Let $y = f(x)$, and
suppose that $f'(c) = 0$, so that there is a stationary point at $x = c$. If $f''(c) > 0$,
the curve is concave upwards in the neighbourhood of $x = c$, and so the
stationary point
is a local mini-
mum (Fig. 5).
If $f''(c) < 0$,
the curve is
concave down-
wards, and the
stationary
point is a local
maximum
(Fig. 6). There-
fore we have

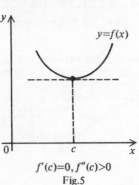

$f'(c)=0, f''(c)>0$
Fig.5

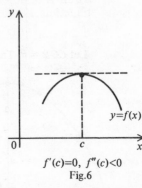

$f'(c)=0,\ f''(c)<0$
Fig.6

the following second-derivative test to distinguish between maxima and minima.

> $f'(c) = 0$ and $f''(c) > 0 \Rightarrow y = f(x)$ has a local minimum at $x = c$
>
> $f'(c) = 0$ and $f''(c) < 0 \Rightarrow y = f(x)$ has a local maximum at $x = c$.

Qu. 6.   Check that this test works in Example 1.

Qu. 7.   Sketch the graphs of $y = f(x)$ when
        (i) $f(x) = x^4$;   (ii) $f(x) = -x^4$; (iii) $f(x) = x^3$.
        Show that $f'(0) = f''(0) = 0$ for all three functions.

Notice that this test does *not* say what happens when $f''(c) = 0$. Qu. 7 shows
that the stationary point may be a minimum, a maximum, or a point of
inflexion when $f''(c) = 0$. The second-derivative test does not help in this
case, and we have to revert to the original method of investigating the sign of
$f'(x)$ for x just less than and just greater than c. It is always worth bearing this
original method in mind anyway, since it is sometimes easier to find the sign
of $f'(x)$ near $x = c$ than to carry out the extra differentiation needed to find
$f''(c)$.

Qu. 8.   Show that $f'(x) = (x - c)\,g(x) \Rightarrow f''(c) = g(c)$. This can save much
        hard work when using the second derivative test.

Example 2: Find the greatest possible area of a trapezium inscribed in a circle of radius a, when one of the parallel sides of the trapezium is a diameter of the circle.

Solution: With the notation of the figure,
$A\hat{B}C + B\hat{C}D = \pi$, since $AB\|DC$, and
$A\hat{B}C + C\hat{D}A = \pi$, since $ABCD$ is cyclic. Therefore $B\hat{C}D = C\hat{D}A$ so $ABCD$ is an isosceles trapezium.

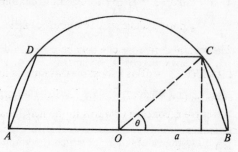

Let $C\hat{O}B = \theta$. Then the distance between AB and CD is $a \sin \theta$ and $DC = 2a\cos\theta$. The area of the trapezium, S, is therefore given by

$$S = \tfrac{1}{2}a \sin \theta \,(2a + 2a \cos \theta)$$
$$= a^2 \sin \theta \,(1 + \cos \theta) \qquad \left(0 \leqslant \theta \leqslant \frac{\pi}{2}\right).$$

Then $\dfrac{dS}{d\theta} = a^2 \cos \theta \,(1 + \cos \theta) - a^2 \sin^2 \theta$

$$= a^2 \,(\cos \theta + \cos^2 \theta - \sin^2 \theta)$$
$$= a^2 \,(2 \cos^2 \theta + \cos \theta - 1) \quad \text{since } \sin^2 \theta = 1 - \cos^2 \theta$$

and $\dfrac{d^2S}{d\theta^2} = -a^2 \,(4 \cos \theta \sin \theta + \sin \theta) < 0$ for $0 < \theta \leqslant \dfrac{\pi}{2}$.

$$\frac{dS}{d\theta} = 0 \Leftrightarrow \quad 2 \cos^2 \theta + \cos \theta - 1 = 0$$
$$\Leftrightarrow (2 \cos \theta - 1)(\cos \theta + 1) = 0$$
$$\Leftrightarrow \qquad\qquad \cos \theta = \tfrac{1}{2} \text{ or } -1.$$

The only solution for which $0 \leqslant \theta \leqslant \dfrac{\pi}{2}$ is $\theta = \dfrac{\pi}{3}$. Since $\dfrac{d^2S}{d\theta^2} < 0$,

$\theta = \dfrac{\pi}{3}$ gives a local maximum, and

$\theta = \dfrac{\pi}{3} \Rightarrow S = a^2 \dfrac{\sqrt{3}}{2}(1 + \tfrac{1}{2}) = \dfrac{3\sqrt{3}a^2}{4}$. Since θ is confined to the

interval $0 \leqslant \theta \leqslant \dfrac{\pi}{2}$ we check the values of S at the endpoints:

$\theta = 0 \Rightarrow S = 0, \theta = \dfrac{\pi}{2} \Rightarrow S = a^2 < \dfrac{3\sqrt{3}a^2}{4}$. Therefore the greatest

possible area is $\dfrac{3\sqrt{3}a^2}{4}$. $\qquad\qquad\qquad\qquad\qquad\qquad\quad \square$

Exercise 2A

1–5. Find $\dfrac{dy}{dx}, \dfrac{d^2y}{dx^2}, \dfrac{d^3y}{dx^3}$ for each of the following.

1. $y = 2x^3 + 7x^2 - 5x + 3$.   2. $y = \sqrt{x} + \dfrac{1}{x^2}$.

3. $y = \sin x + \frac{1}{2}\cos 2x + \frac{1}{3}\sin 3x$.   4. $y = \tan x$.   5. $y = \ln(1 + x^2)$.

6. If $y = 5\sin 4x + 7\cos 4x$ prove that $\dfrac{d^2y}{dx^2} + 16y = 0$.

7. Prove that $f(x) = e^{2x}\cos 3x \Rightarrow f''(x) + 13f(x) = 4f'(x)$.

8. Find the values of x for which the graph of $y = 2x^3 + 3x^2 - 12x + 5$ is
   (a) rising; (b) concave upwards. Sketch the graph.

9. Find the stationary points on $y = 7x^6 + 6x^7$, and say what type of point each one is. Do likewise for $y = 6x^5 + 5x^6$.

10. Find the maximum and minimum points on $y = 3x^4 - 8x^3 - 18x^2 + 7$, and distinguish between them.

11. Find the values of x for which the function $f(x) = e^x(2x^2 - 3x + 2)$ has
    (i) a maximum; (ii) a minimum; (iii) an inflexion.
    Draw a rough sketch of the graph of the function.   [O]

12. Find the coordinates of the points of inflexion of the curve $y = \dfrac{1}{x^2 + 12}$.
    Sketch the curve.

13. Show that $y = x + \sin x$ has an infinite number of stationary points of inflexion, but no maximum or minimum. Sketch the curve.

14. (i) Prove that the graph of a cubic polynomial $y = ax^3 + bx^2 + cx + d$ always has just one point of inflexion.
    (ii) What can be said about the number of points of inflexion of the graph of a fourth-degree polynomial? Give an example of each type which can occur.

15. The cost of manufacturing x articles of a certain type is $C(x)$. Although in practice x is an integer, we assume here and in the next two questions that x may take all real values between certain limits. Explain why the extra cost of manufacturing one additional article is approximately $C'(x)$. For this reason C' is called the marginal cost function. Write the statement 'marginal cost is positive but decreases with x' in calculus notation, and sketch the graph of $C(x)$ in this case.

16. The revenue from selling the x articles of Qu. 15 is $R(x)$. Explain the meaning of the marginal revenue, $R'(x)$. What is happening on the graph of $R(x)$ at a point of maximum marginal revenue?

17. The profit $P(x)$ from selling the x articles of Qu. 15 is $R(x) - C(x)$. Prove that the graph of $P(x)$ has a local maximum at a point where
    (a) marginal revenue equals marginal cost and (b) marginal revenue is increasing slower than marginal cost.

18. Prove that all the maximum points of the curve $y = e^{-x} \sin x$ lie on $y = Ae^{-x}$ and that all the minimum points lie on $y = -Ae^{-x}$, where A is a constant to be found. Prove also that the points of inflexion lie alternately on the curves $y = e^{-x}$ and $y = -e^{-x}$. Show all the curves $y = \pm e^{-x}$, $y = \pm Ae^{-x}$, $y = e^{-x} \sin x$ on a single diagram.

19. The cross section of a rectangular beam has breadth x and depth y.
    (i) The strength S of the beam is given by the formula $S = kxy^2$, where k is a constant. Prove that the strongest beam that can be cut from a circular log of radius a has $y = \sqrt{2}.x$.
    (ii) The stiffness of the beam is proportional to the product of the breadth and the cube of the depth. Find the relation between depth and breadth for the stiffest beam which can be cut from a circular log.

20. A right-circular cone is inscribed in a sphere of fixed radius a and centre O. Taking θ as indicated in the diagram, find the volume of the cone in terms of a and θ. Show that the volume is greatest when $\cos \theta = \frac{1}{3}$, and find what fraction of the sphere is then occupied by the cone.

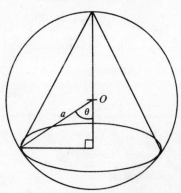

21. Find in terms of k the minimum distance from the point $(0, k)$ to the curve $y = x^2$, considering the various cases that arise as k takes all real values.

22. In a circle of radius a and centre O parallel chords AB, CD subtend angles 2α, 2θ at O, as shown in the diagram. Prove that the area of trapezium $ABCD$ is $a^2 (\sin \alpha + \sin \theta)(\cos \alpha - \cos \theta)$, and check that this formula still holds when θ is obtuse. If α remains fixed but θ varies $(\alpha \leqslant \theta \leqslant \pi)$ find in terms of α the value of θ which gives the greatest area $ABCD$.

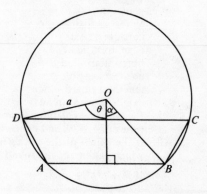

23. A rectangular sheet of paper $ABCD$
    with $AB = 20$ cm is folded so that
    the corner D just reaches the edge
    BC. If D' is the original position of
    D the acute angle between the
    crease and AD' is θ; $\theta = \alpha$ when the
    crease passes through A, and $\theta = \beta$
    when D is at B. The sheet is long,
    so that α is small and β is near
    $\dfrac{\pi}{2}$.

    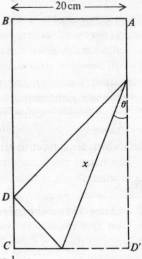

    (i) Show that if $\alpha \leqslant \theta \leqslant \dfrac{\pi}{4}$ the
        length of the crease x cm is
        given by $x = \dfrac{20}{\sin\theta\,(1 + \cos 2\theta)}$.
        Find the minimum length of
        crease for $\alpha \leqslant \theta \leqslant \dfrac{\pi}{4}$.

        [It is easier to maximize the denominator.]

    (ii) Find x in terms of θ for $\dfrac{\pi}{4} \leqslant \theta \leqslant \beta$.

    (iii) Sketch the graph of x against θ for $\alpha \leqslant \theta \leqslant \beta$.

24. (i) Two corridors of
        widths a and b with
        vertical walls meet at
        right angles. A rec-
        tangular board is to
        be carried round the
        corner in a vertical
        plane; the board is
        almost as high as the
        corridors, so its top
        edge must be kept
        horizontal. Find the
        greatest possible
        length of board which
        can be moved round

        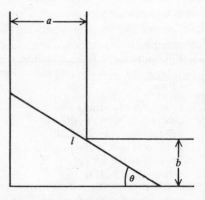

        the corner, explaining your reasoning carefully. [Hint: find l in
        terms of a, b, θ, and hence find the minimum length of l.]

    (ii) If the height of the ceiling is c show that the longest straight stick
        which can be moved round the corner is of length
        $$\left[\left(a^{\frac{2}{3}} + b^{\frac{2}{3}}\right)^3 + c^2\right]^{\frac{1}{2}}.$$

2.2 Locating the roots of an equation

The number α is said to be a *solution* or *root* of the equation $f(x) = 0$ if and only if $f(\alpha) = 0$. The formulae for the roots of linear or quadratic equations are simple and familiar; there are also formulae for solving cubic and quartic (fourth degree) equations, but they are so cumbersome that they are not often used. It is usually impossible to solve other equations exactly, so approximate methods must be used to find the roots to the required accuracy. In what follows we assume that the numerical coefficients in the equations are known exactly, though in practice they may have to be estimated from data which are subject to error.

The first step in finding the roots of an equation is to get some idea of their whereabouts, in order to be able to say, for example, that there is a root between -4 and -5, or that there are two positive roots less than 10. If we know the number of roots and have some idea of their values we say that we have *located* the roots. The precision with which this can be done depends on the particular equation. We now describe some common methods for locating roots.

1. Graphical method

The roots of $f(x) = 0$ can be located by finding where the graph $y = f(x)$ meets the x-axis. In some cases it is better to use a rearrangement of the equation and find the intersections of two simpler graphs, as in Vol. 1 §3.10. For example, to locate the roots of $\sec x + x = 2$ we sketch on the same axes $y = \sec x$ and $y = 2 - x$, and find the x-coordinates of the points where these graphs meet.

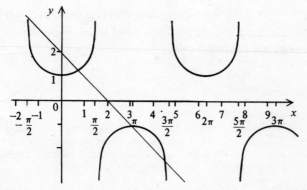

This shows that there are infinitely many roots, including ones between $-\dfrac{\pi}{2}$ and -1, 0 and 1, 3 and π, 4 and $\dfrac{3\pi}{2}$, and so on.

2. Sign-change search

If $f(x) \equiv x^4 - 3x^3 - 4x - 9$ the graph of $y = f(x)$ cannot be sketched without a good deal of preliminary work, so in this case we locate the roots by tabulating values of $f(x)$ against x:

x	-2	-1	0	1	2	3	4
$f(x)$	39	-1	-9	-15	-25	-21	39

When $x = -2$, $f(x)$ is positive and the graph is above the x-axis; when $x = -1$, $f(x)$ is negative and the graph is below the x-axis. Since this is the graph of a polynomial it is continuous,† and must therefore cross the x-axis somewhere between -2 and -1. Similarly the graph must cross the x-axis between 3 and 4.

Qu. 1. Sketch a graph to show that if $F(a)$ and $F(b)$ are of opposite signs then there may be more than one root of $F(x) = 0$ between a and b.

Qu. 2. Taking $f(x)$ as above, show that $f'(x) \equiv x^2(4x - 9) - 4$, and that $f'(x)$ is positive if $x > 3$ and negative if $x < 2$. Deduce that there is just one root of $f(x) = 0$ between -2 and -1, just one root between 3 and 4, and no other roots for $x < -2$ or $x > 4$.

Qu. 3. Sketch a graph to show that there may be roots of $F(x) = 0$ between a and b even if $F(a)$ and $F(b)$ have the same sign.

Qu. 4. By writing $f(x) \equiv x^3(x - 3) - 4x - 9$ show that $f(x) = 0$ has no root between 2 and 3.

Qu. 5. Let $g(x) \equiv \sec x + x - 2$. Show that $g(1) > 0$ and $g(2) < 0$ but that there is no root of $g(x) = 0$ between 1 and 2. Why does the change of sign test fail in this case?

---

† This is an example of a property which is taken for granted here, but would be defined and proved in a more advanced course.

The general method of locating a root
of $f(x) = 0$ between consecutive integers
by searching for a change of sign of $f(x)$
is summarized by this flow diagram.
Capital letters stand for the contents
of storage locations, and the symbol
$:=$ means 'is set equal to'. Thus
$A := a$ means that the number a is
placed in store A; $F := f(A)$ means
that the function f is evaluated for
the number in location A and the result
is placed in location F; $A := B$ means
that the number in location B is copied
into location A. Whenever a number is
put into a store, the previous contents
of that store are obliterated.

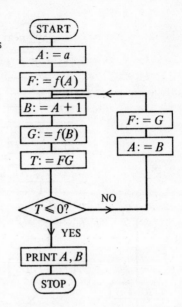

Notice the convention of using a
rectangle for an instruction box and a
diamond for a decision box.

Qu. 6.   Work through this flow diagram, taking $a = 0$, $f(x) \equiv x^4 - 3x^3 - 4x - 9$,
and tabulating the contents of A, F, B, G, T at each step.

If $f(x)$ is a *polynomial* in which the coefficient of the term of highest degree
is 1, then all the roots of $f(x) = 0$ lie between $\pm(|M| + 1)$, where M is the
coefficient of greatest magnitude; this is proved in Ex. 13B, Qu. 18. For
example, the equation $4x^6 - 3x^7 = 8 - 18x^3$ can be rearranged as $f(x) = 0$,
where $f(x) \equiv x^7 - \frac{4}{3}x^6 - 6x^3 + \frac{7}{3}$ is in the right form, with $M = -6$, and therefore all its roots lie between 7 and -7. This is worth knowing because it defines
the maximum extent of the sign-change search.

3. Decimal search

Having discovered that one of the roots of $f(x) \equiv x^4 - 3x^3 - 4x - 9 = 0$ lies
between 3 and 4, we can find a better approximation by evaluating $f(x)$ for
$x = 3.0, 3.1, 3.2$ and so on until the sign of $f(x)$ changes.
  We find that $f(3.5) = -1.56$ and $f(3.6) = 4.59$ (working to 2 D.P.), so the
root is between 3.5 and 3.6. We can then work from 3.50 at intervals of 0.01
to locate the root more precisely, and so on. This is called a *decimal search*
for the root.

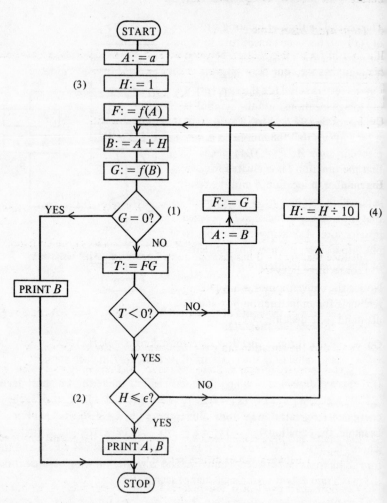

The flow diagram for a decimal search is a modification of the previous flow diagram. Decision box (1) takes account of the possibility that we find the root exactly, and decision box (2) ensures that we keep going until the root is located between numbers differing by not more than a previously fixed number ϵ. This method is slow and laborious, but it is simple to program for a computer.

Qu. 7. Work through this flow diagram, taking $a = 0$, $f(x) \equiv 7x - 10$, $\epsilon = 0.01$, tabulating the contents of the stores.

4. Repeated bisection

If a root of $f(x) = 0$ is located between a and b, the method known as *repeated bisection* has the same flow diagram as above except that box (3) is

$\boxed{H: = (b-a)/2}$ and box (4) is $\boxed{H: = H/2}$.

Qu. 8.   Tabulate the results using repeated bisection with
$f(x) \equiv x^4 - 3x^3 - 4x - 9, a = 3, b = 4, \epsilon = \frac{1}{4}$.

Exercise 2B

1. Locate each of the three roots of $x^3 - 5x^2 + 5 = 0$ between two consecutive integers.

2. If $f(x) \equiv x^3 - 6x^2 + 15x - 25$, show that $f'(x)$ is always positive, and deduce that $f(x) = 0$ has just one root. Locate this root between consecutive integers.

3. Locate the roots of $3x^4 + 4x^3 - 12x^2 + 1 = 0$.

4. Locate each of the roots of the equations (i) $\sin x = \frac{1}{2}x$; (ii) $\cos x = \frac{1}{2}x$ with an error less than $\pi/2$.

5. Show that the equation $\tan x = \dfrac{1}{x}$ has one root between $n\pi$ and $(n + \frac{1}{2})\pi$

   for each positive integer n. State the corresponding result about the negative roots.

6. Sketch on the same axes $y = \log_{10} x$ and $y = \frac{1}{5}(x - 1)$. Use a decimal search to find the larger root of $5 \log_{10} x = x - 1$ correct to 1 D.P.

7. Use tables and the decimal search method to locate the positive root of

   $e^x = \dfrac{1}{x} + 1$ between values differing by 0.01.

8. The flow chart overleaf is designed to find the square root of the number x. Carry out the instructions given in the flow chart for the pairs of numbers

x	t
4	1
0.36	0.01
−0.0001	2

   tabulating the values of a, b, c at each stage.
   Explain the purpose of the number t and the reason for the instruction $c: = x$.
   Explain how the flow chart works, and indicate in what circumstances the flow chart will fail to find the square root of the number x.
   Explain how the flow chart may be modified to find the cube root of the number x.   [MEI]

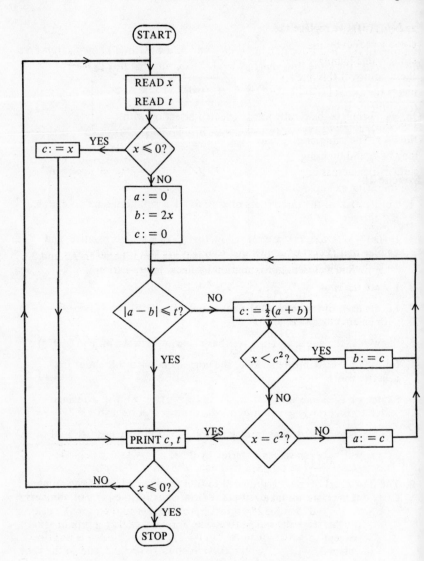

2.3 Linear interpolation

The search methods of §2.2 concentrate on the sign of $f(x)$, but make no use of its magnitude. One way to avoid this waste of information is to use *linear interpolation*. Having found that $f(x)$ changes sign between the points $(a, f(a))$

and $(b, f(b))$, we replace the curve $y = f(x)$ by the straight line joining these points. If this line meets the x-axis at $(c, 0)$ then c is taken as the approximation to the root. The value of c may be found by using similar triangles, as in the following example.

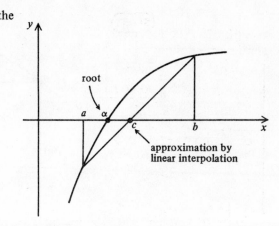

root

approximation by
linear interpolation

Example 1: Given that $x^4 - 3x^3 - 4x - 9 = 0$ has a root between 3.5 and 3.6, find a closer approximation by linear interpolation.

Solution: We know that $f(3.5) = -1.56, f(3.6) = 4.59$.
Let the straight line joining $(3.5, -1.56)$ and $(3.6, 4.59)$ meet the x-axis at $(3.5 + h, 0)$. Then by similar triangles

$$\frac{h}{1.56} = \frac{0.1 - h}{4.59}$$

So that
$4.59h = 0.156 - 1.56h$
and
$$h = \frac{0.156}{6.15} \approx 0.025.$$

Therefore we take
$3.5 + 0.025 = 3.525$ as our approximation to the root. We cannot at this stage say much about the accuracy of this approximation, except to note that since $f''(x) = 12x^2 - 18x$, which is positive when $3.5 < x < 3.6$, the curve is concave upward, and so the root is actually greater than 3.525. □

Linear interpolation can also be used when we know the values of $f(a), f(b)$ and want to find $f(c)$, where c lies between a and b. If the chord joining $(a, f(a))$ and $(b, f(b))$ is used instead of the curve $y = f(x)$ then, in the notation of the figure,

$$f(c) \approx f(a) + RQ.$$

Since triangles PQR, PST are similar,

$$\frac{RQ}{PQ} = \frac{TS}{PS},$$

i.e. $\dfrac{RQ}{c-a} = \dfrac{f(b)-f(a)}{b-a}$,

so that $RQ = \dfrac{c-a}{b-a}\,[f(b)-f(a)].$

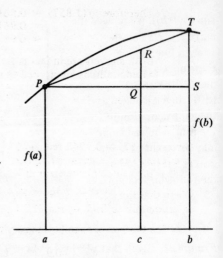

Therefore

$$f(c) \approx f(a) + \frac{c-a}{b-a}\,[f(b)-f(a)]$$

$$= f(a) + \theta\,\Delta f,$$

where θ is the fraction $\dfrac{c-a}{b-a}$ and Δf is the increment $f(b)-f(a)$.

Example 2: Given the following values

x	0.85	0.86	1.85	1.86
$\sin x$	0.7513	0.7578	0.9613	0.9585

use linear interpolation to estimate (i) sin 0.857; (ii) sin 1.857.

Solution: (i) $a = 0.85, b = 0.86, c = 0.857; \theta = \dfrac{0.007}{0.01} = 0.7.$

$$\begin{aligned} f(b) &= 0.7578 \\ f(a) &= 0.7513 \\ \hline \Delta f &= 0.0065 \end{aligned}$$

Therefore $f(0.857) \approx 0.7513 + 0.7 \times 0.0065 = 0.7513 + 0.004$
$$\approx 0.7559.$$

(ii) $a = 1.85, b = 1.86, c = 1.857; \theta = \dfrac{0.007}{0.01} = 0.7.$

$$\begin{aligned} f(b) &= 0.9585 \\ f(a) &= 0.9613 \\ \hline \Delta f &= -0.0028 \end{aligned}$$ (notice the sign).

Therefore $f(1.857) \approx 0.9613 + 0.7 \times (-0.0028)$
$= 0.9613 - 0.001\,96$
$\approx 0.9593.$

Both answers are in fact correct to 4 D.P., though more sophisticated methods are needed to show this.   □

Exercise 2C

1. Given that $57° = 0.9948$ rad and $58° = 1.0123$ rad, use linear interpolation to express 1 rad in degrees to 3 D.P. In fact 1 rad $= 57.296°$ (to 3 D.P.); what is the source of the error in your answer?

2. Given that $\tan 71.5° = 2.9887$ and $\tan 71.6° = 3.0061$, use linear interpolation to find (i) $\tan 71.54°$; (ii) the angle whose tangent is 3.

3. (i) The Normal probability density function ϕ takes the following values:

x	0.9	1.0	1.1
$\phi(x)$	0.2661	0.2420	0.2179

   Use linear interpolation to estimate $\phi(0.96)$ and $\phi(1.02)$, giving each answer to 4 D.P.

   (ii) Given that $\phi(x) \equiv \dfrac{1}{\sqrt{(2\pi)}}\, e^{-\frac{1}{2}x^2}$, prove that $y = \phi(x)$ has a point of inflexion at $x = 1$. Sketch the curve in the neighbourhood of $x = 1$, and use it to decide whether your estimates in (i) are likely to be too large or too small.

4. Find out and describe how the difference columns in four-figure logarithm tables are constructed. Explain why no values are given in the difference columns for four-figure tables of tangents when the angle is near $90°$. Use numerical examples to illustrate your answers.

5. Show that the equation $x^4 - x^3 + x - 2 = 0$ has a root between 1 and 2. Use one bisection and then one linear interpolation to locate this root more precisely.

6. Use linear interpolation to find a closer approximation to each of the roots in Ex. 2B Qu. 1. State whether each approximation is greater or less than the corresponding root.

7. Show that the equation $6 \cos x = e^x$ has a root between 1 and 2. Find a closer approximation by linear interpolation.

8. The shaded area under the curve $y = \dfrac{1}{x}$ from $x = 1$ to $x = t$ equals half the area of the rectangle shown in the diagram. Find an equation for t. Use a decimal search to locate the

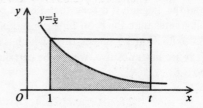

relevant root between values differing by 0.1. Find a closer approximation
by linear interpolation.

9. Show that the approximation to the root of $f(x) = 0$ between a and b
   given by linear interpolation is $\dfrac{a\,f(b) - b\,f(a)}{f(b) - f(a)}$. This is known as the
   inverse linear interpolation formula.

2.4 Newton-Raphson approximations

When finding a root of an equation approximately by linear interpolation, we
work with a chord as a replacement for the curve near the root. Instead of a
chord we could use a tangent to the curve as a local approximation (see
Vol. 1 §5.15).

For example, we know (Ex. 2B, Qu. 2) that the equation
$f(x) \equiv x^3 - 6x^2 + 15x - 25 = 0$ has just one root, which lies between 3 and 4
since $f(3) = -7$ and $f(4) = 3$.
Because $f'(x) \equiv 3x^2 - 12x + 15$ which is always positive,
and $f''(x) \equiv 6x - 12$ which is positive for $x > 2$,
the shape of the graph of $y = f(x)$
between 3 and 4 is as shown
in the diagram (not to scale).
Replace the curve by its
tangent at $(4, 3)$, and let
this tangent meet the
x-axis at $(4 - p, 0)$.
The gradient of the
tangent is $f'(4) = 15$.

Therefore $\dfrac{3}{p} = 15$

and so $p = 0.2$.

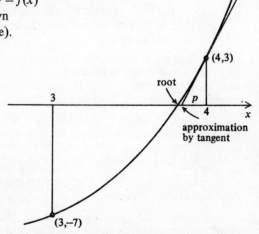

The approximation to the root given by this method is $4 - 0.2 = 3.8$, and
from the diagram it is clear that 3.8 is a better approximation than 4, though
still greater than the root. Linear interpolation between 3 and 4 gives the
approximation 3.7 (check this), which is less than the root since $f'(x)$ and
$f''(x)$ are both positive. We can be certain then that the root lies between 3.7
and 3.8.

To obtain a closer approximation, we can use the tangent at $x = 3.8$. The arithmetic begins to get heavy, so it is advisable to use a calculator. The best way to evaluate polynomials is then by *nested multiplication*. For example, rearrange $x^3 - 6x^2 + 15x - 25$ as $x(x(x - 6) + 15) - 25$, and evaluate by starting with x and successively adding the next coefficient and then multiplying by x. This is particularly convenient if the value of x can be stored and recalled.

We find that $f(3.8) = 0.232$ and $f'(3.8) = 12.72$.
Therefore (from the figure)

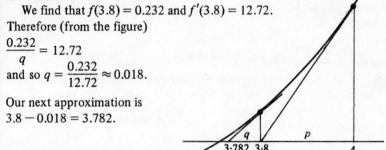

$$\frac{0.232}{q} = 12.72$$
and so $q = \dfrac{0.232}{12.72} \approx 0.018$.

Our next approximation is
$3.8 - 0.018 = 3.782$.

The process of replacing the curve by a tangent uses the linear approximation $\delta y \approx \dfrac{dy}{dx} \delta x$ (Vol. 1 §5.15), which in function notation states that $f(a + h) - f(a) \approx f'(a) . h$. In this application we are thinking of a as an approximation to a root α of the equation $f(x) = 0$.

If $\qquad \alpha = a + h \qquad$ then $\quad f(a + h) = f(\alpha) = 0,$

and so $\qquad -f(a) \approx f'(a) . h, \qquad$ i.e. $\qquad h \approx -\dfrac{f(a)}{f'(a)}.$

Therefore $\quad \alpha = a + h \approx a - \dfrac{f(a)}{f'(a)}.$

This shows that

> if a is an approximation to a root of $f(x) = 0$
>
> then $a - \dfrac{f(a)}{f'(a)}$ is in general a better approximation.

This idea was first used by Newton, and was published in its present form by Joseph Raphson in 1690; it is called the Newton-Raphson method (or Newton's method). It is important to notice the words 'in general', since $a - \dfrac{f(a)}{f'(a)}$ will *not* always be a better approximation. These figures show the sort of difficulty which may arise; see also Ex. 2D, Qu. 5.

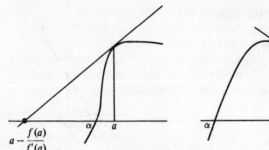
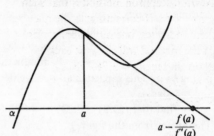

$a - \dfrac{f(a)}{f'(a)}$ α   a   α   a $a - \dfrac{f(a)}{f'(a)}$

Example: Find the root of $\sec x + x = 2$ near $x = 1$ correct to 2 D.P.

Solution: Let $f(x) \equiv \sec x + x - 2$. Then $f'(x) \equiv \sec x \tan x + 1$.

Take $a_1 = 1$ as first approximation.

$$f(1) = 1.851 + 1 - 2 = 0.851; \quad f'(1) = 3.882.$$

The second approximation is

$$a_2 = 1 - \frac{0.851}{3.882} = 0.781.$$

$$f(0.781) = 0.189; \qquad\qquad\qquad f'(0.781) = 2.396.$$

The third approximation is

$$a_3 = 0.781 - \frac{0.189}{2.396} = 0.702.$$

$$f(0.702) = 0.012; \qquad\qquad\qquad f'(0.702) = 2.108.$$

The fourth approximation is

$$a_4 = 0.702 - \frac{0.012}{2.108} = 0.696.$$

The fact that a_3 and a_4 agree to 2 D.P. does not guarantee that the root is 0.70 to this accuracy, for although the successive approximations are now changing slowly we may still be further from the root than we want to be. So to make sure, we evaluate $f(0.695) = -0.003$ and $f(0.705) = 0.018$.
Since there is a change of sign we can be certain that the root lies between 0.695 and 0.705, and hence that the root is 0.70 to 2 D.P.   □

One further complication that can arise is illustrated in our original example where $f(x) \equiv x^3 - 6x^2 + 15x - 25$. Taking $a_1 = 4$ we found $a_2 = 3.8$ and $a_3 = 3.782$, and because of the shape of the curve we expect all these to be bigger than the root. If we work to 2 D.P. instead of 3 D.P. when finding a_3, we obtain $a'_3 = 3.78$. But $f(3.78) = -0.020\,248$, and so rounding to 2 D.P. has carried us across the root. One of the great advantages of the

Newton-Raphson method is that such errors are self-correcting; as the diagram indicates, one more application of the method will carry us back again. In fact $a_3 = 3.782$ gives $a_4 = 3.781\,618$, while $a'_3 = 3.78$ gives $a'_4 = 3.781\,619$.

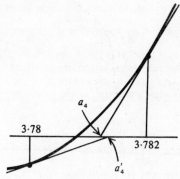

Exercise 2D

1–4. Apply the Newton-Raphson method to find a better approximation a_2 to the root of the given equation near the given first approximation a_1.

1. $x^4 - 10x + 6 = 0, a_1 = 2.$ **2.** $5x^3 + 3x^2 + 11 = 0, a_1 = -1.5.$

3. $10x^{\frac{1}{3}} = x + 4, a_1 = 27.$ **4.** $3\sin x + x = 3, a_1 = \pi.$

5. Let a_1 be an approximation to a root α of $f(x) = 0$. Draw diagrams to show that if $f''(x)$ has the same sign as $f(a_1)$ for x between a_1 and α, then the next approximation a_2 given by the Newton-Raphson method is certainly better than a_1, but that otherwise it may be worse.

6. (i) Show that $2x^3 + 6x - 7 = 0$ has a root between 0 and 1. Starting from (a) $a_1 = 1$, (b) $a_1 = 0$, find a_2 by the Newton-Raphson method in each case. In which case can you be certain *in advance* that a_2 is closer to the root than a_1? Which case actually gives a better approximation in this example?
   (ii) Repeat (i) using the equation $2x^3 + 6x - 1 = 0$.

7. The equation $x^3 - x^2 + 1 = 0$ has a root between 0 and -1. Why is it impossible to apply the Newton-Raphson method with $a_1 = 0$? Taking $a_1 = -1$, find a_2. Then use linear interpolation to obtain closer bounds for the root.

8. Tabulate values of $p(x) = x^3 + x^2 - 2x - 1$ for the range 1.1 (0.1) 1.4 of x.† Use linear interpolation to obtain an approximation to the zero of the polynomial $p(x)$ in this interval to two decimal places, and make one application of Newton's method to obtain a further approximation.
                                                                              [MEI]

9. Show that $2x^4 - 5x^2 - 1 = 0$ has a root between 1 and 2, and use the substitution $u = x^2$ to find this root to 3 D.P. Show that taking $a_1 = 1$ and using the Newton-Raphson method produces the oscillating sequence $a_2 = -1, a_3 = 1, a_4 = -1, \ldots$. Use a sketch graph of $y = 2x^4 - 5x^2 - 1$ to illustrate what is happening.

† i.e. for x from 1.1 to 1.4 inclusive at intervals of 0.1.

10. The tangents at points A and B on a circle with centre O meet at T. TO produced meets the circle again at C, and angle AOC is θ radians $(\frac{\pi}{2} < \theta < \pi)$. Prove that if the area of quadrilateral $OATB$ equals the area of sector $OACB$, then $\tan \theta + \theta = 0$.

    The relevant solution of this equation is close to an integer. Using this integer as the first approximation, find the second approximation by Newton's method, giving your answer to 2 decimal places.   [NI]

11. (i) Use sketch graphs of $y = \ln x$ and $y = 2 - x$ to show that $\ln x + x - 2 = 0$ has just one root.

    (ii) The flow chart describes the calculation of this root by the Newton-Raphson method. Show that in this case

    $$p(x) = \frac{x(3 - \ln x)}{x + 1}.$$

    (iii) Carry out the calculation when $k = 0.01$.

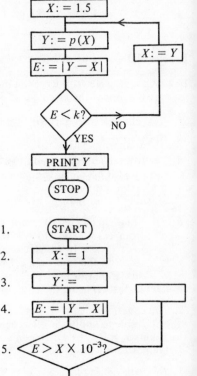

12. It is required to find the positive root of the equation $2\cos x - x = 0$, using Newton's method. The figure shows part of a flow diagram for this purpose. Complete the diagram and explain the reasons for boxes 2, 5 and 6. Calculate the required root, using this flow diagram.

    1. START

    2. $X := 1$

    3. $Y :=$

    4. $E := |Y - X|$

    5. $E > X \times 10^{-3}?$

    6. PRINT Y

    7. STOP   [MEI]

13. The area between the coordinate axes and the arc $y = \cos x$ from $x = 0$ and $x = \frac{\pi}{2}$ is bisected by the line OP drawn from the origin O to a suitable point P of the arc. Show that the x-coordinate of P satisfies

$2 \sin x - x \cos x - 1 = 0$. Show also that the solution of this equation lies between $x = \frac{\pi}{4}$ and $x = \frac{\pi}{3}$, and use linear interpolation to estimate it.

Obtain a further approximation by one application of Newton's method.

[MEI]

14. Show that the equation $y^3 - 3y = 0.90$ has one positive and two negative roots. Obtain the positive root to 4 significant figures by Newton's method. Write a flow chart for the determination of one of the negative roots of the equation to 6 significant figures, supposing that this is justified by the accuracy of the given coefficients. [MEI]

15. Taking the value 2 as a first approximation (x_1), a second approximation (x_2) to $\sqrt[3]{2}$ is found by the use of Newton's method applied to the function f defined by $f : x \to x^{k+3} - 2x^k$. Show that $x_2 = \frac{2(k+3)}{k+4}$. Select that integer value of k (> -4) for which $|x_2{}^3 - 2|$ has its smallest value. Using this value for k, obtain a rational approximation x_3 to $\sqrt[3]{2}$. [JMB]

2.5 Maclaurin and Taylor approximations

The linear approximation $f(x) \approx f(0) + x f'(0)$ for small values of x is obtained by using the tangent at the point $A (0, f(0))$ to replace the curve $y = f(x)$. The tangent is the best-fitting straight line near $x = 0$: it passes through A and its gradient equals the gradient of the curve at A. But curves curve, and straight lines do not, so to improve on this approximation we seek to replace $y = f(x)$ by a simple curve which touches it at A and has the same rate of change of gradient there.

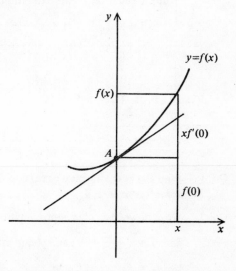

For example, suppose that $f(x) \equiv e^x$. Then $f(0) = 1, f'(x) \equiv e^x, f'(0) = 1$, so the linear approximation is $e^x \approx 1 + x$. To improve on this, we consider the rate of change of gradient, or second derivative: $f''(x) \equiv e^x, f''(0) = 1$. The quadratic function $q(x)$ which touches $y = e^x$ at $(0, 1)$ and has the same second derivative there must satisfy $q(x) = f(0) = 1, q'(0) = f'(0) = 1$, $q''(0) = f''(0) = 1$.

Now if $q(x) \equiv a_0 + a_1 x + a_2 x^2$ then $q(0) = a_0$

$$q'(x) \equiv \quad\quad a_1 + 2a_2 x, \quad\quad q'(0) = a_1$$
$$q''(x) \equiv \quad\quad\quad 2a_2, \quad\quad q''(0) = 2a_2.$$

Therefore $a_0 = 1, a_1 = 1, a_2 = \frac{1}{2}$, and we have the quadratic approximation

$$e^x \approx 1 + x + \frac{x^2}{2}.$$

Similarly, to improve on this we find the cubic polynomial
$c(x) \equiv a_0 + a_1 x + a_2 x^2 + a_3 x^3$ such that $c(0) = f(0), c'(0) = f'(0)$,
$c''(0) = f''(0), c'''(0) = f'''(0)$. This gives $a_0 = 1, a_1 = 1, 2a_2 = 1, 6a_3 = 1$, so
that $a_0 = 1, a_1 = 1, a_2 = \frac{1}{2}, a_3 = \frac{1}{6}$, and we have the cubic approximation

$$e^x \approx 1 + x + \frac{x^2}{2} + \frac{x^3}{6}.$$

The extent of these improvements is shown by the following table and graphs.

x	linear approximation $1 + x$	quadratic approximation $1 + x + \dfrac{x^2}{2}$	cubic approximation $1 + x + \dfrac{x^2}{2} + \dfrac{x^3}{6}$	e^x
-2	-1	1	-0.333	0.135
-1	0	0.5	0.333	0.368
0	1	1	1	1
1	2	2.5	2.667	2.718
2	3	5	6.333	7.389

Qu. 1. Find the next (fourth degree) approximation to e^x, and evaluate it for $x = -2, -1, 0, 1, 2$.

Qu. 2. Write down the cubic approximation for e^{-x}. Find the product of this and $1 + x + \dfrac{x^2}{2} + \dfrac{x^3}{6}$, and comment on your answer.

We can now deal with the general case. Suppose that the function f is such that $f(0), f'(0), f''(0), \ldots, f^{(n)}(0)$ all exist, and that we wish to find a polynomial $p(x)$ which agrees with f and its first n derivatives at $x = 0$, so that
$p(0) = f(0), p'(0) = f'(0), \ldots, p^{(n)}(0) = f^{(n)}(0)$.   (A)
This gives $n + 1$ conditions to be satisfied, so we try a polynomial with $n + 1$ coefficients, i.e. a polynomial of degree n.

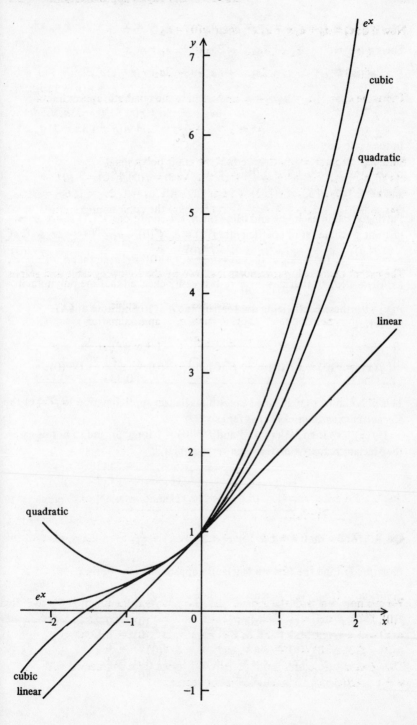

Let $p(x) \equiv a_0 + a_1 x + a_2 x^2 + a_3 x^3 + \ldots + a_r x^r + \ldots + a_n x^n$.

Then $p'(x) \equiv a_1 + 2a_2 x + 3a_3 x^2 + \ldots + r a_r x^{r-1} + \ldots + n a_n x^{n-1}$

$p''(x) \equiv 2a_2 + 6a_3 x + \ldots + r(r-1) a_r x^{r-2} + \ldots + n(n-1) a_n x^{n-2}$

$p'''(x) \equiv 6a_3 + \ldots + r(r-1)(r-2) a_r x^{r-3} + \ldots + n(n-1)(n-2) a_n x^{n-3}$

In the same way

$$p^{(r)}(x) \equiv r! a_r + \ldots + n(n-1) \ldots (n-r+1) a_n x^{n-r}$$

and $p^{(n)}(x) \equiv n! a_n$.

Putting $x = 0$ in each line and using the $n + 1$ conditions at (A) above gives just one possible set of coefficients: $f(0) = a_0, f'(0) = a_1, f''(0) = 2a_2$ so that $a_2 = \dfrac{f''(0)}{2!}, f'''(0) = 6a_3$ so that $a_3 = \dfrac{f'''(0)}{3!}, f^{(r)}(0) = r! a_r$ so that $a_r = \dfrac{f^{(r)}(0)}{r!}$, $f^{(n)}(0) = n! a_n$ so that $a_n = \dfrac{f^{(n)}(0)}{n!}$. It is easily checked that the polynomial $p(x)$ with these coefficients does satisfy the $n + 1$ conditions at (A). Therefore

$$f(x) \approx f(0) + x f'(0) + \frac{x^2}{2!} f''(0) + \frac{x^3}{3!} f'''(0) + \ldots + \frac{x^n}{n!} f^{(n)}(0).$$

This polynomial $p(x)$ is called the nth Maclaurin approximation to $f(x)$ or the *Maclaurin expansion* of $f(x)$ as far as x^n.

If $f(x) \equiv e^x$ then $f^{(n)}(x) \equiv e^x$ and $f^{(n)}(0) = 1$ for all n, and so in this case the Maclaurin approximations are very simple:

$$e^x \approx 1 + x + \frac{x^2}{2!} + \frac{x^3}{3!} + \ldots + \frac{x^n}{n!}.$$

Qu. 3. Show that $e \approx 1 + 1 + \dfrac{1}{2!} + \dfrac{1}{3!} + \ldots + \dfrac{1}{n!}$.

Example 1: Find the first six Maclaurin approximations to $\sin x$.

Solution: If
$$
\begin{aligned}
f(x) &\equiv \sin x & \text{then} && f(0) &= 0 \\
f'(x) &\equiv \cos x & & & f'(0) &= 1 \\
f''(x) &\equiv -\sin x & & & f''(0) &= 0 \\
f'''(x) &\equiv -\cos x & & & f'''(0) &= -1 \\
f^{(4)}(x) &\equiv \sin x & & & f^{(4)}(0) &= 0 \\
f^{(5)}(x) &\equiv \cos x & & & f^{(5)}(0) &= 1 \\
f^{(6)}(x) &\equiv -\sin x & & & f^{(6)}(0) &= 0
\end{aligned}
$$

The Maclaurin approximations are

$$p_1(x) \equiv p_2(x) \equiv x$$

$$p_3(x) \equiv p_4(x) \equiv x - \frac{x^3}{3!} \equiv x - \frac{x^3}{6}$$

$$p_5(x) \equiv p_6(x) \equiv x - \frac{x^3}{3!} + \frac{x^5}{5!} \equiv x - \frac{x^3}{6} + \frac{x^5}{120}. \qquad \square$$

Qu. 4.   How do these approximations reflect the fact that $\sin x$ is an odd function, i.e. $\sin(-x) \equiv -\sin x$?

Qu. 5.   Write down the first six Maclaurin approximations to $\cos x$.
The patterns established in Example 1 and Qu. 5 suggest the following Maclaurin approximations, which can be proved formally by induction.

$$\sin x \approx x - \frac{x^3}{3!} + \frac{x^5}{5!} - \frac{x^7}{7!} + \ldots + \frac{(-1)^k x^{2k+1}}{(2k+1)!}$$

$$\cos x \approx 1 - \frac{x^2}{2!} + \frac{x^4}{4!} - \frac{x^6}{6!} + \ldots + \frac{(-1)^k x^{2k}}{(2k)!}$$

Qu. 6.   If these Maclaurin approximations are called $S_k(x)$ and $C_k(x)$ respectively, show that $S_k'(x) = C_k(x)$ and $C_k'(x) = -S_{k-1}(x)$.

Example 2: Find the nth Maclaurin approximation to $(1+x)^\alpha$.

Solution: If $f(x) \equiv (1+x)^\alpha$ then $f(0) = 1$

$$f'(x) \equiv \alpha(1+x)^{\alpha-1} \qquad\qquad f'(0) = \alpha$$

$$f''(x) \equiv \alpha(\alpha-1)(1+x)^{\alpha-2} \qquad\qquad f''(0) = \alpha(\alpha-1)$$

$$f'''(x) \equiv \alpha(\alpha-1)(\alpha-2)(1+x)^{\alpha-3} \qquad f'''(0) = \alpha(\alpha-1)(\alpha-2)$$

$$f^{(n)}(x) \equiv \alpha(\alpha-1)(\alpha-2)\ldots(\alpha-n+1)(1+x)^{\alpha-n}$$

$$f^{(n)}(0) = \alpha(\alpha-1)(\alpha-2)\ldots(\alpha-n+1).$$

Therefore the nth Maclaurin approximation to $(1+x)^\alpha$ is

$$1 + \alpha x + \frac{\alpha(\alpha-1)}{2!}x^2 + \frac{\alpha(\alpha-1)(\alpha-2)}{3!}x^3 + \ldots$$

$$+ \frac{\alpha(\alpha-1)(\alpha-2)\ldots(\alpha-n+1)}{n!}x^n.$$

Notice that these are precisely the first $n+1$ terms of the expansion of $(1+x)^\alpha$ by the Binomial Theorem (Vol. 1 §4.13). $\square$

We cannot at this stage say much about the accuracy of these Maclaurin approximations. We know that if α is not a positive integer, the Binomial expansion of $(1+x)^\alpha$ is not valid for $|x| > 1$, and so the successive Maclaurin approximations to $(1 + x)^\alpha$ will not approach $(1 + x)^\alpha$ if $|x| > 1$. On the other hand, direct calculation shows that $\pi - \dfrac{\pi^3}{3!} + \dfrac{\pi^5}{5!} - \dfrac{\pi^7}{7!} + \dfrac{\pi^9}{9!} \approx 0.0069$, which is remarkably close to $\sin \pi \, (= 0)$, and it can be shown that for $\sin x$, $\cos x$ and e^x the values of successive Maclaurin approximations tend to the value of the function however large $|x|$ is. We shall consider this in more detail in §11.4.

Finally, what happens if we want approximations to $f(x)$ centred on $x = a$ instead of $x = 0$? These follow directly from what we have already done, for

$$f(x) \equiv f(a + (x - a)) \equiv f(a + h), \text{ where } h = x - \mathrm{a}.$$

Since a is constant, we can let $f(a + h) \equiv g(h)$.

Then $f(a + h) \equiv g(h) \approx g(0) + h\,g'(0) + \dfrac{h^2}{2!}\,g''(0) + \ldots + \dfrac{h^n}{n!}\,g^{(n)}(0).$

But $g'(h) \equiv f'(a + h), g''(h) \equiv f''(a + h), \ldots, g^{(n)}(h) \equiv f^{(n)}(a + h)$, and so $g(0) = f(a), g'(0) = f'(a), g''(0) = f''(a), \ldots, g^{(n)}(0) = f^{(n)}(a)$. Therefore

$$f(a + h) \approx f(a) + h\,f'(a) + \dfrac{h^2}{2!}\,f''(a) + \ldots + \dfrac{h^n}{n!}\,f^{(n)}(a)$$

or, equivalently,

$$f(x) \approx f(a) + (x - a)\,f'(a) + \dfrac{(x - a)^2}{2!}\,f''(a) + \ldots + \dfrac{(x - a)^n}{n!}\,f^{(n)}(a).$$

This is called the nth Taylor approximation to $f(x)$ centred at $x = a$.

The history of the Taylor and Maclaurin approximations is rather muddled. The Scottish mathematician James Gregory had used them forty years before Brook Taylor published his account in 1715. In 1742 Colin Maclaurin, who was Gregory's successor as professor at Edinburgh, gave his expansion. He stated that it was a special case $(a = 0)$ of Taylor's result, but for some reason it has been credited to him as a separate theorem.

Example 3: Find the nth Taylor approximation to $\ln x$ centred at $x = 1$.

Solution: Since $\ln x$ is undefined if $x \leqslant 0$, it is impossible to find approximations centred on $x = 0$. That is why we change the centre to the next simplest value, $x = 1$.

Let $f(x) \equiv \ln x$. Then $f'(x) \equiv x^{-1}, f''(x) \equiv -1 . x^{-2}$, $f'''(x) \equiv (-1)(-2)\,x^{-3}, \ldots, f^{(n)}(x) \equiv (-1)(-2) \ldots (-n + 1)\,x^{-n}$. (This can be proved formally by induction if required.) Thus $f(1) = 0, f'(1) = 1, f''(1) = -1, f'''(1) = 2, \ldots$, $f^{(n)}(1) = (-1)^{n-1}(n - 1)!$, and so the Taylor approximation centred at $x = 1$ is

$$\ln(1 + h) \approx 0 + h - \frac{1}{2}h^2 + \frac{2}{3!}h^3 - \frac{6}{4!}h^4 + \dots$$

$$+ (-1)^{n-1}\frac{(n-1)!}{n!}h^n,$$

i.e. $$\ln(1 + h) \approx h - \frac{h^2}{2} + \frac{h^3}{3} - \frac{h^4}{4} + \dots + (-1)^{n-1}\frac{h^n}{n}.$$

In §11.4 we shall show that successive approximations tend to $\ln(1 + h)$ if and only if $-1 < h \leqslant 1$.   □

Exercise 2E

1. This flow diagram gives a method for calculating e. Explain why. Carry out this calculation, working to 8 D.P., and show that $e = 2.718282$ (to 6 D.P.).

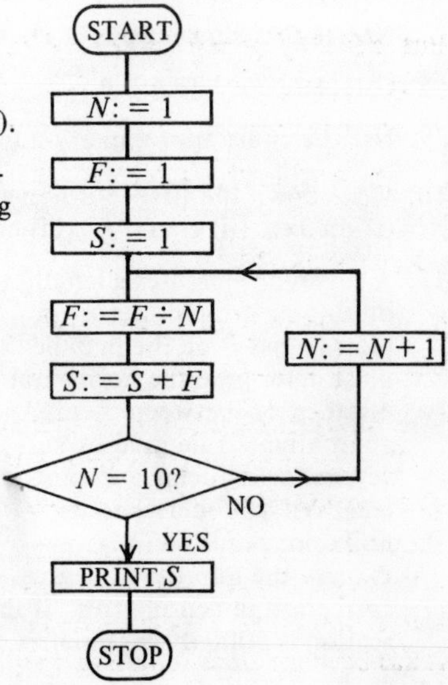

2. Write down the nth Maclaurin approximation for each of the following, giving the first four non-zero terms and the last term.
   (i) e^{-x}; (ii) e^{2x}; (iii) $e^x + e^{-x}$;
   (iv) $\dfrac{e^{3x} - e^{-x}}{e^{2x}}$.

3. Use a Maclaurin approximation to calculate $\dfrac{1}{\sqrt{e}}$ to 5 D.P.

4. Write down the Maclaurin expansion of $e^{-\frac{1}{2}x^2}$ (i) as far as x^6; (ii) as far as x^8. It can be shown that $e^{-\frac{1}{2}x^2}$ always lies between these two approximations. Use them to estimate $\displaystyle\int_0^1 e^{-\frac{1}{2}x^2}\,dx$ indicating the accuracy of your answer. [Integrals of $e^{-\frac{1}{2}x^2}$ are important in the theory of the Normal probability distribution.]

5. If $E_n(x) = \displaystyle\sum_{r=0}^{n} \frac{x^r}{r!}$ show that   (i) $E_n'(x) = E_{n-1}(x)$;
   (ii) $\int E_n(x)\,dx = E_{n+1}(x) + c$.

6. [$E_n(x)$ is as in Question 5.] (i) Simplify $e^x \times e^y$.
   (ii) Expand $E_3(x) \times E_3(y)$, and show that if terms of degree 4 or more (i.e. terms in $x^a y^b$ with $a + b \geqslant 4$) are neglected, then $E_3(x) \times E_3(y) = E_3(x + y)$.
   (iii) Show that the coefficient of $x^r y^s$ in the expansion of $E_n(x) \times E_n(y)$.

Flow diagram (right side):

START

$N := 1$

$F := 1$

$S := 1$

$F := F \div N$

$S := S + F$

$N := N + 1$

$N = 10?$ — NO

YES

PRINT S

STOP

is $\dfrac{1}{r!} \times \dfrac{1}{s!}$, and that if $r + s = m$ this coefficient is $\dfrac{1}{m!}$ $^{m}C_r$. Deduce that $E_n(x) \times E_n(y) = E_n(x + y)$ if terms of degree $n + 1$ or more are neglected.

7. Sketch the graphs of $y = \sin x$ and its successive Maclaurin approximations up to the one in x^9 as accurately as you can, using the following values.

$\sin x$	0	0.84	0.91	0.14	-0.76
x	0	1.00	2.00	3.00	4.00
$x^3/3!$	0	0.17	1.33	4.50	10.67
$x^5/5!$	0	0.01	0.27	2.02	8.53
$x^7/7!$	0	0.00	0.03	0.43	3.25
$x^9/9!$	0	0.00	0.00	0.05	0.72

Notice that $\sin x$ always lies between successive Maclaurin approximations.

8. Calculate $\cos 1$ to 4 D.P.

9. Use the cubic approximation to $\sin x$ to calculate $\sin 18°$ $(= \sin \dfrac{\pi}{10})$ to 4 D

10. Write down the first four non-zero terms in the Maclaurin expansions of
(i) $\sin 3x$; (ii) $\cos(x^2)$; (iii) $\cos^2 x \left[= \tfrac{1}{2}(1 + \cos 2x) \right]$.

11. A surveyor measures a length AB on sloping ground. Before he plots A and B on the map he must determine the horizontal distance AC between them. An approximate rule used by surveyors for reducing a sloping length of 100 metres to its horizontal equivalent is 'Square the number of degrees in the slope, multiply by $1\tfrac{1}{2}$ and obtain the *correction* in centimetres.' If the slope $\theta°$ is α radians, prove that the correction is $5000\,\alpha^2$ centimetres. Show that the rule is approximately correct

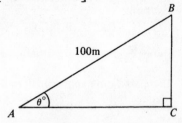

12. An approximate rule used by builders to find the length, c, of a circular arc ABC is $c = \dfrac{8b - a}{3}$, where a and b are as shown in the sketch. If O is the centre of the circle, show that $b = 2r \sin \dfrac{\theta}{2}$ and $a = 2r \sin \theta$. Using the cubic approximation to $\sin x$, show that $8b - a = 6r\theta$. Hence verify the rule. Find the percentage error caused by using this rule when $\theta = \dfrac{\pi}{3}$.

13. Sketch the graph of $y = f(x)$ for $0 \leqslant x \leqslant 1$, where

$$f(x) = \begin{cases} \dfrac{\sin x}{x} & \text{when } x \neq 0 \\ 1 & \text{when } x = 0. \end{cases}$$

Prove that $\displaystyle\int_0^1 f(x)\, dx$ differs from $\frac{17}{18}$ by less than $\frac{1}{600}$.

14. Use the cubic approximation to $\sin x$ to show that the positive root of $\sin x = x^2$ is approximately $\sqrt{15} - 3$.

15. The expression $1 - \frac{3}{2}x^2 + \frac{5}{2}x^3$ is the Taylor approximation to $f(x)$ for small values of x. Write down the values of $f'(0), f''(0), f'''(0)$. Sketch the graph of f near $x = 0$.   [SMP]

16. Obtain the expansion of $e^{\sin x}$ in ascending powers of x up to and including the term in x^4,
    (i) by calculating the values of the first four derivatives of $e^{\sin x}$ at $x = 0$,
    (ii) by taking e^y, where $y = \sin x$, and substituting the expansion of $\sin x$ into the expansion of e^y.   [O]

17. Why would you expect the Maclaurin expansion of $\tan x$ to contain only odd powers? Obtain this expansion as far as the term in x^3. Check that the expansions satisfy $\cos x \tan x = \sin x$ to this accuracy.

18. Find the Maclaurin expansion of $\sec x$ up to and including the term in x^4. Hence, or otherwise, find the Maclaurin expansion of $\ln(\sec x + \tan x)$, up to and including the term in x^5.

19. Napier's approximate formula for the calculation of logarithms was
$\ln \dfrac{a}{b} = \frac{1}{2}(a - b)\left(\dfrac{1}{a} + \dfrac{1}{b}\right)$. Assuming that $\dfrac{a}{b} = 1 + x$, where x is small enough for x^4 to be neglected, show that the error in the formula is $\frac{1}{6}x^3$.   [L]

20. Find the Maclaurin expansion of $\ln \cos x$ as far as the term in x^4. What is the connection with Question 17? Sketch the graph of $y = \ln \cos x$.

21. By means of the expansions of e^x and $\ln(1 + x)$, or otherwise, prove that $\left(1 + \dfrac{1}{n}\right)^n = e\left(1 - \dfrac{1}{2n} + \dfrac{11}{24n^2} - \dfrac{7}{16n^3} + \dots\right)$. Hence show that e is given by the formula $2e = (1.1)^{10} + (0.9)^{-10}$ with an error of approximately 0.46%.   [MEI]

22. Find the first and second Taylor approximations to $x^{\frac{2}{3}}$ centred at $x = 8$, and compare the values they give for $10^{\frac{2}{3}}$.

23. Let $f(x) \equiv 35 - 56x + 36x^2 - 10x^3 + x^4$. Show that $f'(2) = f''(2) = 0$ (so the second derivative test fails to give the nature of the stationary point at $x = 2$). Use Taylor's method to express $f(x)$ in the form $a_0 + a_1(x - 2) + a_2(x - 2)^2 + a_3(x - 2)^3 + a_4(x - 2)^4$. Hence sketch the graph of $y = f(x)$ near $x = 2$, and state the nature of the stationary point.

24. A set of tables of $\ln x$ gives the values of the function at intervals of 0.01 for $1 \leqslant x < 10$. Values at intermediate points are calculated from the approximation $\ln(p + \alpha) \approx \ln p + \dfrac{\alpha}{p}$. Give a simple justification for this rule, and explain why $|\alpha|$ need never exceed 0.005. State, giving a reason, whether the value given by the approximation is too high or too low.

[SMP]

25. (i) The diagram shows part of the graph of $y = f'(x)$ with the tangent at B. For a graph of this shape what are the signs of $f'(a), f''(a), f'''(a)$?

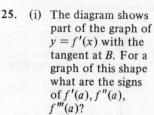

(ii) Find expressions for area $ABDF$ and area $ABEF$ in terms of f, f', a, h, and deduce that area $BDE = f(a + h) - f(a) - h f'(a)$.

(iii) By using triangle BCE, deduce that for this curve the error in the linear approximation $f(a + h) \approx f(a) + h f'(a)$ is less than $\frac{1}{2}h^2 f''(a)$.

(iv) By taking $a = 49, h = 1$ show that $353.5 < 50^{\frac{3}{2}} < 353.6$.

2.6 Simpson's rule

We have already met two common methods for estimating the value of a definite integral $\displaystyle\int_a^b f(x)\,dx$ by using straight line approximations (Vol. 1 §6.11). Both methods start by dividing the area under $y = f(x)$ from a to b into strips

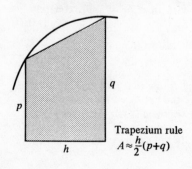

Trapezium rule
$$A \approx \frac{h}{2}(p+q)$$

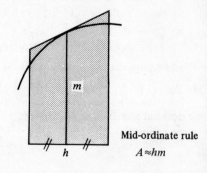

Mid-ordinate rule
$$A \approx hm$$

of equal width h. The curve is then replaced either by its chord, giving the

trapezium rule $$A \approx \frac{h}{2}(p+q),$$

or by its tangent at the mid-interval point, giving the mid-ordinate rule $A \approx hm$. In both cases the accuracy can be improved by using more strips, though of course this involves more calculation.

It turns out to be more efficient, however, to improve our estimate for each strip by using a curve rather than a straight line as a replacement. As usual, the simplest curve for this purpose is a parabola of the form $y = ax^2 + bx + c$. Since this has three coefficients to be chosen, we can fit a parabola through three points of the curve. Therefore in this method we deal with *pairs* of strips of equal width h. We now need a formula for the area under the parabola in terms of h and the three ordinates p, q, r, corresponding to the trapezium rule $\frac{h}{2}(p+q)$. Since this

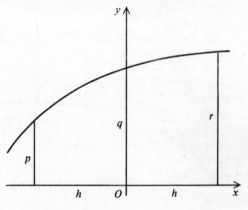

formula will depend on the width of the strips but not on the position of the y-axis, we can choose the origin so that the three points through which the parabola passes are $(-h, p), (0, q), (h, r)$. If the parabola through these points is $y = ax^2 + bx + c$, then the area we require is

$$A = \int_{-h}^{h} (ax^2 + bx + c)\,dx$$

$$= \left[\frac{ax^3}{3} + \frac{bx^2}{2} + cx \right]_{-h}^{h}$$

$$= \frac{2ah^3}{3} + 2ch$$

$$= \frac{h}{3}(2ah^2 + 6c).$$

We now have to express this in terms of p, q, r. Since $(-h, p), (0, q), (h, r)$ lie on the parabola,

$$p = ah^2 - bh + c$$
$$q = c$$
$$r = ah^2 + bh + c.$$

Therefore $p + r = 2ah^2 + 2c$ and $q = c$, so that $2ah^2 + 6c = p + 4q + r$.

$$\text{Therefore} \quad A = \frac{h}{3}(p + 4q + r).$$

This result is called *Simpson's rule*, after Thomas Simpson who published it in 1743. It had been used as early as 1668 by James Gregory, who as usual got no credit for his discovery.

When estimating $\int_a^b f(x)\,dx$ using Simpson's rule, we divide the interval

from a to b into n equal parts, where n must be *even*.

Let $\dfrac{b-a}{n} = h$, and

let $f(a + rh) = y_r$,
so that $y_0 = f(a)$
   and $y_n = f(b)$;
notice that there are an odd number $(n + 1)$ of ordinates.

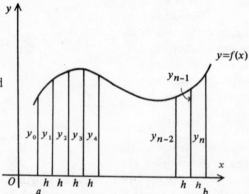

Applying Simpson's rule to each double strip gives

$$\int_a^b f(x)\,dx \approx \frac{h}{3}(y_0 + 4y_1 + y_2) + \frac{h}{3}(y_2 + 4y_3 + y_4) + \ldots$$
$$+ \frac{h}{3}(y_{n-2} + 4y_{n-1} + y_n),$$

i.e.
$$\boxed{\int_a^b f(x)\,dx \approx \frac{h}{3}\{y_0 + y_n + 4(y_1 + y_3 + \ldots + y_{n-1}) + 2(y_2 + y_4 + \ldots + y_{n-2})\}.}$$

This can be remembered as

$$\text{Integral} \approx \frac{\text{strip width}}{3} \times \{\text{ends} + 4 \times \text{odds} + 2 \times \text{other evens}\}.$$

The following example gives a convenient way to set out the calculation.

Example: Use Simpson's rule with six strips to estimate $\int_{40}^{100} \operatorname{cosec}(x°)\,dx$.

Solution:

$x°$	$\operatorname{cosec}(x°)$	(×4)	(×2)
40	1.5557		
50		1.3054	
60			1.1547
70		1.0642	
80			1.0154
90		1.0000	
100	1.0154		
	2.5711	3.3696	2.1701
	13.4784	× 4	× 2
	4.3402	13.4784	4.3402
	20.3897		

$$h = \frac{100 - 40}{6} = 10.$$

$$\int_{40}^{100} \operatorname{cosec}(x°)\,dx \approx \frac{10 \times 20.3897}{3} = 68.0 \text{ to 3 S.F.} \qquad \square$$

Simpson's rule is exact for areas under parabolas. It is surprising to find that it is also exact for areas under cubic curves.
For let $y = C(x)$ be a cubic curve through $(-h, p)$, $(0, q)$, (h, r), and let $y = Q(x)$ be the parabola through the same points. The difference between

$\int_{-h}^{h} C(x)\,dx$ and $\int_{-h}^{h} Q(x)\,dx$ is

$$\Delta = \int_{-h}^{h} (C(x) - Q(x)), \text{ shaded in the diagram.}$$

But if $D(x) \equiv C(x) - Q(x)$ then $D(x)$ is a cubic polynomial which is zero at $x = -h$, $x = 0$, $x = h$. Therefore $D(x) \equiv k(x + h)\,x\,(x - h) \equiv k(x^3 - h^2 x)$,

and $\Delta = \int_{-h}^{h} D(x)\,dx = k\left[\frac{x^4}{4} - \frac{h^2 x^2}{2}\right]_{-h}^{h} = 0.$

Therefore $\int_{-h}^{h} C(x)\,dx = \int_{-h}^{h} Q(x)\,dx = \frac{h}{3}(p + 4q + r).$

This indicates that to improve on Simpson's rule we should need to take not a cubic but a fourth-degree approximating curve, (but see Ex. 2F, Qu. 16).

Exercise 2F

1. Use Simpson's rule with four strips to estimate $\int_0^2 \frac{1}{1+x^3}\, dx$. Give your answer to 3 S.F.

2. Use Simpson's rule with six strips to estimate $\int_1^4 \log_{10} x\, dx$. Give your answer to 4 S.F.

3. Find, to 3 D.P., the error involved in estimating $\int_0^4 \frac{1}{1+x}\, dx$ by using Simpson's rule with 5 ordinates.   [SMP]

4. The table gives the velocity v m s^{-1} of an electric milk float t seconds after starting from rest.

t	0	2	4	6	8	10	12
v	0	3.0	4.7	5.8	6.6	7.1	7.3

Use Simpson's rule to estimate the distance travelled in 12 seconds.

5. Tabulate values of $f(x) \equiv \sqrt{\{27 + (x-3)^3\}}$ for integral values of x from 0 to 6 inclusive and sketch the graph of $y = f(x)$ for the interval

$0 \leqslant x \leqslant 6$. Given that $F(t) \equiv \int_0^t f(x)\, dx$, use Simpson's rule and the

calculate values of $f(x)$ to estimate $F(2)$ and $F(6)$.   [L]

6. Tabulate to 3 D.P. the value of the function $f(x) \equiv \sqrt{(1+x^2)}$ for values of x from 0 to 0.8 at intervals of 0.1. Use these values to estimate

$\int_0^{0.8} f(x)\, dx$ by Simpson's method, (i) using all the ordinates, and

(ii) using only ordinates at intervals of 0.2. Draw any conclusions you can about the accuracy of the results.   [MEI]

7. Use Simpson's rule with four strips to estimate the volume of the solid of

revolution generated by rotating $y = \frac{1}{2+\sqrt{x}}$ from $x = 5$ to $x = 9$ about the x-axis.

8. The table shows the diameter d cm of a vase at height h cm above its base.

h	0	3	6	9	12	15	18	21	24
d	4.6	7.5	4.6	3.1	2.8	3.2	4.0	4.6	6.2

Estimate the volume of the vase, which may be treated as a solid of revolution.

9. On board the clipper Cutty Sark at Greenwich one can see the tonnage calculation done in 1869 by the ship's designer Hercules Linton. This involves working out the areas of eleven equally spaced cross sections of the hull. For each one Linton divides half the cross section into six

horizontal strips of
equal width, and
then uses Simpson's
rule. At the largest
cross section the
depth of the hull is
20.4 feet, and its
half-widths at the
seven levels are

17.15, 17.40, 17.45,
17.30, 16.10, 10.65,
0.60 feet.

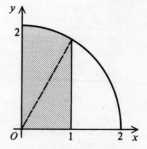

Show that the area
of this cross section
is just over 600 square feet.
If you knew all eleven cross-sectional areas, how would you calculate the
volume of the hull, given that the length of the Cutty Sark's hull is
213 feet?

10. Given that it is required to use Simpson's rule to evaluate $\int_0^6 f(x)\,dx$,

where $f(x) \equiv \sqrt{(400 + x^2)} - 20$, show that it is preferable, when using
tables to obtain values of $f(x)$, to use $f(x)$ in the equivalent form

$$f(x) \equiv \frac{x^2}{\sqrt{(400 + x^2)} + 20}.$$

Use Simpson's rule with three ordinates to evaluate the integral, giving
the result to 4 S.F.
An alternative method uses the approximation $f(x) \approx \dfrac{x^2}{40}$ and direct

integration. Show that the two methods give results differing by less than
2%.   [MEI]

11. Given that $x \geqslant 4$, show that $e^{-\frac{1}{2}x^2} \leqslant e^{-2x}$ and hence show that

$$\int_4^8 e^{-\frac{1}{2}x^2}\,dx < 0.0002. \quad \text{[Take } e^{-8} \text{ to be 0.0003.]}$$

Use Simpson's rule with 5 ordinates to estimate the value of $\int_0^4 e^{-\frac{1}{2}x^2}\,dx$,

and hence obtain an estimate of $\int_{-8}^8 e^{-\frac{1}{2}x^2}\,dx$.   [JMB]

12. The shaded area is bounded by the lines
$x = 0$, $x = 1$, $y = 0$ and the circle centre O
radius 2. Prove that this shaded area is
$\dfrac{\pi}{3} + \dfrac{\sqrt{3}}{2}$. (The broken line suggests a method.)

Find the approximation to this area given by
Simpson's rule with two strips, and deduce
that $\pi \approx 1 + \sqrt{15} - \sqrt{3}$. Find the percentage error in this approximation.

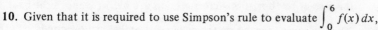

13. By writing a^x as $e^{x \ln a}$ $(a > 0)$ find $\int_0^1 a^x dx$ in terms of a. Find also the approximation to $\int_0^1 a^x dx$ which is obtained by using Simpson's rule with three ordinates, and deduce that $\ln a \approx \dfrac{6(a-1)}{a+4\sqrt{a}+1}$. Show that this leads to $\ln 3 \approx \frac{3}{2}(\sqrt{3}-1)$, and use tables to estimate the error in this approximation.

14. (i) The top and bottom of a trough are horizontal rectangles, not necessarily of the same shape but with the top edges parallel to the bottom edges. The vertical height of the trough is h, and 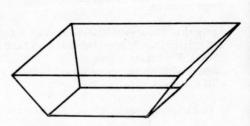 the area of the horizontal cross section at a height x above the base is $A(x)$. Explain why $A(x)$ is a quadratic function of x, and deduce that the volume of the trough is

$$\frac{h}{6}\left\{ A(0) + 4A\left(\frac{h}{2}\right) + A(h) \right\}.$$

  (ii) Derive from this the formula for the volume of a pyramid with a rectangular base.

  (iii) The roof space of a house has a rectangular base of width a and length b with a centrally placed ridge of length c at a height h above it, as in the diagram. Prove that the volume of the roof space is $\frac{1}{6}ha(2b+c)$.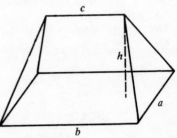

15. State the trapezium rule for finding an approximation I_h to the integral $I = \int_a^b f(x)\,dx$, using n intervals of length h. If I_k is another approximation obtained by using intervals of length k, and if the errors $I - I_h$, $I - I_k$ are taken as ch^2, ck^2 respectively, show that a better approximation is given by $I \approx I_h + \dfrac{h^2}{k^2 - h^2}(I_h - I_k)$. If n is even and $k = 2h$, obtain Simpson's rule in the form

$$I \approx \frac{h}{3}\{f_0 + 4f_1 + 2f_2 + 4f_3 + \ldots + 2f_{n-2} + 4f_{n-1} + f_n\},$$

where $f_r = f(a + rh), r = 0, 1, \ldots, n$.

Use the trapezium rule and Simpson's rule with $n = 4$ to find approximations to

$$I = \int_{-1}^{1} e^{-2x^2} dx.$$ [MEI]

16. An approximation to $\int_{-1}^{1} f(x)\, dx$ is given by

$$\frac{2\,[p\,f(-\alpha) + q\,f(0) + p\,f(\alpha)]}{2p + q}, \quad \text{where } \alpha > 0.$$

Prove that this approximation is exact if $f(x)$ is a polynomial of degree not greater than 5, provided that α and the ratio of p to q are suitably chosen. Find these values.

What modifications are necessary to the result if the integral is taken from a to b, instead of from -1 to 1?   [O & C]

3 Vectors I

3.1 Scalars and vectors: basic ideas

In the physical world many quantities are fully specified by a real number together with the name of the unit being used (if any); for example, the volume of a sphere, the temperature of the bath water, or the kinetic energy of a cricket ball. Such quantities are known as *scalar quantities* and the numbers which are their magnitudes are called *scalars*. Other quantities have a directional property as well as a magnitude; for example, the position of a ship relative to a lighthouse, or the velocity of a rocket, or the force on an object due to an attached spring.

We define a non-zero *vector* as an entity which has three characteristics: a positive number known as the magnitude of the vector, a direction in space, and a sense (defining the idea of direction more precisely). The *zero vector* has magnitude zero, but its direction and sense are not defined.

direction   direction with sense   direction with opposite sense

In 1586 Simon Stevin of Bruges, following the work of Archimedes (250 B.C.), published papers showing how forces could be represented in magnitude and direction by directed line segments, and could be compounded by the triangle of forces. By the early nineteenth century many vector quantities were known. The word 'vector' (meaning 'carrier') was introduced by William Hamilton, who presented the first mathematical treatment of the subject in Dublin in 1843. Independently, Hermann Grassmann in Stettin published similar work in 1844.

We use bold print $\mathbf{a}$, $\mathbf{b}$, $\mathbf{F}$, ... to denote vectors; in script this is indicated by a wavy underline: $\underset{\sim}{a}$, $\underset{\sim}{b}$, $\underset{\sim}{F}$, ... The zero vector is denoted by $\mathbf{0}$. The magnitude of a vector is shown by using either the modulus sign or italic print; thus the magnitudes of $\mathbf{u}$, $\mathbf{F}$ are denoted by $|\mathbf{u}|$, $|\mathbf{F}|$ or u, F respectively.

If $\mathbf{a}$, $\mathbf{b}$ are vectors which are parallel and have the same magnitude and sense, we say that $\mathbf{a}$ and $\mathbf{b}$ are *equal*, and write $\mathbf{a} = \mathbf{b}$.

Vectors are represented diagrammatically by directed line segments, such as $\overrightarrow{PQ}$. The length of line segment PQ represents the magnitude of the vector, according to some scale, and the direction and sense of $\overrightarrow{PQ}$ represent the direction and sense of the vector. If P and Q coincide, the directed line segment $\overrightarrow{PQ}$ degenerates to the point P representing the zero vector. Each directed line segment represents one and only one vector. Each vector may be represented by infinitely many directed line segments, all starting at different points, but if $\overrightarrow{AB}$ represents vector **v**, then once we have chosen the position of A, the position of B is uniquely defined. We refer to the vector represented by $\overrightarrow{PQ}$ as **PQ** (in script PQ).

In this text we reserve the notation $\overrightarrow{PQ}$ for the directed line segment from P to Q; if P, R are distinct points, and $\overrightarrow{PQ}$, $\overrightarrow{RS}$ are parallel and of the same magnitude and sense, then $\overrightarrow{PQ}$, $\overrightarrow{RS}$ are distinct directed line segments, both representing the same vector which may be referred to as **PQ** or **RS**. The statement **AB** = **CD** implies that the directed line segments $\overrightarrow{AB}$, $\overrightarrow{CD}$ are parallel and have the same magnitude and sense, though they may well be situated in different

positions. There is an analogy here with rational numbers: the two fractions $\frac{12}{33}$ and $\frac{4}{11}$ are different expressions for the same rational number, and we write $\frac{12}{33} = \frac{4}{11}$. By convention we usually reduce a fraction to its lowest terms, but we can choose the most convenient form for any particular application, such as adding $\frac{2}{33}$ or making an order comparison with $\frac{1}{3}$. So with vectors, we can choose the most convenient representative line segment for the work we are doing.

If $\overrightarrow{TH}$, $\overrightarrow{DM}$ are parallel directed line segments, with opposite senses, $\overrightarrow{DM}$ being shorter than $\overrightarrow{TH}$, then **DM** is a vector which has a smaller magnitude than **TH**, and the vectors **DM**, **TH** are parallel, but have opposite senses.

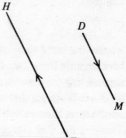

Directed line segments are also used to represent translations.

If $\overrightarrow{AA_1}$ represents the translation T which takes shape S to S_1, and $\overrightarrow{A_1A_2}$ represents the translation T' which takes S_1 to S_2, then the single translation which takes S to S_2 is represented by $\overrightarrow{AA_2}$.

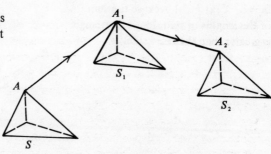

This leads us to define the addition of two vectors as follows:

(i)   take an arbitrary point P,
(ii)   take the unique directed line segment $\overrightarrow{PQ}$ which represents **a**,
(iii) take the unique directed line segment $\overrightarrow{QR}$ which represents **b**,
(iv)   we define **a** + **b** to be the unique vector represented by $\overrightarrow{PR}$.

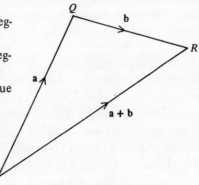

This is summarized as the *Triangle Law*:

> if vectors **a** and **b** are represented by $\overrightarrow{PQ}$ and $\overrightarrow{QR}$ respectively, their sum **a** + **b** is represented by $\overrightarrow{PR}$.

Notice that **a** + **0** = **a** and **0** + **a** = **a**, so **0** is the identity element for vector addition.

Qu. 1. Prove that the result **a** + **b** is independent of the position of P.

We use the same sign + for the addition of vectors as for the addition of numbers because these two different operations have many similar properties. The sum of two or more vectors is often called their *resultant*.

If **a** + **b** are parallel and in the same sense, then $|\mathbf{a} + \mathbf{b}| = |\mathbf{a}| + |\mathbf{b}|$. This relation is also true if either **a** or **b** is the zero vector. In all other cases, $|\mathbf{a} + \mathbf{b}| < |\mathbf{a}| + |\mathbf{b}|$ because the sum of the lengths of two sides of a triangle is greater than the third side.

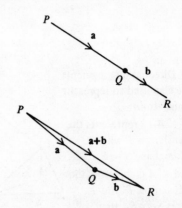

A quantity such as a force or a velocity which has magnitude, direction and sense may be represented by a directed line segment; two such quantities

may be combined together by the Triangle Law, forming a third quantity. If
this third (resultant) quantity is usefully related to the original two quantities,
they are called *vector quantities*. For example, if the original two quantities
are the velocity of an aeroplane relative to the air and the velocity of the
wind, then the resultant is the velocity of the aeroplane relative to the ground,
which is clearly worth finding. If the original two quantities are two forces
acting on a rigid body then the Triangle Law gives the magnitude, direction
and sense of the resultant force, but does not give the line of action of this
force. Vector addition is useful here but needs to be treated with care. As
another example, consider an airport's passenger traffic: on a certain day
1000 northbound passengers and 1000 eastbound passengers pass through
the airport; these quantities have magnitude, direction and sense, but the
vector sum is approximately 1414 passengers in the direction north-east, a
result which serves no practical purpose, so these quantities are not vector
quantities.

Physicists, in particular, often prefer to use a parallelogram rule for adding
vectors, stating that
the representations
$\vec{PQ}, \vec{PS}$ of **a** and **b**
respectively must
start at the same
point P. Point R
completes paral-
lelogram $PQRS$;

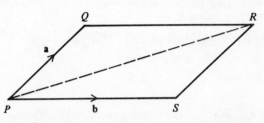

then **a** + **b** is represented by $\vec{PR}$. This is particularly useful when adding vectors
representing forces as it helps to emphasize that the resultant of two forces acts
through the common point of their lines of action.

Using the top triangle in the diagram of the parallelogram we see that
PR = **PQ** + **QR**; using the lower triangle we see that **PR** = **PS** + **SR**. But
a = **PQ** = **SR** because $\vec{PQ}, \vec{SR}$ have the same magnitude, direction and sense;
similarly **b** = **PS** = **QR**, so that **a** + **b** = **b** + **a**. This is the commutative property
of vector addition.

Notice what happens if we add two vectors which are equal in magnitude,
equal in direction, but opposite in sense:

$$\textbf{AB} + \textbf{BA} = \textbf{0}.$$

So we define −**u**, called the *negative* of **u**, as the vector which has the same
magnitude and direction as **u**, but the opposite sense; therefore −**AB** = **BA**.

Qu. 2.   What is −(−**u**)?

We define subtraction of vectors, denoted by −, in this way:
a − b means a + (−b). Again notice
that the same sign − is used for the
subtraction of vectors and for the
subtraction of numbers; this is
because these two different
operations share many common
properties.

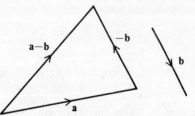

Qu. 3. Show that in parallelogram *PQRS* on page 57 **QS** = **b** − **a** and express
   SQ in terms of **a** and **b**.

Sums of more than two vectors can be formed in various ways.

Let **a**, **b**, **c** be represented by $\overrightarrow{PQ}$, $\overrightarrow{QR}$, $\overrightarrow{RS}$ respectively.
 Then (**a** + **b**) + **c** = **PR** + **RS** = **PS**
  and **a** + (**b** + **c**) = **PQ** + **QS** = **PS**,
so that (**a** + **b**) + **c** = **a** + (**b** + **c**)
for any vectors **a**, **b**, **c**.

This is the *associative* property of vector addition.

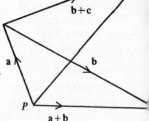

Qu. 4. Why is it unnecessary to put brackets in the expression **a** + **b** + **c** + **d**?

Suppose **a**, **b** are parallel vectors with the same
sense with $|\mathbf{b}| = \frac{1}{2}|\mathbf{a}|$. Then **b** + **b** = **a**, and it
would be convenient if we could say $\mathbf{b} = \frac{1}{2}\mathbf{a}$.
Similarly, if **p**, **q** are parallel, but with opposite
sense, and $|\mathbf{p}| = 3|\mathbf{q}|$, then −**q** − **q** − **q** = **p** and
it would be convenient if we could write
p = −3 **q**. Generalizing this, we make the
following definition of multiplication of a vector by a scalar:

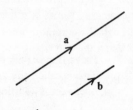

> *m***a** is a vector of magnitude $|m|\,a$;
> *m***a** and **a** are parallel, with the same sense if *m* is positive,
> but with opposite sense if *m* is negative;
> if *m* = 0, or **a** = **0** then *m***a** = **0**.

Division by a non-zero scalar is also possible:

$$\frac{\mathbf{a}}{m} = \frac{1}{m}\mathbf{a}, m \neq 0.$$

Example 1: Show that $\mathbf{AB} + \mathbf{DC} = \mathbf{AD} + \mathbf{CB}$ implies that C and D are coincident.

Solution: $\mathbf{AB} + \mathbf{DC} = \mathbf{AD} + \mathbf{CB} \Rightarrow \mathbf{DA} + \mathbf{AB} + \mathbf{BC} + \mathbf{DC} = \mathbf{0}$

$\Rightarrow \qquad\qquad\qquad 2\,\mathbf{DC} = \mathbf{0}$

$\Rightarrow \qquad\qquad\qquad\quad \mathbf{DC} = \mathbf{0}$

$\Rightarrow C$ and D coincide.   □

Example 2: Show that $3(\mathbf{a} + \mathbf{b}) = 3\mathbf{a} + 3\mathbf{b}$.

Solution: Let $\overrightarrow{PQ}$ represent $\mathbf{a}$, $\overrightarrow{QR}$ represent $\mathbf{b}$, so that $\overrightarrow{PR}$ represents $\mathbf{a} + \mathbf{b}$.
Let triangle $P'Q'R'$ be an enlargement of triangle PQR from some centre O, scale factor 3.

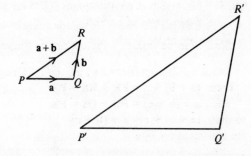

Then each side of triangle $P'Q'R'$ is parallel to the corresponding side of triangle PQR, and each side of triangle $P'Q'R'$ is 3 times as long as the corresponding side of triangle PQR. Moreover, as the scale factor is positive, $\overrightarrow{P'R'}$ has the same sense as $\overrightarrow{PR}$, and similarly for the other directed line segments.
Thus $\overrightarrow{P'Q'}, \overrightarrow{Q'R'}, \overrightarrow{P'R'}$ represent $3\mathbf{a}, 3\mathbf{b}, 3(\mathbf{a} + \mathbf{b})$ respectively; and from triangle $P'Q'R'$, we see that $\overrightarrow{P'R'}$ also represents $3\mathbf{a} + 3\mathbf{b}$.
Therefore $3(\mathbf{a} + \mathbf{b}) = 3\mathbf{a} + 3\mathbf{b}$ □

Qu. 5. Prove that $m(\mathbf{a} + \mathbf{b}) = m\mathbf{a} + m\mathbf{b}$.
This is called the *distributive* property of multiplication by a scalar over vector addition.

Qu. 6. Prove that $(mn)\mathbf{a} = m(n\mathbf{a})$.
Notice that this resembles the associative property of multiplication; but on the left we have a multiplication of scalars, followed by multiplication of a vector by a scalar, while on the right we twice multiply a vector by a scalar.

Qu. 7. Prove that $(m + n)\mathbf{a} = m\mathbf{a} + n\mathbf{a}$.
Again notice that the two + signs have different meanings: on the left side + refers to addition of scalars, and on the right + refers to addition of vectors, two different operations.

Summary

1. $\mathbf{a} + \mathbf{b} = \mathbf{b} + \mathbf{a}$

2. $(\mathbf{a} + \mathbf{b}) + \mathbf{c} = \mathbf{a} + (\mathbf{b} + \mathbf{c})$ 3. $(mn)\mathbf{a} = m(n\mathbf{a})$

3. $m(\mathbf{a} + \mathbf{b}) = m\mathbf{a} + m\mathbf{b}$ 5. $(m + n)\mathbf{a} = m\mathbf{a} + n\mathbf{a}$

Exercise 3A

1. Which of the following are vector quantities and which are scalar quantities? Velocity, mass, weight, area, density, volume, force, time, acceleration, electrical resistance, energy, momentum, temperature.

2. Find the sum of $\mathbf{DC}, -\mathbf{DE}, \mathbf{BE}, -\mathbf{EC}, \mathbf{AB}$.

3. Prove that P and Q are the same point if $\mathbf{PR} + \mathbf{ST} = \mathbf{QT} + \mathbf{SR}$.

4. Prove that $|\mathbf{a} - \mathbf{b}| \geqslant |\mathbf{a}| - |\mathbf{b}|$. Under what conditions does equality occur?

5. The Mid-point Theorem states that in any triangle the line joining the mid-points of two of the sides is parallel to the third side and half its length. Use vectors to prove this theorem.

6. Let O be the centre of the regular hexagon $ABCDEF$. Forces represented by $\mathbf{OA}, 3\mathbf{ED}, \mathbf{OF}, \mathbf{CD}$ act at a point. Find a vector which represents the resultant of these forces.

7. Let $ABCDEFGH$ be a cube, labelled as shown. Find the resultant of four concurrent forces represented by $\mathbf{AG}, \mathbf{BE}, \mathbf{CF}, 3\mathbf{HD}$.

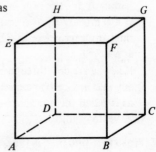

8. Given that $\mathbf{a}$ and $\mathbf{b}$ are not zero vectors, nor are they parallel, when does $\lambda\mathbf{a} = \mu\mathbf{b}$? Solve the equation $6\mathbf{a} = (x + y)\mathbf{a} + (x - 2y)\mathbf{b}$ to find scalars x and y.

9. If $\mathbf{p} = 3\mathbf{x} - \mathbf{y}$ and $\mathbf{q} = \mathbf{x} - 3\mathbf{y}$ find $\mathbf{x}$ and $\mathbf{y}$ in terms of $\mathbf{p}$ and $\mathbf{q}$, stating the vector properties that have been used.

10. If $\mathbf{a}$ is a vector of magnitude 2 units due east, $\mathbf{b}$ is a vector of magnitude 3 units due north, $\mathbf{c}$ is a vector of magnitude 3 units vertically upwards, find the magnitude of the vector $\mathbf{a} + \mathbf{b} + 2\mathbf{c}$, and the angle it makes with the horizontal.

11. If $PQRS$ is a skew quadrilateral, i.e. P, Q, R, S are not coplanar and $K, L,$ M, N are the mid-points of PQ, QR, RS, SP respectively, find $\mathbf{KN}, \mathbf{LM}$ in terms of $\mathbf{QS}$. What does this prove about quadrilateral $KLMN$?

12. Rotations of a rigid body about a fixed point have magnitude (the size of the angle through which the rotation takes place), direction (the axis of rotation) and sense (when you look along the axis in the sense of the arrow, the rotation is clockwise). By considering a half turn about an axis pointing east, and a half turn about an axis pointing north, or otherwise, show that finite rotations cannot be added by the triangle law, and are thus not vector quantities.

3.2 Vectors in 2 and 3 dimensions

Before we proceed, we need to define a few more terms: we will describe three or more vectors as *coplanar* if their representations as directed line segments could all be in the same plane. In particular the zero vector is coplanar with any two vectors.

Vectors which have magnitude 1 are called *unit vectors*; the unit vector which is parallel to $\mathbf{a}$ and has the same sense will be denoted by $\hat{\mathbf{a}}$. Notice that $\mathbf{a} = a\hat{\mathbf{a}}$.

We are now able to prove the *Unique Components Theorem*:

> (i) If $\mathbf{a}, \mathbf{b}$ are any two non-parallel, non-zero vectors in a plane, then any vector $\mathbf{v}$ in that plane can be expressed uniquely as $\alpha\mathbf{a} + \beta\mathbf{b}$.
>
> (ii) If $\mathbf{a}, \mathbf{b}, \mathbf{c}$ are any three non-coplanar, non-zero vectors, then any vector $\mathbf{v}$ can be expressed uniquely as $\alpha\mathbf{a} + \beta\mathbf{b} + \gamma\mathbf{c}$.

(i) We prove separately that $\mathbf{v}$ can be expressed as $\alpha\mathbf{a} + \beta\mathbf{b}$, and that the expression so obtained is unique; the uniqueness part of the proof will, however, be left as an exercise for the reader, who should not attempt it until he has read the proof of part (ii).

Let $\overrightarrow{PQ}, \overrightarrow{PR}, \overrightarrow{PS}$ represent $\mathbf{v}, \mathbf{a}, \mathbf{b}$ respectively. Let R' be on PR (possibly PR produced), such that $R'Q$ is parallel to PS; then $\overrightarrow{PR'}$ represents a vector $\alpha\mathbf{a}$, where α is some scalar; $R'Q$ is parallel to $PS \Rightarrow \overrightarrow{R'Q}$ represents a vector $\beta\mathbf{b}$ for some scalar β. Thus $\mathbf{v} = \alpha\mathbf{a} + \beta\mathbf{b}$.

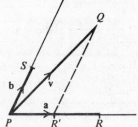

(ii) Again we divide the proof into two stages, and prove (a) that the required expression exists, and (b) that it is unique.

(a) *Existence*: Let $\mathbf{v}, \mathbf{a}, \mathbf{b}, \mathbf{c}$ be represented by $\vec{PQ}, \vec{PR}, \vec{PS}, \vec{PT}$ respectively.

Let U be a point in the plane PRS such that UQ is parallel to PT; then $\vec{UQ}$ represents a vector $\gamma\mathbf{c}$, for some scalar γ, and $\vec{PU}$ represents some vector $\mathbf{v}_1$, coplanar with $\mathbf{a}, \mathbf{b}$.

Therefore $\mathbf{v} = \mathbf{v}_1 + \gamma\mathbf{c}$
$$= \alpha\mathbf{a} + \beta\mathbf{b} + \gamma\mathbf{c}$$
by part (i) of this theorem.

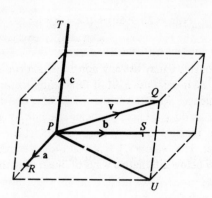

(b) *Uniqueness*: Suppose $\mathbf{v}$ can be expressed in two supposedly different ways.

That is $\mathbf{v} = \alpha_1\mathbf{a} + \beta_1\mathbf{b} + \gamma_1\mathbf{c}$

and $\mathbf{v} = \alpha_2\mathbf{a} + \beta_2\mathbf{b} + \gamma_2\mathbf{c}$ are alternative expressions for $\mathbf{v}$.

Subtracting: $\mathbf{0} = (\alpha_1 - \alpha_2)\mathbf{a} + (\beta_1 - \beta_2)\mathbf{b} + (\gamma_1 - \gamma_2)\mathbf{c}$

$\Leftrightarrow -(\alpha_1 - \alpha_2)\,\mathbf{a} = (\beta_1 - \beta_2)\mathbf{b} + (\gamma_1 - \gamma_2)\mathbf{c}$ \hfill (1)

The vector $-(\alpha_1 - \alpha_2)\mathbf{a}$ is either $\mathbf{0}$ or is parallel to $\mathbf{a}$ (which is not coplanar with $\mathbf{b}$ and $\mathbf{c}$); the vector $(\beta_1 - \beta_2)\mathbf{b} + (\gamma_1 - \gamma_2)\mathbf{c}$ is either $\mathbf{0}$ or is coplanar with $\mathbf{b}$ and $\mathbf{c}$. Thus the only possibility is that $-(\alpha_1 - \alpha_2)\mathbf{a} = \mathbf{0}$, which means that $\alpha_1 = \alpha_2$.

But equation (1) can be rearranged with $-(\beta_1 - \beta_2)\mathbf{b}$ on the left, and a similar argument shows that $\beta_1 = \beta_2$; similarly $\gamma_1 = \gamma_2$. This completes the proof that the expression for $\mathbf{v}$ is unique.

Qu. 1.   Complete the uniqueness part of the proof of part (i) of this theorem.

In the three-dimensional case, we are in effect forming a parallelepiped which has $\mathbf{v}$ as a diagonal, with all its edges parallel to $\mathbf{a}, \mathbf{b}, \mathbf{c}$. We have just proved that given $\mathbf{a}, \mathbf{b}, \mathbf{c}, \mathbf{v}$ this can be done in one and only one way, provided that $\mathbf{a}, \mathbf{b}, \mathbf{c}$ are not coplanar, nor zero; $\mathbf{a}, \mathbf{b}, \mathbf{c}$ can be of any magnitude (except zero), and inclined to each other at any angle (except zero and π).

The vectors $\alpha\mathbf{a}, \beta\mathbf{b}, \gamma\mathbf{c}$ are called *component vectors* of $\mathbf{v}$; the scalars α, β, γ are called the *components* of $\mathbf{v}$ relative to the *base vectors* $\mathbf{a}, \mathbf{b}, \mathbf{c}$; we say that $\mathbf{v}$ has been *resolved* into its component vectors.

The various properties summarized on page 60 allow us to add, subtract, or multiply vectors by scalars by adding, subtracting or multiplying by scalars the component vectors. For vectors to be equal, their respective components must be equal. For example

$$(3a - 2b + 5c) + (6a + b - 7c) = 9a - b - 2c$$
$$4(3a - 2b + 5c) = 12a - 8b + 20c.$$

Although we may use any non-zero, non-coplanar vectors as the base vectors, we usually use a set of three units vectors **i**, **j**, **k** which are mutually perpendicular. We often think of **i** as being one unit long going from left to right across the page, and **j** as being one unit long pointing towards the top of the page; then **k** is one unit long, perpendicular to the page, coming up out of the paper. The vectors **i**, **j**, **k** in that order are said to form a right-handed triad. We show below alternative ways of illustrating these vectors.

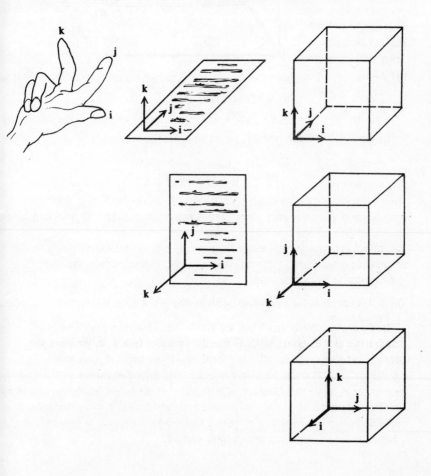

In three dimensions, the Cartesian system of coordinates (x, y, z) requires three perpendicular axes; the x-axis, the y-axis and the z-axis (in that order) also form a right-handed triad. It is conventional to take $\mathbf{i}$, $\mathbf{j}$, $\mathbf{k}$ along the

positive x, y, z axes respectively, as shown in the diagram. Any point in space is specified by its own (unique) coordinates. If, for example, the point P has coordinates $(2, 4, 3)$ and O is the origin, then

$$\mathbf{OP} = \mathbf{OA} + \mathbf{AB} + \mathbf{BP}$$
$$= 2\mathbf{i} + 4\mathbf{j} + 3\mathbf{k}.$$

This can be written more compactly using column vector notation:

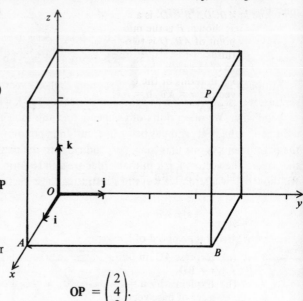

$$\mathbf{OP} = \begin{pmatrix} 2 \\ 4 \\ 3 \end{pmatrix}.$$

In general, if P is the point (x, y, z) and O is the origin

$$\mathbf{OP} = x\mathbf{i} + y\mathbf{j} + z\mathbf{k} = \begin{pmatrix} x \\ y \\ z \end{pmatrix}.$$

In particular $\mathbf{i} = \begin{pmatrix} 1 \\ 0 \\ 0 \end{pmatrix}$, $\mathbf{j} = \begin{pmatrix} 0 \\ 1 \\ 0 \end{pmatrix}$, $\mathbf{k} = \begin{pmatrix} 0 \\ 0 \\ 1 \end{pmatrix}$. Column vector notation is used only

when the base vectors are $\mathbf{i}$, $\mathbf{j}$, $\mathbf{k}$.

As before, we can add, subtract, or multiply a vector by a scalar by adding, subtracting or multiplying by a scalar the component vectors. Equally, we can carry out the same operations by dealing only with the components.

For example

$$(3\mathbf{i} + 7\mathbf{j} - 2\mathbf{k}) + (4\mathbf{i} - 12\mathbf{j} - 8\mathbf{k}) = 7\mathbf{i} - 5\mathbf{j} - 10\mathbf{k},$$

$$\begin{pmatrix} 3 \\ 7 \\ -2 \end{pmatrix} + \begin{pmatrix} 4 \\ -12 \\ -8 \end{pmatrix} = \begin{pmatrix} 7 \\ -5 \\ -10 \end{pmatrix}, \begin{pmatrix} p \\ q \\ r \end{pmatrix} - \begin{pmatrix} x \\ y \\ z \end{pmatrix} = \begin{pmatrix} p-x \\ q-y \\ r-z \end{pmatrix},$$

$$\tfrac{1}{3} \begin{pmatrix} 6 \\ -2 \\ 9 \end{pmatrix} = \begin{pmatrix} 2 \\ -\tfrac{2}{3} \\ 3 \end{pmatrix}.$$

Qu. 2. Prove that $a = \begin{pmatrix} x \\ y \\ x \end{pmatrix} \Rightarrow a = \sqrt{(x^2+y^2+z^2)}$ and $\hat{a} = \begin{pmatrix} x/\sqrt{(x^2+y^2+z^2)} \\ y/\sqrt{(x^2+y^2+z^2)} \\ z/\sqrt{(x^2+y^2+z^2)} \end{pmatrix}$.

Example 1: $ABCDA'B'C'D'$ is a cuboid as shown. P is the mid point of AB; Q is two-thirds of the way from D' to C'. Express the vector **PQ** in terms of the base vectors $\mathbf{a} = \mathbf{AB}$, $\mathbf{b} = \mathbf{AD}$, $\mathbf{c} = \mathbf{AA'}$.

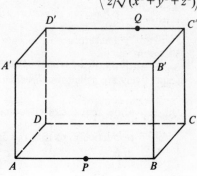

Solution: $\mathbf{PQ} = \mathbf{PB} + \mathbf{BB'} + \mathbf{B'C'} + \mathbf{C'Q}$

$\qquad = \frac{1}{2}\mathbf{a} + \mathbf{c} + \mathbf{b} + (-\frac{1}{3}\mathbf{a})$

$\qquad = \frac{1}{6}\mathbf{a} + \mathbf{b} + \mathbf{c}.$ $\qquad\qquad\qquad$ □

Example 2: Using the cuboid of Example 1:

   (a) Express $\mathbf{AC'}$ in terms of the base vectors $\mathbf{p} = \mathbf{PA}$, $\mathbf{q} = \mathbf{PB'}$, $\mathbf{r} = \mathbf{BD}$.

   (b) Explain why $\mathbf{u} = \mathbf{PA'}$, $\mathbf{v} = \mathbf{CD'}$, $\mathbf{w} = \mathbf{B'B}$ cannot be used as a set of base vectors.

Solution: (a) Find a route from A to C' by directions parallel to the base vectors. One such route is $A \to P \to B' \to D' \to C'$.

$\qquad$ Therefore $\mathbf{AC'} = \mathbf{AP} + \mathbf{PB'} + \mathbf{B'D'} + \mathbf{D'C'}$

$\qquad\qquad\qquad\quad = -\mathbf{p} + \mathbf{q} + \mathbf{r} + (-2\mathbf{p}) = -3\mathbf{p} + \mathbf{q} + \mathbf{r}.$

   (b) $\mathbf{u} = \mathbf{PA'}$, $\mathbf{v} = \mathbf{CD'} = \mathbf{BA'}$, $\mathbf{w} = \mathbf{B'B}$

$\qquad \overrightarrow{PA'}, \overrightarrow{BA'}, \overrightarrow{B'B}$ all lie in plane $ABB'A' \Rightarrow \mathbf{u}, \mathbf{v}, \mathbf{w}$ are coplanar.

$\qquad\qquad\qquad\qquad\qquad\qquad\qquad\qquad\qquad\qquad\qquad\qquad\quad$ □

Example 3: Let $\mathbf{a} = 2\mathbf{i} + 2\mathbf{j} - \mathbf{k}$. Find the magnitude of $\mathbf{a}$; find $\hat{\mathbf{a}}$; find the cosines of the angles $\mathbf{a}$ makes with the positive x, y, z axes respectively.

Solution: $a = \sqrt{(2^2 + 2^2 + (-1)^2)} = \sqrt{9} = 3$

$\hat{\mathbf{a}} = \dfrac{\mathbf{a}}{a} = \frac{2}{3}\mathbf{i} + \frac{2}{3}\mathbf{j} - \frac{1}{3}\mathbf{k}.$

The required angles are $P\hat{O}x, P\hat{O}y, P\hat{O}z$ respectively, of which the first two are acute, and the third obtuse. Since $OP = 3$ and the **i**-component of **OP** is 2,

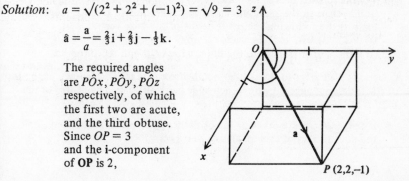

$$\cos P\hat{O}x = \tfrac{2}{3}.$$

Similarly $$\cos P\hat{O}y = \tfrac{2}{3}.$$

OP makes an acute angle with the negative z-axis; the cosine of this angle is $\tfrac{1}{3}$.

Therefore $$\cos P\hat{O}x = -\tfrac{1}{3}. \qquad \square$$

The cosines of the angles between the vector **a** and the unit vectors **i, j, k** respectively are called the *direction cosines* of **a**.

Qu. 3. Show that the direction cosines of $\mathbf{a} = x\mathbf{i} + y\mathbf{j} + z\mathbf{k}$ are $\dfrac{x}{a}, \dfrac{y}{a}, \dfrac{z}{a}$

respectively, so that the direction cosines of **a** are the components of $\hat{\mathbf{a}}$.

Exercise 3B

Questions 1—4 refer to the parallelepiped *ABCDEFGH* as shown in the diagram.

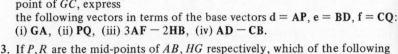

1. Express the following vectors in terms of the base vectors **a** = **AC**, **b** = **AD**, **c** = **DH**:
   (i) **AF**, (ii) **DF**, (iii) **GA**, (iv) **HB**.

2. If P is the mid-point of AB, Q is the mid-point of GC, express the following vectors in terms of the base vectors **d** = **AP**, **e** = **BD**, **f** = **CQ**:
   (i) **GA**, (ii) **PQ**, (iii) $3\mathbf{AF} - 2\mathbf{HB}$, (iv) $\mathbf{AD} - \mathbf{CB}$.

3. If P, R are the mid-points of AB, HG respectively, which of the following sets of vectors could not be used as a set of base vectors?
   (i) {**AB, FB, GD**}, (ii) {**FG, DH, PR**}, (iii) {**GB, FG, PR**}.

4. Let the lengths of AB, AD, AE be 6, 2, 3 units respectively. If **p, q, r** are unit vectors in the same direction and sense as **BC, GC, EF** respectively, express the following vectors in terms of the base vectors **p, q, r**:
   (i) **AG**, (ii) **DB**, (iii) the vector from G to the mid-point of AB,
   (iv) the vector from H to the point where DB and AC intersect.

Questions 5—8 refer to the regular octahedron *ABCDEF*, as shown at the top of the following page, in which each edge is 1 unit long. The centre of the solid is at O.

5. If OA, OB, OE are taken as the positive x, y, z axes, respectively, find the coordinates of each of the six vertices.

6. If $\mathbf{AD} = \mathbf{i}$, $\mathbf{DC} = \mathbf{j}$,
   express $\mathbf{k}$ in terms of
   a directed line seg-
   ment from the dia-
   gram, and express
   the following vectors
   in terms of $\mathbf{i}$, $\mathbf{j}$, $\mathbf{k}$:
   (i) $\mathbf{DF}$, (ii) $\mathbf{AC}$,
   (iii) $\mathbf{OB}$, (iv) $\mathbf{EF}$.

7. Use the letters of the
   diagram to answer
   this question:
   (a) If $\mathbf{i} = \sqrt{2}\ \mathbf{OB}$,
       $\mathbf{j} = \sqrt{2}\ \mathbf{OF}$, name
       a vector equal to $\mathbf{k}$.
   (b) If $\mathbf{k} = \mathbf{AB}$, $\mathbf{j} = \mathbf{DA}$
       name a vector
       equal to $\mathbf{i}$.
   (c) If $\mathbf{i} = \mathbf{CD}$,
       $\mathbf{k} = \sqrt{2}\ \mathbf{OE}$, name
       a vector equal to $\mathbf{j}$.

8. Using $\mathbf{a} = \mathbf{DA}$, $\mathbf{c} = \mathbf{DC}$,
   $\mathbf{e} = \mathbf{DE}$ as the base
   vectors, resolve the
   following into their
   component vectors:
   (i) $\mathbf{EF}$, (ii) $\mathbf{AC}$,
   (iii) $\mathbf{AP}$ where $\mathbf{FP} = \mathbf{PC}$, (iv) $\mathbf{BR}$ where $\mathbf{CR} = -3\ \mathbf{BE}$.

9. If $\mathbf{a} = 3\mathbf{i} - 7\mathbf{j} + 5\mathbf{k}$, $\mathbf{b} = 4\mathbf{i} + 11\mathbf{j} - 2\mathbf{k}$, $\mathbf{c} = 6\mathbf{i} - 2\mathbf{j} + 5\mathbf{k}$ find
   (i) $\mathbf{a} + \mathbf{b} - \mathbf{c}$, (ii) $\mathbf{c} - 2\mathbf{a}$, (iii) $\hat{\mathbf{a}}$, (iv) $\hat{\mathbf{c}}$ in terms of $\mathbf{i}$, $\mathbf{j}$, $\mathbf{k}$.

10. If $\mathbf{d} = \begin{pmatrix} 9 \\ -12 \\ 20 \end{pmatrix}$, find $\hat{\mathbf{d}}$, and the angles $\mathbf{d}$ makes with the positive x, y, z axes
    respectively.

11. What are the direction cosines of $\mathbf{e} = 3\mathbf{i} - 4\mathbf{j} + 12\mathbf{k}$ and $\mathbf{f} = \begin{pmatrix} -12 \\ 16 \\ 15 \end{pmatrix}$?

12. Show that the sum of the squares of the direction cosines of any vector
    is 1.

13. Show that the following sets of vectors are not suitable vectors to be used
    as a base set:
    (i) $\mathbf{p} = \begin{pmatrix} 3 \\ 7 \\ 1 \end{pmatrix}$, $\mathbf{q} = \begin{pmatrix} 4 \\ -2 \\ -6 \end{pmatrix}$, $\mathbf{r} = \begin{pmatrix} 17 \\ 17 \\ -9 \end{pmatrix}$;
    (ii) $\mathbf{p} = \begin{pmatrix} 4 \\ 1 \\ 3 \end{pmatrix}$, $\mathbf{q} = \begin{pmatrix} 6 \\ -1 \\ -1 \end{pmatrix}$, $\mathbf{r} = \begin{pmatrix} -11 \\ 1 \\ 0 \end{pmatrix}$;

(iii) $\mathbf{p} = \begin{pmatrix} 3 \\ 2 \\ 7 \end{pmatrix}$, $\mathbf{q} = \begin{pmatrix} 1 \\ -1 \\ -1 \end{pmatrix}$, $\mathbf{r} = \begin{pmatrix} 1 \\ 1 \\ 3 \end{pmatrix}$.

14. The non-zero vectors $\mathbf{a} = \begin{pmatrix} x_1 \\ y_1 \\ z_1 \end{pmatrix}$, $\mathbf{b} = \begin{pmatrix} x_2 \\ y_2 \\ z_2 \end{pmatrix}$ are represented by $\overrightarrow{PQ}, \overrightarrow{QR}$

respectively. Use Pythagoras' Theorem to show that

$$\mathbf{a}, \mathbf{b} \text{ are perpendicular} \Rightarrow x_1 x_2 + y_1 y_2 + z_1 z_2 = 0.$$

Is the converse also true?

15. Prove that $\mathbf{a} = \begin{pmatrix} \frac{2}{7} \\ \frac{3}{7} \\ -\frac{6}{7} \end{pmatrix}$, $\mathbf{b} = \begin{pmatrix} \frac{6}{7} \\ \frac{2}{7} \\ \frac{3}{7} \end{pmatrix}$, $\mathbf{c} = \begin{pmatrix} -\frac{3}{7} \\ \frac{6}{7} \\ \frac{2}{7} \end{pmatrix}$ are mutually perpendicular

unit vectors (hint: see Qu. 14).
Do they form a right-handed triad, taken in the order $\mathbf{a}$, $\mathbf{b}$, $\mathbf{c}$?

3.3 Lines in three dimensions

If O and P are two points in space, then the vector **OP** is called the *position vector of P relative to O*. If there is no possibility of ambiguity we often drop the phrase 'relative to O'. We normally denote points by capital letters and we will adopt the convention that the position vector of a point is denoted by the corresponding small letter, so that **p** is the position vector of P, **a** is the position vector of A, and so on, all relative to the same point O, known as the origin. (The one exception to this rule is that **0** is the position vector of O.) We will often refer to 'the point **p**' meaning 'the point with position vector **p**'.

By the rule for addition

$$\mathbf{PQ} = \mathbf{PO} + \mathbf{OQ} = -\mathbf{p} + \mathbf{q} = \mathbf{q} - \mathbf{p}.$$

Similarly $\mathbf{DM} = \mathbf{m} - \mathbf{d}$
and $\mathbf{TH} = \mathbf{h} - \mathbf{t}$.

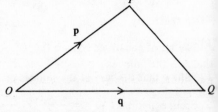

A particular straight line in space can be precisely specified in a variety of ways: (i) by means of two points on the line,
        (ii) by means of one point on the line and the direction of the line,
        (iii) as the intersection of two planes.
We will use method (ii): we want to find an expression for the position vector of any point on the straight line through the point A with position vector **a**, the line being parallel to the vector **d**.

If R with position vector $\mathbf{r}$ is an arbitrary
point on the line, then
$\mathbf{AR}$ is parallel to $\mathbf{d}$ (or else $\mathbf{AR} = \mathbf{0}$)

    $\Leftrightarrow$ $\mathbf{AR} = \lambda\mathbf{d}$ where λ is a scalar

    $\Leftrightarrow$ $\mathbf{r} - \mathbf{a} = \lambda\mathbf{d}$

    $\Leftrightarrow$ $\mathbf{r} = \mathbf{a} + \lambda\mathbf{d}$.

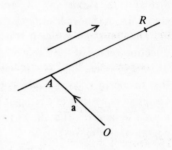

This is the *vector equation* of the line
through A parallel to $\mathbf{d}$. The scalar λ is a
parameter, and may take any real value,
including zero. Each value of λ corresponds
to a distinct point on the line, distant $|\lambda|d$
from A. Positive values of λ give points on
one side of A, while negative values correspond to points on the other side of
A. Conversely, each point on the line corresponds to a unique real value of λ.

The vector $\mathbf{d}$ is called a *direction vector* of the line; the direction cosines
of $\mathbf{d}$ are known as the *direction cosines* of the line.

Suppose the arbitrary point R has Cartesian coordinates (x, y, z) with
respect to axes through O. Let the corresponding coordinates for A be

(x_1, y_1, z_1). Let $\mathbf{d}$ be the column vector $\begin{pmatrix} l \\ m \\ n \end{pmatrix}$. Then $\mathbf{r} = \begin{pmatrix} x \\ y \\ z \end{pmatrix}$ and $\mathbf{a} = \begin{pmatrix} x_1 \\ y_1 \\ z_1 \end{pmatrix}$,

so that the vector equation of the line becomes

$$\begin{pmatrix} x \\ y \\ z \end{pmatrix} = \begin{pmatrix} x_1 \\ y_1 \\ z_1 \end{pmatrix} + \lambda \begin{pmatrix} l \\ m \\ n \end{pmatrix}.$$

Equating components we get
$$\left. \begin{array}{l} x = x_1 + \lambda l \\ y = y_1 + \lambda m \\ z = z_1 + \lambda n \end{array} \right\} \quad \text{(A)}$$

The set of equations (A) are the *parametric equations* of the line; if we say
that R is the point $(x_1 + \lambda l, y_1 + \lambda m, z_1 + \lambda n)$, we are expressing the coordinates of R in *terms of the parameter* λ.

From equations (A) $\lambda = \dfrac{x - x_1}{l} = \dfrac{y - y_1}{m} = \dfrac{z - z_1}{n}$, assuming for the

moment that none of l, m, n is zero.

$\dfrac{x - x_1}{l} = \dfrac{y - y_1}{m} = \dfrac{z - z_1}{n}$ are the *Cartesian equations* of the straight

line through (x_1, y_1, z_1) in the direction of the vector $\begin{pmatrix} l \\ m \\ n \end{pmatrix}$.

If any of l, m, n is zero, the Cartesian equations of the straight line do not seem to make sense. Suppose $l = 0$; then the line we are thinking of is perpendicular to the x-axis, and all points on that line have the same x-coordinate; equations (A) above lead to the same result, $x = x_1$. Similar results hold if m or n is zero, but notice that we cannot have $l = m = n = 0$, for then the direction vector of the line would be the zero vector (which has no direction) and the so-called 'line' would have only one point. We adopt the convention that the occurrence of a zero as the denominator of an expression in the Cartesian equations of a straight line implies that the corresponding numerator must also be zero. This allows us to say that *all* straight lines have Cartesian equations of the form

$$\frac{x - x_1}{l} = \frac{y - y_1}{m} = \frac{z - z_1}{n}.$$

Qu. 1. Write the Cartesian equations of the following lines in the above form:

   (i) $x = 3 + 2\lambda, y = 4, z = 5 - \lambda$
   (ii) $x = 5 + 4\lambda, y = 3, z = -2$
   (iii) the line through $(1, -4, 5)$ with direction vector $2\mathbf{j} - 3\mathbf{k}$
   (iv) the line through $(-6, 7, 10)$ parallel to the x-axis
   (v)   the z-axis.

Qu. 2. Give the coordinates of the arbitrary point on these lines in terms of the parameter λ:

   (i) $\dfrac{x - 1}{2} = \dfrac{y + 3}{4} = \dfrac{z - 5}{0}$

   (ii) $\dfrac{x + 4}{0} = \dfrac{y - 8}{0} = \dfrac{z - 2}{9}$

   (iii) $\dfrac{x}{0} = \dfrac{y}{1} = \dfrac{z}{0}.$

Example 1: Find the Cartesian equations of the line through $(6, 3, -5)$ in the specified directions: (a) in the direction of the vector $\begin{pmatrix} 4 \\ -8 \\ 7 \end{pmatrix}$

   (b) parallel to the line $\dfrac{x}{3} = \dfrac{y - 10}{-2} = \dfrac{z + 8}{13}$

   (c) perpendicular to the y-axis, but inclined at $60°$, $30°$ to the positive x and z axes respectively.

Solution: (a) By inspection the equations are

$$\frac{x - 6}{4} = \frac{y - 3}{-8} = \frac{z + 5}{7}.$$

(b) The given line and the line we are to find must have the same direction, specified by the direction vector $\begin{pmatrix} 3 \\ -2 \\ 13 \end{pmatrix}$. So the equations we want are

$$\frac{x-6}{3} = \frac{y-3}{-2} = \frac{z+5}{13}.$$

(c) Since the line we are to find is inclined at $60°, 90°, 30°$ to the positive x, y, z axes respectively, it has direction cosines $\frac{1}{2}, 0, \frac{\sqrt{3}}{2}$ respectively. Therefore a unit direction vector for the line is $\begin{pmatrix} \frac{1}{2} \\ 0 \\ \frac{\sqrt{3}}{2} \end{pmatrix}$; another direction vector (not a unit vector) is $\begin{pmatrix} 1 \\ 0 \\ \sqrt{3} \end{pmatrix}$.

There are infinitely many possible direction vectors, all scalar multiples of these.

So the equations of the straight line through $(6, 3, -5)$ are

$$\frac{x-6}{1} = \frac{y-3}{0} = \frac{z+5}{\sqrt{3}}. \qquad \square$$

Note Some writers prefer to give the equations obtained at the end of part (c) as the pair of equations $\dfrac{x-6}{1} = \dfrac{z+5}{\sqrt{3}}$, $y = 3$.

Our next example illustrates various ways of finding the coordinates of points on a line with stated Cartesian equations.

Example 2: Find the coordinates of three points on the line

$$\frac{x-5}{4} = \frac{y+7}{3} = \frac{z-2}{-2}.$$

Solution (i):   By inspection $x = 5, y = -7, z = 2$ satisfy the equations, so $(5, -7, 2)$ is a point on the line. $\qquad \square$

Solution (ii):   Choose any value for x (or y or z) and calculate the other corresponding coordinates. For example, if $x = 9$ then $y = -4$ and $z = 0$. That is $(9, -4, 0)$ is a point on the line.
Notice that if the equation had contained the expression $\dfrac{x-5}{0}$ then 5 is the x-coordinate of all points on the line, so we could not 'choose' a value for x, and substitution of $x = 5$ would not lead us anywhere. $\qquad \square$

Solution (iii):   Let $\lambda = \dfrac{x-5}{4} = \dfrac{y+7}{3} = \dfrac{z-2}{-2}$. Choose values for λ and calculate the corresponding values for x, y, z.
For example $\lambda = -3 \Rightarrow x = -7, y = -16, z = 8$.
Therefore $(-7, -16, 8)$ is a point on the line. $\qquad \square$

Example 3: Describe the line $\dfrac{x+4}{0} = \dfrac{y - \frac{3}{4}}{5} = \dfrac{z - \frac{3}{2}}{0}$.

Solution: A direction vector for this line is $\begin{pmatrix} 0 \\ 5 \\ 0 \end{pmatrix}$, so the direction cosines are

0, 1, 0 respectively. The line is perpendicular to both the x and z axes, and is parallel to the y-axis. It goes through the point $(-4, \frac{3}{4}, \frac{3}{2})$. A general point on the line is $(-4, \lambda, \frac{3}{2})$ where λ may take any value.   □

Example 4: Find the vector equation and the Cartesian equations of the line through A $(3, -7, -4)$ and B $(5, 2, 2)$.

Solution: $\mathbf{AB} = \mathbf{b} - \mathbf{a} = \begin{pmatrix} 2 \\ 9 \\ 6 \end{pmatrix}$. This vector is in the direction of the line and

may be used as a direction vector of the line. The line through A

with direction $\begin{pmatrix} 2 \\ 9 \\ 6 \end{pmatrix}$ has

the vector equation $\qquad \mathbf{r} = \begin{pmatrix} 3 \\ -7 \\ -4 \end{pmatrix} + \lambda \begin{pmatrix} 2 \\ 9 \\ 6 \end{pmatrix}$,

and Cartesian equations $\qquad \dfrac{x-3}{2} = \dfrac{y+7}{9} = \dfrac{z+4}{6}$.

These are the required equations.   □

Neither the vector equation nor the Cartesian equations of a straight line are unique. Two sets of Cartesian equations represent the same straight line if and only if both the following conditions are satisfied:
- (i) a set of coordinates which satisfies one of the sets of equations also satisfies the other set of equations;
- (ii) a direction vector of the line represented by one set of equations is a non-zero scalar multiple of a direction vector of the line represented by the other set of equations.

Similar conditions apply to the vector equations of straight lines. Where possib we use direction vectors which have integral components with no common factors.

Two distinct lines in a plane are either parallel or intersecting; in three dimensions there are three possibilities: they may be parallel, or intersecting, or *skew*—that is neither parallel nor intersecting.

We now illustrate how to find whether two lines intersect, and if they do, the coordinates of the intersection point. A similar method can be used if the equations are given in vector form.

Example 5: $\dfrac{x+9}{4} = \dfrac{y+5}{1} = \dfrac{z+1}{-2}$ (1)

$\dfrac{x-8}{-5} = \dfrac{y-2}{-4} = \dfrac{z-5}{8}$ (2)

$\dfrac{x-8}{-5} = \dfrac{y-2}{-4} = \dfrac{z+15}{8}$ (3)

(i) Find where the lines represented by equations (1) and (2) intersect.

(ii) Find where the lines represented by equations (1) and (3) intersect.

Solution: (i) If there is a point (x_1, y_1, z_1) lying on both lines, then

$$\frac{x_1+9}{4} = \frac{y_1+5}{1} = \frac{z_1+1}{-2} = \lambda \text{ say}$$

and $\dfrac{x_1-8}{-5} = \dfrac{y_1-2}{-4} = \dfrac{z_1-5}{8} = \mu \text{ say}.$

Therefore $x_1 = \quad 4\lambda - 9 = -5\mu + 8$ (4)

$y_1 = \quad \lambda - 5 = -4\mu + 2$ (5)

$z_1 = -2\lambda - 1 = \quad 8\mu + 5.$ (6)

Solving equations (4) and (5) simultaneously: $\lambda = 3, \mu = 1$. These values do not satisfy equation (6), so we conclude that the lines represented by equations (1) and (2) do not intersect.

(ii) As in part (i), suppose that $\lambda = \dfrac{x_1+9}{4} = \dfrac{y_1+5}{1} = \dfrac{z_1+1}{-2}$

and $\mu = \dfrac{x_1-8}{-5} = \dfrac{y_1-2}{-4} = \dfrac{z_1+15}{8}$

Then $x_1 = \quad 4\lambda - 9 = -5\mu + 8$ (7)

$y_1 = \quad \lambda - 5 = -4\mu + 2$ (8)

$z_1 = -2\lambda - 1 = \quad 8\mu - 15$ (9)

Solving equations (7) and (8) simultaneously: $\lambda = 3, \mu = 1$. These values do satisfy equation (9). The values of x_1, y_1, z_1 are those obtained from equations (7), (8), (9) using $\lambda = 3$ (or $\mu = 1$). Thus the lines represented by equations (1) and (3) intersect at the point $(3, -2, -7)$.   □

Qu. 3. Example 5 has shown that the same method can be used whether or not two lines intersect. Use this method with the lines

$$\frac{x+5}{-12} = \frac{y+4}{-3} = \frac{z+3}{6} \quad \text{and} \quad \frac{x+9}{4} = \frac{y+5}{1} = \frac{z+1}{-2}.$$

Comment on your results.

Exercise 3C

1. Find the Cartesian equations of the following lines:

   (i) through $(5, -2, 6)$ in the direction $\begin{pmatrix} 3 \\ 1 \\ 4 \end{pmatrix}$

   (ii) through $(2, -3, -7)$ parallel to $\dfrac{x+5}{7} = \dfrac{y-13}{3} = \dfrac{z-4}{-2}$

   (iii) through $(-4, 1, 0)$ parallel to the z-axis
   (iv) through $(5, -1, -3)$ perpendicular to the x-axis, but equally inclined to the positive y and z axes
   (v) through $(-3, 0, 0)$ inclined at $60°$ to the positive x-axis, at $45°$ to the positive y-axis, and at an acute angle to the positive z-axis
   (vi) the y-axis.

2. Find the Cartesian equations of the lines joining the following pairs of points:

   (i) $(3, 2, -7), (5, -13, -4)$ (ii) $(-8, -13, -9), (12, 7, 1)$
   (iii) $(3, 0, 0), (0, 0, 5)$ (iv) $(0, 0, 0), (-10, 4, -6)$
   (v) $(a, 3a, 2a), (2a, a, -a)$ (vi) $(a, a^2, a^2 + b^2), (b, b^2, 2ab)$

3. Find the vector equations of the lines of Question 2.

4. Let $ABCD$ be a parallelogram. Find the vector equations of the following lines in terms of $\mathbf{a}$, $\mathbf{b}$ and $\mathbf{c}$:
   (i) AB (ii) AD (iii) AC (iv) BD.

5. Describe the inclination of the following lines to the axes.

   (i) $\dfrac{x}{0} = \dfrac{y}{1} = \dfrac{z}{0}$ (ii) $\dfrac{x}{1} = \dfrac{y}{0} = \dfrac{z}{1}$

   (iii) $\dfrac{x}{1} = \dfrac{y}{0} = \dfrac{z}{0}$ (iv) $\dfrac{x}{1} = \dfrac{y}{-1} = \dfrac{z}{1}$

   (v) $\dfrac{x}{-1} = \dfrac{y}{-1} = \dfrac{z}{1}$.

6. For each of the following pairs of lines state whether the lines intersect, are parallel, are skew, or are coincident. If the lines intersect, find the coordinates of the intersection point.

   (i) $\dfrac{x+7}{5} = \dfrac{y+3}{-2} = \dfrac{z+12}{8}, \dfrac{x-17}{7} = \dfrac{y+1}{3} = \dfrac{z+4}{-4}$

   (ii) $\dfrac{x+5}{3} = \dfrac{y+30}{7} = \dfrac{z-1}{1}, \dfrac{x-8}{4} = \dfrac{y-6}{-2} = \dfrac{z+5}{3}$

   (iii) $\dfrac{x-11}{5} = \dfrac{y+7}{4} = \dfrac{z-2}{-3}, \dfrac{x-1}{-5} = \dfrac{y-1}{-4} = \dfrac{z+1}{3}$

(iv) $\dfrac{x+3}{2} = \dfrac{y+5}{-1} = \dfrac{z-7}{9}, \dfrac{x-11}{-2} = \dfrac{y+12}{1} = \dfrac{z-70}{-9}$

(v) $\dfrac{x-15}{3} = \dfrac{y+15}{-2} = \dfrac{z-30}{6}$, the y-axis.

7. The line with Cartesian equations $\dfrac{x+5}{2} = \dfrac{y-14}{-10} = \dfrac{z+13}{11}$ meets

$\dfrac{x-3}{2} = \dfrac{y+5}{-3} = \dfrac{z+17}{-5}$ at A and meets $\dfrac{x}{1} = \dfrac{y+5}{1} = \dfrac{z-7}{-2}$ at B.

Calculate the length of AB.

8. Show that the three lines $\dfrac{x-3}{5} = \dfrac{y-2}{0} = \dfrac{z+5}{12}$

$$\dfrac{x}{2} = \dfrac{y-18}{-4} = \dfrac{z+9}{4}$$

$$\dfrac{x-2}{1} = \dfrac{y+6}{8} = \dfrac{z+9}{4}$$

form the sides of a triangle, and find the lengths of all three sides.

9. (a) Show that $\mathbf{r} = \mathbf{a} + \lambda\,(\hat{\mathbf{p}} + \hat{\mathbf{q}})$ is the vector equation of the internal bisector of angle BAC where $\mathbf{p} = \mathbf{AB}$ and $\mathbf{q} = \mathbf{AC}$.

   (b) What is the vector equation of the external bisector of angle BAC?

10. Find the Cartesian equation of the internal bisector of angle BAC where A, B, C have the given coordinates:
    (i) $A\,(2, 3, 7)$, $B\,(4, 3, 7)$, $C\,(0, 1, 8)$
    (ii) $A\,(4, -5, 3)$, $B\,(-3, -1, -1)$, $C\,(1, 1, -3)$
    (iii) $A\,(3, -9, -5)$, $B\,(5, 5, 0)$, $C\,(1, 2, 5)$
    (iv) $A\,(0, 1, 6)$, $B\,(3, 2, 6)$, $C\,(1, 0, 5)$.

11. Find the incentre of triangle ABC where A, B, C have the coordinates $(3, 4, -5)$, $(3, 4, 6)$, $(7, 12, -13)$ respectively. (The *incentre* is the point where the internal angle bisectors meet.)

12. Given that $\dfrac{x+4}{3} = \dfrac{y-7}{-2} = \dfrac{z-b}{a}$

    and $\dfrac{x}{2} = \dfrac{y+5}{1} = \dfrac{z-b-4}{2a}$

intersect at P, find a and show that b may take any real value. Find also the coordinates of P, in terms of b.

13. The lines $\dfrac{x-3}{-2} = \dfrac{y}{2} = \dfrac{z}{3}, \dfrac{x}{3} = \dfrac{y-7}{-1} = \dfrac{z}{9}$ are called l_1, l_2 respectively.

Show that l_1 and l_2 are skew.
A line through the origin O meets l_1, l_2 at P, Q respectively. By using general coordinates for P and Q and a condition that O, P, Q are collinear, find the Cartesian equations of OPQ, and the coordinates of P and Q.

3.4 Ratio Theorem

If we say that R *divides PQ in the ratio* $m:n$, we mean that R is on the straight line through P and Q with $n\,PR = m\,RQ$, with the following additional convention: if m and n have the same sign, R is between P and Q;
    if m and n have opposite signs, R is not between P and Q.
It will be noticed that if R is between P and Q, **PR**, **RQ** have the same sense, but not otherwise, so $n\mathbf{PR} = m\mathbf{RQ}$ in all cases. Each value of the ratio (except -1) corresponds to a unique point R.

For example: A divides PQ in the ratio $1:2$

B divides PQ in the ratio $4:-1$

C divides PQ in the ratio $-5:2$

D divides PQ in the ratio $1:-8$

Qu. 1.   Explain why R cannot divide PQ in the ratio $-1:1$.

Example 1: The point R divides PQ in the ratio $m:n$.

Prove that $\mathbf{r} = \dfrac{n\mathbf{p} + m\mathbf{q}}{m + n}$. This is known as the *Ratio Theorem*.

Solution:

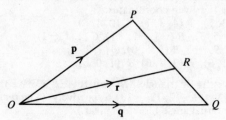

R divides PQ in the ratio $m:n$

$\Leftrightarrow \qquad n\,\mathbf{PR} = m\,\mathbf{RQ}$

$\Leftrightarrow \quad n\,(\mathbf{r} - \mathbf{p}) = m\,(\mathbf{q} - \mathbf{r})$

$\Leftrightarrow \quad (m + n)\mathbf{r} = n\,\mathbf{p} + m\,\mathbf{q}$

$\Leftrightarrow \qquad\qquad \mathbf{r} = \dfrac{n\,\mathbf{p} + m\,\mathbf{q}}{m + n}$ provided that $m + n \neq 0$. $\square$

The expression $\dfrac{n\mathbf{p} + m\mathbf{q}}{m + n}$ is a weighted mean, with 'weights' m, n. Notice the order in which m and n occur; this can be remembered by taking an extreme case. For example, if $m = 99$ and $n = 1$ then R is very close to Q; by the Ratio Theorem $\mathbf{r} = \frac{1}{100}\mathbf{p} + \frac{99}{100}\mathbf{q}$, which correctly shows that the $\mathbf{q}$ component is much more important than the $\mathbf{p}$ component.

Qu. 2.   Show that the mid-point of PQ has position vector $\frac{1}{2}(\mathbf{p} + \mathbf{q})$.

Example 2: Prove that the medians of a triangle are concurrent.

Solution (*i*):   Let ABC be the triangle; let L, M, N be the mid-points of BC, CA, AB respectively, and take an origin O not in the plane of the triangle.

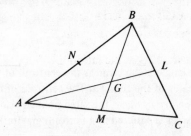

Then $\mathbf{l} = \frac{1}{2}(\mathbf{b} + \mathbf{c})$, $\mathbf{m} = \frac{1}{2}(\mathbf{c} + \mathbf{a})$, $\mathbf{n} = \frac{1}{2}(\mathbf{a} + \mathbf{b})$. Let G be the intersection of AL and BM. Suppose G divides AL in the ratio $\lambda : 1$; then $\mathbf{g} = \dfrac{\mathbf{a} + \lambda\mathbf{l}}{1 + \lambda} = \dfrac{2\mathbf{a} + \lambda\mathbf{b} + \lambda\mathbf{c}}{2(1 + \lambda)}$. Suppose G divides BM in the ratio $\mu : 1$; then

$\mathbf{g} = \dfrac{\mathbf{b} + \mu\mathbf{m}}{1 + \mu} = \dfrac{\mu\mathbf{a} + 2\mathbf{b} + \mu\mathbf{c}}{2(1 + \mu)}$. Since vectors $\mathbf{a}, \mathbf{b}, \mathbf{c}$ are not coplanar we can use the Unique Components Theorem of page 61 to equate the coefficients of $\mathbf{c}$:

$$\frac{\lambda}{2 + 2\lambda} = \frac{\mu}{2 + 2\mu} \Leftrightarrow \lambda = \mu.$$

Using the Unique Components Theorem to equate the coefficients of $\mathbf{a}$:

$$\lambda = \mu = 2.$$

It can easily be seen that these values give equal coefficients for $\mathbf{b}$. Therefore $\mathbf{g} = \dfrac{\mathbf{a} + \mathbf{b} + \mathbf{c}}{3}$ and by the symmetry of $\mathbf{a}, \mathbf{b}, \mathbf{c}$ in this expression, G must also divide CN in the same $2 : 1$ ratio. That is, G is on each median, so the medians are concurrent.   □

Solution (*ii*):   The proof is much shorter if we know the answer! Using the same triangle ABC, with L, M, N defined as before, let G divide AL in the ratio $2 : 1$; then $\mathbf{g} = \frac{1}{3}(\mathbf{a} + 2\mathbf{l})$

$$= \frac{\mathbf{a} + \mathbf{b} + \mathbf{c}}{3}.$$

By the symmetry of $\mathbf{a}, \mathbf{b}, \mathbf{c}$ in this expression, G must also divide BM and CN in the same $2 : 1$ ratio. That is, G is on each median, so the medians are concurrent.   □

The point G of Example 2, with position vector $\dfrac{\mathbf{a}+\mathbf{b}+\mathbf{c}}{3}$ is known as the *centroid* of triangle ABC.

The *centroid* of any system of masses $m_1, m_2, m_3, \ldots, m_k$ situated at points with position vectors $\mathbf{p}_1, \mathbf{p}_2, \mathbf{p}_3, \ldots, \mathbf{p}_k$ respectively is defined as the point with position vector

$$\frac{m_1\mathbf{p}_1 + m_2\mathbf{p}_2 + m_3\mathbf{p}_3 + \ldots + m_k\mathbf{p}_k}{m_1 + m_2 + m_3 + \ldots + m_k} = \frac{\displaystyle\sum_{i=1}^{k} m_i\,\mathbf{p}_i}{\displaystyle\sum_{i=1}^{k} m_i}.$$ It will be noticed that the

the centroid of a triangle is the same as the centroid of three equal masses situated at the vertices of the triangle. The centroid of a system of masses is also known as the *centre of mass*; the same point is also the *centre of gravity* if the system of masses is situated in a uniform gravitational field.

Example 2: Show that the position of the centroid of a system of masses is independent of the origin chosen.

Solution: We need to show that applying the definition of centroid when all position vectors are relative to some point A leads us to the the same point as we get from the definition when the position vectors are relative to O.

Let $m_1, m_2, \ldots, m_k$ be the masses situated at $\mathbf{p}_1, \mathbf{p}_2, \ldots \mathbf{p}_k$ respectively relative to O. The position vector of A relative to O is $\mathbf{a}$.

Then relative to O, the centroid is at $\mathbf{g} = \dfrac{\displaystyle\sum_{i=1}^{k} m_i\,\mathbf{p}_i}{\displaystyle\sum_{i=1}^{k} m_i}$.

Relative to A, the position vector of the mass m_i is $\mathbf{p}_i - \mathbf{a}$ (for $i = 1$ to k) so that the position vector of the centroid is

$$\frac{\displaystyle\sum_{i=1}^{k} m_i(\mathbf{p}_i - \mathbf{a})}{\displaystyle\sum_{i=1}^{k} m_i} = \frac{\displaystyle\sum_{i=1}^{k} m_i\,\mathbf{p}_i - \mathbf{a}\sum_{i=1}^{k} m_i}{\displaystyle\sum_{i=1}^{k} m_i} = \mathbf{g} - \mathbf{a}, \text{ which is the}$$

same point as we obtained earlier when position vectors originated at O.   □

Qu. 3.   System S_1 of masses has centroid G_1 and total mass M_1; system S_2 has centroid G_2 and total mass M_2. Show that the combined system S containing all the masses of S_1 and S_2 has the same centroid as the system of two masses, M_1 at G_1 and M_2 at G_2.

The Ratio Theorem also gives us a test which may be applied to see if three points are collinear. We prove below that the following three statements are equivalent:

(i) points P, Q, R are collinear;
(ii) $\mathbf{PQ} = \lambda\, \mathbf{PR}$ for some scalar λ;
(iii) $l\mathbf{p} + m\mathbf{q} + n\,\mathbf{r} = \mathbf{0}$ for some l, m, n (not all zero) where $l + m + n = 0$.

Statement (i) $\Rightarrow$ R divides PQ in some ratio, say $m_1 : n_1$, where m_1, n_1 are not both zero.

$$\Rightarrow \mathbf{r} = \frac{n_1\mathbf{p} + m_1\mathbf{q}}{m_1 + n_1}$$

$$\Rightarrow n_1\mathbf{p} + m_1\mathbf{q} - (m_1 + n_1)\mathbf{r} = \mathbf{0}$$

$\Rightarrow$ statement (iii) where $l = n_1, m = m_1, n = -(m_1 + n_1)$.

Statement (iii) $\Rightarrow$ $l\mathbf{p} + m\mathbf{q} = (l + m)\mathbf{r}$ by substituting for n. We are given that not all l, m, n are zero; suppose, for example, that n is not zero. Then $l + m \neq 0$ and $\mathbf{r} = \dfrac{l\mathbf{p} + m\mathbf{q}}{l + m}$, so that R divides PQ in the ratio $m : l$ and P, Q, R are collinear. (If n is zero, we could proceed by the same method using l or m instead of n.)

Statements (i) and (ii) are seen to be equivalent from the definition of multiplication of a vector by a scalar, thus completing the proof.

Example 3: Find the centroid C of masses 1, 2, 3 at $(3, 1, 5)$, $(3, 4, 8)$, $(5, 1, 7)$ respectively and prove that it is collinear with $A(0, -3, 8)$ and $B(12, 12, 5)$.

Solution: $\mathbf{c} = \tfrac{1}{6}\left[\begin{pmatrix} 3 \\ 1 \\ 5 \end{pmatrix} + 2\begin{pmatrix} 3 \\ 4 \\ 8 \end{pmatrix} + 3\begin{pmatrix} 5 \\ 1 \\ 7 \end{pmatrix} \right] = \tfrac{1}{6}\begin{pmatrix} 24 \\ 12 \\ 42 \end{pmatrix} = \begin{pmatrix} 4 \\ 2 \\ 7 \end{pmatrix}.$

i.e. centroid C is at $(4, 2, 7)$.

$$\mathbf{AC} = \mathbf{c} - \mathbf{a} = \begin{pmatrix} 4 \\ 5 \\ -1 \end{pmatrix}; \quad \mathbf{AB} = \mathbf{b} - \mathbf{a} = \begin{pmatrix} 12 \\ 15 \\ -3 \end{pmatrix} = 3\,\mathbf{AC}.$$

Therefore A, B, C are collinear. $\square$

Exercise 3D

1. Prove that the points with position vectors $\mathbf{a}, \mathbf{b}, 4\mathbf{b} - 3\mathbf{a}$ are collinear, and draw a diagram to illustrate the situation.

2. Let $ABCD$ be a parallelogram. Find $\mathbf{d}$ in terms of $\mathbf{a}, \mathbf{b}$ and $\mathbf{c}$.

3. Use vector methods to prove that the diagonals of a parallelogram bisect each other.

4. If $ABCD$ is a parallelogram and P is the mid-point of BC, prove that AP, BD trisect each other.

5. If $PQRS$, $P'Q'R'S'$ are parallelograms, prove that the mid-points of PP', QQ', RR', SS' are the vertices of a parallelogram. Is it necessary for $PQRS$, $P'Q'R'S'$ to be coplanar?

6. The points L, M, N are the mid-points of BC, CA, AB respectively. Prove that $\mathbf{OA} + \mathbf{OB} + \mathbf{OC} = \mathbf{OL} + \mathbf{OM} + \mathbf{ON}$ where O is any point. What does this tell you about the centroids of triangles ABC, LMN?

7. The three points A, B, C are the vertices of a triangle. Prove that

   G is the centroid of triangle $ABC \Rightarrow \mathbf{GA} + \mathbf{GB} + \mathbf{GC} = \mathbf{0}$.

   Also state and prove the converse.

8. Triangles ABC, $A'B'C'$ have centroids G, G' respectively. Prove that $\mathbf{AA'} + \mathbf{BB'} + \mathbf{CC'} = 3\mathbf{GG'}$.

9. Triangle ABC has centroid G; points M, N bisect CA, AB respectively. Prove that MN quadrisects AG.

10. Prove that the centroids of the four triangles ABP, ABQ, CDP, CDQ form the vertices of a parallelogram. Is it necessary for the four triangles to be coplanar?

11. (a) Draw one diagram to illustrate the vectors $\mathbf{a}, \mathbf{b}, -\mathbf{b}, \mathbf{a} + \mathbf{b}, \mathbf{a} - \mathbf{b}$. Using the geometry of a semi-circle, or otherwise, prove that $|\mathbf{a}| = |\mathbf{b}| \Rightarrow \mathbf{a} + \mathbf{b}$ and $\mathbf{a} - \mathbf{b}$ are perpendicular.
    (b) Use vector methods to prove that the diagonals of a rhombus bisect each other at right angles.

12. Triangle ABC has circumcentre O. Using O as origin, and with the usual convention about position vectors, the point H has position vector $\mathbf{h} = \mathbf{a} + \mathbf{b} + \mathbf{c}$. Prove the following.
    (i) AH is perpendicular to BC. [Hint: use the test of Question 11(a).]
    (ii) The altitudes of triangle ABC are concurrent at H. The point H is known as the *orthocentre* of triangle ABC.
    (iii) The points O, G, H are collinear with $3OG = OH$, where G is the centroid of triangle ABC. The line OGH is known as the *Euler line* of triangle ABC.

13. Prove that the lines joining the mid-points of opposite edges of a parallelepiped and the four diagonals are concurrent, all bisecting each other.

14. (a) Prove that the three lines joining mid-points of opposite edges of a tetrahedron bisect each other. Call the point of concurrency F.
    (b) Prove that the four line segments joining the vertices of a tetrahedron to the centroids of the opposite faces are conconcurrent. Call the point of concurrency G, and show that G divides each line segment in the ratio $3 : 1$.
    (c) Prove that F and G are the same point. This point is the *centroid* of the tetrahedron.

15. Let G be the centroid of the tetrahedron $ABCD$. (See Question 14.)
Prove that
$$\mathbf{GA} + \mathbf{GB} + \mathbf{GC} + \mathbf{GD} = \mathbf{0}.$$

16. Triangle ABC has sides $BC = p$, $CA = q$ and $AB = r$.
  (i) If the internal bisector of angle A meets BC at D, prove that
  $BD : DC = r : q$.
  Hint: use the Sine Formula on triangles ABD, ACD.
  (ii) Deduce that the position vector of the incentre of the triangle is
  $$\frac{p\mathbf{a} + q\mathbf{b} + r\mathbf{c}}{p + q + r}.$$

17. (a) Prove that the centre of mass of masses $\tan A$, $\tan B$, $\tan C$ placed
  respectively at the vertices A, B, C of a triangle is at the orthocentre
  of the triangle.
  (b) Find the masses which must be placed at A, B, C if the centre of
  mass is to be at the circumcentre of triangle ABC.

18. The points A, B, C (no two of which lie on a line through the origin O)
are represented by position vectors $\mathbf{a}$, $\mathbf{b}$, $\mathbf{c}$ and the points A', B', C' by
vectors $\alpha\mathbf{a}$, $\beta\mathbf{b}$, $\gamma\mathbf{c}$ where α, β, γ are distinct real numbers. If BC meets
$B'C'$ at the point L with position vector $\mathbf{l}$, show that
$$\mathbf{l} = \frac{\beta(1-\gamma)}{(\beta-\gamma)}\mathbf{b} + \frac{\gamma(1-\beta)}{(\gamma-\beta)}\mathbf{c}.$$

Write down similar expressions for the vectors $\mathbf{m}$, $\mathbf{n}$ corresponding to
the points M (where AC meets $A'C'$) and N (where AB meets $A'B'$).
Prove that L, M, N are collinear.
[This is the theorem of Desargues (1648)].   [SMP]

4 Calculus I

4.1 Integration by parts

In Vol. 1 §8.4 the rule for differentiating products was proved:

$$\frac{d}{dx}(uv) = u\frac{dv}{dx} + v\frac{du}{dx}; \text{ this can be rearranged as } u\frac{dv}{dx} = \frac{d}{dx}(uv) - v\frac{du}{dx}.$$

Integrating both sides with respect to x we obtain

$$\int u\frac{dv}{dx}\,dx = uv - \int v\frac{du}{dx}\,dx.$$

The process of integration which utilizes this formula is known as *integration by parts*. It is frequently used when the integrand is a product. The first step is to integrate one of the factors of the integrand; we shall see in Qu. 2 that the arbitrary constant of integration which we would normally require may be taken as zero at this stage.

Example 1: Find $\int x \cos x \, dx$.

Solution: Let $u = x, \dfrac{dv}{dx} = \cos x$.

Then $\dfrac{du}{dx} = 1, v = \sin x +$ an arbitrary constant which we take as zero.

Therefore $\int x \cos x \, dx = x \cdot \sin x - \int \sin x \cdot 1 \cdot dx$

$$= x \sin x + \cos x + c. \qquad \square$$

Integration by parts permits us to replace one integral by another, but whether the new integration is easier than the original depends on how we have chosen u and $\dfrac{dv}{dx}$. In Example 1 we could have taken $u = \cos x, \dfrac{dv}{dx} = x$,

obtaining $\dfrac{du}{dx} = -\sin x, v = \dfrac{x^2}{2} +$ an arbitrary constant which we again take as zero. This gives us

$$\int x \cos x \, dx = \cos x \cdot \frac{x^2}{2} - \int \frac{x^2}{2}(-\sin x)\,dx$$

$$= \tfrac{1}{2}x^2 \cos x + \tfrac{1}{2}\int x^2 \sin x \, dx.$$

This is a true result, but one which does not help us perform the integration.

Qu. 1.   Find (i) $\int x \sin x \, dx$ (ii) $\int x \ln x \, dx$ (iii) $\int x \sec^2 x \, dx$.

Qu. 2.   Prove that $u(v + c) - \int (v + c) \dfrac{du}{dx} \, dx = uv - \int v \dfrac{du}{dx} \, dx$, where c is a

constant and u, v depend on x. This shows that we may safely take as zero the constant of integration when carrying out the first step of integration by parts.

Definite integration can be carried out in much the same way.

Example 2: Evaluate $\int_0^1 x \, e^x \, dx$.

Solution: Let $u = x, \dfrac{dv}{dx} = e^x$. Then $\dfrac{du}{dx} = 1, v = e^x$.

Therefore $\int_0^1 x \, e^x \, dx = \left[x \, e^x - \int e^x . 1 . dx \right]_0^1$

$$= \left[x \, e^x - e^x \right]_0^1 = (e - e) - (0 - 1) = 1.$$

An alternative method of setting out the first part of the

integration is $\int_0^1 x \, e^x dx = \left[x \, e^x \right]_0^1 - \int_0^1 e^x . 1 . dx$, and then the

evaluation of the first term may proceed at once. $\square$

A number of integrands which at first sight do not appear to be products may be dealt with following the method of the next example where we take one of the factors to be 1.

Example 3: Find $\int \ln x \, dx$.

Solution: Let $u = \ln x, \dfrac{dv}{dx} = 1$. Then $\dfrac{du}{dx} = \dfrac{1}{x}, v = x$.

Therefore $\int \ln x \, dx = \ln x . x - \int x . \dfrac{1}{x} . dx$

$$= x \ln x - \int dx = x \ln x - x + c.   \square$$

Sometimes we have to use integration by parts more than once, as illustrated in the following example.

Example 4: Find $\int x^2 \cos x \, dx$.

Solution: Let $u = x^2, \dfrac{dv}{dx} = \cos x$. Then $\dfrac{du}{dx} = 2x, v = \sin x$.

Therefore $\int x^2 \cos x \, dx = x^2 . \sin x - \int \sin x . 2x . dx$

$$= x^2 \sin x - 2 \int x \sin x \, dx.$$

To find $\int x \sin x \, dx$, let $u_1 = x, \dfrac{dv_1}{dx} = \sin x$, so that $\dfrac{du_1}{dx} = 1, v_1 = -\cos x$. Then

$$\int x \sin x \, dx = x \cdot (-\cos x) - \int (-\cos x) \cdot 1 \cdot dx$$

$$= -x \cos x + \int \cos x \, dx$$

$$= -x \cos x + \sin x + c_1, \text{ where } c_1 \text{ is an arbitrary}$$
$$\text{constant.}$$

Therefore $\int x^2 \cos x \, dx = x^2 \sin x + 2x \cos x - 2 \sin x + c.$ □

Qu. 3. Investigate what would happen in the solution of Example 4 if we had used $u_1 = \sin x, \dfrac{dv_1}{dx} = x$ when finding $\int x \sin x \, dx$.

As a final illustration we include an example where integrating by parts twice brings us back to the original integral.

Example 5: Find $\int e^x \sin x \, dx$.

Solution: Let I be one integral of $e^x \sin x$ with respect to x.

$$\text{Then } I = \int e^x \sin x \, dx \qquad\qquad \left[u = e^x, \dfrac{dv}{dx} = \sin x \right]$$

$$= e^x \cdot (-\cos x) - \int (-\cos x) \cdot e^x \, dx$$

$$= -e^x \cos x + \int e^x \cos x \, dx \qquad \left[u_1 = e^x, \dfrac{dv_1}{dx} = \cos x \right]$$

$$= -e^x \cos x + e^x \cdot \sin x - \int \sin x \cdot e^x \, dx$$

$$= e^x (\sin x - \cos x) - (I + c_1) \text{ where } c_1 \text{ is an arbitrary}$$
$$\text{constant, since } I \text{ was defined as one inte-}$$
$$\text{gral of } e^x \sin x \text{ and all such integrals differ}$$
$$\text{by only a constant.}$$

Therefore $2I = e^x (\sin x - \cos x) - c_1$ and so
$$\int e^x \sin x \, dx = \tfrac{1}{2} e^x (\sin x - \cos x) + c.$$ □

Exercise 4A

1–15. Integrate with respect to x.

1. $x \cos 2x$ 2. $x \sin 2x$ 3. $x (1 + x)^{19}$
4. $\ln 2x$ 5. $x^2 \ln x$ 6. $x e^{2x}$
7. $x^2 e^x$ 8. $x^2 \sin x$ 9. $(\ln x)^2$

10. $(1 + x^2) \ln x$ 11. $e^{-x} \sin x$ 12. $x \tan^2 x$

13. $x^3 e^{2x}$ 14. $e^x \cos x$ 15. $x \cos^2 x.$

16. Evaluate   (i) $\int_0^{\pi} x \sin x \, dx$ (ii) $\int_0^{\pi} (x^2 + 3x + 2) \cos x \, dx.$

17. The region R is bounded by the x-axis, the curve $y = \ln x$, and the line $x = e$. Calculate   (i) the area of R (ii) the volume of the solid of revolution formed by rotating R about the x-axis.

18. Find the mistake in the following:

$$\begin{aligned} `I = \int \tan x \, dx &= \int \sin x \sec x \, dx \\ &= -\cos x \sec x + \int \cos x \sec x \tan x \, dx \\ &= -1 + \int \tan x \, dx = -1 + I. \end{aligned}$$

Therefore $0 = -1.$'

19. Find $\int x^3 e^{x^2} \, dx$. (Hint: investigate $\dfrac{d}{dx}(e^{x^2})$.)

20. Find $\int e^{x^{\frac{1}{2}}} \, dx.$

21. (i) Show that $\int \sin^2 x \, dx = -\sin x \cos x + \int \cos^2 x \, dx$. By writing $\cos^2 x \equiv 1 - \sin^2 x$ deduce that $\int \sin^2 x \, dx = \dfrac{x}{2} - \dfrac{\sin 2x}{4} + c.$

   (ii) Prove the result of (i) again using $\cos 2x \equiv 1 - 2 \sin^2 x.$

   (iii) Find $\int \cos^2 x \, dx.$

22. Find $\int \sec^3 x \, dx$. (Hint: consider $\int \sec x \sec^2 x \, dx.$)

23. Find $\int e^{2x} \sin 3x \, dx.$

24. Find $\int e^x (x + 2) \ln (x + 1) \, dx.$

25. (i) By evaluating $\int_0^x e^{-t} \, dt$ directly and also by parts, show that
$$1 = e^{-x} + xe^{-x} + \int_0^x t \, e^{-t} \, dt.$$

   (ii) Integrate by parts again to show that
$$1 = e^{-x} + x \, e^{-x} + \frac{x^2}{2} \, e^{-x} + \int_0^x \frac{t^2}{2} \, e^{-t} \, dt.$$

   (iii) Prove by induction that
$$1 = e^{-x}\left(1 + x + \frac{x^2}{2!} + \ldots + \frac{x^n}{n!}\right) + \int_0^x \frac{t^n}{n!} \, e^{-t} \, dt$$
   and deduce that
$$e^x = 1 + x + \frac{x^2}{2!} + \ldots + \frac{x^n}{n!} + E_n(x),$$

   where $E_n(x) = e^x \int_0^x \frac{t^n}{n!} e^{-t} \, dt.$

4.2 Inverse functions

A *function* was defined in Vol. 1 §1.1 as a relation which maps objects to images in such a way that each object has one and only one image. The *domain* is the set of possible objects, and the *range* is the set of images. The *codomain* is a set containing all the images, but possibly containing other elements which are not images. We say that a function maps the domain *onto* the range but *into* the codomain. In this section, except where otherwise stated, we take the domain to be the largest possible set of real numbers.

Suppose function f has domain D and range R; suppose also that $y_1 = f(x_1)$ and $y_2 = f(x_2)$. From the definition of a function, $x_1 = x_2 \Rightarrow y_1 = y_2$ for all $x_1, x_2 \in D$. But in general $y_1 = y_2 \not\Rightarrow x_1 = x_2$. By representing the function f graphically, we see that a vertical line can cut the graph at most once, while a horizontal line may cut it any number of times. If f is also such that $y_1 = y_2 \Rightarrow x_1 = x_2$ whenever $y_1, y_2 \in R$, then f is described as a *one-to-one* function; otherwise f is a *many-to-one* function. A horizontal line will cut the graph of a one-to-one function at most once (as in Fig. 1), while the graph of a many-to-one function will be cut more than once by at least one horizontal line (as in Fig. 2).

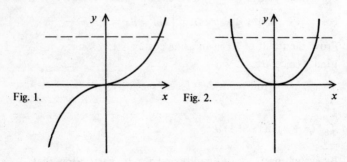

Fig. 1.   Fig. 2.

By restricting the domain of a function f to the set D_1 (a subset of the original domain D) we may obtain a function which is one-to-one. If this is so, we say that f is one-to-one over domain D_1. In Fig. 2, if we restrict the domain to non-negative values then we have created a function which is one-to-one over the domain $\{x : x \geqslant 0\}$.

Qu. 1.   Refer to the diagrams (a) to (e). State which of the diagrams illustrate one-to-one functions, and which illustrate many-to-one functions. In the case of many-to-one functions, state how the domain could be restricted to create a one-to-one function which has the same range as the original function.

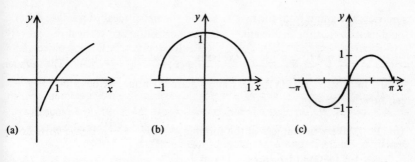

(a)

(b)

(c)

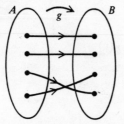

(d)

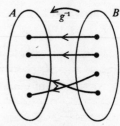

(e)

Consider the function g, with domain A, and range B. If g is a one-to-one function over A, each element of A corresponds to just one element of B, and *vice versa*, so that for any y in B there is one and only one x in A such that $g(x) = y$. This defines another function, which maps B onto A; this new function is called the *inverse* of g, and is denoted by g^{-1}. We can illustrate g, g^{-1} by mapping diagrams. Notice that all the members of sets A and B are used, and that the domain of g^{-1} is the range of g, and the range of g^{-1} is the domain of g.

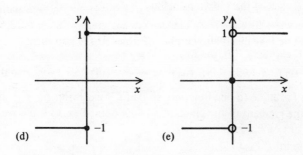

Notice further that (x, y) is a point on the graph of g if and only if (y, x) is a point on the graph of g^{-1}, showing that the graph of g^{-1} is the result of reflecting the graph of g in the line $y = x$. The equation of the graph of g^{-1} may be obtained from the equation of the graph of g merely by replacing x by y and y by x; we may then want to make y the subject of the equation, but that is not always possible. This suggests an alternative way of obtaining the graph of the inverse function: exchange the x and y labels on the axes, and view the graph from the reverse side of the paper (against a bright light), holding it so that the new x and y axes are horizontal and vertical respectively.

Be careful not to confuse $g^{-1}(x)$ with $\dfrac{1}{g(x)}$; when using index form, $\dfrac{1}{g(x)}$ should be written as $[g(x)]^{-1}$. If f and f^{-1} are the same, we describe f as a *self-inverse* function.

Qu. 2.   Make x the subject of the formula $y = \sqrt{\left(\dfrac{x-1}{x}\right)}$. Hence write down the inverse of $f : x \mapsto \sqrt{\left(\dfrac{x-1}{x}\right)}$.

Qu. 3.   For each of the functions below
   (a) sketch the curve $y = f(x)$;
   (b) sketch the graph of $y = f^{-1}(x)$;
   (c) express f^{-1} in the form $f^{-1} : x \mapsto \ldots$
      (i) $f : x \mapsto 3x + 5$;   (ii) $f : x \mapsto e^x$;   (iii) $f : x \mapsto x^2, x \geqslant 0$.

Suppose f is a one-to-one function, so that f^{-1} exists; let $g = f^{-1}$. Then by symmetry the gradient of the graph of f at the point $P(x, y)$ is the reciprocal of the gradient of the graph of g at the point $P'(y, x)$.

With the usual notation, the derivative of f

at $P = f'(x)$

$= $ limit of $\dfrac{\delta y}{\delta x}$ as $\delta x \to 0$

$= \dfrac{dy}{dx}$;

the derivative of g
at $P' = g'(y)$

$= $ limit of $\dfrac{g(y + \delta y) - g(y)}{\delta y}$ as $\delta y \to 0$

$= $ limit of $\dfrac{\delta x}{\delta y}$ as $\delta y \to 0$

$= \dfrac{dx}{dy}$.

Thus, for a one-to-one function

$$\dfrac{dy}{dx} = 1 \bigg/ \dfrac{dx}{dy} \text{ provided that } \dfrac{dy}{dx}, \dfrac{dx}{dy} \neq 0.$$

Example: If $y = \sqrt{x}$ find $\dfrac{d^2x}{dy^2}$ without making x the subject of the equation.

Solution: $y = \sqrt{x} \Rightarrow \dfrac{dy}{dx} = \dfrac{1}{2\sqrt{x}} \Rightarrow \dfrac{dx}{dy} = 2\sqrt{x}.$

Therefore $\dfrac{d^2x}{dy^2} = \dfrac{d}{dy}(2\sqrt{x}) = \dfrac{d}{dx}(2\sqrt{x}) \cdot \dfrac{dx}{dy}$

$$= \frac{1}{\sqrt{x}} \cdot 2\sqrt{x} = 2. \qquad \square$$

Exercise 4B

Except where otherwise stated, the domain of a function is the largest possible set of real numbers.

1. For each of the following, sketch the graph of $y = f(x)$; if f^{-1} exists find an expression for $f^{-1}(x)$.
   (i) $f(x) \equiv x^3 + 2$;   (ii) $f(x) \equiv \sqrt{(1 - x^2)}, 0 \leqslant x \leqslant 1$;
   (iii) $f(x) \equiv x^3 - 3x + 2$;   (iv) $f(x) \equiv \sqrt{\left(1 + \dfrac{1}{x}\right)}.$

2. Find two linear functions and one non-linear function which are self-inverse.

3. If $y = \dfrac{1}{(x + 1)^2}$, find $\dfrac{dy}{dx}, \dfrac{dx}{dy}, \dfrac{d^2y}{dx^2}, \dfrac{d^2x}{dy^2}$ without first expressing x in terms of y.

4. If $y = \sqrt{\left(\dfrac{x}{1 + x}\right)}$, express x in terms of y.
   Find $\dfrac{dy}{dx}$ (i) in terms of x, (ii) in terms of y.

5. If $f(x) \equiv \dfrac{1 + x}{2 + x}$ find $f'(x)$ and deduce that f is one-to-one over domain $\{x : x \neq -2\}$. Find an expression for $f^{-1}(x)$.

6. If $f(x) \equiv x^5 + x + 1$ show that f is one-to-one over the set of real numbers. Find the range of f, the domain and range of f^{-1}, and $f^{-1}(3)$, $f^{-1}(-1), f^{-1}(35)$. This is an example of a function f which has an inverse although there is no explicit formula for $f^{-1}(x)$.

7. Show that $f : x \mapsto \dfrac{3x + 2}{5x - 3}$ is a self-inverse function.

4.3 Inverse trigonometric functions

To find inverses of the trigonometric functions we have to restrict their domains since they are not one-to-one over the set of real numbers. We choose

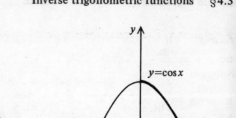

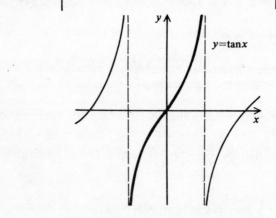

values as close to the origin as possible, and restrict the domains as shown in the table.

Function	Restricted domain
Sine	$\{ x : -\frac{\pi}{2} \leqslant x \leqslant \frac{\pi}{2} \}$
Cosine	$\{ x : \ 0 \ \leqslant x \leqslant \pi \}$
Tangent	$\{ x : -\frac{\pi}{2} < x < \frac{\pi}{2} \}$

The inverse of the sine function (known as the *inverse sine* function or *arcsine* function) is defined as the function which maps x onto y such that $\sin y = x$, $-\frac{\pi}{2} \leqslant y \leqslant \frac{\pi}{2}$. We write $y = \sin^{-1} x$, or $y = \arcsin x$.

Similarly the *inverse cosine* function is defined as the function which maps x onto y such that $\cos y = x$, $0 \leqslant y \leqslant \pi$; we write $y = \cos^{-1} x$, or

$y = \arccos x$. The *inverse tangent* function maps x onto y such that $\tan y = x, -\dfrac{\pi}{2} < y < \dfrac{\pi}{2}$; we write $y = \tan^{-1}x$, or $y = \arctan x$.

It is important to remember that the sine function maps numbers to numbers, (see Vol. 1 §8.3). Therefore the inverse sine function also maps numbers to numbers, and so do the inverse cosine and inverse tangent functions. This means that $\sin^{-1}\tfrac{1}{2}$, $\cos^{-1}\left(-\dfrac{\sqrt{3}}{2}\right)$, $\tan^{-1}1$ are the numbers $\dfrac{\pi}{6}, \dfrac{5\pi}{6}, \dfrac{\pi}{4}$ respectively.

Qu. 1. Sketch the graphs of the functions $f : x \mapsto \sin^{-1}x$, $g : x \mapsto \cos^{-1}x, h : x \mapsto \tan^{-1}x$, and state their domains and ranges.

Qu. 2. Show that $\sin^{-1}x + \cos^{-1}x \equiv \dfrac{\pi}{2}$.

Methods of solving trigonometric equations have already been dealt with in Vol. 1, Part 7, but there most equations were restricted to the domain $0° \leqslant \theta \leqslant 360°$. If we are looking for the *general solution* of an equation, we want a rule or formula which will give us all the solutions (and no other values). For example, consider the equation $\sin x = \tfrac{1}{2}$. Since $\sin x$ is positive, all the solutions will be in the first and second quadrants; we also notice that $\sin^{-1}\tfrac{1}{2} = \dfrac{\pi}{6}$. Thus the general solution is

$$x = 2n\pi + \frac{\pi}{6} \quad \text{or} \quad (2n+1)\pi - \frac{\pi}{6} \quad \text{where } n \text{ is any integer,}$$

which may also be written as $x = m\pi + (-1)^m \dfrac{\pi}{6}$ where m is any integer.

Qu. 3. Find the general solution of the equations

(a) $\cos x = \dfrac{1}{\sqrt{2}}$, (b) $\tan x = -\sqrt{3}$.

We conclude this section by mentioning that the inverse secant, cosecant and cotangent functions, denoted in the obvious way, are defined by

$$\sec^{-1}x \equiv \cos^{-1}\left(\frac{1}{x}\right), \ \operatorname{cosec}^{-1}x \equiv \sin^{-1}\left(\frac{1}{x}\right), \ \cot^{-1}x \equiv \tan^{-1}\left(\frac{1}{x}\right) \text{ with } \cot^{-1}0 = \frac{\pi}{2}.$$

Exercise 4C

1. Find the exact value of the following, leaving answers in terms of π.

(i) $\cos^{-1}\tfrac{1}{2}$ (ii) $\tan^{-1}\dfrac{1}{\sqrt{3}}$ (iii) $\sin^{-1}(-1)$

(iv) $\cos^{-1}0$ (v) $\sec^{-1}(-2)$ (vi) $\cot^{-1}\left(-\dfrac{1}{\sqrt{3}}\right)$.

2. Evaluate the following to 3 S.F.

   (i) $\sin^{-1}\frac{1}{3}$ (ii) $\tan^{-1}1.5$ (iii) $\cos^{-1}(-0.6)$
   (iv) $\operatorname{cosec}^{-1}(-4)$ (v) $\cos^{-1}\frac{2}{5}$ (vi) $\sec^{-1}5$.

3. Sketch the graphs of the following functions and state their domains and ranges:

   (i) $f : x \mapsto \sec^{-1}x$ (ii) $g : x \mapsto \operatorname{cosec}^{-1}x$ (iii) $h : x \mapsto \cot^{-1}x$.

4. If $f(x) \equiv \cos x + \sin x, -\dfrac{\pi}{4} \leqslant x \leqslant \dfrac{\pi}{4}$, find $f^{-1}(x)$.

5. Show that $\sin^{-1}(-x) = -\sin^{-1}x$, and $\tan^{-1}(-x) = -\tan^{-1}x$. State and prove a relationship between $\cos^{-1}(-x)$ and $\cos^{-1}x$.

6. Show that $\sin^{-1}(\sin \pi) \neq \pi$. Under what conditions is $\sin^{-1}(\sin x) = x$?

7. Give the general solution of (a) $\sin x = k$, (b) $\sec x = k$, (c) $\tan x = k$, in each case stating any necessary restrictions on k.

8–13. Find the general solution of the equation; where possible give the answer as a rational multiple of π, otherwise leave the solution in a form involving an inverse trigonometric function.

   8. $\sin 2x = \sin x$ 9. $\cos x - \sin x = \sqrt{2}$

   10. $3\cos x + 4\sin x = 2.5$ 11. $\tan 2x = 4\tan x$

   12. $\cos x = \cos \frac{1}{2}x$ 13. $2\sin x - \cos x = 1$

4.4 Integration using inverse trigonometric functions

Suppose $y = \sin^{-1}x$; then $\sin y = x$. By differentiating with respect to x, using the chain rule for the derivative of $\sin y$, we obtain $\cos y \cdot \dfrac{dy}{dx} = 1$.

If $-1 < x < 1$, then $-\dfrac{\pi}{2} < y < \dfrac{\pi}{2}$ so that $\cos y > 0$, and

$$\frac{dy}{dx} = \frac{1}{\cos y} = \frac{1}{\sqrt{(1 - \sin^2 y)}} = \frac{1}{\sqrt{(1 - x^2)}}.$$

At $x = 1$ and at $x = -1$, the tangents to $y = \sin^{-1}x$ are vertical, and $\dfrac{dy}{dx}$ is undefined.

We have thus shown that $\dfrac{d}{dx}(\sin^{-1}x) = \dfrac{1}{\sqrt{(1 - x^2)}}$

Similarly $\dfrac{d}{dx}(\cos^{-1}x) = -\dfrac{1}{\sqrt{(1 - x^2)}}$

and $\dfrac{d}{dx}(\tan^{-1}x) = \dfrac{1}{1 + x^2}$

Qu. 1. Prove these last two results.

Qu. 2. Find the derivatives of (i) $\sin^{-1} 5x$ (ii) $\sin^{-1}(3x-4)$ (iii) $\tan^{-1}\dfrac{3x}{2}$ (iv) $\tan^{-1}(2-3x)$.

The inverse trigonometric functions are particularly useful in enabling us to perform a number of integrations. In the examples which follow we make frequent use of the identities $\sin^2\theta + \cos^2\theta \equiv 1$, $\tan^2\theta + 1 \equiv \sec^2\theta$.

Example 1: Find $\displaystyle\int \frac{1}{\sqrt{(9-x^2)}}\, dx$.

Solution: Let $u = \sin^{-1}\dfrac{x}{3}$, then $\qquad 3\sin u = x$,

$$\text{so that} \quad 3\cos u \,.\, du = dx$$
$$\text{and} \quad \sqrt{(9-x^2)} = \sqrt{(9 - 9\sin^2 u)} = 3\cos u.$$

Therefore $\displaystyle\int \frac{1}{\sqrt{(9-x^2)}}\, dx = \int \frac{3\cos u \,.\, du}{3\cos u} = u + c = \sin^{-1}\frac{x}{3} + c.$

$\square$

Example 2: Find $\displaystyle\int \frac{1}{\sqrt{(a^2-x^2)}}\, dx, a > 0$.

Solution: Let $u = \sin^{-1}\dfrac{x}{a}$, then $\qquad a\sin u = x$,

$$\text{so that} \quad a\cos u \,.\, du = dx$$
$$\text{and} \quad \sqrt{(a^2-x^2)} = \sqrt{(a^2 - a^2\sin^2 u)} = a\cos u.$$

Therefore $\displaystyle\int \frac{1}{\sqrt{(a^2-x^2)}}\, dx = \int \frac{a\cos u \,.\, du}{a\cos u} = u + c = \sin^{-1}\frac{x}{a} + c.$

$\square$

Example 3: Find $\displaystyle\int \frac{1}{16 + x^2}\, dx$.

Solution: Let $u = \tan^{-1}\dfrac{x}{4}$, then $\qquad 4\tan u = x$

$$\text{so that} \quad 4\sec^2 u \,.\, du = dx$$
$$\text{and} \quad 16 + x^2 = 16 + 16\tan^2 u$$
$$= 16\sec^2 u.$$

Therefore $\displaystyle\int \frac{1}{16 + x^2}\, dx = \int \frac{4\sec^2 u \,.\, du}{16\sec^2 u} = \tfrac{1}{4}u + c$

$$= \tfrac{1}{4}\tan^{-1}\frac{x}{4} + c. \qquad \square$$

Qu. 3. Find (i) $\displaystyle\int \frac{1}{\sqrt{(4-x^2)}}\, dx$ (ii) $\displaystyle\int \frac{1}{\sqrt{(4-9x^2)}}\, dx$ (iii) $\displaystyle\int \frac{1}{4+x^2}\, dx$.

Qu. 4.   Show that $\int \dfrac{1}{a^2 + x^2}\, dx = \dfrac{1}{a} \tan^{-1} \dfrac{x}{a} + c, a \neq 0.$

Sometimes it is necessary to remove a constant factor, or to complete a square before using one of the formulae of this section.

Example 4: Find $\int \dfrac{dx}{4 + 9x^2}.$

Solution (i): $\displaystyle \int \frac{dx}{4 + 9x^2} = \frac{1}{9} \int \frac{dx}{\frac{4}{9} + x^2}$

$$= \frac{1}{9}\left(\frac{3}{2} \tan^{-1} \frac{3x}{2} \right) + c \qquad \text{using Qu. 4.}$$

$$= \frac{1}{6} \tan^{-1} \frac{3x}{2} + c. \qquad\qquad \square$$

Solution (ii): $4 + 9x^2 = 4\left(1 + \frac{9}{4}x^2\right).$

Let $u = \tan^{-1} \dfrac{3x}{2}$, then $\qquad \dfrac{2}{3} \tan u = x,$

so that $\dfrac{2}{3} \sec^2 u \,.\, du = dx$

and $\qquad 4 + 9x^2 = 4 \sec^2 u.$

Therefore $\displaystyle \int \frac{dx}{4 + 9x^2} = \int \frac{\frac{2}{3} \sec^2 u \,.\, du}{4 \sec^2 u} = \frac{1}{6} u + c$

$$= \frac{1}{6} \tan^{-1} \frac{3x}{2} + c. \qquad \square$$

Example 5: Find $\int \dfrac{dx}{x^2 + 2x + 5}.$

Solution: $\displaystyle \int \frac{dx}{x^2 + 2x + 5} = \int \frac{dx}{2^2 + (x + 1)^2}$

$$= \int \frac{du}{2^2 + u^2} \qquad \text{where } u = x + 1$$

$$= \frac{1}{2} \tan^{-1} \frac{u}{2} + c = \frac{1}{2} \tan^{-1} \frac{x + 1}{2} + c.$$

Many people will find it unnecessary to use the substitution $u = x + 1$, and will be able to go straight from $\int \dfrac{dx}{2^2 + (x + 1)^2}$ to the answer. $\qquad\qquad \square$

If we are evaluating a definite integral, we may quote the formula for the indefinite integral and then substitute the limits. For example

$$\int_{-\frac{2}{3}}^{\frac{2}{3}} \frac{dx}{4+9x^2} = \left[\frac{1}{6}\tan^{-1}\frac{3x}{2}\right]_{-\frac{2}{3}}^{\frac{2}{3}} \qquad \text{see Example 4}$$

$$= \frac{1}{6}\tan^{-1}1 - \frac{1}{6}\tan^{-1}(-1) = \frac{\pi}{12}.$$

Alternatively, we may use a substitution provided we remember to change the limits of integration. For example

$$\int_{-\frac{2}{3}}^{\frac{2}{3}} \frac{dx}{4+9x^2} = \int_{-\frac{\pi}{4}}^{\frac{\pi}{4}} \frac{\frac{2}{3}\sec^2 u . du}{4\sec^2 u} \qquad \text{see Example 4, solution (ii);}$$

$$\text{when } x=\frac{2}{3}, u=\frac{\pi}{4}; \text{ when } x=-\frac{2}{3}, u=-\frac{\pi}{4}.$$

$$= \left[\frac{1}{6}u\right]_{-\frac{\pi}{4}}^{\frac{\pi}{4}} = \frac{\pi}{12}.$$

Summary

$$\int \frac{dx}{\sqrt{(a^2-x^2)}} = \sin^{-1}\frac{x}{a} + c, a > 0$$

$$\int \frac{dx}{a^2+x^2} = \frac{1}{a}\tan^{-1}\frac{x}{a} + c, a \neq 0.$$

Exercise 4D

1. Differentiate the following with respect to x:
   (i) $\sin^{-1}2x$ (ii) $\tan^{-1}5x$ (iii) $\sin^{-1}(3x^2)$ (iv) $\sec^{-1}2x$
   (v) $\tan^{-1}(e^x)$ (vi) $\text{cosec}^{-1}(1-x^2)$

2. Integrate the following with respect to x:

   (i) $\dfrac{1}{\sqrt{(36-x^2)}}$ (ii) $\dfrac{4}{25+4x^2}$ (iii) $\dfrac{1}{\sqrt{(9-2x^2)}}$ (iv) $\dfrac{1}{\sqrt{(1-(x+4)^2)}}$

   (v) $\dfrac{1}{3+2x^2}$ (vi) $\dfrac{1}{1+4(x+2)^2}$ (vii) $\dfrac{1}{\sqrt{(3+2x-x^2)}}$

   (vii) $\dfrac{3}{9x^2+6x+5}$.

3. Show that $\displaystyle\int \frac{dx}{-\sqrt{(a^2-x^2)}}$ may be expressed as $\cos^{-1}\dfrac{x}{a} + c_1$ and as

   $-\sin^{-1}\dfrac{x}{a} + c_2$ where c_1, c_2 are arbitrary constants. Explain how the two

   results are compatible, and express c_2 in terms of c_1.

4. Evaluate the following:

    (i) $\int_0^{\frac{1}{4}} \dfrac{dx}{\sqrt{(1-4x^2)}}$ (ii) $\int_{-\frac{1}{2}}^{\frac{1}{2}} \dfrac{dx}{\sqrt{(3-6x^2)}}$ (iii) $\int_{-1}^{1} \dfrac{dx}{1+9x^2}$

    (iv) $\int_{\sqrt{\frac{5}{6}}}^{\sqrt{\frac{5}{2}}} \dfrac{dx}{5+2x^2}$.

5. Use integration by parts to find:

    (i) $\int \sin^{-1}x\,dx$ (ii) $\int \cos^{-1}x\,dx$ (iii) $\int \tan^{-1}x\,dx$

    (iv) $\int \cot^{-1}x\,dx$.

6. Find the Maclaurin expansions of (i) $\tan^{-1}x$, (ii) $\sin^{-1}x$ as far as the term in x^5. From your answer to part (ii) deduce the expansion for $\cos^{-1}x$ as far as the term in x^5.

7. Use the substitution $u = \sin^{-1}\dfrac{x}{a}$ to find $\int_0^b \sqrt{(a^2-x^2)}\,dx, a > b > 0$.

    Sketch the curve $y = \sqrt{(a^2-x^2)}$, and give a geometrical interpretation of each term of your integral.

8. Integrate each of the following with respect to x:

    (i) $\dfrac{1}{x^2-6x+13}$ (ii) $\dfrac{1}{\sqrt{(7-12x-4x^2)}}$ (iii) $\dfrac{1}{4x^2+20x+29}$

    (iv) $\dfrac{1}{x^2-6x+9}$ (v) $\dfrac{1}{\sqrt{(4x-x^2)}}$.

9. Integrate each of the following with respect to x:

    (i) $\sqrt{(4-3x^2)}$ (ii) $\sqrt{(3-4x-4x^2)}$ (iii) $(x+1)\sqrt{(1-x^2)}$.

10. (a) Find $\dfrac{dx}{\sqrt{(x(1-x))}}$ (i) using the substitution $u = \cos^{-1}(1-2x)$,

                                              (ii) using the substitution $u = \sin^{-1}\sqrt{x}$.

    (b) Prove that $\cos^{-1}(1-2x) \equiv 2\sin^{-1}\sqrt{x}$.

11. (a) Sketch the graph of $y = \sec^{-1}x$ and deduce that $\dfrac{dy}{dx}$ is always positive.

    (b) Show that $\dfrac{d}{dx}\sec^{-1}x = \dfrac{1}{|x|\sqrt{(x^2-1)}}$ and find $\dfrac{d}{dx}\operatorname{cosec}^{-1}x$.

    (c) Find $\int \dfrac{dx}{x\sqrt{(x^2-a^2)}}, 0 < a < x$.

12. (a) Show that $\int \sec x\,dx = \ln|\sec x + \tan x| + c$.

    (b) Use integration by parts followed by the substitution $u = \sec^{-1}x$ to find

      (i) $\int \sec^{-1}x\,dx, x > 1$;   (ii) $\int \operatorname{cosec}^{-1}x\,dx, x > 1$.

4.5 Differentiation of implicit functions

If a function $f : x \mapsto y$ is specified by an equation in x and y, and y is not the subject of the equation, then f is known as an *implicit* function. Frequently it is necessary to restrict the range of values of x, y in some way so that f is a function. For example,

$$x^2 + y^2 = 1, y \geqslant 0$$

defines the function illustrated in the diagram, while

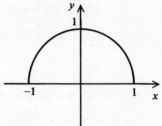

$$x^2 + y^2 = 1, y \leqslant 0$$

clearly defines a different function. Often we do not mention the restrictions that are necessary to make the function unambiguous, but tacitly assume that there are such restrictions. Example 1 below illustrates the usefulness of this apparent laxity. By using the chain rule, the derivative $\dfrac{dy}{dx}$ may be found without making y the subject of the equation.

Example 1: If $x^3 + y^3 = x^3 y^2 + 2x - 1$, find $\dfrac{dy}{dx}$ in terms of x and y.

Solution: $x^3 + y^3 = x^3 y^2 + 2x - 1 \Rightarrow \dfrac{d}{dx}(x^3 + y^3) = \dfrac{d}{dx}(x^3 y^2 + 2x - 1)$

$$\Rightarrow 3x^2 + \frac{d}{dx}(y^3) = \frac{d}{dx}(x^3 y^2) + 2.$$

By the chain rule, $\dfrac{d}{dx}(y^3) = 3y^2 \cdot \dfrac{dy}{dx}$.

Using the product formula and the chain rule

$$\frac{d}{dx}(x^3 y^2) = 3x^2 y^2 + 2x^3 y \cdot \frac{dy}{dx}.$$

Therefore $3x^2 + 3y^2 \dfrac{dy}{dx} = 3x^2 y^2 + 2x^3 y \dfrac{dy}{dx} + 2$

$\Leftrightarrow \qquad \dfrac{dy}{dx}(3y^2 - 2x^3 y) = 3x^2 y^2 - 3x^2 + 2$

$\Leftrightarrow \qquad\qquad \dfrac{dy}{dx} = \dfrac{3x^2 y^2 - 3x^2 + 2}{y(3y - 2x^3)},$

provided $y(3y - 2x^3) \neq 0$. □

Qu. 1. What can be said about the tangents to the curve of Example 1 at points where $y(3y - 2x^3) = 0$?

Example 2: Find a formula for the gradient of the curve

$$(x-3)^2 + (y-2)^2 = 25.$$

Solution: The equation represents a circle, centre $(3,2)$, radius 5. Differentiating with respect to x, we obtain

$$2(x-3) + 2(y-2) \cdot \frac{dy}{dx} = 0$$

$$\Leftrightarrow \qquad \frac{dy}{dx} = -\frac{x-3}{y-2} \text{ provided } y \neq 2.$$

If $y = 2$, the gradient of the circle is undefined.

Therefore at (x,y) on the circle $(x-3)^2 + (y-2)^2 = 25$, the gradient is $-\dfrac{x-3}{y-2}$.

Notice that we are dealing with the two distinct functions

$$(x-3)^2 + (y-2)^2 = 25, y \geqslant 2$$

and $\qquad (x-3)^2 + (y-2)^2 = 25, y \leqslant 2$

and that the gradient formula is the same for both.   □

Qu. 2.   Find $\dfrac{dy}{dx}$ in terms of x and y when

(i) $x^2 + y^2 = 4$ (ii) $xy^2 = 2x + 3y$ (iii) $\sin x + \sin y = 1$.

Sometimes it is easier to find the derivative of products or quotients by using logarithms, as shown below.

Example 3:   If $y = \dfrac{(x^2+5)^3}{(4x^3+1)^2}$, find $\dfrac{dy}{dx}$.

Solution: $y = \dfrac{(x^2+5)^3}{(4x^3+1)^2}$

$$\Leftrightarrow \quad \ln y = 3\ln(x^2+5) - 2\ln(4x^3+1)$$

$$\Rightarrow \frac{1}{y} \cdot \frac{dy}{dx} = \frac{3 \cdot 2x}{x^2+5} - \frac{2 \cdot 12x^2}{4x^3+1} = 6x\left(\frac{1}{x^2+5} - \frac{4x}{4x^3+1}\right)$$

$$\Rightarrow \quad \frac{dy}{dx} = 6xy\left(\frac{1}{x^2+5} - \frac{4x}{4x^3+1}\right)$$

$$= \frac{6x(x^2+5)^3}{(4x^3+1)^2}\left(\frac{1}{x^2+5} - \frac{4x}{4x^3+1}\right)$$

$$= \frac{6x(1-20x)(x^2+5)^2}{(4x^3+1)^3} \qquad\qquad □$$

Exercise 4E

1. Find $\dfrac{dy}{dx}$ in terms of x and y:

   (i) $x^3 + y^3 = 1$ (ii) $\ln(xy) = x + y$ (iii) $(x - 3)(y + 4) = 2$
   (iv) $y = k^x$, where k is constant   (v) $y = x^x$.

2. Find the gradient of the curve $x^2 + 3xy + y^2 = x + y + 8$ at the point $(1, 2)$.

3. A cylinder of radius r and height h is closed at both ends. Its total surface area is 15. Find $\dfrac{dr}{dh}$.

4. For mercury between $15°C$ and $270°C$, the vapour pressure p and the absolute temperature T are related by the equation

$$\ln(pT^c) = a - \frac{b}{T},$$

   where a, b, c are constants. Find $\dfrac{dp}{dT}$.

5. The pressure p and volume V of a gas expanding adiabatically are related by the equation $pV^\gamma = c$, where γ, c are constants. Show that $V^2 \dfrac{d^2p}{dV^2} = \gamma(\gamma + 1)p$ and find $p^2 \dfrac{d^2V}{dp^2}$ in terms of γ and V only.

6. Van der Waal's equation $\left(p + \dfrac{a}{V^2}\right)(V - b) = RT$ connects the pressure p, the volume V, and the temperature T of a gas, where a, b, R are constants. In this question take T as constant also. The critical point of a substance is when its gaseous and liquid phases have the same density: it occurs when $\dfrac{d^2p}{dV^2} = \dfrac{dp}{dV} = 0$. Find the values of p, V, T at the critical point in terms of a, b, R.

7. Find the gradient of the curve $x^2 + xy + 4y^2 = 10$ at the point $(2, 1)$. Find also the coordinates of the points at which the tangent to the curve is parallel to the y-axis.   [MEI]

8. If $x^2 + y^2 = 2y$, find $\dfrac{dy}{dx}$ in terms of x and y without first finding y in terms of x. Prove that $\dfrac{d^2y}{dx^2} = \dfrac{1}{(1 - y)^3}$.   [L]

9. Find the values of d^2y/dx^2 at the points on the curve $x^2 + xy + y^2 = 1$ at which the gradient is 1.   [L]

10. If $y = \arctan x$, prove that $\dfrac{d^2y}{dx^2} + 2x\left(\dfrac{dy}{dx}\right)^2 = 0$ and deduce that

$$\frac{d^3y}{dx^3} - 8x^2\left(\frac{dy}{dx}\right)^3 + 2\left(\frac{dy}{dx}\right)^2 = 0.$$ [L]

11. The sides BC, CA, AB of triangle ABC are of lengths $b + y, b, b - y$ respectively. Use the Sine Rule to show that $\sin A - 2 \sin B + \sin C = 0$, and find $\dfrac{dA}{dB}$ in terms of cosines of A, B, C.

12. (M) A particle of mass m hangs from a fixed point by a string of length l. At time t the string makes a small angle θ with the vertical. Show that the potential energy of the mass is $mgl (1 - \cos \theta)$ and the kinetic energy is $\frac{1}{2} m l^2 \left(\dfrac{d\theta}{dt}\right)^2$. Assuming that m, g, l, and the sum of the two energies are constant, show that either $\dfrac{d\theta}{dt} = 0$ or $\dfrac{d^2\theta}{dt^2} \approx -\dfrac{g\theta}{l}$.

13. Given that $y^3 = e^{2x} y + e^{-2x}$, by differentiating the equation with respect to x, express $\dfrac{dy}{dx}$ in terms of x and y. If $x = h$ and $y = k$ when $\dfrac{dy}{dx} = 0$, show that $ke^{4h} = 1$ and hence find the values of h and k. [MEI]

14. If $f(x) = \sqrt{(1 - x^2)} \cdot \cos^{-1} x$, where the inverse cosine is taken to be in the range $0 \leqslant \cos^{-1} x \leqslant \pi$, prove that $(1 - x^2) f'(x) + x f(x) = x^2 - 1$. Differentiate this result three times and prove, by induction or otherwise, that
$$(1 - x^2) f^{(n)}(x) - (2x - 3)x f^{(n-1)}(x) - (n - 1)(n - 3) f^{(n-2)}(x) = 0$$
when $n = 4, 5, 6, \ldots$
Obtain expressions for $f^{(2n)}(0)$ and $f^{(2n+1)}(0)$ when $n \geqslant 1$.   [O & C]

4.6 Differentiation with a parameter

Suppose C is the locus of the point (x, y) such that $x = X(t), y = Y(t)$, where X, Y are functions of a variable t; then the equations $x = X(t)$, $y = Y(t)$ are the *parametric* equations of C; the variable t is known as the *parameter*. It is common to talk about 'the point t' when we mean 'the point with parameter t'. At times, but not always, we may be able to eliminate t from the parametric equations to obtain a single equation in x and y only (see Vol. 1 §3.12). Parametric equations for a curve are particularly useful when they are simpler than the single equation representing the same curve. As with implicit functions, by applying certain restrictions to the values of t we may obtain a function $f : x \mapsto y$.

If we restrict t such that X is one-to-one, then X^{-1} exists, and we have $y = Y(t) = Y(X^{-1}(x))$.

Using the chain rule, $\dfrac{dy}{dx} = \dfrac{dy}{dt} \cdot \dfrac{dt}{dx}$. But $\dfrac{dt}{dx} = 1 \Big/ \dfrac{dx}{dt}$ (see §4.2), so that

$$\frac{dy}{dx} = \frac{dy}{dt} \bigg/ \frac{dx}{dt}, \quad \frac{dx}{dt} \neq 0.$$

The second derivative may be found by using the chain rule again:

$\frac{d^2y}{dx^2} = \frac{d}{dx}\left(\frac{dy}{dx}\right) = \frac{d}{dt}\left(\frac{dy}{dx}\right) \cdot \frac{dt}{dx} = \frac{d}{dt}\left(\frac{dy}{dx}\right) \bigg/ \frac{dx}{dt}$ provided all the

derivatives exist.

Qu. 1.   (i) At the point P on the curve C, $\frac{dx}{dt} = 0$, $\frac{dy}{dx} \neq 0$. By considering

the shape of C near to P, show that the tangent to C at P is
parallel to the y-axis.

(ii) By considering $x = t^3$, $y = kt^3$ where k is some constant, show

that $\frac{dy}{dx}$ may take any value at points where $\frac{dx}{dt} = \frac{dy}{dt} = 0$.

Example: (a) Sketch the graph of the curve given parametrically as $x = t^4$,
$y = t^3 - t + 1$.

(b) Find the angle between the tangents at the point where the
curve crosses itself.

Solution: (a) We first sketch the graphs of $x = t^4$, $y = t^3 - t + 1$, using
the same vertical axis for both graphs. By the usual methods

it may be established that $\left(-\frac{1}{\sqrt{3}}, 1 + \frac{2}{3\sqrt{3}}\right)$ and

$\left(\frac{1}{\sqrt{3}}, 1 - \frac{2}{3\sqrt{3}}\right)$ are the maximum and minimum respectively

of the curve $y = t^3 - t + 1$.

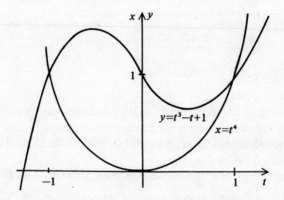

As t increases from 0 to $\frac{1}{\sqrt{3}}$, x increases from 0 to $\frac{1}{9}$, and

y decreases from 1 to $1 - \dfrac{2}{3\sqrt{3}} \approx 0.615$; initially x increases

very slowly while y decreases more rapidly, but as t
approaches $1/\sqrt{3}$, y decreases more slowly. This enables us
to sketch the arc AB.

As t increases from $1/\sqrt{3}$ to 1, x and y increase to 1;
when $t \approx 1/\sqrt{3}$, y is changing slowly. This enables us to add
the arc BC to the sketch.

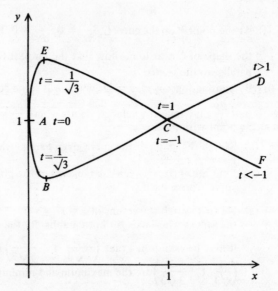

As t increases from 1, x and y increase from 1, x increasing more rapidly than y. This enables us to add the arc CD
to the sketch.

Similar arguments as t decreases from 0 allow us to complete the sketch; the point C corresponds to both $t = 1$ and
$t = -1$.

Solution: (b) $x = t^4 \qquad\Rightarrow \dfrac{dx}{dt} = 4t^3;$

$y = t^3 - t + 1 \Rightarrow \dfrac{dy}{dt} = 3t^2 - 1.$ Therefore $\dfrac{dy}{dx} = \dfrac{3t^2 - 1}{4t^3}.$

$t = 1 \;\;\Rightarrow \dfrac{dy}{dx} = \dfrac{1}{2};$ this is the gradient of arc BCD at C;

$t = -1 \Rightarrow \dfrac{dy}{dx} = -\dfrac{1}{2};$ this is the gradient of arc ECF at C.

Therefore both tangents to the curve at C make an angle of
$\tan^{-1}\tfrac{1}{2}$ with the x-axis. Thus the angle between the tangents
at C is $2\tan^{-1}\tfrac{1}{2} \approx 53°$.   □

Qu. 2.   Find $\dfrac{d^2y}{dx^2}$ in terms of t. Hence show that on both parts of the curve

through C there are points of inflexion at C.

Exercise 4F

1. Eliminate t from the following parametric equations, and also find
$\dfrac{dy}{dx}$ in terms of t:

   (i) $x = a \cos t, y = b \sin t$ (ii) $x = at^2, y = 2at$

   (iii) $x = ct, y = \dfrac{c}{t}$ (iv) $x = a \sec t, y = b \tan t$.

2. Sketch the *semi-cubical parabola*: $x = at^2, y = at^3$. Find the equation of
the tangent at the point t_1, and show that this tangent intersects the curve
again at the point with parameter $-\dfrac{t_1}{2}$.

3. Sketch the curve $x = e^{kt} \cos t, y = e^{kt} \sin t$ for (i) $k > 0$, (ii) $k = 0$,
(iii) $k < 0$. Show that $\dfrac{dy}{dx} = \tan(\alpha + t)$ where $\alpha = \cot^{-1} k$.

4. Sketch the curve with parametric equations $x = \cos t, y = \cos 2t$. A cup-
like vessel is formed by rotating this curve about the y-axis; find the
greatest volume of liquid that this vessel can hold.

5. The curve $x = a \cos^3 t, y = a \sin^3 t$ is known as an *astroid*. Sketch the
curve and find its equation in terms of x, y and a only.
The point P is on the astroid in the first quadrant; if P is the point with
parameter p, find the equation of the tangent to the astroid at P. This
tangent meets the x-axis at S and the y-axis at T. Show that ST is of
constant length. Use this property to draw an astroid.

6. A circular wheel with centre C, radius r is rolling without slipping along
horizontal ground, T being the point of contact of wheel and ground.
The locus of T is the straight line OX; Y is vertically above O. The point
P is fixed to the circumference of the wheel; initially P is at O; at time t
angle PCT is θ. Taking OX, OY as the axes, show that P is the point
(x,y) where
$$x = r(\theta - \sin\theta), \quad y = r(1 - \cos\theta).$$

Sketch the locus of P. This curve is a *cycloid*. Prove that the tangent to
the cycloid at P passes through the highest point of the wheel. Also show
that $y^2 \dfrac{d^2y}{dx^2} = -r$.

7. Given that $x(1 + t^2) = 1, y(1 + t^2) = t$, prove that $\dfrac{dy}{dx} = \dfrac{1}{2}\left(t - \dfrac{1}{t}\right)$ and
find an expression for $\dfrac{d^2y}{dx^2}$ in terms of t.   [MEI]

8. Sketch the curve given by the parametric equations $x = t + 1, y = 4 - t^2$. Find the equation of the normal to this curve at the point P at which $t = 1$. If this normal meets the curve again at Q and the x-axis at R, show that $PQ : QR = 5 : 7$.   [L]

9. A curve C has the parametric form $x = t^2, y = t^3$. Find the equation of the tangent to C at the point with parameter t.
   If the tangents to the curve at the points with parameters t_1, t_2, t_3 meet at the point (X, Y), prove that $t_1 + t_2 + t_3 = 0$, and find X and Y in terms of t_1, t_2, t_3 (or any two of them).   [O & C]

10. Sketch the curve given parametrically by the equations $x = 3t^2, y = t^3$. Show that the tangent at the point P, where $t = 2$, cuts the curve at the point Q where $t = -1$. Find the area of the finite region bounded by the curve and the line PQ.   [L]

11. Find the equation of the tangent to the curve $x = f(\theta), y = g(\theta)$ at the point P which has the parameter θ. Given that the equation of the tangent at P is $y \cos \theta = x \sin \theta - a\theta$, where a is a non-zero constant, prove that

$$f(\theta) \sin \theta - g(\theta) \cos \theta = a\theta,$$

and that

$$f(\theta) = a(\cos \theta + \theta \sin \theta).$$

Find also the corresponding expression for $g(\theta)$.
Find the equation of the normal at P and prove that it touches a fixed circle with centre at the origin. Find the radius of this circle, and the coordinates of the point at which the normal touches it.   [O & C]

4.7 Hyperbolic functions

The hyperbolic cosine and hyperbolic sine functions are defined by the identities

$$\boxed{\cosh x \equiv \frac{e^x + e^{-x}}{2}, \ \sinh x \equiv \frac{e^x - e^{-x}}{2}.}$$

We pronounce cosh as it is spelled, while sinh is pronounced as 'sine-ch' or 'shine'. (See Ex. 10D Qu. 16 for an explanation of the term 'hyperbolic function.')

Qu. 1.   Sketch the graphs of $y = e^x, y = e^{-x}, y = -e^{-x}, y = \cosh x$ and $y = \sinh x$ on the same axes. State the domain and range of the cosh and sinh functions.

The functions have many properties which are similar to those of the cosine and sine functions, most of which can be proved without difficulty from the definitions.

Qu. 2.   Prove that

$$\cosh(-x) \equiv \cosh x, \quad \sinh(-x) \equiv -\sinh x$$

$$\cosh^2 x - \sinh^2 x \equiv 1$$

$$\sinh(x+y) \equiv \sinh x \cosh y + \cosh x \sinh y$$

$$\cosh(x+y) \equiv \cosh x \cosh y + \sinh x \sinh y$$

$$\frac{d}{dx}(\cosh x) = \sinh x, \quad \frac{d}{dx}(\sinh x) = \cosh x.$$

Qu. 3.   Simplify $\cosh x - \sinh x$ and deduce that $\cosh x > \sinh x$. Also prove that $\cosh x > |\sinh x|$.

The definitions of the remaining four hyperbolic functions are similar to the definitions of the corresponding circular functions:

$$\tanh x \equiv \frac{\sinh x}{\cosh x}, \quad \coth x \equiv \frac{1}{\tanh x}, \quad \operatorname{sech} x \equiv \frac{1}{\cosh x}, \quad \operatorname{cosech} x \equiv \frac{1}{\sinh x}.$$

Suppose $y = \tanh x$. Then $y = \dfrac{e^x - e^{-x}}{e^x + e^{-x}}$ [from the definitions]

$$= \frac{1 - e^{-2x}}{1 + e^{-2x}} \quad \begin{array}{l}\text{[dividing numerator and} \\ \text{denominator by } e^x]\end{array}$$

$$\to 1 \text{ as } x \to +\infty$$

Qu. 4.   (a) Show that   (i) $\tanh x \to -1$ as $x \to -\infty$

(ii) $-1 < \tanh x < 1$

(iii) $\tanh(-x) = -\tanh x$.

(b) Show that $\dfrac{d}{dx}(\tanh x) = \operatorname{sech}^2 x$ and deduce that

$$0 < \frac{d}{dx}(\tanh x) \leqslant 1.$$

(c) Sketch the graph of $y = \tanh x$.

A convenient way of remembering many of the formulae involving hyperbolic functions is known as *Osborn's rule*: to change a trigonometric formula into a hyperbolic formula replace cos by cosh, and sin by sinh, etc., and reverse the sign of the product of two sines, tangents, cotangents, and cosecants. Thus $\cos^2 x + \sin^2 x \equiv 1$ becomes $\cosh^2 x - \sinh^2 x \equiv 1$ and

$$\tan 2x \equiv \frac{2\tan x}{1 - \tan^2 x} \text{ becomes } \tanh 2x = \frac{2\tanh x}{1 + \tanh^2 x}. \text{ An explanation of}$$

why this rule works is given in §13.3. The rule is helpful, but should be used with caution as it only applies to the basic 'usual' formulae, and not to modified identities, nor to many of the calculus formulae. For example

$$\frac{\cos^2 x}{1 - \sin x} = 1 + \sin x \quad \text{but} \quad \frac{\cosh^2 x}{1 - \sinh x} \neq 1 + \sinh x$$

and $\quad \dfrac{d}{dx}(\cos x) = -\sin x \quad$ but $\dfrac{d}{dx}(\cosh x) \neq -\sinh x.$

When solving equations involving hyperbolic functions it is generally easier to use the definitions of the functions than to use the various identities.

Example 1: Solve $\qquad \cosh x = 2 \sinh x - 1$

Solution: $\qquad\qquad\qquad \cosh x = 2 \sinh x - 1$

$\Leftrightarrow \qquad\quad \dfrac{e^x + e^{-x}}{2} = e^x - e^{-x} - 1$

$\Leftrightarrow \quad e^x - 3e^{-x} - 2 = 0$

$\Leftrightarrow \quad (e^x)^2 - 2e^x - 3 = 0$

$\Leftrightarrow \quad (e^x - 3)(e^x + 1) = 0$

$\Leftrightarrow \qquad\qquad\quad e^x = 3 \quad [e^x \text{ is positive for real } x \Rightarrow e^x \neq -1]$

$\Leftrightarrow \qquad\qquad\quad x = \ln 3.$ $\qquad\qquad\qquad\qquad\qquad$ □

Example 2: Solve for t in terms of k: $\cosh t = k, \ k \geqslant 1.$

Solution: From the graph we expect the unique solution

$\qquad\qquad t = 0$ when $k = 1,$

and a pair of solutions $t = t_1$ or t_2 (say) when $k > 1,$ with $t_1 + t_2 = 0.$

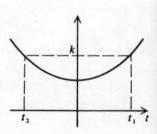

$\qquad\qquad\qquad\quad \cosh t = k$

$\Leftrightarrow \qquad\quad e^t + e^{-t} = 2k$

$\Leftrightarrow (e^t)^2 - 2ke^t + 1 = 0$

$\Leftrightarrow \qquad\qquad\qquad e^t = \dfrac{2k \pm \sqrt{(4k^2 - 4)}}{2} \quad$ [using formula for the solution of quadratic equations]

$\qquad\qquad\qquad\qquad = k \pm \sqrt{(k^2 - 1)}$

$\Leftrightarrow \qquad\qquad\qquad t = \ln(k + \sqrt{(k^2 - 1)}) = t_1$ say

$\qquad\qquad\qquad$ or $t = \ln(k - \sqrt{(k^2 - 1)}) = t_2$ say.

If $k = 1, t_1 = t_2 = \ln 1 = 0.$

If $k \geqslant 1, t_1 + t_2 = \ln\{(k + \sqrt{(k^2 - 1)})(k - \sqrt{(k^2 - 1)})\}$

$\qquad\qquad\qquad\qquad = \ln\{k^2 - (k^2 - 1)\} = \ln 1 = 0.$

It is clear from the expressions that $t_1 > t_2$ if $k > 1.$ $\qquad\qquad$ □

The hyperbolic cosine function is not one-to-one over the domain of real numbers. However, if we restrict the domain to non-negative real numbers, the cosh function is one-to-one and therefore has an inverse function.

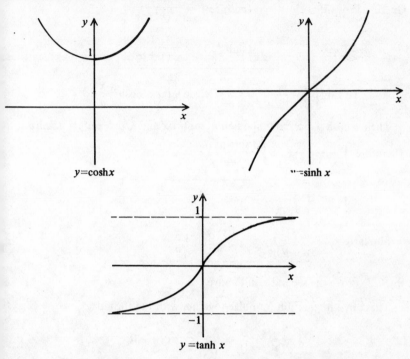

$y = \cosh x$

$y = \sinh x$

$y = \tanh x$

The sinh and tanh functions are one-to-one over the domain of real numbers, so it is unnecessary to restrict their domain when we create their inverse functions.

The inverse of the cosh function is defined as the function which maps x onto y such that $\cosh y = x$, $y \geqslant 0$; we write $y = \cosh^{-1} x$. Similarly, the inverse sinh function maps x onto y such that $\sinh y = x$; we write $y = \sinh^{-1} x$. The inverse tanh function maps x onto y such that $\tanh y = x$ and we write $y = \tanh^{-1} x$.

Perhaps surprisingly the inverse hyperbolic functions can be expressed explicitly in terms of other functions with which we are already familiar. For example, if $y = \cosh^{-1} x$, then $\cosh y = x$

$$\Leftrightarrow \qquad y = \ln(x + \sqrt{(x^2 - 1)}) \geqslant 0$$

$$\text{or} \quad y = \ln(x - \sqrt{(x^2 - 1)}) \leqslant 0$$

as in Example 2.

But $\cosh^{-1} x$ is non-negative.

Therefore $\cosh^{-1} x = \ln(x + \sqrt{(x^2 - 1)})$.

Qu. 5.   Prove that $\sinh^{-1}x = \ln(x + \sqrt{(x^2 + 1)})$.

Qu. 6.   Prove that
$$\frac{d}{dx}(\cosh^{-1}x) = \frac{1}{\sqrt{(x^2-1)}}$$
$$\frac{d}{dx}(\sinh^{-1}x) = \frac{1}{\sqrt{(1+x^2)}}$$

To find $\displaystyle\int \frac{dx}{\sqrt{(x^2-a^2)}}, 0 < a < x$, we let $u = \cosh^{-1}\dfrac{x}{a}$.

Then $a\cosh u = x$, so that $a\sinh u . du = dx$ and $\sqrt{(x^2-a^2)} = a\sinh u$.

Therefore $\displaystyle\int \frac{dx}{\sqrt{(x^2-a^2)}} = \int \frac{a\sinh u . du}{a\sinh u} = u + c$,

and
$$\int \frac{dx}{\sqrt{(x^2-a^2)}} = \cosh^{-1}\frac{x}{a} + c, a > 0.$$

Similarly
$$\int \frac{dx}{\sqrt{(a^2+x^2)}} = \sinh^{-1}\frac{x}{a} + c, a > 0.$$

Qu. 7.   Prove the second result above.

These two integration formulae together with the formulae

$$\int \frac{dx}{\sqrt{(a^2-x^2)}} = \sin^{-1}\frac{x}{a} + c, a > 0,$$

$$\int \frac{dx}{a^2+x^2} = \frac{1}{a}\tan^{-1}\frac{x}{a} + c, a \neq 0$$

should be memorized; confusion as to which inverse function is required can be avoided by asking oneself the question 'Which substitution would simplify the denominator?'

Example 3: Find (i) $\displaystyle\int \frac{dx}{\sqrt{(3 + 4x^2)}}$ (ii) $\displaystyle\int \frac{dx}{\sqrt{(4x^2 + 12x - 7)}}$.

Solution: (i) $\displaystyle\int \frac{dx}{\sqrt{(3 + 4x^2)}} = \int \frac{dx}{2\sqrt{(\frac{3}{4} + x^2)}} = \frac{1}{2}\sinh^{-1}\frac{2x}{\sqrt{3}} + c$

(ii) $\displaystyle\int \frac{dx}{\sqrt{(4x^2 + 12x - 7)}} = \int \frac{dx}{\sqrt{((2x + 3)^2 - 16)}}$

$$= \frac{1}{2}\cosh^{-1}\frac{2x + 3}{4} + c, \ x > \frac{1}{2}. \qquad \square$$

Exercise 4G

1. State and prove the identity concerning hyperbolic functions corresponding to
   (i) $\cos 2A \equiv 1 - 2 \sin^2 A$
   (ii) $\cos A - \cos B \equiv -2 \sin \dfrac{A+B}{2} \sin \dfrac{A-B}{2}$
   (iii) $\tan(A+B) \equiv \dfrac{\tan A + \tan B}{1 - \tan A \tan B}$
   (iv) $\sin 3A \equiv 3 \sin A - 4 \sin^3 A$.

2. Sketch the graphs of the following functions and state their domains and ranges:
   (i) $f : x \mapsto \operatorname{sech} x$,   (ii) $g : x \mapsto \operatorname{cosech} x$,   (iii) $h : x \mapsto \coth x$.

3. Sketch the graphs of the following:
   (i) $y = \sinh^{-1} x$,   (ii) $y = \cosh^{-1} x$,   (iii) $y = \tanh^{-1} x$.

4. Solve the equations:
   (i) $2 \sinh x + \cosh x + 1 = 0$ (ii) $10 \cosh x = 2 \sinh x + 11$
   (iii) $4 \tanh x = \coth x$ (iv) $3 \tanh x = 4(1 - \operatorname{sech} x)$.

5. Find the real values of x and y which satisfy the equations
   $$\sinh x + \sinh y = \tfrac{25}{12}$$
   $$\cosh x - \cosh y = \tfrac{5}{12}.$$ [MEI]

6. Prove that: (i) $\alpha = \ln \tan \beta \Leftrightarrow \tanh \alpha = -\cos 2\beta$
   (ii) $\alpha = \ln \tan \left(\dfrac{\pi}{4} + \dfrac{\beta}{2} \right) \Leftrightarrow \tanh \alpha = \sin \beta$.

7. Prove that if $p = \tfrac{1}{2} \ln(2 + \sqrt{5})$ and $q = \ln(1 + \sqrt{2})$ then
   $\tanh x < \sinh x < \operatorname{sech} x < \cosh x < \operatorname{cosech} x < \coth x$ if $0 < x < p$, and
   $\tanh x < \operatorname{sech} x < \sinh x < \operatorname{cosech} x < \cosh x < \coth x$ if $p < x < q$.
                                                                          [MEI]

8. Differentiate each of the following with respect to x:
   (i) $\sinh 3x$ (ii) $\cosh(5x - 2)$ (iii) $\sinh^2 x$
   (iv) $\sinh(\ln x)$ (v) $\tanh^{-1} x$ (vi) $\tanh^{-1}(\sin x)$
   (vii) $\cosh^{-1}(\sec x)$ (viii) $\tan^{-1}(\sinh x)$.

9. Integrate each of the following with respect to x:
   (i) $\sinh^2 x$ (ii) $e^x \cosh x$ (iii) $x \sinh x$
   (iv) $\sinh^{-1} x$ (v) $\cosh^{-1} x$ (vi) $\cosh 2x \cosh 4x$
   (vii) $\sinh^3 x$ (viii) $\coth^2 x$.

10. Find: (i) $\int \operatorname{sech} x \, dx$ Hint: use the substitution $u = e^x$.
    (ii) $\int e^x \tanh x \, dx$.

11. Integrate the following with respect to x:
    (i) $\dfrac{1}{\sqrt{(4 + x^2)}}$ (ii) $\dfrac{1}{\sqrt{(x^2 - 9)}}$ (iii) $\dfrac{1}{\sqrt{(9x^2 + 16)}}$

(iv) $\dfrac{1}{\sqrt{(7x^2 - 5)}}$ (v) $\dfrac{1}{\sqrt{(x^2 - 6x + 13)}}$ (vi) $\dfrac{1}{\sqrt{(9x^2 + 6x - 8)}}$

(vii) $\dfrac{1}{\sqrt{(25x^2 + 20x + 4)}}$.

12. Find the Maclaurin expansions for the following up to and including the term in x^4: (i) $\cosh x$ (ii) $\sinh x$ (iii) $\tanh^{-1} x$.

13. Show that the Maclaurin expansions for $\ln(x + \sqrt{(1 + x^2)})$ and $\sin x$ are identical as far as the term involving x^4.

14. Show that $\displaystyle\int_0^x \sqrt{(1 + p^2)}\, dp = \tfrac{1}{2}x\sqrt{(1 + x^2)} + \tfrac{1}{2}\ln(x + \sqrt{(1 + x^2)})$.   [MEI]

15. If $-\dfrac{\pi}{2} < x < \dfrac{\pi}{2}$ and a is any real constant, show that the equation

$\sin x = \tanh a$ has just one solution, and prove that $\tan x = \sinh a$ and $\sec x = \cosh a$ for this value of x.

16. (i) Find $\dfrac{d}{dx}(\tanh^{-1} x)$ and deduce $\displaystyle\int \dfrac{dx}{1 - x^2}$.

   (ii) Show that $\dfrac{1}{1 - x^2} \equiv \dfrac{1}{2}\left(\dfrac{1}{1 + x} + \dfrac{1}{1 - x}\right)$ and hence find an alterna-

   tive form for $\displaystyle\int \dfrac{dx}{1 - x^2}$.

   (iii) Prove that $\tanh^{-1} x \equiv \dfrac{1}{2}\ln\left(\dfrac{1 + x}{1 - x}\right)$

17. Show that the function $y = A\cosh nx + B\sinh nx$, where A, B, n are constants, satisfies the equation $d^2y/dx^2 = n^2 y$.
    If $n = \tfrac{1}{2}$, find the values of A, B so that $y = 2$, $dy/dx = \tfrac{1}{2}$ when $x = 0$.
    With these values of A, B, n, find and identify the stationary value of y and sketch the graph of y against x.

18. Find the area enclosed by the curves $y = \cosh x$, $y = \sinh x$, $x = 0$, $x = 1$.

19. If $\sinh a = \tfrac{3}{4}$ show that $\cosh a = \tfrac{5}{4}$ and $a = \ln 2$. Obtain the first five terms in the Taylor series for $\cosh x$ in powers of $(x - \ln 2)$.   [L]

20. Any heavy, uniform, perfectly flexible cable suspended from two points (not in the same vertical line) takes the shape of a *catenary*, a curve

whose equation is $y = c\cosh\dfrac{x}{c}$. Let P be any point on this curve; let M

be the foot of the perpendicular from P to the x-axis. The point Q is on the tangent to the curve at P, such that MQ is perpendicular to PQ.
Prove that   (i) $MQ = c$
              (ii) the x-coordinate of P is $c\ln\tan\left(\dfrac{\pi}{4} + \dfrac{\theta}{2}\right)$, where $\tan\theta$ is
                    the gradient of PQ
              (iii) the product of the y-coordinates of P and Q is c^2.

5 Complex numbers I

5.1 The development of the number system

As our number system has evolved, the reason for each stage in its development has been the desire to solve some new practical or theoretical problem. For many thousands of years the counting numbers $1, 2, 3, \ldots$ met all the needs of primitive man.

Qu. 1. Show that the set of counting numbers is closed under the operations of addition and multiplication, but not under subtraction and division. Does the set contain an identity for addition or for multiplication? Are there additive or multiplicative inverses?

As society grew more complex and trading increased, the concept of a fraction arose; the earliest record we have of a systematic way of writing fractions is in an Egyptian papyrus of about 1650 B.C. The Greeks were the first to study mathematics for its intellectual interest rather than its usefulness. By 500 B.C. Pythagoras and his followers had developed the belief that everything in geometry and practical mathematics could be explained in terms of whole numbers or their ratios (or, as we would put it, in terms of positive *rational* numbers).

Qu. 2. Under which of the operations $+, -, \times, \div$ is the set of positive rational numbers closed? What is the multiplicative inverse of $\dfrac{p}{q}$?

Qu. 3. Prove that between any two positive rational numbers there is another positive rational number. This property is described by saying that the positive rational numbers are *dense*.

Qu. 4. Prove that $\dfrac{p+r}{q+s}$ lies between $\dfrac{p}{q}$ and $\dfrac{r}{s}$.

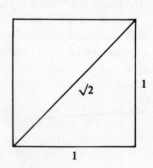

The discovery about a century later that even such a simple length as the diagonal of a unit square (which is $\sqrt{2}$ by Pythagoras' Theorem) is not a rational number came as a great shock, for it showed that although the rational numbers are everywhere dense on the number line, there are still points on the line which do not represent any rational number.

One of the major achievements of Greek mathematics is the theory of proportion in which Eudoxus ($c.$ 370 B.C.) managed to include these *irrational* numbers in a satisfactory logical framework, so that by this time the number system consisted of the positive rational and irrational numbers.

Qu. 5.   Proof that $\sqrt{2}$ is irrational.
   (i)   Suppose that $\sqrt{2}$ is rational, i.e. that $\sqrt{2} = \dfrac{p}{q}$ where p and q are integers with no common factor.
   (ii) Show that $p^2 = 2q^2$, so that p^2 is even.
   (iii) By considering $(2k + 1)^2$, show that the square of an odd integer is odd, and deduce that p is even. Let $p = 2m$, where m is an integer.
   (iv) Deduce that $q^2 = 2m^2$, and hence that q is also even. Thus p and q have the common factor 2, which contradicts (i). So we conclude that $\sqrt{2}$ is irrational.

Qu. 6.   Adapt Qu. 5 to show that $\sqrt{3}$ is irrational, and see how this method breaks down for $\sqrt{4}$.

Qu. 7.   Prove that $\sqrt[3]{2}$ is irrational.

Qu. 8.   Prove that the positive irrational numbers are dense.

Qu. 9.   Prove that if α and β are rational, then $\alpha + \beta\sqrt{2}$ is irrational.

Qu. 10. Give either a proof or a counterexample for each of the following statements about positive numbers:
   (i)   rational + irrational   = irrational;
   (ii) irrational + irrational = irrational;
   (iii) rational $\times$ irrational   = irrational;
   (iv) irrational $\times$ irrational = irrational.

The development of negative numbers took a surprisingly long time. About a thousand years after Eudoxus, the Hindu mathematician Brahmagupta ($c.$ 630 A.D.) gave the first definite mention of negative numbers, and stated the usual rules of signs. It was even later, in 876 in India, that the earliest undoubted zero occurs in writing, though it had probably been known earlier. This completed the set of *real* numbers, which consists of all positive and negative rational and irrational numbers and zero, and which can be represented by the complete real number line.

Qu. 11. Draw a single Venn diagram taking {real numbers} as the universal set, and showing the following subsets: {integers}, {rational numbers}, {irrational numbers}, {positive numbers}, {negative numbers}, {zero}. Mark the numbers 3, $\frac{22}{7}$, π (which is irrational), $-1, -1.4142, -\sqrt{2}$ in your diagram.

These Hindu mathematicians knew how to solve quadratic equations, and soon realized that some equations have no real solutions. For example there is no real number x such that $x^2 = -1$, and if we try to use 'the formula' to solve

$$3x^2 + 4x + 5 = 0$$

we get
$$x = \frac{-4 \pm \sqrt{(16 - 60)}}{6} = \frac{-4 \pm \sqrt{(-44)}}{6};$$

since -44 has no real square root we can go no further. For many years there was no reason to worry about this; it was accepted that such quadratic equations had no solution, just as the Greeks had accepted that $x + 5 = 3$ had no solution. But by the middle of the 16th century, the Italians Tartaglia and Cardan had discovered how to solve cubic and quartic (fourth degree) equations. Their method for solving cubic equations forced mathematicians to take the square roots of negative numbers seriously, since the real solutions of some cubic equations could not be reached without some understanding of the non-real solutions of a related quadratic equation (see Ex. 5A Qu. 28). From then on the square roots of negative numbers were found increasingly useful, and became an essential tool in mathematics. For nearly four hundred years however there was considerable unease about their respectability, and it was not until 1831 that Gauss published the first really satisfactory account of what he called *complex* numbers.

5.2 Working with complex numbers

Following the historical pattern, we shall first learn how to use complex numbers, and then in §13.4, when they are more familiar, we shall look again at the question of their formal definition.

So let us assume that we can add to the set of real numbers an element we call j which combines with itself and with the real numbers according to the usual laws of algebra, but which has the extra property that $j^2 = -1$. The notation i instead of j is also commonly used. The square roots of other negative numbers can be expressed in terms of j, for example
$$\sqrt{(-484)} = \sqrt{(484 \times (-1))} = \sqrt{(484)} \times \sqrt{(-1)} = 22j.$$

The solutions of the quadratic equation $t^2 - 4t + 125 = 0$ are given by the formula as
$$t = \frac{4 \pm \sqrt{-484}}{2} = \frac{4 \pm 22j}{2} = 2 \pm 11j,$$

and we can easily check that these do satisfy the equation. For

$$t = 2 + 11j \Rightarrow t^2 - 4t + 125 = (2 + 11j)^2 - 4(2 + 11j) + 125$$
$$= 4 + 44j + 121j^2 - 8 - 44j + 125$$
$$= 4 + 44j - 121 - 8 - 44j + 125$$
$$\text{(since } j^2 = -1\text{)}$$
$$= 0.$$

Qu. 1. What laws of algebra are being used here? Check that $2 - 11j$ is also a solution.

A number of the form $x + yj$, where x and y are real, is called a complex number. We call x the *real part* of the complex number, and y the *imaginary part*. The terms 'real' and 'imaginary' were introduced by Descartes in 1637, and are now too well established to be changed, though Gauss' work has made it clear that complex numbers are just as real as, say, rational numbers. (Or should we say that rational numbers are just as imaginary as complex numbers?) Notice in particular that the imaginary part y of $x + yj$ is a real number!

If $x_1 = x_2$ and $y_1 = y_2$, then the complex numbers $x_1 + y_1 j$ and $x_2 + y_2 j$ are equal. The addition, subtraction and multiplication of complex numbers is straightforward:

$$(x_1 + y_1 j) + (x_2 + y_2 j) = (x_1 + x_2) + (y_1 + y_2)j$$
$$(x_1 + y_1 j) - (x_2 + y_2 j) = (x_1 - x_2) + (y_1 - y_2)j$$
$$(x_1 + y_1 j)(x_2 + y_2 j) = x_1 x_2 + x_1 y_2 j + y_1 x_2 j + y_1 y_2 j^2$$
$$= (x_1 x_2 - y_1 y_2) + (x_1 y_2 + y_1 x_2)j$$
$$\text{since } j^2 = -1.$$

From these results we can make a useful deduction:

$$x_1 + y_1 j = x_2 + y_2 j \Rightarrow x_1 - x_2 = (y_2 - y_1)j$$
$$\text{(subtracting } x_2 + y_1 j \text{ from both sides)}$$
$$\Rightarrow (x_1 - x_2)^2 = ((y_2 - y_1)j)^2$$
$$\Rightarrow (x_1 - x_2)^2 = -(y_2 - y_1)^2.$$

Now $x_1 - x_2$ and $y_2 - y_1$ are real numbers, so their squares cannot be negative. Unless both sides of the final equation are zero we have a contradiction. Thus $\quad x_1 + y_1 j = x_2 + y_2 j \Rightarrow x_1 - x_2 = 0$ and $y_2 - y_1 = 0$
$$\Rightarrow x_1 = x_2 \text{ and } y_1 = y_2.$$

This shows that equality of real and imaginary parts is a necessary as well as sufficient condition for equality of complex numbers. (In case this seems entirely trivial, remember that when rational numbers are constructed from pairs of integers $\dfrac{x_1}{y_1} = \dfrac{x_2}{y_2} \not\Rightarrow x_1 = x_2$ and $y_1 = y_2$.)

In order to divide $x_1 + y_1 j$ by $x_2 + y_2 j$, we have to find real numbers p and q such that

$$p + qj = \frac{x_1 + y_1 j}{x_2 + y_2 j}$$

or $(p + qj)(x_2 + y_2 j) = x_1 + y_1 j.$

Expanding the left-hand side and equating real and imaginary parts gives the simultaneous equations

$$px_2 - qy_2 = x_1$$
$$py_2 + qx_2 = y_1$$

which have the solutions $p = \dfrac{x_1 x_2 + y_1 y_2}{x_2{}^2 + y_2{}^2}$, $q = \dfrac{y_1 x_2 - x_1 y_2}{x_2{}^2 + y_2{}^2}$.

Thus

$$\boxed{\frac{x_1 + y_1 j}{x_2 + y_2 j} = \frac{x_1 x_2 + y_1 y_2}{x_2{}^2 + y_2{}^2} + \frac{y_1 x_2 - x_1 y_2}{x_2{}^2 + y_2{}^2}\, j}$$

unless $x_2{}^2 + y_2{}^2 = 0$, i.e. unless $x_2 = y_2 = 0$, in which case the division is impossible since the equations have no solution (or arbitrary solutions if $x_1 = y_1 = 0$ also).

Qu. 2. Express $\dfrac{1}{x + yj}$ in the form $p + qj$.

In practice it is easier to obtain the same result by noting that $(x_2 + y_2 j)(x_2 - y_2 j) = x_2{}^2 - y_2{}^2 j^2 = x_2{}^2 + y_2{}^2$, which is real.

For example
$$\frac{7 + 4j}{3 + 5j} = \frac{7 + 4j}{3 + 5j} \times \frac{3 - 5j}{3 - 5j}$$

$$= \frac{21 - 35j + 12j - 20j^2}{3^2 + 5^2}$$

$$= \frac{41}{34} - \frac{23}{34}\, j.$$

The complex number $x - yj$ is called the *complex conjugate* of $x + yj$, and the first step in dividing by a complex number is usually to multiply the numerator and denominator by the conjugate of the denominator. Note that $x + jy$ and $x - jy$ are factors of $x^2 + y^2$, which until now we could not factorize.

Exercise 5A

1–14. Express the following in the form $x + yj$.

1. $(3 + 4j) + (8 - 2j)$ 2. $(4 + 9j) - (1 + 3j)$ 3. $(8 - 7j) - (5 + 3j)$

4. $(1 + 3j)(5 + 2j)$ 5. $(3 - 6j)(2 + 9j)$ 6. $2j(8 - 5j)$

7. $(7 + 6j)^2$ 8. $(3 + 3j)(4 - 4j)$ 9. $(6 + j)(5 - 7j)(2 - 3j)$

10. $(3 - 4j)^3$ 11. $\dfrac{1 + 5j}{1 + j}$ 12. $\dfrac{5 - j}{6 + 2j}$ 13. $\dfrac{4 - 7j}{3 - j}$

14. $\dfrac{47 + 33j}{5 - 3j}$

15. There are two square roots of 1, $+1$ and -1. Show that $1, -\dfrac{1}{2} + \dfrac{\sqrt{3}}{2}j$, $-\dfrac{1}{2} - \dfrac{\sqrt{3}}{2}j$ are three cube roots of 1. Find four fourth roots of 1.

16. Simplify (i) j^3; (ii) j^4; (iii) j^{29}; (iv) $\dfrac{1}{j}$; (v) $\dfrac{1}{j^{46}}$.

17. Evaluate (i) $\displaystyle\sum_{n=1}^{100} j^n$; (ii) $\displaystyle\sum_{n=-17}^{33} j^n$.

18. Solve the equations   (i) $z^2 - 4z + 13 = 0$ (ii) $9z^2 + 12z + 29 = 0$
    (iii) $5z^2 + 6z + 7 = 0$ (iv) $z^2 + 2jz - 5 = 0$.

19. Find quadratic equations in the form $az^2 + bz + c = 0$ whose roots are

    (i) $\dfrac{3j}{2}, \dfrac{-3j}{2}$ (ii) $5 + 2j, 5 - 2j$

    (iii) $-7 + j, -7 - j$ (iv) $4 + \sqrt{6} \cdot j, 4 - \sqrt{6} \cdot j$
    (v) $-3, 1 + 2j$ (vi) $4 + j, 5 - 3j$.

20. Find real numbers a and b with $a > 0$ such that
    (i) $(a + bj)^2 = -3 + 4j$ (ii) $(a + bj)^2 = 9 + 40j$
    (iii) $(a + bj)^2 = j$.

21. Given that $z = 2 + 3j$ is a solution of the equation

$$z^2 + (a - j)z + 16 + bj = 0,$$

    where a and b are real, find a, b and the other solution of the equation.

22. Solve the equation $z^2 + (-5 + j)z + 6 - 2j = 0$, giving the solutions in the form $x + yj$.

23. One solution of the cubic equation $z^3 - 3z^2 - 6z - 20 = 0$ is a positive integer. Find all three solutions.

24. Find all three solutions of $z^3 - 2z^2 - 8z + 21 = 0$.

25. Prove that if the complex number $p + qj$ is a solution of a quadratic or cubic equation with *real* coefficients, then so is the conjugate complex number $p - qj$.

26. (i) Prove that $(1 + j)^2 = 2j$ and $(1 - j)^2 = -2j$.
   (ii) By writing $z^4 + 4$ as $z^4 - (2j)^2$ and using (i), express $z^4 + 4$ as the product of four complex linear factors.
   (iii) By taking the product of conjugate pairs of these linear factors, express $z^4 + 4$ as the product of two real quadratic factors, and check your answer by expanding these.

27. Use Question 26 to solve the equations (i) $z^4 + 4 = 0$; (ii) $z^4 = -1$.

28. The Cardan-Tartaglia method for solving $x^3 = px + q$.

   (i) Let $x = u + v$, where $uv = \dfrac{p}{3}$. Show that $u^3 + v^3 = q$ and deduce that $u^3 + \dfrac{p^3}{27u^3} = q$.
   (ii) Put $t = u^3$ and obtain the quadratic equation $t^2 - qt + \dfrac{p^3}{27} = 0$. Let the solutions of this be $t = t_1$ and $t = t_2$.
   (iii) Show that if $u^3 = t_1$ then $v^3 = t_2$ and *vice versa*. Deduce that one solution of $x^3 = px + q$ is $x = \sqrt[3]{t_1} + \sqrt[3]{t_2}$.
   (iv) Use this method to find one solution of $x^3 = 6x + 6$ to 3 D.P. Check your answer by substitution.

29. Verify that the method of Qu. 28 applied to the equation $x^3 = 15x + 4$ leads to the quadratic equation $t^2 - 4t + 125 = 0$, and hence to the solution $x = \sqrt[3]{(2 + 11j)} + \sqrt[3]{(2 - 11j)}$.

When Rafael Bombelli (1526–1573) considered this solution he had 'a wild thought': that the cube roots of conjugate complex numbers might themselves be conjugate, so that if $\sqrt[3]{(2 + 11j)} = a + bj$ then $\sqrt[3]{(2 - 11j)} = a - bj$. Then $x = a + bj + a - bj = 2a$, which is real. Bombelli was trying to obtain the obvious solution $x = 4$, so he naturally took $a = 2$. Find b, and check that Bombelli's wild thought was correct.

5.3 Representing complex numbers geometrically

The fact that the complex number $x + yj$ is specified by the ordered pair (x, y) of real numbers leads naturally to a very useful way of picturing complex numbers. Referred to a set of rectangular Cartesian axes in a plane, the complex number $x + yj$ is represented by the point (x, y).

Thus $2 + j$ is represented by $(2, 1)$
        $-4 - 2j$ is represented by $(-4, -2)$
        $2j$ is represented by $(0, 2)$
        3 is represented by $(3, 0)$, as shown on the next page.

All real numbers are represented by points on the x-axis, which is therefore called the *real axis*. Complex numbers of the form $0 + yj$ are called *pure imaginary*, and are represented by points on the y-axis, which is also called

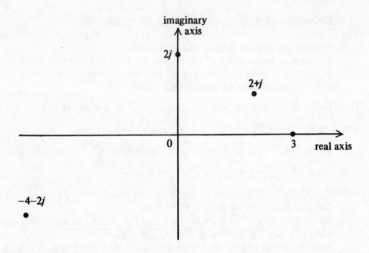

the *imaginary axis*. This visual interpretation of complex numbers is called the *complex plane* or the *Argand diagram* after J. R. Argand who published an account of it in 1806, though the Norwegian surveyor Caspar Wessel really deserves the credit for the idea, which he described in 1797 in a paper to the Danish Academy of Sciences.

The Argand diagram represents each complex number by a single point, so it is useful to use a single letter (often z or w) to denote a complex number. The real and imaginary parts of z are denoted by $\mathrm{Re}(z)$ and $\mathrm{Im}(z)$ respectively: for example, if $z = 5 - 3j$ then $\mathrm{Re}(z) = 5$ and $\mathrm{Im}(z) = -3$.

The complex conjugate of z is denoted by z^*, so that if $a = x + yj$ then $z^* = x - yj$. The point representing z^* is the reflection in the real axis of the point representing z. The notation $\bar{z}$ instead of z^* is also common.

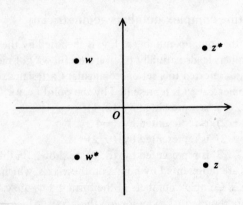

Qu. 1.   How are the points representing z and $-z$ related?

The following properties of conjugates should be noted.

1. $(z^*)^* = z$

2. $(z_1 + z_2)^* = z_1{}^* + z_2{}^*$

3. $(z_1 z_2)^* = z_1{}^* z_2{}^*$

4. $\left(\dfrac{z_1}{z_2}\right)^* = \dfrac{z_1{}^*}{z_2{}^*}$ $(z_2 \neq 0)$

5. $\operatorname{Re}(z) = \dfrac{1}{2}(z + z^*)$, $\operatorname{Im}(z) = \dfrac{1}{2j}(z - z^*)$

6. z is real $\Leftrightarrow z = z^*$
   z is pure imaginary $\Leftrightarrow z = -z^*$.

The proofs follow immediately from the definitions. For example, to prove 3, let $z_1 = x_1 + y_1 j, z_2 = x_2 + y_2 j$.

Then
$$z_1 z_2 = (x_1 x_2 - y_1 y_2) + (x_1 y_2 + y_1 x_2)j$$
so
$$(z_1 z_2)^* = (x_1 x_2 - y_1 y_2) - (x_1 y_2 + y_1 x_2)j$$
$$= (x_1 - y_1 j)(x_2 - y_2 j)$$
$$= z_1{}^* z_2{}^*.$$

Qu. 2.   Prove properties $1, 2, 5, 6$.

Qu. 3.   Let $w = \dfrac{z_1}{z_2}$ so that $w z_2 = z_1$. Use property 3 to prove property 4.

If $z_1 = x_1 + y_1 j$ and $z_2 = x_2 + y_2 j$ then $z_1 + z_2 = (x_1 + x_2) + (y_1 + y_2)j$. So to add complex numbers, we add the real and imaginary parts separately. The points representing $0, z_1, z_1 + z_2, z_2$ form a parallelogram. This immediately reminds us of vector addition: $\begin{pmatrix} x_1 \\ y_1 \end{pmatrix} + \begin{pmatrix} x_2 \\ y_2 \end{pmatrix} = \begin{pmatrix} x_1 + x_2 \\ y_1 + y_2 \end{pmatrix}$, and suggests

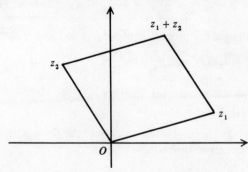

another useful geometrical interpretation, in which $z = x + yj$ is represented by the column vector $\begin{pmatrix} x \\ y \end{pmatrix}$. This is the position vector of the point (x, y) which represents z in the Argand diagram, though any other directed line segment with the same magnitude, direction and sense will represent z equally well.

For example, $z_1 - z_2$ can be thought of in two ways

    (A) the complex number which must be added to z_2 to give z_1

    (B) the sum $z_1 + (-z_2)$.

These give rise to the distinct vector diagrams (A) and (B), and in (A) the vector representing $z_1 - z_2$ is not a position vector.

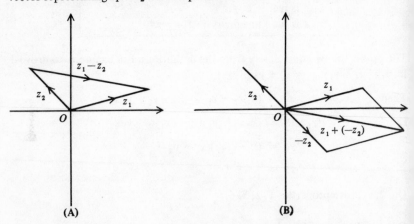

             (A)   (B)

The length of the vector representing z is called the *modulus* of z, and is denoted by $|z|$. Thus $z = x + yj \Rightarrow |z| = \sqrt{(x^2 + y^2)}$. If we are using the Argand diagram point representation, then $|z|$ is the distance from the origin of the point representing z. If z is real, $z = x$ say, then $|z| = \sqrt{x^2}$, so that $|z|$ is the absolute value of x, which fits in with our previous use of the modulus sign.

Notice that
$$\boxed{|z|^2 = x^2 + y^2 = (x + yj)(x - yj) = zz^*}$$

Qu. 4.   Use this and the properties of conjugates to show that
$$|z_1 z_2| = |z_1||z_2| \quad \text{and} \quad \left| \frac{z_1}{z_2} \right| = \frac{|z_1|}{|z_2|} \qquad (z_2 \neq 0).$$

Qu. 5.   Find particular complex numbers z_1, z_2 for which
$$|z_1 + z_2| \neq |z_1| + |z_2|.$$

Qu. 6.   Show that $|z_1 - z_2|$ is the distance of the point representing z_1 from the point representing z_2.

The point and vector representations are so closely linked that they are sometimes used simultaneously. For example in the diagram below, $2 + j$ and $3 + 5j$ are represented by points, while $1 + 4j$ is represented by the directed line segment joining them.

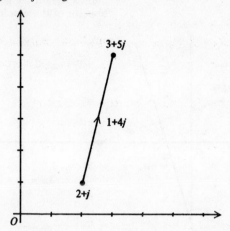

From now on, for brevity, we shall use 'the point $2 + j$' for 'the point representing $2 + j$', 'the vector $1 + 4j$' for 'the vector representing $1 + 4j$', and so on.

Example: Draw separate Argand diagrams showing the sets of points z where
   (i) $|z - 2| \leqslant 3$;
   (ii) $|z - 2| = |z + 1 - j|$.

Solution: (*i*) $|z - 2| \leqslant 3$ ⟺ the vector $z - 2$ from the point 2 to the point z has length $\leqslant 3$
   ⟺ the point z lies on or within the circle with centre at the point 2 and radius 3.

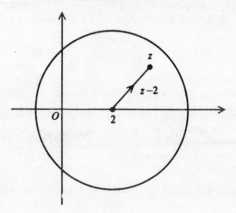

(ii) $|z-2| = |z+1-j|$ ⇔ vectors $z-2$ and $z+1-j$ have
equal lengths
 ⇔ point z is equidistant from the
points 2 and $-1+j$
 ⇔ point z lies on the perpendicular
bisector of these points.

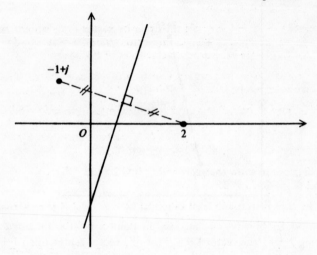

□

Exercise 5B

1. If $z = 1 + 3j$ and $w = 4 + 2j$, plot the following points on a single
Argand diagram:
  (i) z (ii) w (iii) $z + w$ (iv) $-w$ (v) $z - w$
  (vi) $w - z$ (vii) w^* (viii) $-z^*$ (ix) $4z - w$ (x) $3w - 2z$

  (xi) wz (xii) w^2 (xiii) $\dfrac{1}{z}$ (xiv) $\dfrac{z}{w}$.

2. If $z = 1 + j$, plot the points z^n on a single Argand diagram for
$n = -1, 0, 1, 2, 3, 4, 5$. Join each point to its predecessor and to the
origin.

3. If $z = -4 + 2j$, plot the following points on a single Argand diagram:
  (i) z (ii) $3z$ (iii) $\frac{1}{2}z$ (iv) $0z$ (v) $-\frac{1}{4}z$
  (vi) $-z$ (vii) $-2z$.
What is the relation between the points z and λz, where λ is real?

4. (i) Represent the complex numbers $2 + 3j$ and $j(2 + 3j)$ by position
vectors on a single diagram.
  (ii) Do likewise for (a) $-4 - j$ and $j(-4 - j)$; (b) $5 - 2j$ and $j(5 - 2j)$.
  (iii) What is the relation between the vectors z and jz?
  (iv) Interpret the statement $j^2 = -1$ geometrically.

5. Prove that the point $\frac{1}{2}(z_1 + z_2)$ is halfway between the points z_1 and z_2.

6. Prove that points z_1, z_2, z_3, z_4 are the vertices in this order round a parallelogram $\Leftrightarrow z_1 + z_3 = z_2 + z_4$, provided that the term 'parallelogram' is taken to include certain degenerate cases, which should be described.

7. If the points $-3 + 2j, -1 - j, 4 - 3j$ are three vertices of a parallelogram, find all the possible fourth vertices. Draw all the parallelograms on a single Argand diagram.

8. The points $6 + 2j, -4 + 8j$ are the opposite vertices of a square. What is the centre of the square? What is the vector from the centre to $6 + 2j$? Use Qu. 4 (iii) to find the other two vertices of the square.

9. The points $2z, 2w$ are the opposite vertices of a square. Prove that the other two vertices are $(1 \pm j)z + (1 \mp j)w$.

10. Draw separate Argand diagrams showing the sets of points z where
    (i) $|z| = 3$ (ii) $|z - 3| \leqslant 1$ (iii) $|z - 2j| = 2$
    (iv) $|z + 2| > 4$ (v) $|z + 3 + 2j| = 3$ (vi) $|4 - 3j - z| < 5$
    (vii) $|z - 5 + j| = 0$ (viii) $1 \leqslant |z + 4j| \leqslant 5$.

11. Use a diagram to show that if $|z - 2 - j| \leqslant 2$ then
    $\sqrt{5} - 2 \leqslant |z| \leqslant \sqrt{5} + 2$.

12. What are the greatest and least values of $|z + 3 - 2j|$ if $|z - 5 + 4j| \leqslant 3$?

13. Prove that $\{z : |z - 2 + 7j| \geqslant 10\} \cap \{z : |z + 4 + 2j| \leqslant 2\} = \emptyset$.

14. Draw separate Argand diagrams showing the sets of points z where
    (i) $|z| = |z - 2|$ (ii) $|z| \geqslant |z + 3j|$ (iii) $\mathrm{Re}(z) = 5$
    (iv) $|z + 1 - j| = |z - 1 + j|$ (v) $\mathrm{Im}(z + 4 - 3j) < 0$
    (vi) $|4 + 4j - z| = |3 - j + z|$ (vii) $\mathrm{Re}(z + 5 + 7j) = \mathrm{Im}(z - 2 - 6j)$
    (viii) $|z + 8 + 2j| = |z - 6 + 2j| = |z - 4 - 4j|$.

15. Prove that for all integers n (positive or negative)
    (i) $(z^n)^* = (z^*)^n$ (ii) $|z^n| = |z|^n$.

16. Prove that if $(7 + 2j)^n = x_n + y_n j$ (where x_n and y_n are real) then
    $x_n^2 + y_n^2 = 53^n$.
    Hence express 53^3 as the sum of two perfect squares.

17. Use conjugates to prove that $|z - w|^2 + |z + w|^2 = 2|z|^2 + 2|w|^2$. Deduce that the sum of the squares of the sides of a parallelogram equals the sum of the squares of the diagonals.

18. Prove that $zw^* + z^*w$ is real and that $zw^* - z^*w$ is pure imaginary.

19. (i) The points Z, W represent the complex numbers z, w in the Argand diagram, where $zw \neq 0$. Explain why

    OW is perpendicular to **OZ** $\Leftrightarrow w = \lambda jz$ where λ is real.

    Deduce that **OW** is perpendicular to **OZ** $\Leftrightarrow zw^* + z^*w = 0$.
    (ii) The points A, B, C, D represent the complex numbers a, b, c, d. Prove that

    AB is perpendicular to **CD**
    $$\Leftrightarrow ac^* + a^*c + bd^* + b^*d = ad^* + a^*d + bc^* + b^*c.$$

20. Use the result of Qu. 19 (i) to prove that if $|z| = |w|$ then the vectors $z + w$ and $z - w$ are perpendicular.
    Interpret this geometrically (i) as a statement about a rhombus
                                       (ii) as a statement about a circle.

21. Show by using conjugates that if k is positive and $k \neq 1$ then
$$|z - k^2| = k|z - 1| \; \Leftrightarrow \; |z| = k,$$
    and give a geometrical interpretation.
    Examine and comment upon the exceptional case, $k = 1$.

22. Prove that if $|a| = |b|$, then the general point on the line through the point p perpendicular to the line joining points a and b can be written as $p + \lambda(a + b)$, where λ is real.
    $ABCD$ is a cyclic quadrilateral. Through the midpoint of each side or diagonal a line is drawn perpendicular to the opposite side or other diagonal. Prove that these six lines are concurrent.

5.4 The triangle inequalities

One important difference between the set of real numbers and the set of complex numbers is that the real numbers can be arranged in order, whereas the complex numbers can not. This is not really surprising, since real numbers can be represented by points in order along a line, but complex numbers need two dimensions (points or vectors in a plane) for their representation. It is impossible to define a satisfactory order relation (such as 'less than') in the set of complex numbers, and statements like '$4 + 3j > 2 - 7j$' or '$5 + 8j$ is positive' are meaningless. But since the modulus of a complex number is real we can have inequalities involving moduli, and some of these are dealt with now.

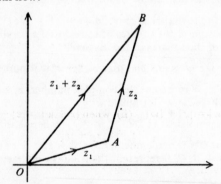

If the complex numbers z_1, z_2 are represented by vectors **OA**, **AB** then their sum $z_1 + z_2$ is represented by **OB**, using the triangle law of addition.

Since in any triangle OAB

$$OB \leqslant OA + AB$$

it follows that

$$|z_1 + z_2| \leqslant |z_1| + |z_2|, \tag{1}$$

with equality if and only if the vectors z_1, z_2 have the same direction and sense (or $z_1 = 0$ or $z_2 = 0$) in which case the 'triangle' OAB is a straight line segment.

Replacing z_1 by $z_1 + z_2$ and z_2 by $-z_2$ in (1) gives

$$|(z_1 + z_2) + (-z_2)| \leqslant |z_1 + z_2| + |-z_2|$$

or

$$|z_1| \leqslant |z_1 + z_2| + |z_2|$$

so that

$$|z_1 + z_2| \geqslant |z_1| - |z_2|. \tag{2}$$

Similarly, by interchanging z_1 and z_2,

$$|z_1 + z_2| \geqslant |z_2| - |z_1| \tag{3}$$

Statements (2) and (3) show that $|z_1 + z_2|$ is greater than or equal to the positive difference between $|z_1|$ and $|z_2|$, i.e. that

$$|z_1 + z_2| \geqslant ||z_1| - |z_2||.$$

Qu. 1. Give a geometrical interpretation of this last statement. When will equality occur?

Thus we have the following bounds on the modulus of $z_1 + z_2$:

$$\boxed{||z_1| - |z_2|| \leqslant |z_1 + z_2| \leqslant |z_1| + |z_2|.}$$

These are called the *triangle inequalities*. They can also be proved directly without using geometry. (See Ex. 5C, Qu. 5.)

Exercise 5C

1. If $z_1 = 7 + 5j$, $z_2 = 12 + 7j$ find $|z_1|$, $|z_2|$ and $|z_1 + z_2|$, and check that $|z_1 + z_2| \leqslant |z_1| + |z_2|$.

2. If $z = -9 + 12j$ and $|w| = 5$ find w
   (i) when $|z + w| = |z| + |w|$ (ii) when $|z + w| = |z| - |w|$.

3. Prove that $||z_1| - |z_2|| \leqslant |z_1 - z_2| \leqslant |z_1| + |z_2|$.

4. (i) Prove that $|z_1 + z_2 + z_3| \leqslant |z_1| + |z_2| + |z_3|$, and state the circumstances in which there is equality.
   (ii) Prove by induction that
   $$|z_1 + z_2 + \ldots + z_n| \leqslant |z_1| + |z_2| + \ldots + |z_n|.$$

5. (i) Prove that $-|z| \leqslant \mathrm{Re}(z) \leqslant |z|$
    (ii) Use the fact that $|z|^2 = zz^*$ to prove that
    $$|z_1 + z_2|^2 = |z_1|^2 + 2\,\mathrm{Re}\,(z_1 z_2{}^*) + |z_2|^2$$
    (iii) Deduce the triangle inequalities without using geometry.

6. If $|z| < 1$ prove that $1 - |z| \leqslant \left| \dfrac{1}{1-z} \right| \leqslant \dfrac{1}{1-|z|}$. Verify this when
    $z = 0.1 + 0.1j$.

7. (i) Prove that if $|z| < 1$ then $|1 + z + z^2 + \ldots + z^n| \leqslant \dfrac{1}{1-|z|}$.

    (ii) Prove that if $|w| > 1$ then $|1 + w + w^2 + \ldots + w^n| \leqslant \dfrac{|w|^{n+1}}{|w| - 1}$.
    Hint: put $z = \dfrac{1}{w}$ and use (i).

8. Verify that $(a-c)(b-d) = (a-b)(c-d) + (a-d)(b-c)$. Prove
    that if A, B, C, D are any four points in a plane, then
    $$AC \cdot BD \leqslant AB \cdot CD + AD \cdot BC.$$

9. In the complex polynomial equation
    $$z^n + a_{n-1} z^{n-1} + a_{n-2} z^{n-2} + \ldots + a_1 z + 1 = 0,$$
    it is given that the complex numbers $a_{n-1}, a_{n-2}, \ldots, a_1$ satisfy
    $$|a_{n-1}| \leqslant 1, \; |a_{n-2}| \leqslant 1, \ldots, \; |a_1| \leqslant 1.$$
    Show that any root of the equation must lie in the annular region
    $\tfrac{1}{2} < |z| < 2$.   [CS]

5.5 The polar form of complex numbers

The position of the point representing the complex number z in the Argand diagram can be described by means of its distance r from O and its bearing θ measured from the direction of the positive real axis; r and θ are called the *polar coordinates* of the point.

The distance r is $|z|$, the modulus of z as defined in the previous section. The angle θ is not uniquely defined, since adding or subtracting any whole

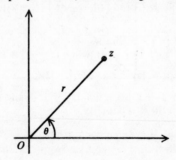

multiple of 2π to it will not alter the bearing. But there is just one value of θ for which $-\pi < \theta \leqslant \pi$; this is called the *principal argument* of z, and is denoted by arg z.

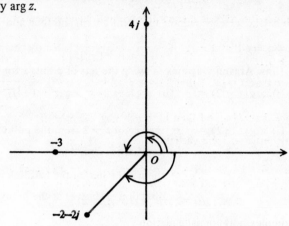

For example, $\arg 4j = \dfrac{\pi}{2}$

$$\arg(-3) = \pi$$

$$\arg(-2 - 2j) = -\frac{3\pi}{4}.$$

The argument of zero is not defined.

Qu. 1. Prove that in general $\arg z^* = -\arg z$. What are the exceptions?

Qu. 2. Prove that $\arg(-z) = \arg z - \pi$ for $0 < \arg z \leqslant \pi$. Find $\arg(-z)$ in terms of $\arg z$ for $-\pi < \arg z \leqslant 0$.

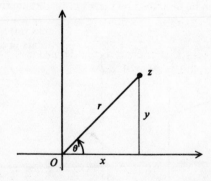

It is clear from the diagram that if $z = x + yj$, $|z| = r$, $\arg z = \theta$ then $x = r \cos \theta$, $y = r \sin \theta$, so that $z = r(\cos \theta + j \sin \theta)$. This is called the *polar* or *modulus-argument* form of z. Given x and y, we can easily find r and θ, since $r = \sqrt{(x^2 + y^2)}$ and $\cos\theta : \sin\theta : 1 = x : y : r$. (It is tempting to say more simply that $\theta = \tan^{-1}\dfrac{y}{x}$, but this would always give θ between $\pm\dfrac{\pi}{2}$. It is necessary to decide which quadrant the point z is in, and then add $\pm\pi$ to $\tan^{-1}\dfrac{y}{x}$ if necessary.)

Example: Draw Argand diagrams showing the sets of points z for which

    (i) $\arg(z - 2) = \dfrac{\pi}{6}$ (ii) $\arg(z - 2) = \arg(z + 1 - j)$.

Solution: (i) $\arg(z - 2) = \dfrac{\pi}{6}$ $\Leftrightarrow$ the vector $z - 2$ from the point 2 to the point z has direction $\dfrac{\pi}{6}$

                    $\Leftrightarrow$ the point z lies on the half line with direction $\dfrac{\pi}{6}$ starting at the point 2.

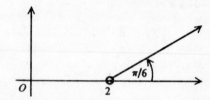

(Note that the point 2 does not belong to this set, since $\arg 0$ is undefined.)

(ii) $\arg(z - 2) = \arg(z + 1 - j)$ $\Leftrightarrow$ the vectors from points 2 and $-1 + j$ to the point z are in the same direction and sense

                                     $\Leftrightarrow$ z lies on the line joining points 2 and $-1 + j$, but does not lie between these points or at either of them.

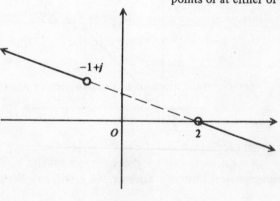

$\square$

Exercise 5D

1–15. Find the modulus and principal argument of the following. If an argument is not a simple multiple of π, give it in radians to 2 D.P.

1. 1 2. -2 3. $3j$ 4. $-4j$

5. $1 + j$ 6. $-\sqrt{8} + \sqrt{8}j$ 7. $1 - \sqrt{3}j$ 8. $-\sqrt{3} + j$

9. $7 \cos \dfrac{\pi}{7} + 7j \sin \dfrac{\pi}{7}$ 10. $\frac{1}{12} \cos 2 + \frac{1}{12} j \sin 2$

11. $6(\cos 4 + j \sin 4)$ 12. $3 - 4j$ 13. $-5 - 12j$

14. $-6 + 11j$ 15. $-7 - 3j$

16. Given that $\arg(3 + 2j) = \alpha$, express the following in terms of α.
    (i) $\arg(-3 - 2j)$ (ii) $\arg(2 + 3j)$ (iii) $\arg(-3 + 2j)$
    (iv) $\arg(2 - 3j)$.

17. Write these complex numbers in polar form.
    (i) $5 - 5j$ (ii) $3 + \sqrt{3}j$ (iii) $\cos\theta - j\sin\theta$
    (iv) $7\sin\theta + 7j\cos\theta$ (v) $1 + j\tan\theta$.

18. Use a vector diagram to find the modulus and argument of $1 + \cos\theta + j\sin\theta$.

19. Draw separate Argand diagrams showing the sets of points z where
    (i) $\arg z = \dfrac{2\pi}{3}$ (ii) $\arg(z - 3j) = 0$ (iii) $\arg(z + 4) \leqslant -\dfrac{\pi}{2}$

    (iv) $\arg(z - 2 - 5j) = \dfrac{\pi}{4}$ (v) $\arg(z + 3 + j) > 0$

    (vi) $\dfrac{\pi}{4} \leqslant \arg(z + 4 - 2j) \leqslant \dfrac{\pi}{3}$.

20. Illustrate on the Argand diagram
$$\{z : |z - 4 - 2j| \leqslant 3\} \cap \left\{z : \arg(z + 1 + 2j) \geqslant \dfrac{\pi}{4}\right\}.$$

21. What is the greatest value of $\arg z$ if $|z - 8j| \leqslant 4$?

22. If k is real and $|z| \leqslant k$, prove that $0 \leqslant |z + k| \leqslant 2k$ and
$-\dfrac{\pi}{2} < \arg(z + k) < \dfrac{\pi}{2}$. Find the greatest and least values of $|z + 2k|$
and $\arg(z + 2k)$.

23. Draw separate Argand diagrams showing the sets of points z where
    (i) $\arg z = \arg(z + 1)$ (ii) $\arg(z + j) = \arg(z - 3)$
    (iii) $\arg(z + j) = -\arg(z - 3)$
    (iv) $\arg(z - 2 - 5j) + \arg(z + 4 - 2j) = 0$.

24. Prove that in the Argand diagram $\left\{z : \arg(z - 1) = \arg(z + 1) + \dfrac{\pi}{4}\right\}$
is part of $\{z : |z - j| = \sqrt{2}\}$, and draw a diagram to show which part.

5.6 Multiplication in the Argand diagram

The polar form provides a very convenient way of interpreting the multiplication of complex numbers.

For if

$$z_1 = r_1(\cos\theta_1 + j\sin\theta_1) \quad \text{and} \quad z_2 = r_2(\cos\theta_2 + j\sin\theta_2)$$

then

$$
\begin{aligned}
z_1 z_2 &= r_1 r_2 (\cos\theta_1 + j\sin\theta_1)(\cos\theta_2 + j\sin\theta_2) \\
&= r_1 r_2 (\cos\theta_1\cos\theta_2 - \sin\theta_1\sin\theta_2 + j(\sin\theta_1\cos\theta_2 + \cos\theta_1\sin\theta_2)) \\
&= r_1 r_2 (\cos(\theta_1 + \theta_2) + j\sin(\theta_1 + \theta_2)).
\end{aligned}
$$

Therefore $\qquad |z_1 z_2| = |z_1||z_2|$

and $\quad \arg(z_1 z_2) = \arg z_1 + \arg z_2 \quad (\pm 2\pi$ if necessary, to give the principal argument).

So to multiply complex numbers in polar form we multiply their moduli and *add* their arguments. This can be interpreted geometrically in two ways.

(A)   To obtain the vector $z_1 z_2$, enlarge the vector z_1 by scale factor $|z_2|$, and rotate it through $\arg z_2$ anticlockwise about O. (Or alternatively enlarge the vector z_2 by scale factor $|z_1|$ and rotate it through $\arg z_1$.)

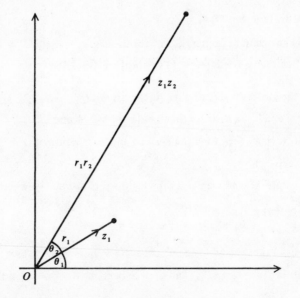

(B) To obtain the point $z_1 z_2$, draw the triangle formed by the points
     $0, 1, z_1$. Enlarge this triangle by scale factor $|z_2|$ and rotate it through
     $\arg z_2$ anticlockwise about O, so that the vertex at the point 1 moves
     to the point z_2. Then the vertex at the point z_1 moves to the point
     $z_1 z_2$. The initial and final triangles are similar.

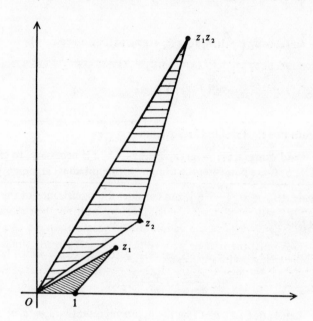

Qu. 1.   Draw an accurate diagram of (B) for the case $z_1 = 2 + j,\, z_2 = 1 + 4j$.
            Draw another diagram interchanging the roles of z_1 and z_2.

To obtain the corresponding results for division, let $w = \dfrac{z_1}{z_2}$. Then $z_1 = wz_2$ so that $|z_1| = |w|\,|z_2|$ and $\arg z_1 = \arg w + \arg z_2$ ($\pm 2\pi$ if necessary).

> Thus $\left| \dfrac{z_1}{z_2} \right| = \dfrac{|z_1|}{|z_2|}$
>
> and $\arg \left(\dfrac{z_1}{z_2} \right) = \arg z_1 - \arg z_2$ ($\pm 2\pi$ if necessary).

In particular, since $\arg 1 = 0$, $\arg \left(\dfrac{1}{z} \right) = -\arg z$.

Qu. 2. Prove this final result directly from $\dfrac{1}{z} = \dfrac{z^*}{|z|^2}$.

Qu. 3. If A, B, C are points representing a, b, c, show that angle $ABC = \arg \left(\dfrac{a-b}{c-b} \right)$, and consider the significance of the sign of this argument.

Example: Prove that the triangle with vertices a, b, c is directly similar to the triangle with vertices a', b', c' if and only if
$$\frac{c-a}{b-a} = \frac{c'-a'}{b'-a'}.$$

Solution: Let A, B, C, A', B', C' be the points representing a, b, c, a', b', c'.

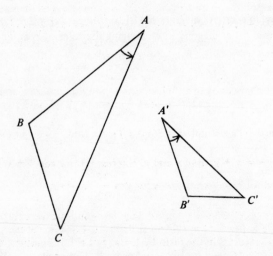

Then ABC, A', B', C' are directly similar if and only if

$$\frac{AC}{AB} = \frac{A'C'}{A'B'}$$

and angle BAC = angle $B'A'C'$, both angles being in the same sense because the similarity is direct.
Therefore

$$\frac{|c-a|}{|b-a|} = \frac{|c'-a'|}{|b'-a'|} \quad \text{and} \quad \arg\left(\frac{c-a}{b-a}\right) = \arg\left(\frac{c'-a'}{b'-a'}\right)$$

and so

$$\frac{c-a}{b-a} = \frac{c'-a'}{b'-a'}. \qquad \qquad \square$$

Qu. 4.   What is the condition for opposite similarity of these triangles?

Exercise 5E

1. Draw an Argand diagram showing the points $z, w, wz, \dfrac{z}{w}, \dfrac{w}{z}$, where

   (i) $z = 4\left(\cos\dfrac{\pi}{3} + j\sin\dfrac{\pi}{3}\right)$, $w = 2\left(\cos\dfrac{\pi}{6} + j\sin\dfrac{\pi}{6}\right)$

   (ii) $z = \cos\dfrac{2\pi}{3} + j\sin\dfrac{2\pi}{3}$, $w = 3\left(\cos\dfrac{\pi}{2} + j\sin\dfrac{\pi}{2}\right)$

   (iii) $z = 3\left(\cos\left(-\dfrac{5\pi}{6}\right) + j\sin\left(-\dfrac{5\pi}{6}\right)\right)$, $w = 2\left(\cos\dfrac{5\pi}{6} + j\sin\dfrac{5\pi}{6}\right)$.

2. Given the point z on an Argand diagram, show how to construct the points: (i) $2z$ (ii) $|z|$ (iii) jz (iv) $1+jz$ (v) $(1+j)z$ (vi) z^2.

3. The point z moves along the imaginary axis from the point $-j$ to the point j. Describe the motion of the points   (i) $3z$ (ii) $2jz$ (iii) $z + 3 + 2j$ (iv) $(3 + 2j)z$ (v) z^* (vi) z^2.

4. The point z moves once anticlockwise round the circle $|z| = 1$, starting at $z = 1$. Describe the motion of the points of Qu. 3 in this case.

5. If $z = \cos\theta + j\sin\theta$ and $w = \cos\phi + j\sin\phi$ find $\dfrac{z}{w} + \dfrac{w}{z}$ in terms of θ, ϕ.

6. If $\dfrac{-1+j}{1+\sqrt{3}j} = x + yj$ find x and y. Express $-1 + j$ and $1 + \sqrt{3}j$ in polar form. Deduce exact expressions for $\cos\dfrac{5\pi}{12}$ and $\sin\dfrac{5\pi}{12}$.

7. The points A, B, C in the Argand diagram represent the complex numbers a, b, c and $c = (1 - k)a + kb$. Prove that if k is real then C lies on AB and divides AB in the ratio $k : 1 - k$, but if k is complex then, in triangle ABC, $AC : AB = |k| : 1$ and angle $BAC = \arg k$.

8. If z_1, z_2, z_3 are the vertices of a triangle, prove that the point $\frac{1}{3}(z_1 + z_2 + z_3)$ is the centroid of the triangle.

9. Directly similar triangles BCL, CAM, ABN are drawn on the sides of a triangle ABC. Prove that triangles ABC, LMN have the same centroid.

10. Let $\omega = \cos \frac{2}{3}\pi + j \sin \frac{2}{3}\pi$. Show that

   (i) $\omega = \frac{1}{2}(-1 + \sqrt{3}j)$ (ii) $\omega^3 = 1$ (iii) $\omega^* = \dfrac{1}{\omega} = \omega^2$
   (iv) $1 + \omega + \omega^2 = 0$.

11. (i) If ω is as in Qu. 10, how are the vectors z and ωz related geometrically?
   (ii) If $1 + 2j$ and $5 + 4j$ are two vertices of an equilateral triangle, find both possible positions for the third vertex.

12. If the points a and b are two vertices of an equilateral triangle, prove that the third vertex is either $b + \omega (b - a)$ or $b + \omega^2 (b - a)$, and that these can be written as $-\omega a - \omega^2 b$ and $-\omega^2 a - \omega b$ respectively. Deduce that the triangle with vertices z_1, z_2, z_3 is equilateral if and only if
$$z_1 + \omega z_2 + \omega^2 z_3 = 0 \quad \text{or} \quad z_1 + \omega^2 z_2 + \omega z_3 = 0.$$

13. Prove that it is impossible to have an equilateral triangle with vertices whose Cartesian coordinates are all rational.

14. Prove that a necessary and sufficient condition for the points z_1, z_2, z_3 to form an equilateral triangle is
$$z_1^2 + z_2^2 + z_3^2 = z_2 z_3 + z_3 z_1 + z_1 z_2.$$

15. (i) On the sides of any triangle, equilateral triangles are drawn, all pointing outward. Using Qu. 12, prove that the centroids of these equilateral triangles form another equilateral triangle. This is *Napoleon's theorem*.
   (ii) Prove that the theorem is still true if the equilateral triangles are drawn inwards rather than outwards.
   (iii) Prove that the triangle of centroids in (i), the corresponding triangle in (ii), and the original triangle all have the same centroid.

5.7 De Moivre's Theorem

In §5.6 we saw that, with the usual notation,
$$z_1 z_2 = r_1 r_2 (\cos (\theta_1 + \theta_2) + j \sin (\theta_1 + \theta_2)).$$

In particular if $z = \cos \theta + j \sin \theta$, so that $|z| = 1$, then
$$z^2 = \cos (\theta + \theta) + j \sin (\theta + \theta) = \cos 2\theta + j \sin 2\theta,$$
$$z^3 = z^2 . z = \cos (2\theta + \theta) + j \sin (2\theta + \theta) = \cos 3\theta + j \sin 3\theta, \text{ and so on.}$$

These are simple cases of a very useful result, called *De Moivre's Theorem*:

> If n is any integer, then
> $$(\cos\theta + j\sin\theta)^n = \cos n\theta + j\sin n\theta.$$

The proof of this is in three parts.

(i) When n is a positive integer we use induction.
The theorem is obviously true when $n = 1$, and

$$(\cos\theta + j\sin\theta)^k = \cos k\theta + j\sin k\theta$$
$$\Rightarrow (\cos\theta + j\sin\theta)^{k+1} = (\cos k\theta + j\sin k\theta)(\cos\theta + j\sin\theta)$$
$$= \cos(k\theta + \theta) + j\sin(k\theta + \theta)$$
$$= \cos(k+1)\theta + j\sin(k+1)\theta$$

so by induction the theorem is true for all positive integers n.

(ii) We define $z^0 = 1$ for all complex numbers $z \neq 0$. Therefore

$$(\cos\theta + j\sin\theta)^0 = 1$$
$$= \cos 0 + j\sin 0.$$

(iii) When n is a negative integer, let $n = -m$. As a preliminary step we note that

$$(\cos\alpha + j\sin\alpha)(\cos(-\alpha) + j\sin(-\alpha)) = \cos(\alpha - \alpha) + j\sin(\alpha - \alpha) = 1,$$

so that

$$\frac{1}{\cos\alpha + j\sin\alpha} = \cos(-\alpha) + j\sin(-\alpha).$$

Now

$$(\cos\theta + j\sin\theta)^n = (\cos\theta + j\sin\theta)^{-m}$$
$$= \frac{1}{(\cos\theta + j\sin\theta)^m}$$
$$= \frac{1}{\cos m\theta + j\sin m\theta} \quad \text{using the result from part (i) of the proof}$$
$$= \cos(-m\theta) + j\sin(-m\theta) \quad \text{using the preliminary result}$$
$$= \cos n\theta + j\sin n\theta.$$

This completes the proof of the theorem for all integers n. In §13.1 we shall consider how the theorem can be extended to include rational powers. Abraham de Moivre (1667–1754) was born in France, came to England as a refugee when he was eighteen, and spent the rest of his long life in London. He never explicitly published the theorem which is named after him, though he showed in a paper of 1707 that he was familiar with its use.

Example 1: Express $\cos 5\theta$ in terms of $\cos \theta$.

Solution: By de Moivre's Theorem

$$\cos 5\theta + j \sin 5\theta = (\cos \theta + j \sin \theta)^5$$
$$= c^5 + 5j\,c^4 s - 10\,c^3 s^2 - 10j\,c^2 s^3 + 5\,cs^4 + j\,s^5$$
$$\text{(where } c = \cos \theta, s = \sin \theta).$$

Equating real parts

$$\cos 5\theta = c^5 - 10\,c^3 s^2 + 5\,cs^4.$$

But $s^2 = 1 - c^2$, so

$$\cos 5\theta = c^5 - 10\,c^3 (1 - c^2) + 5c\,(1 - c^2)^2$$
$$= c^5 - 10\,c^3 + 10\,c^5 + 5c - 10\,c^3 + 5c^5$$

i.e. $\cos 5\theta = 16 \cos^5\theta - 20 \cos^3\theta + 5 \cos \theta.$ □

Qu. 1. Check this by putting $\theta = 0$.

Qu. 2. By equating imaginary parts find $\sin 5\theta$ in terms of $\sin \theta$.

If $z = \cos \theta + j \sin \theta$ then $z^n = \cos n\theta + j \sin n\theta$

and $z^{-n} = \cos(-n\theta) + j \sin(-n\theta) = \cos n\theta - j \sin n\theta.$

Therefore

$$\boxed{\begin{aligned} z = \cos \theta + j \sin \theta \Rightarrow \cos n\theta &= \tfrac{1}{2}(z^n + z^{-n}) \\ \sin n\theta &= \frac{1}{2j}(z^n - z^{-n}). \end{aligned}}$$

Example 2: Express $\cos^4\theta \sin^2\theta$ in terms of multiple angles.

Solution: Let $z = \cos \theta + j \sin \theta$.

Then since $\cos \theta = \frac{1}{2}(z + z^{-1})$ and $\sin \theta = \frac{1}{2j}(z - z^{-1})$,

$$\cos^4\theta \sin^2\theta = \frac{1}{2^4}(z + z^{-1})^4 \cdot \frac{-1}{2^2}(z - z^{-1})^2$$

So $-2^6 \cos^4\theta \sin^2\theta = (z + z^{-1})^4 (z - z^{-1})^2$

$$= (z + z^{-1})^2 [(z + z^{-1})(z - z^{-1})]^2 \qquad \text{(A)}$$

$$= (z + z^{-1})^2 (z^2 - z^{-2})^2$$

$$= (z^2 + 2 + z^{-2})(z^4 - 2 + z^{-4})$$

$$= z^6 - 2z^2 + z^{-2} + 2z^4 - 4 + 2z^{-4} + z^2 - 2z^{-2} +$$

$$= (z^6 + z^{-6}) + 2(z^4 + z^{-4}) - (z^2 + z^{-2}) - 4$$

$$= 2\cos 6\theta + 4\cos 4\theta - 2\cos 2\theta - 4.$$

Therefore $\cos^4\theta \sin^2\theta = 2^{-5}(2 + \cos 2\theta - 2\cos 4\theta - \cos 6\theta).$

Notice in particular step (A), which saves quite a lot of work. □

Qu. 3.   Check the result of Example 2
(i) by putting $\theta = 0$ (ii) by putting $\theta = \dfrac{\pi}{4}$.

Qu. 4.   Use Example 2 to find $\int \cos^4\theta \sin^2\theta \, d\theta$.

Example 3: Sum the series
$$1 + {}^nC_1 \cos\theta + {}^nC_2 \cos 2\theta + {}^nC_3 \cos 3\theta + \ldots + \cos n\theta.$$

Solution: At first sight this series suggests the binomial expansion $(1 + \cos\theta)^n$: the coefficients $1, {}^nC_1, {}^nC_2, {}^nC_3, \ldots$ are right, but we have multiple angles, $\cos r\theta$, instead of powers of cosines, $\cos^r\theta$. This indicates that De Moivre's Theorem can be used.

Let $C = 1 + {}^nC_1 \cos\theta + {}^nC_2 \cos 2\theta + \ldots + \cos n\theta$

and $S = \quad {}^nC_1 \sin\theta + {}^nC_2 \sin 2\theta + \ldots + \sin n\theta.$

Then $C + jS = 1 + {}^nC_1(\cos\theta + j\sin\theta) + {}^nC_2(\cos 2\theta + j\sin 2\theta) +$
$$\ldots + (\cos n\theta + j\sin n\theta)$$

$$= 1 + {}^nC_1 z + {}^nC_2 z^2 + \ldots + z^n,$$
$$\text{where } z = \cos\theta + j\sin\theta$$

$$= (1 + z)^n.$$

This expresses $C + jS$ compactly; now we have to separate real and imaginary parts.

Now $1 + z = 1 + \cos\theta + j\sin\theta$

$$= 2\cos^2\frac{\theta}{2} + 2j\sin\frac{\theta}{2}\cos\frac{\theta}{2}$$

$$= 2\cos\frac{\theta}{2}\left(\cos\frac{\theta}{2} + j\sin\frac{\theta}{2}\right).$$

Therefore $(1 + z)^n = 2^n \cos^n\frac{\theta}{2}\left(\cos\frac{\theta}{2} + j\sin\frac{\theta}{2}\right)^n$

$$= 2^n \cos^n\frac{\theta}{2}\left(\cos\frac{n\theta}{2} + j\sin\frac{n\theta}{2}\right).$$

Taking the real part,

$$C = 2^n \cos^n\frac{\theta}{2}\cos\frac{n\theta}{2}. \qquad\qquad □$$

Qu. 5.   State the result obtained by equating imaginary parts in Example 3.

Notice that in all three examples we have used De Moivre's Theorem to establish results which involve only real numbers, and which could be obtained without using complex numbers at all, though usually this would be more difficult. Also, in Examples 1 and 3, by equating real and imaginary parts we get two results for the price of one.

Exercise 5F

1–4. Evaluate

1. $\left(\cos \dfrac{\pi}{4} + j \sin \dfrac{\pi}{4}\right)^8$

2. $\left(\cos \dfrac{\pi}{6} + j \sin \dfrac{\pi}{6}\right)^{-3}$

3. $\left(\cos \dfrac{5\pi}{18} + j \sin \dfrac{5\pi}{18}\right)^6$

4. $\left(3 \cos\left(\dfrac{-\pi}{12}\right) + 3j \sin\left(\dfrac{-\pi}{12}\right)\right)^{-4}$

5–8. Simplify

5. $(\cos(-\alpha) + j \sin(-\alpha))^7$

6. $(\cos^2 \theta + j \sin \theta \cos \theta)^{10}$

7. $(\cos \alpha + j \sin \alpha)^5 (\cos \beta + j \sin \beta)^3$

8. $(\cos \phi - j \sin \phi)^4 (\cos \phi + j \sin \phi)^{-8}$

9. Simplify $(1 + j)^6$ (i) by means of the binomial expansion
   (ii) by first expressing $1 + j$ in polar form
   Check that your answers agree.

10. Prove that $\cos 4\theta = \cos^4 \theta - 6 \cos^2 \theta \sin^2 \theta + \sin^4 \theta$ and that
    $\sin 4\theta = 4 \cos^3 \theta \sin \theta - 4 \cos \theta \sin^3 \theta$.
    Use these results to find $\tan 4\theta$ in terms of $\tan \theta$.

11. Express $\cos 7\theta$ and $\dfrac{\sin 7\theta}{\sin \theta}$ in terms of $\cos \theta$.

12. Prove that $\cos 6\theta = 32 \cos^6 \theta - 48 \cos^4 \theta + 18 \cos^2 \theta - 1$. Use this to solve the equations
    (i) $32x^6 - 48x^4 + 18x^2 - 1 = 0$ (ii) $32x^6 - 48x^4 + 18x^2 = \tfrac{3}{2}$.

13. If $t = \tan \theta$ show that $\cos n\theta = \cos^n \theta \,(1 - {}^nC_2 t^2 + {}^nC_4 t^4 - \ldots)$
    and $\sin n\theta = \cos^n \theta \,({}^nC_1 t - {}^nC_3 t^3 + \ldots)$
    Find $\tan n\theta$ in terms of t.

14. Express in terms of cosines of multiple angles
    (i) $\cos^4 \theta$ (ii) $\sin^6 \theta$ (iii) $\cos^3 \theta \sin^4 \theta$

15. Express in terms of sines of multiple angles
    (i) $\sin^3 \theta$ (ii) $\sin^5 \theta$ (iii) $\cos^4 \theta \sin^3 \theta$

16. Prove that $\cos^m \theta \sin^n \theta$ can be expressed in terms of the cosines of multiple angles if n is even, and in terms of the sines of multiple angles if n is odd.

17. Use Qu. 14 and Qu. 15 to find
    (i) $\int \sin^6 \theta \, d\theta$ (ii) $\displaystyle\int_0^{\pi/2} \cos^3 \theta \sin^4 \theta \, d\theta$ (iii) $\displaystyle\int_0^{\pi} \cos^4 \theta \sin^3 \theta \, d\theta$

18. Express $\cos^5\theta$ in the form $a\cos 5\theta + b\cos 3\theta + c\cos\theta$, and hence find all the solutions of the equation $16\cos^5\theta = \cos 5\theta$.

19. By expressing $\cos^{2n}\theta$ in terms of the cosines of multiple angles prove that
$$\int_0^\pi \cos^{2n}\theta \, d\theta = \frac{(2n)!\,\pi}{2^{2n}(n!)^2}.$$

20. Use $\cos n\theta = \frac{1}{2}(z^n + z^{-n})$ to find
$$\cos\theta + \cos 3\theta + \cos 5\theta + \ldots + \cos(2n-1)\theta.$$

21. Express $2 + \cos\dfrac{2\pi}{3} + j\sin\dfrac{2\pi}{3}$ in polar form. Deduce that
$$2^n + {}^nC_1\,2^{n-1}\cos\frac{2\pi}{3} + {}^nC_2\,2^{n-2}\cos\frac{4\pi}{3} + \ldots + \cos\frac{2n\pi}{3} = 3^{n/2}\cos\frac{n\pi}{6}.$$

State the corresponding result for sines.

22. Sum the series
$$\cos\alpha + {}^nC_1\cos(\alpha+\beta) + {}^nC_2\cos(\alpha+2\beta) + \ldots + \cos(\alpha+n\beta).$$

23. Let $C = 1 + \cos\theta + \cos 2\theta + \ldots + \cos(n-1)\theta$
    and $S = \quad\ \sin\theta + \sin 2\theta + \ldots + \sin(n-1)\theta$.
    Show that $C + jS$ is a geometric progression with common ratio
    $z = \cos\theta + j\sin\theta$. By summing this progression and taking real and
    imaginary parts, show that $C = \dfrac{1 - \cos\theta + \cos(n-1)\theta - \cos n\theta}{2 - 2\cos\theta}$
    and find S.

24. Adapt the method of Qu. 23 to find $\displaystyle\sum_{r=0}^{n-1} k^r\cos r\theta$ and $\displaystyle\sum_{r=0}^{n-1} k^r\sin r\theta$,

    where k is constant. If $|k| < 1$ what are $\displaystyle\sum_{r=0}^{\infty} k^r\cos r\theta$ and $\displaystyle\sum_{r=0}^{\infty} k^r\sin r\theta$?

25. Prove that if $\theta \neq (n+\frac{1}{2})\pi$, $\displaystyle\sum_{r=1}^{\infty} \sin^r\theta \sin r\theta = (1 - 2\cot\theta + \csc^2\theta)^{-1}$.

6 Rational functions

6.1 Polynomials

A polynomial of degree n in x is an expression which can be put in the form $c_0 x^n + c_1 x^{n-1} + c_2 x^{n-2} + \ldots + c_{n-1} x + c_n$, where $c_0, c_1, c_2, \ldots, c_{n-1}, c_n$ are numbers, called the *coefficients* of the polynomial, with $c_0 \neq 0$; c_0 is the *leading* coefficient, and c_n is the *constant term*.

If $P(x)$ is a polynomial, then the function $x \mapsto P(x)$ is a *polynomial function*. Polynomial functions are easy to evaluate, differentiate and integrate, so it is often useful to use them as approximations to more complicated functions, as we did in §2.5.

Any non-zero number can be considered as a polynomial of degree 0; it is also convenient to regard 0 as a polynomial, though the degree of this *zero polynomial* is not defined.

Qu. 1. Show that the product of two polynomials of degrees m, n is a polynomial of degree $m + n$. Show that this would no longer hold if any finite degree were assigned to the zero polynomial.

Two polynomials $P_1(x)$, $P_2(x)$ are said to be *identically equal* if $P_1(x) = P_2(x)$ for every value of x; in this case we write $P_1(x) \equiv P_2(x)$. The set of polynomials is closed under addition, subtraction and multiplication (Vol. 1 §3.1), but not under division since for example $(x + 1)/(x + 2)$ is not a polynomial. However, if $D(x)$ and $P(x)$ are polynomials with $D(x) \not\equiv 0$ and degree of $D(x) \leqslant$ degree of $P(x)$, it is possible to find by algebraic long division unique polynomials $Q(x)$ and $R(x)$, with degree of $R(x) <$ degree of $D(x)$ or $R(x) \equiv 0$, such that $P(x) \equiv D(x) . Q(x) + R(x)$, (Vol. 1 §3.3). This follows the pattern dividend = divisor $\times$ quotient + remainder which is familiar from the division of numbers.

Qu. 2. Find $Q(x)$ and $R(x)$ when $P(x) \equiv 4x^4 - 5x + 3, D(x) \equiv 2x^2 - x + 3$.

In particular if $D(x)$ is the linear polynomial $x - a$, then $R(x)$ is a constant R and we have $P(x) \equiv (x - a) Q(x) + R$. Putting $x = a$ in this identity shows that $R = P(a)$, a result known as the *Remainder Theorem*.

Qu. 3. Deduce from the Remainder Theorem that $x - a$ is a factor of $P(x) \Leftrightarrow P(a) = 0$.

Qu. 4. Show that the remainder when $P(x)$ is divided by $hx + k$ is $P\left(-\dfrac{k}{h}\right)$.

Alternatively we can use the important identity

$$x^n - a^n \equiv (x - a)(x^{n-1} + x^{n-2}a + x^{n-3}a^2 + \ldots + x\,a^{n-2} + a^{n-1}).$$

This identity can be established by expanding the right-hand side.

If $\qquad P(x) \equiv c_0 x^n + c_1 x^{n-1} + \ldots + c_{n-1} x + c_n$

then $\qquad P(a) = c_0 a^n + c_1 a^{n-1} + \ldots + c_{n-1}a + c_n$

and $\; P(x) - P(a) \equiv c_0(x^n - a^n) + c_1(x^{n-1} - a^{n-1}) + \ldots + c_{n-1}(x - a)$.

Each term on the right has $(x - a)$ as a factor, and so

$$P(x) - P(a) \equiv (x - a)\,Q(x),$$

where
$Q(x) \equiv c_0(x^{n-1} + x^{n-2}a + \ldots + a^{n-1}) + c_1(x^{n-2} + \ldots + a^{n-2}) + \ldots + c_{n-1}$,
a polynomial of degree $n - 1$ which also has leading coefficient c_0.

Thus $\qquad P(x) \equiv (x - a)\,Q(x) + P(a)$, as before.

Qu. 5. Prove that if n is odd then

$$x^n + a^n \equiv (x + a)(x^{n-1} - x^{n-2}a + x^{n-3}a^2 - \ldots - x\,a^{n-2} + a^{n-1}).$$

We can now establish the following very useful *Factor Theorem*:

> If $P(x)$ is a polynomial of degree $n\ (\geqslant 1)$ with leading coefficient c_0 and if $P(a_1) = P(a_2) = P(a_3) = \ldots = P(a_n) = 0$ for n distinct numbers $a_1, a_2, a_3, \ldots, a_n$ then
> $$P(x) \equiv c_0(x - a_1)(x - a_2)(x - a_3) \ldots (x - a_n).$$

The proof is by induction.
(i) If $P(x)$ is of degree 1 and $P(a_1) = 0$ then $P(x) \equiv c_0 x + c_1$ with $P(a_1) = c_0 a_1 + c_1 = 0$. Therefore $c_1 = -c_0 a_1$ and $P(x) \equiv c_0 x - c_0 a_1 \equiv c_0(x - a_1)$, as required.
(ii) Suppose that for some particular $k > 1$ the theorem holds for all polynomials of degree $k - 1$. Let $P(x)$ be a polynomial of degree k with leading coefficient c_0, such that $P(x) = 0$ when x takes any one of the k distinct values $a_1, a_2, \ldots, a_k$. Then by the Remainder Theorem, since $P(a_1) = 0$, $P(x) \equiv (x - a_1)\,Q(x)$, where $Q(x)$ is of degree $k - 1$ and also has leading coefficient c_0.
Putting $x = a_r\ (r = 2, 3, \ldots, k)$ we have $(a_r - a_1)\,Q(a_r) = P(a_r) = 0$. But $a_r - a_1 \neq 0$, and so $Q(a_r) = 0$ for $r = 2, 3, \ldots, k$.

Since $Q(x)$ is of degree $k-1$ we may apply the theorem:

$$Q(x) \equiv c_0(x - a_2)(x - a_3) \dots (x - a_k).$$

Therefore $P(x) \equiv (x - a_1)\, c_0(x - a_2)(x - a_3) \dots (x - a_k)$

$$\equiv c_0(x - a_1)(x - a_2)(x - a_3) \dots (x - a_k),$$

so that the theorem holds for $n = k$ also.

From (i) and (ii) by induction the theorem holds for all $n \geqslant 1$.

An immediate deduction from the Factor Theorem is that a polynomial $P(x)$ of degree n cannot take the value zero for more than n distinct values of x. For if $P(x) = 0$ for $x = a_1, a_2, \dots, a_n$, then $P(x) \equiv c_0(x - a_1)(x - a_2) \dots (x - a_n)$, and so if b is any value distinct from the a_r then $P(b) = c_0(b - a_1)(b - a_2) \dots (b - a_n) \neq 0$.

Now suppose that $P(x) \equiv c_0 x^n + c_1 x^{n-1} + \dots + c_n$ and $Q(x) \equiv d_0 x^n + d_1 x^{n-1} + \dots + d_n$ take equal values for more than n distinct values of x. Then the difference

$$P(x) - Q(x) \equiv (c_0 - d_0)x^n + (c_1 - d_1)x^{n-1} + \dots + (c_n - d_n)$$

is zero for more than n distinct values of x, and so cannot be a polynomial of degree n or less. The only remaining possibility is that $P(x) - Q(x)$ is the zero polynomial. Thus all the coefficients of $P(x) - Q(x)$ must vanish: $c_0 - d_0 = c_1 - d_1 = \dots = c_n - d_n = 0$, and hence $c_0 = d_0$, $c_1 = d_1$, $\dots$, $c_n = d_n$. We deduce that

> if $P(x)$, $Q(x)$ are polynomials of degree $\leqslant n$ and $P(x) = Q(x)$ for more than n distinct values of x, then the coefficients of x^r in $P(x)$ and $Q(x)$ are equal for $r = 0, 1, 2, \dots, n$, and $P(x) \equiv Q(x)$.

The process of using $c_0 = d_0$, $c_1 = d_1$, $\dots$ is called *equating coefficients*.

Example 1: Find, if possible, constants a, b, c
      (i) such that $n^2 \equiv an(n + 2) + b(n + 1) + c(n + 2)$;
      (ii) such that $n^2 \equiv an(n + 2) + b(n + 1)^2 + c$.

Solution: (i) Equating coefficients of n^2, n^1, n^0 gives $a = 1$, $2a + b + c = 0$, $b + 2c = 0$, which gives the unique solution $a = 1$, $b = -4$, $c = 2$.
         (ii) Equating coefficients of n^2, n^1, n^0 gives $a + b = 1$, $2a + 2b = 0$, $2a + b + c = 0$. The first two equations are inconsistent, so no such a, b, c can be found.   □

Example 2: Prove that if a, b, c are distinct, then

$$x^2 \equiv \frac{a^2(x-b)(x-c)}{(a-b)(a-c)} + \frac{b^2(x-c)(x-a)}{(b-c)(b-a)} + \frac{c^2(x-a)(x-b)}{(c-a)(c-b)}$$

Solution: When $x = a$, R.H.S. $= \dfrac{a^2(a-b)(a-c)}{(a-b)(a-c)} + 0 + 0 = a^2 =$ L.H.S.

Similarly both sides are equal when $x = b$ and when $x = c$. Thus the two sides are quadratic polynomials which are equal for three distinct values of x. Therefore they are identically equal.

□

These general results about factors do not depend on the type of numbers we are using, rational, real or complex. But this does become important when we deal with particular cases: thus $x^2 - 2$ has no factors if we are restricted to rational numbers, but using real numbers, $x^2 - 2 \equiv (x - \sqrt{2})(x + \sqrt{2})$. Similarly $x^2 - 4x + 125$ has no real factors, but with complex numbers $x^2 - 4x + 125 \equiv (x - 2 - 11j)(x - 2 + 11j)$.

Notice the one-way connection between algebraic and arithmetic factors:

$$6x^2 + 7x - 3 \equiv (3x - 1)(2x + 3) \Rightarrow 667 = 29 \times 23 \text{ (putting } x = 10)$$

but $x^2 + 1$ has no real factors $\not\Rightarrow 3^2 + 1$ is prime.

Each factor $x - a$ of a polynomial $P(x)$ corresponds to a root $x = a$ of the polynomial equation $P(x) = 0$. Even when it is difficult to find these roots, we may be able to say something about them from graphical considerations, as in the following example.

Example 3: Investigate the roots of $x^4 - 8x^3 + 10x^2 = k$ for various k.

Solution: Let $P(x) \equiv x^4 - 8x^3 + 10x^2$.

Then $P'(x) \equiv 4x^3 - 24x^2 + 20x \equiv 4x\,(x - 1)(x - 5)$

and $P''(x) \equiv 12x^2 - 48x + 20$.

$$P'(x) = 0 \Leftrightarrow x = 0, 1, \text{ or } 5$$
$$P(0) = 0, \qquad P''(0) = 20$$
$$P(1) = 3, \qquad P''(1) = -16$$
$$P(5) = -125, \quad P''(5) = 80.$$

The curve $y = P(x)$ has minimum points at $(0, 0)$ and $(5, -125)$, and a maximum point at $(1, 3)$; its shape is indicated (not to scale) in the diagram overleaf.

Considering the lines $y = k$ we see that the number of real roots is 0 for $k < -125$, 1 for $k = -125$, 2 for $-125 < k < 0$ or $k > 3$, 3 for $k = 0$ or $k = 3$, and 4 for $0 < k < 3$.

For $k = -125, 0, 3$ we can factorize and find all the roots; this enables us to locate the roots more accurately in the other cases.

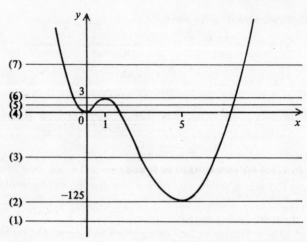

Qu. 6. Complete the investigation in Example 3 along these lines.

Exercise 6A

1. If $x^4 + px^3 - x^2 + qx - 12$ has factors $x + 1$ and $x + 2$, find the constants p, q and the remaining factors.

2. The polynomial $P(x) \equiv x^3 + ax^2 + bx + c$ leaves remainders $-36, -20, 0$ on division by $x + 1, x + 2, x + 3$ respectively. Solve the equation $P(x) = 0$.

3. Prove that $x^n - a^n \equiv (x - a)(x^{n-1} + x^{n-2}a + \ldots + a^{n-1})$ by summing a suitable geometric progression with common ratio a/x.

4. Show that $x - a, x^2 - a^2, x^3 - a^3$, are all factors of $x^5 + 3x^4a + 2x^3a^2 - x^2a^3 - 3xa^4 - 2a^5$, and find three other factors.

   [MEI]

5. Given that $P(x) \equiv (x - a)^2 Q(x)$, where $Q(x)$ is a polynomial, show that $P'(a) = 0$. Given that the polynomial $2x^3 + 11x^2 + 12x - 9$ has a factor $(x - a)^2$, find the value of a. Express the polynomial as a product of linear factors.

   [JMB]

6. A polynomial $P(x)$ has remainder $Rx + S$ on division by $(x - a)(x - b)$. If $a \neq b$, express the constants R, S in terms of $a, b, P(a), P(b)$. Find the remainder when $P(x)$ is divided by $(x - a)^2$, and explain the connection with the Taylor expansion of $P(x)$ centred at $x = a$.

7. Prove that
   (i) there are infinitely many a, b, c for which
        $$x \equiv a(x - 1) + b(x - 2) + c(x - 3);$$
   (ii) there are unique a, b, c for which
        $$x^2 \equiv a(x - 1)^2 + b(x - 2)^2 + c(x - 3)^2;$$

(iii)   there are no a, b, c for which
$$x^3 \equiv a(x-1)^3 + b(x-2)^3 + c(x-3)^3.$$

8. Express $\dfrac{y^7}{y^3+1}$ in the form $P(y) + \dfrac{R(y)}{1+y^3}$ where $P(y)$ and $R(y)$ are

polynomials in y and $R(y)$ is of degree less than three. By putting

$y = 1/x$ in this, or otherwise, show that $\displaystyle\int_1^2 \dfrac{dx}{x^4(1+x^3)} = \tfrac{7}{24} + \tfrac{2}{3}\ln\tfrac{3}{4}$.
[L]

9. Express $x^3 - 3x^2 + x + 2$ in the form $p(x-2)^3 + q(x-2)^2 + r(x-2) + s$
where p, q, r, s are constants to be found.
Given a cubic polynomial $P(x) \equiv Ax^3 + Bx^2 + Cx + D$, the following
(incomplete) flow diagram is intended (by dividing $P(x)$ repeatedly by
$x - H$) to find the constants appropriate for expressing $P(x)$ 'in terms
of $x - H$'. It contains an error.

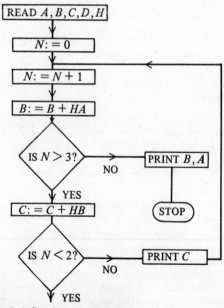

(i)   Describe briefly the purpose of N.
(ii)   Correct the error and complete the diagram.   [SMP]

10. Find constants a, b, c such that $(an+1)^3 - (bn-1)^3 \equiv 24n^2 + c$.
Hence show that $1^2 + 2^2 + \ldots + n^2 = \tfrac{1}{6}n(n+1)(2n+1)$.

11. (i)   If $f(r) \equiv r(r+1)(r+2)$, show that $f(r) - f(r-1) \equiv 3r(r+1)$.

Deduce that $\displaystyle\sum_{r=1}^{n} r(r+1) = \tfrac{1}{3}n(n+1)(n+2)$.

(ii)   Starting with $f(r) \equiv r(r+1)(r+2)(r+3)$, use the method of (i)

to find $\displaystyle\sum_{r=1}^{n} r(r+1)(r+2)$.

(iii) Prove that $\sum\limits_{r=1}^{n} r(r+1)(r+2) \ldots (r+k)$

$= \dfrac{1}{k+2}\, n(n+1)(n+2) \ldots (n+k)(n+k+1).$

12. Find a, b, c, d such that

$r^4 \equiv r(r+1)(r+2)(r+3) + ar(r+1)(r+2) + br(r+1) + cr + d.$

Use the results of Qu. 11 to find $\sum\limits_{r=1}^{n} r^4.$

13. (i) Show that the equation of the straight line through $(a, A), (b, B)$
    can be written in the form $y = \dfrac{A(x-b)}{a-b} + \dfrac{B(x-a)}{b-a}.$

   (ii) Show that the equation of the unique quadratic curve through
    $(a, A), (b, B), (c, C)$ is

   $$y = \frac{A(x-b)(x-c)}{(a-b)(a-c)} + \frac{B(x-c)(x-a)}{(b-c)(b-a)} + \frac{C(x-a)(x-b)}{(c-a)(c-b)}.$$

   (iii) Following the pattern of (i) and (ii), write down the equation of
    the unique cubic curve through $(a, A), (b, B), (c, C), (d, D).$

14. Prove that $\dfrac{a^2-x^2}{(a-b)(a-c)} + \dfrac{b^2-x^2}{(b-c)(b-a)} + \dfrac{c^2-x^2}{(c-a)(c-b)} \equiv 1.$

15. (i) Prove that if m has any odd factor then $2^m + 1$ is composite, i.e.
    not prime. [Hint: let $m = ab$, where b is odd. Then
    $2^m + 1 = (2^a)^b + 1$. Now use Qu. 5 on page 141.]

   (ii) Deduce that if $2^m + 1$ is prime then $m = 2^n$ for some $n \geqslant 0$.

    Numbers of the form $2^{2^n} + 1$ are called Fermat numbers;
    see page 8.

16. (i) Prove that if a is an integer greater than 2, then $a^m - 1$ is
    composite.

   (ii) Prove that if m is composite, then so is $2^m - 1$.

   (iii) Deduce that the only primes of the form $a^m - 1$ are of the more
    restricted form $2^p - 1$, where p is prime. Such primes are called
    Mersenne primes after Marin Mersenne (1588–1648).

   (iv) If $M_p = 2^p - 1$ show that M_2, M_3, M_5, M_7 are prime but M_{11} is
    composite. By 1976 twenty-four Mersenne primes had been
    found, including the largest known prime number, M_{19937}.

17. Given that k and h are real, find the ranges of values of h such that
    the equations $x^3 - 3k^2 x + h = 0$ should have (i) three real roots;
    (ii) one positive and two negative real roots; (iii) one negative real
    root and two complex roots. Solve the equation $x^3 - 12x + 11 = 0$.

                                                                      [MEI]

18. Prove that the function $f(x) \equiv x^4 + 4x^3 - 2x^2 - 12x + p$ has two
    minima and one maximum. Deduce that the equation $f(x) = 0$ has
    four distinct roots provided that $-7 < p < 9$.

In the particular case $p = 5$, use the substitution $x = z + k$ and choose k to obtain a quadratic equation for z^2. Hence find the four real roots.

[MEI]

19. Let f be the function which is defined for all real x by

$f(x) \equiv 1 - x + \dfrac{x^2}{2} - \dfrac{x^3}{3} + \dfrac{x^4}{4}$. By considering the derivative of f, or

otherwise, prove that the equation $f(x) = 0$ has no real roots. How

many real roots has the equation $1 - x + \dfrac{x^2}{2} - \dfrac{x^3}{3} + \dfrac{x^4}{4} - \dfrac{x^5}{5} = 0$?

Can you suggest any general result concerning the numbers of real roots of polynomials similar to those above but of higher order?   [SMP]

20. If $f(x) \equiv (x - a)^n g(x)$ where $g(x)$ is a polynomial in x, prove that the first $(n - 1)$ derivatives of $f(x)$ vanish when $x = a$.

If $p_n(x) \equiv \dfrac{d^n}{dx^n} \{(x^2 - 1)^n\}$, use integration by parts to prove that

$\displaystyle\int_{-1}^{1} x^m p_n(x)\, dx = 0$ where m, n are positive integers and $m < n$.

Hence prove that $\displaystyle\int_{-1}^{1} p_M(x)\, p_N(x)\, dx = 0$ if M, N are positive integers

such that $M \neq N$.   [O & C]

6.2 Properties of the roots of polynomial equations

In Vol. 1 §3.9, we proved and used the properties $\alpha + \beta = -\dfrac{b}{a}$, $\alpha\beta = \dfrac{c}{a}$

of the roots α, β of $ax^2 + bx + c = 0$. Here we consider the corresponding properties of the roots of cubic and quartic equations.

Consider first the cubic equation $az^3 + bz^2 + cz + d = 0$, $(a \neq 0)$. We use z instead of x to emphasize that the results of this section apply to complex roots as well as real ones. Suppose that the roots of this equation are α, β, γ. Then by the Factor Theorem of §6.1 we have

$$az^3 + bz^2 + cz + d \equiv a(z - \alpha)(z - \beta)(z - \gamma)$$
$$\equiv az^3 - a(\alpha + \beta + \gamma)z^2 + a(\beta\gamma + \gamma\alpha + \alpha\beta)z - a\,\alpha\beta\gamma.$$

Equating the coefficients of z^2, z^1, z^0 shows that

if α, β, γ are the roots of $az^3 + bz^2 + cz + d = 0$ then

$$\alpha + \beta + \gamma = -\dfrac{b}{a}, \quad \beta\gamma + \gamma\alpha + \alpha\beta = \dfrac{c}{a}, \quad \alpha\beta\gamma = -\dfrac{d}{a}.$$

Qu. 1. The roots of $3z^3 + 5z^2 - 8 = 0$ are α, β, γ. Write down the values of $\alpha + \beta + \gamma, \beta\gamma + \gamma\alpha + \alpha\beta, \alpha\beta\gamma$.

Qu. 2. Find the cubic equations with roots
(i) $2, -4, 7$ (ii) $3, 4 - 2j, 4 + 2j$.

We cannot find the roots directly from these relations, since using them to eliminate β and γ for example will merely lead back to $a\alpha^3 + b\alpha^2 + c\alpha + d = 0$. But if we have additional information about the roots these relations often provide a quick method of solution.

Example 1: Solve the equation $9z^3 - 18z^2 - 4z + 8 = 0$, given that the roots are in arithmetic progression.

Solution: Let the roots be $p - q, p, p + q$. The sum of the roots is $3p$ and the product is $p(p^2 - q^2)$.
Therefore $3p = \frac{18}{9}$, so that $p = \frac{2}{3}$,
and $\frac{2}{3}(\frac{4}{9} - q^2) = -\frac{8}{9}$, so that $q^2 = \frac{4}{9} + \frac{4}{3} = \frac{16}{9}$ and $q = \pm\frac{4}{3}$.
Therefore the roots are $\frac{2}{3} - \frac{4}{3}, \frac{2}{3}, \frac{2}{3} + \frac{4}{3}$, i.e. $-\frac{2}{3}, \frac{2}{3}, 2$.
Notice that we need to use only two of the three relations between the roots in this case.   □

Example 2: The roots of $az^3 + bz^2 + cz + d = 0$ are α, β, γ ($\alpha\beta\gamma \neq 0$).
Find the cubic equation whose roots are $\dfrac{1}{\alpha}, \dfrac{1}{\beta}, \dfrac{1}{\gamma}$.

Solution (i): $\dfrac{1}{\alpha} + \dfrac{1}{\beta} + \dfrac{1}{\gamma} = \dfrac{\beta\gamma + \gamma\alpha + \alpha\beta}{\alpha\beta\gamma} = \dfrac{c/a}{-d/a} = -\dfrac{c}{d}$

$\dfrac{1}{\beta} \cdot \dfrac{1}{\gamma} + \dfrac{1}{\gamma} \cdot \dfrac{1}{\alpha} + \dfrac{1}{\alpha} \cdot \dfrac{1}{\beta} = \dfrac{\alpha + \beta + \gamma}{\alpha\beta\gamma} = \dfrac{-b/a}{-d/a} = \dfrac{b}{d}$

$\dfrac{1}{\alpha} \cdot \dfrac{1}{\beta} \cdot \dfrac{1}{\gamma} = \dfrac{1}{\alpha\beta\gamma} = -\dfrac{a}{d}.$

Therefore the equation with roots $\dfrac{1}{\alpha}, \dfrac{1}{\beta}, \dfrac{1}{\gamma}$ is

$$z^3 - \left(-\frac{c}{d}\right)z^2 + \frac{b}{d}z - \left(-\frac{a}{d}\right) = 0$$

or $dz^3 + cz^2 + bz + a = 0.$

Notice that the original coefficients appear here in reverse order.   □

Solution (ii): Let $w = \dfrac{1}{z}$ so that $z = \dfrac{1}{w}$.

Then $w \in \left\{\dfrac{1}{\alpha}, \dfrac{1}{\beta}, \dfrac{1}{\gamma}\right\} \Leftrightarrow z \in \{\alpha, \beta, \gamma\}$

$\Leftrightarrow az^3 + bz^2 + cz + d = 0$

$$\Leftrightarrow a \cdot \frac{1}{w^3} + b \cdot \frac{1}{w^2} + c \cdot \frac{1}{w} + d = 0$$

$$\Leftrightarrow a + bw + cw^2 + dw^3 = 0. \qquad \square$$

If f is any function with an inverse, the method of the second solution of Example 2 can be used to transform an equation with roots α, β, γ into an equation with roots $f(\alpha), f(\beta), f(\gamma)$ by substituting $z = f^{-1}(w)$ in the original equation and simplifying.

The quartic equation can be treated similarly. If the roots of $az^4 + bz^3 + cz^2 + dz + e = 0$ $(a \neq 0)$ are $\alpha, \beta, \gamma, \delta$ then

$$az^4 + bz^3 + cz^2 + dz + e \equiv a(z-\alpha)(z-\beta)(z-\gamma)(z-\delta)$$
$$\equiv az^4 - a(\alpha + \beta + \gamma + \delta)z^3 + a(\alpha\beta + \alpha\gamma + \alpha\delta + \beta\gamma + \beta\delta + \gamma\delta)z^2$$
$$- a(\beta\gamma\delta + \gamma\delta\alpha + \delta\alpha\beta + \alpha\beta\gamma)z + a\,\alpha\beta\gamma\delta.$$

Equating the coefficients of z^3, z^2, z^1, z^0 shows that

$$\alpha + \beta + \gamma + \delta = -\frac{b}{a}, \quad \alpha\beta + \alpha\gamma + \alpha\delta + \beta\gamma + \beta\delta + \gamma\delta = \frac{c}{a}$$

$$\beta\gamma\delta + \gamma\delta\alpha + \delta\alpha\beta + \alpha\beta\gamma = -\frac{d}{a}, \quad \alpha\beta\gamma\delta = \frac{e}{a}.$$

These sums of roots taken one, two and three at a time are abbreviated to $\Sigma\alpha, \Sigma\alpha\beta, \Sigma\alpha\beta\gamma$ respectively. Other symmetric functions of the roots can be shown similarly; for example

$$\Sigma\alpha^2\beta = \alpha^2\beta + \alpha\beta^2 + \alpha^2\gamma + \alpha\gamma^2 + \alpha^2\delta + \alpha\delta^2 + \beta^2\gamma$$
$$+ \beta\gamma^2 + \beta^2\delta + \beta\delta^2 + \gamma^2\delta + \gamma\delta^2.$$

Example 3: The roots of $z^4 - 2z^3 + 3z^2 - z + 5 = 0$ are $\alpha, \beta, \gamma, \delta$. Find $\alpha^3 + \beta^3 + \gamma^3 + \delta^3$.

Solution: Since $\alpha\beta\gamma\delta = 5$, none of the roots is zero.
Let $S_n = \alpha^n + \beta^n + \gamma^n + \delta^n$; we want to find S_3. Since α is a root, $\alpha^4 - 2\alpha^3 + 3\alpha^2 - \alpha + 5 = 0$. Dividing by α,

$$\alpha^3 - 2\alpha^2 + 3\alpha - 1 + \frac{5}{\alpha} = 0.$$

Similar equations hold for β, γ, δ. Adding these four equations gives
$$S_3 - 2S_2 + 3S_1 - 4 + 5S_{-1} = 0 \qquad \text{(A)}$$

Now $S_1 = \Sigma\alpha = 2$,

$$S_2 = \Sigma\alpha^2 = (\Sigma\alpha)^2 - 2\Sigma\alpha\beta = 2^2 - 2 \times 3 = -2$$

$$S_{-1} = \Sigma\frac{1}{\alpha} = \frac{1}{5}, \text{ using the obvious extension of}$$
$$\text{Example 2 to quartics.}$$

Putting these values in (A) gives
$S_3 - 2 \times (-2) + 3 \times 2 - 4 + 5 \times \frac{1}{5} = 0$ and therefore $S_3 = -7$.
□

Qu. 3. With the notation of Example 3, prove that
$$S_{n+4} - 2S_{n+3} + 3S_{n+2} - S_{n+1} + 5S_n = 0.$$
State the corresponding result for the general quartic equation.

Exercise 6B

1. If α, β, γ are the roots of $z^3 - 2z^2 + 3z + 1 = 0$ find
   (i) $\Sigma \alpha$ (ii) $\Sigma \alpha\beta$ (iii) $\alpha\beta\gamma$ (iv) $\Sigma \alpha^2$ (v) $\Sigma \alpha^3$
   (vi) $\Sigma \alpha^4$ (vii) $\Sigma \dfrac{1}{\alpha}$ (viii) $\Sigma \dfrac{1}{\alpha\beta}$ (ix) $\Sigma \alpha^2\beta$ (x) $\Sigma \dfrac{\alpha + \beta}{\gamma}$.

2. If α, β, γ are the roots of the equation $z^3 - 4z^2 - z + 3 = 0$, find cubic equations whose roots are
   (i) $2\alpha, 2\beta, 2\gamma$ (ii) $\alpha + 2, \beta + 2, \gamma + 2$ (iii) $\beta + \gamma, \gamma + \alpha, \alpha + \beta$.

3. Solve these equations, given that in each case the roots are in arithmetic progression.
   (i) $z^3 + 6z^2 + 3z - 10 = 0$ (ii) $z^3 - 6z^2 + 16 = 0$
   (iii) $z^3 - 9z^2 + 31z - 39 = 0$.

4. The roots of the equation $2z^3 - 12z^2 + kz - 15 = 0$ are in arithmetic progression. Find k and solve the equation.

5. Solve $32z^3 - 14z + 3 = 0$, given that one root is twice another.

6. The roots of $az^3 + bz^2 + cz + d = 0$ are α, β, γ. Form the cubic equations whose roots are

   (i) $\alpha^2, \beta^2, \gamma^2$ (ii) $\dfrac{1}{\alpha^2}, \dfrac{1}{\beta^2}, \dfrac{1}{\gamma^2}$ (iii) $\dfrac{\beta\gamma}{\alpha}, \dfrac{\gamma\alpha}{\beta}, \dfrac{\alpha\beta}{\gamma}$

7. Find the relation between a, b, c, d if the roots of $az^3 + bz^2 + cz + d = 0$ are in geometric progression.
   Solve the equation $8z^3 - 52z^2 + 78z - 27 = 0$.

8. The equation $z^3 + pz^2 + 2pz + q = 0$ has roots $\alpha, 2\alpha, 4\alpha$. Find all the possible values of p, q, α.

9. By considering $(\beta + \gamma)(\gamma + \alpha)(\alpha + \beta)$ show that two of the roots of $z^3 + pz^2 + qz + r = 0$ have equal magnitude but opposite signs if and only if $r = pq$. Find the roots in terms of p and q in this case.

10. Write a flow diagram to find $\alpha^n + \beta^n + \gamma^n$ ($n > 3$) in terms of p, q, r, where α, β, γ are the roots of $z^3 + pz^2 + qz + r = 0$.

11. None of the roots α, β, γ of the equation $x^3 + 3px + q = 0$ is zero. Obtain the equation whose roots are $\beta\gamma/\alpha, \gamma\alpha/\beta, \alpha\beta/\gamma$, expressing its coefficients in terms of p and q. Deduce that $\gamma = \alpha\beta$ if and only if $(3p - q)^2 + q = 0$. Solve the equation $4x^3 - 3x - 9 = 0$.   [L]

12. If $\alpha, \beta, \gamma, \delta$ are the roots of $2z^4 + 5z^3 + 4z^2 - 6z + 2 = 0$ find
    (i) $\Sigma\alpha^2$ (ii) $\Sigma(\alpha + \beta + \gamma)^2$ (iii) $\Sigma(\alpha + \beta)^3$ (iv) $\Sigma\alpha^2\beta$.

13. The product of two of the roots of $z^4 + pz^3 + qz^2 + rz + s = 0$ is equal to the product of the other two. Prove that $s = r^2/p^2$.

14. The roots of $az^4 + bz^3 + cz^2 + dz + e = 0$ are $\alpha, \beta, \gamma, \delta$. Express $\Sigma(\alpha - \beta)^2$ in terms of the coefficients. Hence prove that if all the roots are real then $3b^2 \geqslant 8ac$. Show by an example that the converse is not true.

15. If $\alpha, \beta, \gamma, \delta$ are the roots of the equation $x^4 + 4x^3 + x^2 - 6x + 2 = 0$,
    (a) find the value of $\Sigma\alpha^2$ (b) find the value of $\Sigma\alpha^2\beta$
    (c) obtain the equation whose roots are $\alpha + 1, \beta + 1, \gamma + 1, \delta + 1$ and hence evaluate $\alpha, \beta, \gamma, \delta$.   [L]

6.3 Rational functions and their graphs

A *rational function* is a function $f : x \mapsto \dfrac{P(x)}{Q(x)}$, where P, Q are polynomial functions and Q is not the zero polynomial. Any values of x for which the denominator $Q(x)$ is zero must be excluded from the domain. Throughout this section, except where we state otherwise, the domain of f will be the largest possible, that is $\{x : Q(x) \neq 0\}$; in some cases, for example $x \mapsto \dfrac{1}{1 + x^2}$, this domain is $\mathbb{R}$, the set of all real numbers.

Qu. 1. Show that every polynomial function is a rational function.

When sketching the graph of a rational function f we usually want to know where the graph intersects the axes. The y-intercept is found by evaluating $f(0)$; if 0 is not in the domain of f, there is no y-intercept. The x-intercepts are found by solving $f(x) = 0$.

Qu. 2. Find the points where $y = \dfrac{(x - 2)(x + 5)}{(x + 1)(x - 4)}$ cuts the axes.

We also need to know how the curve behaves near the roots of $Q(x) = 0$. Consider, for example, $y = \dfrac{1}{x - 5}$. Clearly, y is undefined for $x = 5$. When x is slightly greater than 5, $x - 5$ is positive but close to zero, so that y takes a large positive value; when x is slightly less than 5, y is numerically large (i.e. $|y|$ is large), but y is negative; $|y|$ can be made as large as we please by making the difference between x and 5 sufficiently small. We

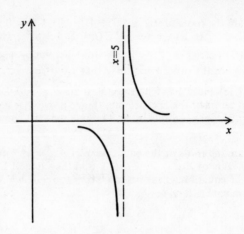

may say that $y \to +\infty$ as $x \to 5$ from the right, and $y \to -\infty$ as $x \to 5$ from the left. The line $x = 5$ is known as a vertical asymptote.

The line $x = a$ is a vertical *asymptote* for the curve $y = \dfrac{P(x)}{Q(x)}$ if $Q(a) = 0$, $P(a) \neq 0$; the signs of $P(x), Q(x)$ when x is close to a enable us to decide whether $y \to +\infty$ or $y \to -\infty$ as $x \to a$ from one side or the other.

Example 1: Sketch the graph of $y = \dfrac{x + 3}{x^2 + 5x - 6}$ for $-7 \leqslant x \leqslant 2$.

Solution: $y = \dfrac{x + 3}{x^2 + 5x - 6} = \dfrac{x + 3}{(x + 6)(x - 1)}$.

(a) Find the intercepts: $x = 0 \Rightarrow y = -\frac{1}{2}$; $y = 0 \Rightarrow x = -3$.
(b) Examine the signs of the numerator and denominator near the vertical asymptotes $x = -6, x = 1$.

	$x \approx -6$		$x \approx 1$	
	$x < -6$	$x > -6$	$x < 1$	$x > 1$
$x + 3$	−	−	+	+
$x + 6$	−	+	+	+
$x - 1$	−	−	−	+
y	−	+	−	+

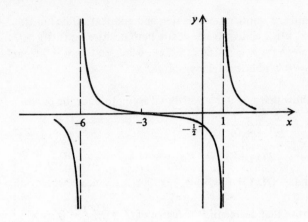

Broken lines are used to indicate the vertical asymptotes.   □

Qu. 3. Describe the behaviour of the following curves near their vertical asymptotes:

(i) $y = \dfrac{x-4}{(x-3)(x-5)}$,   (ii) $y = \dfrac{2-x}{(x+1)^2}$.

Qu. 4. Sketch the curve $y = \dfrac{x^2-5x+4}{x-2}$ for $-1 \leqslant x \leqslant 5$.

Qu. 5. (a) Explain why $P(a) = Q(a) = 0 \Rightarrow \dfrac{P(x)}{Q(x)}$ is not in its simplest form.

(b) How does the graph of $y = \dfrac{x-2}{(x-5)(x-2)}$ differ from the graph of $y = \dfrac{1}{x-5}$?

To complete our sketch graphs we need to examine the behaviour of the curve as $x \to \pm\infty$. For example:

$$y = \frac{1}{x-5} \to 0 \text{ from below as } x \to -\infty \text{ (i.e. } y \to 0 \text{ and } y < 0),$$

$$y = \frac{1}{x-5} \to 0 \text{ from above as } x \to +\infty.$$

As a second example, consider $y = \dfrac{x+3}{x^2+5x-6}$. Dividing numerator and

denominator by x^2 gives $y = \dfrac{\dfrac{1}{x}+\dfrac{3}{x^2}}{1+\dfrac{5}{x}-\dfrac{6}{x^2}} \to 0$ as $x \to \pm\infty$, assuming that the

limits of sum and quotient are the sum and quotient of the limits.
As $x \to +\infty$, $x + 3$ and $x^2 + 5x - 6$ are both positive, so that $y \to 0$ from
above. As $x \to -\infty$, $x + 3 < 0$, $x^2 + 5x - 6 > 0$, so that $y \to 0$ from below.
The line $y = 0$ is a horizontal asymptote.

Qu. 6. Show that $y = 0$ is a horizontal asymptote for the curve

$$y = \frac{3x - 7}{2x^2 - 5x + 21}.$$

Qu. 7. Let $P(x) \equiv a_m x^m + a_{m-1} x^{m-1} + \ldots + a_1 x + a_0$

and $Q(x) \equiv b_n x^n + b_{n-1} x^{n-1} + \ldots + b_1 x + b_0.$

Prove that degree of P < degree of $Q \Rightarrow \dfrac{P(x)}{Q(x)} \to 0$ as $x \to \pm\infty.$

With the notation of Qu. 7, if P and Q are of the same degree (i.e.
$m = n$):

$$\frac{a_n x^n + a_{n-1} x^{n-1} + \ldots + a_0}{b_n x^n + b_{n-1} x^{n-1} + \ldots + b_0} = \frac{a_n + \dfrac{a_{n-1}}{x} + \ldots + \dfrac{a_0}{x^n}}{b_n + \dfrac{b_{n-1}}{x} + \ldots + \dfrac{b_0}{x^n}} \to \frac{a_n}{b_n}$$

as $x \to \pm\infty$, again assuming the necessary limit properties. This means that

the line $y = \dfrac{a_n}{b_n}$ is a horizontal asymptote to the curve $y = \dfrac{P(x)}{Q(x)}.$

Consider the graph of $y = \dfrac{x^2 - 5x + 4}{x - 2}$. In Qu. 4 we investigated its

behaviour when $-1 \leqslant x \leqslant 5$, and now we consider its shape as $x \to \pm\infty$.

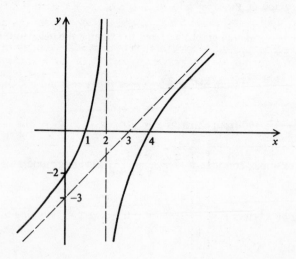

By division the equation may be rewritten in the form $y = x - 3 - \dfrac{2}{x-2}$.

As $x \to +\infty$, $\dfrac{2}{x-2} \to 0$ from above. This means that for large positive x,

the curve $y = \dfrac{x^2 - 5x + 4}{x - 2}$ is below the line $y = x - 3$, the distance between

the two graphs decreasing as x increases. Similarly, if x is negative and $|x|$ is

large, the curve $y = \dfrac{x^2 - 5x + 4}{x - 2}$ is above the line $y = x - 3$, the gap

between them decreasing as x decreases (i.e. as $|x|$ increases.). The line
$y = x - 3$ is known as an oblique asymptote.

If $P(x), Q(x)$ are polynomials, then by polynomial division we may

express $\dfrac{P(x)}{Q(x)}$ in the form $R(x) + \dfrac{S(x)}{Q(x)}$ where $R(x), S(x)$ are polynomials,

and the degree of S is less than the degree of Q. By Qu. 7, $\dfrac{S(x)}{Q(x)} \to 0$ as

$x \to \pm\infty$, so that the curve $y = \dfrac{P(x)}{Q(x)}$ approaches the curve $y = R(x)$

asymptotically.

Example 2: Sketch the graph of $y = \dfrac{x^2 - 5x + 6}{x - 1}$.

Solution: $y = \dfrac{x^2 - 5x + 6}{x - 1} = \dfrac{(x-2)(x-3)}{x-1} = x - 4 + \dfrac{2}{x-1}$.

(a) Find the intercepts: $x = 0 \Rightarrow y = -6$; $y = 0 \Rightarrow x = 2$ or 3.

(b) Examine the behaviour of the curve near the vertical
asymptote $x = 1$:
$$x \approx 1, x < 1 \Rightarrow x^2 - 5x + 6 > 0, x - 1 < 0 \Rightarrow y < 0;$$
$$x \approx 1, x > 1 \Rightarrow x^2 - 5x + 6 > 0, x - 1 > 0 \Rightarrow y > 0.$$

(c) Examine the behaviour of the curve as $x \to \pm\infty$: $y = x - 4$
is an oblique asymptote; as $x \to +\infty$, the curve is above the
asymptote; as $x \to -\infty$, the curve is below the asymptote.

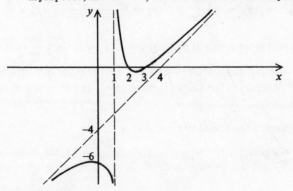

Example 3: Sketch the curve $y = \dfrac{x(x+1)(x-2)}{x-1}$.

Solution: $y = \dfrac{x(x+1)(x-2)}{x-1} = \dfrac{x^3 - x^2 - 2x}{x-1} = x^2 - 2 - \dfrac{2}{x-1}$.

(a) Intercepts: $y = 0 \Rightarrow x = 0$ or -1 or 2.

(b) Near the vertical asymptote $x = 1$:

$x \approx 1, x < 1 \Rightarrow x^3 - x^2 - 2x < 0, x - 1 < 0 \Rightarrow y > 0$;

$x \approx 1, x > 1 \Rightarrow x^3 - x^2 - 2x < 0, x - 1 > 0 \Rightarrow y < 0$.

(c) The curve approaches $y = x^2 - 2$ asymptotically. As $x \to +\infty$, the curve is below the asymptote; as $x \to -\infty$, the curve is above the asymptote.

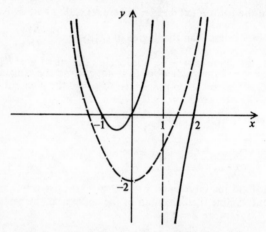

We now consider the problem of sketching the graph of the equation $y^2 = f(x)$. We first notice that the graph exists only for values of x which make $f(x) \geqslant 0$. If (x_1, y_1) satisfies the equation, then so does $(x_1, -y_1)$, so the graph is symmetrical about the x-axis.

Now $y^2 = f(x) \Rightarrow 2y \dfrac{dy}{dx} = f'(x) \Rightarrow \dfrac{dx}{dy} = \dfrac{2y}{f'(x)}, \; f'(x) \neq 0$.

If the curve meets the x-axis at $P(x_1, 0)$ and $f'(x_1) \neq 0$ then $\dfrac{dx}{dy} = 0$ at P, so the tangent at P is parallel to the y-axis. If $f'(x_1) = 0$ the direction of the tangent at P depends on the particular function f, and each case should be treated separately.

Example 4: Sketch the graph of $y^2 = x^2(x+2)$.

Solution: We sketch first the graph of $y = x^2(x+2)$, obtaining curve C_1. We use this to help us sketch C_2, the graph of $y^2 = x^2(x+2)$.

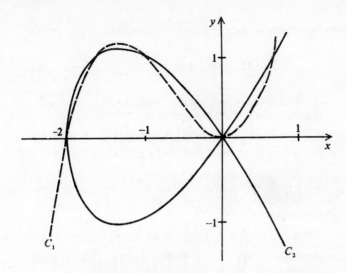

At $x = -2$ the gradient of C_1 is not zero so the tangent to C_2 at $(-2, 0)$ is parallel to the y-axis. At $x = 0$ the gradient of C_1 is zero so we use an alternative method.

Now
$$y^2 = x^2(x + 2)$$
$$\Rightarrow \quad y = \pm x \sqrt{(x + 2)}$$
$$\Rightarrow \frac{dy}{dx} = \pm \left[\sqrt{(x + 2)} + \frac{x}{2\sqrt{(x + 2)}}\right]$$
$$\Rightarrow \frac{dy}{dx} = \pm \sqrt{2} \text{ at } x = 0,$$

so that the sketch may be completed as shown.   □

Qu. 8. Sketch the graph of $y^2 = x^4(x + 2)$.

Exercise 6C

1. Zinc ions in alkaline solution exist in two forms. The total zinc concentration in a solution is given by $ax^2 + \dfrac{b}{x^2}$, where x is the hydroxide concentration, and a, b are positive constants. Sketch the graph of the function for $x > 0$, and state whether it has a maximum or a minimum.

2–18. Sketch the graphs of $y = f(x)$, where $f(x)$ is the given expression.

2. $\dfrac{1}{(x - 3)(x + 1)}$

3. $\dfrac{1}{(x - 3)^2}$

4. $\dfrac{x}{(x + 2)^2}$

5. $\dfrac{2x + 1}{1 - x^2}$

6. $\dfrac{1}{x^2 + 1}$

7. $\dfrac{x(7 + x)}{(1 + x)(2 - x)(5 + x)}$

8. $\dfrac{x^2}{x^2-1}$ 9. $\dfrac{x}{1+x^2}$ 10. $\dfrac{x^2-1}{(x-2)(x-3)}$

11. $\dfrac{1-x^2}{1+x^2}$ 12. $\dfrac{x^3-1}{x}$ 13. $\dfrac{x^2}{5-x}$

14. $\dfrac{1-x^2}{(1+x^2)^2}$ 15. $\dfrac{x(3x+5)}{(x+2)^2}$ 16. $\dfrac{x^3+x^2+4}{x}$

17. $\dfrac{x(3+x)}{(1+x)(2-x)}$ 18. $\dfrac{(1-x)^2(1+2x-x^2)}{(x-2)^2}$

19. Sketch the graphs of $y^2 = f(x)$ where $f(x)$ is as in
(i) Qu. 8   (ii) Qu. 9   (iii) Qu. 10.

20. Sketch the graph of $y^2 = \frac{1}{12}x^3(4-x)$.

21. Sketch the graphs of $y^2 = 3x - x^3 + c$ for $c = 4, 2, 0, -2, -4$.

22. The concentration y of a drug in the blood after x hours is given by
$$y = \frac{x}{20x^2 + 50x + 80}, x \geqslant 0.$$ Sketch the graph of y against x.

23. Sketch the curve $y = \dfrac{x^2 + 2x + 1}{x^4}$. Discuss the dependence on α of the
number of real roots of the equation $\alpha x^4 + x^2 + 2x + 1 = 0$.   [OS]

24. A formula used in the study of lenses is $\dfrac{1}{f} = \dfrac{1}{u} + \dfrac{1}{v}$, where u and v are
the distances from lens to object and lens to image respectively, and f is
the focal length of the lens, regarded as constant in this question. Express
$u + v$ in terms of u, f, mentioning any necessary restrictions, and sketch
the graph of $u + v$ against u, taking f as positive.

25. When a camera is focussed on an object P distant d from the camera,
only objects that distance from the camera are exactly in focus, but
other objects are rendered acceptably sharply; this region of acceptable
sharpness is known as the *depth of field*. The *hyperfocal distance*, h,
is the least value of d such that the depth of field extends to infinity.

The near limit, n, of the depth of field is given by $n = \dfrac{dh}{h+d}$. The far

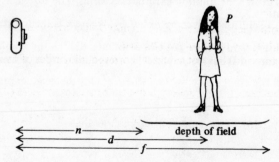

limit, f, of the depth of field is given by $f = \dfrac{dh}{h - d}$, $d < h$; there is no far

limit when $d \geqslant h$. On the same axes sketch graphs of n and f against d for $d > 0$, assuming that h is a positive constant, and show by shading how the depth of field varies with d. Suggest two properties of the hyperfocal distance which would interest photographers.

26. For a current i amp the efficiency e of a transformer is of the form

$\dfrac{bi}{ai^2 + bi + c}$ where a, b, c are positive constants. Sketch the graph of e

against i for $i \geqslant 0$, and show that maximum efficiency occurs when $ai^2 = c$. If $a = 0.64, b = 200, c = 250$, find the maximum efficiency.

6.4 Inequalities

We define the inequality x is greater than y, written $x > y$, to mean that $x - y$ is positive. Then

$x > y \iff x - y$ is positive   by definition

$\quad\ \iff (x + a) - (y + a)$ is positive   because the value of the expression is unchanged

$\quad\ \iff x + a > y + a$ by definition.

Qu. 1. Prove from the definition that
  (a) if a is positive then $x > y \iff ax > ay$;
  (b) if a is negative then $x > y \iff ay > ax$;
  (c) $a > b, x > y \implies a + x > b + y$;
  (d) $x > y, y > z \implies x > z$.

The inequalities $<, \geqslant, \leqslant$ are defined similarly, and equivalent properties may be proved. The basic rules for manipulating inequalities are
  (i)   we may add the same number to both sides of an inequality;
  (ii) we may multiply both sides of an inequality by the same positive number;
  (iii) if both sides of an inequality are multiplied by the same negative number, the inequality is reversed;
  (iv) we may add (but not subtract) corresponding sides of inequalities of the same type (as in Qu. 1(c));
  (v)   inequalities of the same type are transitive (as in Qu. 1 (d)).

Qu. 2. By producing a counter example, show that $\left.\begin{array}{c} a \leqslant b \\ x \leqslant y \end{array}\right\} \not\Rightarrow a - x \leqslant b - y$.

Qu. 3.   Prove that   (a) $x < y \not\Rightarrow x^2 < y^2$;

  (b) $0 < x < y \Rightarrow x^2 < y^2$.

Example 1: Solve $\dfrac{(x-2)(x-3)}{x-1} \leqslant 0$.

Solution (i):   Sketch the curve $y = \dfrac{(x-2)(x-3)}{x-1}$. (This was done in

  Example 2, page 155.) From the graph we see that the solution
  set is $\{x : x < 1 \text{ or } 2 \leqslant x \leqslant 3\}$.   □

Solution (ii):   If we multiply the inequality by $x - 1$, which may be positive
  or negative, we need to consider both possibilities. A better
  method is to multiply by $(x-1)^2$, which is positive, obtaining

$$(x-2)(x-3)(x-1) \leqslant 0, \; x - 1 \neq 0.$$

  Sketch $y = (x-2)(x-3)(x-1), x \neq 1$, from which we see
  that the solution set is $\{x : x < 1 \text{ or } 2 \leqslant x \leqslant 3\}$.

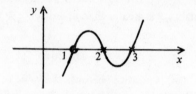

□

Qu. 4.   Sketch graphs to show that $f(x)$ does not necessarily change sign as
x passes through a value for which $f(x)$ is zero or discontinuous.

Qu. 5.   Use each of the two methods just illustrated to solve the inequality
$\dfrac{x^2 - 4x - 5}{x - 4} > 0$.

When solving an inequality such as $f(x) \leqslant g(x)$ it is usually good practice
to rearrange it in the equivalent form $f(x) - g(x) \leqslant 0$, as illustrated in the
next example.

Example 2: Solve $\dfrac{x^2 - 3x + 4}{x - 1} \leqslant 2$.

Solution: $\dfrac{x^2 - 3x + 4}{x - 1} \leqslant 2 \Leftrightarrow \dfrac{x^2 - 3x + 4}{x - 1} - 2 \leqslant 0$

$\Leftrightarrow \dfrac{x^2 - 5x + 6}{x - 1} \leqslant 0$

$\Leftrightarrow \dfrac{(x-2)(x-3)}{x - 1} \leqslant 0$

$\Leftrightarrow x < 1 \text{ or } 2 \leqslant x \leqslant 3$ as in Example 1.   □

Exercise 6D

1–5. Solve the inequalities for real x.

1. $\dfrac{x+2}{(x-2)(x+7)} > 0.$ 2. $\dfrac{(5+x)(3-x)}{(2+x)(1+x)} \geqslant 0$ 3. $\dfrac{x^2-2}{x-3} < 2$

4. $\dfrac{x^3-4}{x-2} \leqslant x+2$ 5. $\dfrac{(2x-3)(x+1)}{x^2} \leqslant -2$

6. Find the set of real x for which   (i) $\left|\dfrac{2x+3}{x-2}\right| \leqslant 1$;   (ii) $\dfrac{2x+3}{x-2} \leqslant 1$.

7. Find the integer value of x which makes $(x+1)^2$ less than $7x-3$ and greater than $5x-1$.   [MEI]

8. Find the set of values of x for which $-3 \leqslant \dfrac{(x-1)(x-5)}{x-3} \leqslant 3$.   [L]

9. Rearrange $y = \dfrac{(x-2)(x+1)}{x^2}$ as a quadratic equation in x. Use the condition for this to have real roots to find, without using calculus, the maximum value of y for real x.

10. If x is real, find the set of possible values of $\dfrac{6x+6}{x^2+3}$.

11. Find constants p, q such that $x^2+6x+10 = (x+p)^2 + q$ for all values of x. Deduce that, if x is real, $0 < \dfrac{1}{x^2+6x+10} \leqslant 1$.   [L]

12. Prove that, when x is real, $1 \leqslant \dfrac{9x^2+8x+3}{x^2+1} \leqslant 11$.   [MEI]

13. Prove that, if a, b, c are positive, then
$(b+c-a)(c+a-b)(a+b-c) \leqslant abc$.   [O & C]

14. (a) Find the ranges of values of the real number x for which the following inequalities hold:

    (i) $\dfrac{1}{x+6} < \dfrac{2}{2-3x}$; (ii) $|5-3x| \leqslant |x+1|$.

    (b) Show by shading on a sketch of the xy-plane, the region for which $x^2+y^2 \leqslant 1, y \geqslant x$ and $y \leqslant x+1$. Hence find
    (i) the greatest value of y,
    (ii) the least value of $x+y$
    for which these inequalities hold.   [MEI]

15. (i) Prove that the sum of a positive number and its reciprocal is at least 2. State and prove a similar relationship concerning a negative number.

    (ii) Prove that $\dfrac{x}{y} + \dfrac{y}{x} + \dfrac{z}{x} + \dfrac{x}{z} + \dfrac{z}{y} + \dfrac{y}{z} \geqslant 6$ for all positive x, y, z.

16. Prove that $a^2+b^2+c^2 \geqslant ab+bc+ca$ for all real a, b, c.

17. The *geometric mean* of a, b is the number c such that a, c, b are consecutive terms of a geometric progression. The *arithmetic mean* is defined similarly. By considering x^2 and y^2, or otherwise, prove that the arithmetic mean of two non-negative numbers is always greater than or equal to their geometric mean. Under what condition does equality occur?

18. The *harmonic mean* of a, b is the number c such that $\dfrac{1}{c}$ is the arithmetic mean of $\dfrac{1}{a}, \dfrac{1}{b}$. Show that the harmonic mean of two positive numbers is less than or equal to their geometric mean. When does equality occur?

19. The *root mean square* of a, b is $\sqrt{\dfrac{a^2 + b^2}{2}}$. Prove that the root mean square of two positive numbers is greater than or equal to their arithmetic mean.

20. The points A, C, E, G are on side PS of trapezium $PQRS$; points B, D, F, H are on QR. Lines PQ, AB, CD, EF, GH, SR are parallel.

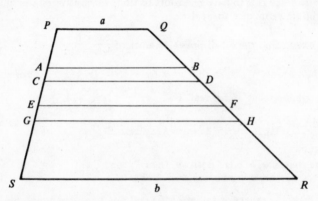

Suppose $PQ = a$, $RS = b$. Show that
  (i) the arithmetic mean of a, b is EF where EF is mid-way between PQ, SR;
  (ii) the geometric mean of a, b is CD where $PQDC, CDRS$ are similar;
  (iii) the harmonic mean of a, b is AB where AB, PR, QS are concurrent;
  (iv) the root mean square of a, b is GH where $PQHG, GHRS$ have the same area.

21. Let S_n stand for the statement
    $$\dfrac{a_1 + a_2 + \ldots a_n}{n} \geqslant \sqrt[n]{a_1 a_2 \ldots a_n},$$ where $\{a_1, a_2, \ldots, a_n\}$ is any set of non-negative real numbers.
  (i) By considering $b_r = \dfrac{a_{2r-1} + a_{2r}}{2}$, use mathematical induction to prove that S_n is true whenever $n = 2^m$ for positive integer m.

(ii) By letting $a_n = \dfrac{a_1 + a_2 + \ldots + a_{n-1}}{n-1}$, prove that $S_n \Rightarrow S_{n-1}$.

(iii) Deduce the truth of S_n for all integral $n \geqslant 2$. Under what condition does equality occur in S_n?

6.5 Partial fractions

When we add together two rational expressions, we start by finding the lowest common multiple of the two denominators. For example

$$\frac{5}{x-3} + \frac{3}{x+2} = \frac{5(x+2) + 3(x-3)}{(x-3)(x+2)} = \frac{8x+1}{(x-3)(x+2)}.$$

The reverse process is known as resolving a rational expression into *partial fractions*.

1. Linear factors

As an example, consider the expression $\dfrac{x+37}{(x-5)(x+1)}$. If we can find constants A, B such that

$$x + 37 \equiv A(x+1) + B(x-5) \tag{1}$$

then by division $\dfrac{x+37}{(x-5)(x+1)} \equiv \dfrac{A}{x-5} + \dfrac{B}{x+1}$. We can find the values of A, B by equating the coefficients of corresponding terms in identity (1):

$$A + B = 1 \quad \text{(equating coefficients of } x\text{)}$$
$$A - 5B = 37 \quad \text{(equating constant terms)}.$$

Thus $A = 7, B = -6$. Therefore $\dfrac{x+37}{(x-5)(x+1)} \equiv \dfrac{7}{x-5} - \dfrac{6}{x+1}$.

An alternative (and usually faster) method of finding the values of A, B uses the fact that identity (1) is true for all values of x. Substituting $x = 5$ gives $42 = 6A \Rightarrow A = 7$, while $x = -1$ gives $36 = -6B \Rightarrow B = -6$ as before.

Qu. 1. Resolve the following expressions into partial fractions:

(i) $\dfrac{x+1}{(x-1)(x+2)}$ (ii) $\dfrac{6x+2}{(x+1)(x-1)(x+3)}$.

Another method of finding the values of A, B in the resolution of $\dfrac{x+37}{(x-5)(x+1)}$ into partial fractions $\dfrac{A}{x-5} + \dfrac{B}{x+1}$ is known as the *cover-up* method. The value of A is found by covering up the factor $x-5$ in the expression $\dfrac{x+37}{(x-5)(x+1)}$ and then putting $x = 5$ in what is left. Similarly,

B is found by covering up $x + 1$, and then evaluating what is left with $x = -1$. It is good practice to include a check to guard against faulty arithmetic. One method of doing this is shown in Example 1.

Qu. 2. By rearranging identity (1), explain why the cover-up method works.

Qu. 3. Verify that $\dfrac{a+x}{(b+x)(c+x)} \equiv \dfrac{a-b}{(b+x)(c-b)} + \dfrac{a-c}{(b-c)(c+x)}$.

Example 1: Use the cover-up method to express $\dfrac{3x^2+13}{(x-3)(x-1)(x+7)}$ in partial fractions.

Solution: The coefficient of $\dfrac{1}{x-3}$ is $\dfrac{3(3)^2+13}{(3-1)(3+7)} = \dfrac{27+13}{20} = 2$.

The coefficient of $\dfrac{1}{x-1}$ is $\dfrac{3(1)^2+13}{(1-3)(1+7)} = \dfrac{3+13}{-16} = -1$.

The coefficient of $\dfrac{1}{x+7}$ is $\dfrac{3(-7)^2+13}{(-7-3)(-7-1)} = \dfrac{147+13}{80} = 2$.

Therefore $\dfrac{3x^2+13}{(x-3)(x-1)(x+7)} \equiv \dfrac{2}{x-3} - \dfrac{1}{x-1} + \dfrac{2}{x+7}$.

(Check: $x = 0 \Rightarrow$ L.H.S. $= \dfrac{13}{(-3)(-1)(7)} = \dfrac{13}{21}$;

R.H.S. $= -\dfrac{2}{3} + 1 + \dfrac{2}{7} = \dfrac{-14+21+6}{21} = \dfrac{13}{21}$.)

$\square$

Qu. 4. Use the cover-up method to express the following as partial fractions:

(i) $\dfrac{x+8}{(x-1)(x+2)}$ (ii) $\dfrac{8x+2}{(x-5)(x-2)(x+2)}$.

2. Quadratic factors

So far we have only considered expressions where the denominator can be factorized into real linear factors, with no repetitions. We consider now the case when one of the factors is a real quadratic expression which cannot be factorized further. Suppose, for example, that we want to resolve $\dfrac{5x^2+6x+1}{(x-1)(x^2+x+2)}$ into partial fractions. Let us guess that one of the partial fractions is (as before)

$$\frac{1}{x-1} \cdot \frac{5+6+1}{1+1+2} \equiv \frac{3}{x-1};$$

then the other is

$$\frac{5x^2+6x+1}{(x-1)(x^2+x+2)} - \frac{3}{x-1} \equiv \frac{5x^2+6x+1-3(x^2+x+2)}{(x-1)(x^2+x+2)}$$

$$\equiv \frac{2x^2+3x-5}{(x-1)(x^2+x+2)}$$

$$\equiv \frac{(x-1)(2x+5)}{(x-1)(x^2+x+2)}$$

$$\equiv \frac{2x+5}{x^2+x+2}.$$

Therefore $\quad \dfrac{5x^2+6x+1}{(x-1)(x^2+x+2)} \equiv \dfrac{3}{x-1} + \dfrac{2x+5}{x^2+x+2}.$

Notice that the second partial fraction has a linear numerator, requiring two constants.

An alternative method is to suppose

$$\frac{5x^2+6x+1}{(x-1)(x^2+x+2)} \equiv \frac{A}{x-1} + \frac{Bx+C}{x^2+x+2}.$$

Then $5x^2+6x+1 \equiv A(x^2+x+2) + (Bx+C)(x-1)$, and we form three equations by equating coefficients of x^2, x^1, x^0. Of course, a combination of methods can be used, such as finding A by the cover-up method (or by substitution), and finding B, C by equating coefficients.

Example 2: Resolve $\dfrac{4x(x-1)}{(x-2)(x^2+4)}$ into partial fractions.

Solution: Suppose $\dfrac{4x(x-1)}{(x-2)(x^2+4)} \equiv \dfrac{A}{x-2} + \dfrac{Bx+C}{x^2+4}.$

Then $\quad 4x(x-1) \equiv A(x^2+4) + (Bx+C)(x-2)$

i.e. $\quad 4x^2-4x \equiv (A+B)x^2+(-2B+C)x+4A-2C.$ (1)

By the cover-up method, $\quad A = \dfrac{4 \cdot 2 \cdot 1}{8} = 1.$

Equating coefficients in identity (1) we find $B = 3, C = 2.$

Therefore $\quad \dfrac{4x(x-1)}{(x-2)(x^2+4)} \equiv \dfrac{1}{x-2} + \dfrac{3x+2}{x^2+4}.$ $\qquad \square$

Qu. 5.   Resolve $\dfrac{x^2 - 6x + 2}{(x - 3)(x^2 - x + 1)}$ into partial fractions.

Qu. 6.   If $f(x) \equiv \dfrac{P(x)}{(x - a)\,Q(x)}$ where $Q(a) \neq 0$, show that the difference between $f(x)$ and its partial fraction $\dfrac{1}{x - a} \cdot \dfrac{P(a)}{Q(a)}$ is $\dfrac{P(x)\,Q(a) - P(a)\,Q(x)}{(x - a)\,Q(a)\,Q(x)}$. Deduce that the numerator of this expression necessarily has the factor $x - a$.

3. Repeated factors

We now look at rational expressions which have repeated linear factors in the denominator. Consider, for example, $\dfrac{5x^2 - x + 3}{(2x - 1)^2 (x + 2)}$. As previously, we guess that one partial fraction is $\dfrac{1}{x + 2} \cdot \dfrac{20 + 2 + 3}{25} = \dfrac{1}{x + 2}$; then the other is

$$\dfrac{5x^2 - x + 3}{(2x - 1)^2 (x + 2)} - \dfrac{1}{x + 2} \equiv \dfrac{5x^2 - x + 3 - (4x^2 - 4x + 1)}{(2x - 1)^2 (x + 2)}$$

$$\equiv \dfrac{x^2 + 3x + 2}{(2x - 1)^2 (x + 2)}$$

$$\equiv \dfrac{(x + 2)(x + 1)}{(2x - 1)^2 (x + 2)} \equiv \dfrac{x + 1}{(2x - 1)^2}.$$

But $x + 1 \equiv \frac{1}{2}(2x - 1) + \frac{3}{2}$, so that $\dfrac{x + 1}{(2x - 1)^2} \equiv \dfrac{\frac{1}{2}}{2x - 1} + \dfrac{\frac{3}{2}}{(2x - 1)^2}$.

Therefore $\dfrac{5x^2 - x + 3}{(2x - 1)^2 (x + 2)} \equiv \dfrac{1}{x + 2} + \dfrac{\frac{1}{2}}{2x - 1} + \dfrac{\frac{3}{2}}{(2x - 1)^2}$. Notice that there is a term involving $2x - 1$, and a term involving $(2x - 1)^2$, both having constant numerators. With this additional information, we could start our solution by supposing $\dfrac{5x^2 - x + 3}{(2x - 1)^2 (x + 2)} \equiv \dfrac{A}{x + 2} + \dfrac{B}{2x - 1} + \dfrac{C}{(2x - 1)^2}$, so that $5x^2 - x + 3 \equiv A(2x - 1)^2 + B(2x - 1)(x + 2) + C(x + 2)$. The values of A, B, C can be found by constructing and solving simultaneous equations, or by a combination of methods; both A and C can be found by the cover-up method.

Example 3: Express $\dfrac{4x + 7}{(x + 2)^2 (x + 1)}$ in partial fractions.

Solution: Suppose $\dfrac{4x + 7}{(x + 2)^2 (x + 1)} \equiv \dfrac{A}{x + 1} + \dfrac{B}{x + 2} + \dfrac{C}{(x + 2)^2}$.

Then $4x + 7 \equiv A(x + 2)^2 + B(x + 2)(x + 1) + C(x + 1)$

i.e. $4x + 7 \equiv (A + B)x^2 + (4A + 3B + C)x + 4A + 2B + C$

$$(1)$$

Using the cover-up method, $A = \dfrac{-4 + 7}{(-1)^2} = 3$.

Equating coefficients in identity (1): $3 + B = 0 \Rightarrow B = -3$;

$$12 - 6 + C = 7 \Rightarrow C = 1.$$

Therefore $\dfrac{4x + 7}{(x + 2)^2 (x + 1)} \equiv \dfrac{3}{x + 1} - \dfrac{3}{x + 2} + \dfrac{1}{(x + 2)^2}$.   □

Qu. 7.   Resolve $\dfrac{4}{(x + 3)^2 (x - 1)}$ into partial fractions.

Qu. 8.   If $f(x) \equiv \dfrac{125}{(x - 1)^3 (x + 4)}$, suggest suitable forms for the four

partial fractions into which $f(x)$ may be resolved, and find them.

So far in this section we have restricted ourselves to real factors. By using complex numbers, every quadratic expression can be expressed in linear factors, which may then be used to form partial fractions with linear denominators.

Qu. 9.   Express $\dfrac{x + 1}{x(x^2 + 1)}$ as the sum of three partial fractions with

complex coefficients.

Exercise 6E

1–12. Resolve the expressions into partial fractions with real coefficients.

1. $\dfrac{3x}{(x - 2)(x + 1)}$

2. $\dfrac{x - 2}{x(x + 2)}$

3. $\dfrac{3}{4x^2 - 25}$

4. $\dfrac{1}{1 - x^2}$

5. $\dfrac{x + 5}{(x - 5)(x + 3)}$

6. $\dfrac{x^2 + x + 1}{x(x - 2)(x + 3)}$

7. $\dfrac{5x^2}{(x^2 + 1)(x + 2)}$

8. $\dfrac{2x^2 + 9x + 7}{(x^2 + 4)(2x - 3)}$

9. $\dfrac{x}{x^3 + 1}$

10. $\dfrac{3}{8 - x^3}$

11. $\dfrac{4}{x^2(x - 1)}$

12. $\dfrac{1}{x^4 - 16}$

13. Use the substitution $y = x + 1$ to express $\dfrac{3x^2 + 1}{(x + 1)^3}$ in the form

$\dfrac{A}{x + 1} + \dfrac{B}{(x + 1)^2} + \dfrac{C}{(x + 1)^3}$.

14. Express $\dfrac{4x(x-4)}{(x-2)(x^2+4)}$ as the sum of partial fractions with linear denominators.

15. Express $(4+5x)/(x-x^2)$ in partial fractions. Hence, or otherwise, find the turning points of the graph of the function $y = (4+5x)/(x-x^2)$, stating about each whether it is a maximum or a minimum or neither.

[O & C]

16. Write $f(x) = \dfrac{x+2}{(x+1)(x-2)}$ in partial fractions. Hence find expressions for $f'(x)$ and $f''(x)$. Find a, b such that $f'(a)$, $f'(b)$ are zero, and evaluate $f''(a)$ and $f''(b)$. What do your answers tell you about the graph of f? Sketch the graph of f, indicating clearly the behaviour of the function for large values of x and near $x = -1$ and $x = 2$. From your graph suggest why you would expect $\displaystyle\int_0^1 f(x)\,dx$ to be negative, and evaluate this integral.

[SMP]

6.6 Uses of partial fractions

In §6.5 we showed that $\dfrac{3x^2+13}{(x-3)(x-1)(x+7)} \equiv \dfrac{2}{x-3} - \dfrac{1}{x-1} + \dfrac{2}{x+7}$. For some purposes the form on the left-hand side is the more useful, but the right-hand side is easier to differentiate†, easier to integrate, easier to expand in ascending powers of x, and easier to evaluate if reciprocals are available.

When integrating $\dfrac{P(x)}{Q(x)}$ where $P(x)$, $Q(x)$ are polynomials, we first check that P is of lower degree than Q, and then resolve $\dfrac{P(x)}{Q(x)}$ into partial fractions, and integrate. If P is not of lower degree than Q we express $\dfrac{P(x)}{Q(x)}$ in the form $R(x) + \dfrac{S(x)}{Q(x)}$ where $R(x)$, $S(x)$ are polynomials, and the degree of S is less than the degree of Q. For example

$$\int \frac{x^3 - x^2 - 3x}{x^2 + 2x + 1}\,dx = \int\left(x - 3 + \frac{2x-3}{(x+1)^2}\right)dx$$

$$= \int\left(x - 3 + \frac{1}{(x+1)^2} + \frac{2}{x+1}\right)dx$$

---

† Strictly speaking we differentiate and integrate functions, not expressions. It is so much easier to speak of differentiating $f(x)$, rather than of differentiating $f: x \mapsto f(x)$, that we tolerate this common 'abuse of language'.

$$= \tfrac{1}{2}x^2 - 3x - \frac{1}{x+1} + 2 \ln |x+1| + c.$$

Notice the form of the integral of fractions like $\dfrac{1}{(x+1)^2}, \dfrac{2}{x+1}$.

Fractions like $\dfrac{4x-7}{x^2 - 6x + 10}$ are more difficult to integrate; if the numerator had been $2x - 6$ (the derivative of the denominator), the integral would have been $\ln |x^2 - 6x + 10|$. So we arrange the numerator as a multiple of the derivative of the denominator plus a constant.

Thus $\displaystyle\int \frac{4x-7}{x^2 - 6x + 10}\, dx = \int \left(\frac{2(2x-6)}{x^2 - 6x + 10} + \frac{5}{x^2 - 6x + 10} \right) dx$

$$= 2 \ln |x^2 - 6x + 10| + \int \frac{5}{(x-3)^2 + 1}\, dx$$

$$= 2 \ln |x^2 - 6x + 10| + 5 \tan^{-1}(x-3) + c.$$

Example 1: Find (i) $\displaystyle\int \frac{3x+7}{x^2 + 2x + 6}\, dx$ (ii) $\displaystyle\int \frac{x^4 - 2}{x^2 - 1}\, dx$.

Solution: (i) As $x^2 + 2x + 6$ does not have real factors, we cannot resolve the integrand into a sum of real partial fractions. We express the numerator as a multiple of the derivative of the denominator plus a constant.

$$3x + 7 \equiv A(2x+2) + B \Rightarrow A = \tfrac{3}{2}, B = 4.$$

Therefore

$$\int \frac{3x+7}{x^2 + 2x + 6}\, dx = \int \left(\frac{3}{2} \cdot \frac{2x+2}{x^2 + 2x + 6} + \frac{4}{x^2 + 2x + 6} \right) dx$$

$$= \tfrac{3}{2} \ln |x^2 + 2x + 6| + \int \frac{4}{(x+1)^2 + 5}\, dx$$

$$= \tfrac{3}{2} \ln |x^2 + 2x + 6| + \frac{4}{\sqrt{5}} \tan^{-1} \frac{x+1}{\sqrt{5}} + c.$$

(ii) $\dfrac{x^4 - 2}{x^2 - 1} \equiv x^2 + 1 - \dfrac{1}{x^2 - 1} \equiv x^2 + 1 + \dfrac{\frac{1}{2}}{x+1} - \dfrac{\frac{1}{2}}{x-1}$.

Therefore

$$\int \frac{x^4 - 2}{x^2 - 1}\, dx = \int \left(x^2 + 1 + \frac{\frac{1}{2}}{x+1} - \frac{\frac{1}{2}}{x-1} \right) dx$$

$$= \tfrac{1}{3}x^3 + x + \tfrac{1}{2} \ln |x+1| - \tfrac{1}{2} \ln |x-1| + c$$

$$= \tfrac{1}{3}x^3 + x + \tfrac{1}{2} \ln \left| \frac{x+1}{x-1} \right| + c. \qquad \square$$

Qu. 1.   Find (i) $\int \dfrac{4x+7}{(x+2)^2(x+1)}\,dx$ (ii) $\int \dfrac{4x(x-1)}{(x-2)(x^2+4)}\,dx.$

The substitution $t = \tan \frac{1}{2}\theta$ often changes the integration of a fraction involving trigonometric functions to the integration of a rational function. We leave it to the reader to prove that

$$\text{if } t = \tan \tfrac{1}{2}\theta, \text{ then } \tan\theta = \frac{2t}{1-t^2},\ \sin\theta = \frac{2t}{1+t^2},$$
$$\cos\theta = \frac{1-t^2}{1+t^2},\quad d\theta = \frac{2dt}{1+t^2}.$$

Example 2: Find $\displaystyle\int \frac{d\theta}{1+\sin\theta}.$

Solution: Let $t = \tan\frac{1}{2}\theta.$ Then $d\theta = \dfrac{2dt}{1+t^2}$ and $\sin\theta = \dfrac{2t}{1+t^2}.$

Therefore $\displaystyle\int \frac{d\theta}{1+\sin\theta} = \int \frac{2dt}{(1+t^2)+2t}$

$$= \int \frac{2dt}{(1+t)^2}$$

$$= -\frac{2}{1+t} + c = -\frac{2}{1+\tan\frac{1}{2}\theta} + c. \quad \square$$

Qu. 2.   Use $t = \tan\frac{1}{2}\theta$ to find (i) $\int \operatorname{cosec}\theta\,d\theta$ (ii) $\int \sec\theta\,d\theta.$

Qu. 3.   By expressing the numerator in the form

$$A\frac{d}{dx}(2\sin x + \cos x) + B(2\sin x + \cos x)$$

find $\displaystyle\int \frac{3\sin x + 7\cos x}{2\sin x + \cos x}\,dx.$

Example 3: Find $\displaystyle\sum_{r=1}^{n} \frac{r+4}{r(r+1)(r+2)}.$

Solution: $\dfrac{r+4}{r(r+1)(r+2)} \equiv \dfrac{2}{r} - \dfrac{3}{r+1} + \dfrac{1}{r+2}.$

Therefore $\displaystyle\sum_{r=1}^{n} \frac{r+4}{r(r+1)(r+2)} = \sum_{r=1}^{n}\left(\frac{2}{r} - \frac{3}{r+1} + \frac{1}{r+2}\right)$

$$= 2 - \frac{3}{2} \; \bigg| + \frac{1}{3}$$

$$+ \frac{2}{2} \bigg| - \frac{3}{3} + \frac{1}{4}$$

$$+ \frac{2}{3} - \frac{3}{4} + \frac{1}{5}$$

$$+ \frac{2}{4} - \frac{3}{5} + \frac{1}{6}$$

$$+ \ldots + \frac{2}{n-2} - \frac{3}{n-1} + \frac{1}{n}$$

$$+ \frac{2}{n-1} - \frac{3}{n} + \frac{1}{n+1}$$

$$+ \frac{2}{n} - \frac{3}{n+1} + \frac{1}{n+2}.$$

Most of the terms cancel out and we are left with

$$\sum_{r=1}^{n} \frac{r+4}{r(r+1)(r+2)} = 2 - \frac{3}{2} + \frac{2}{2} + \frac{1}{n+1} - \frac{3}{n+1} + \frac{1}{n+2}$$

$$= \frac{3}{2} - \frac{2}{n+1} + \frac{1}{n+2}. \qquad \square$$

Qu. 4. Resolve $\dfrac{1+2x}{(1+x)(1-2x)}$ into partial fractions, and hence expand

$\dfrac{1+2x}{(1+x)(1-2x)}$ in ascending powers of x as far as the term in x^3.

Exercise 6F

1–17. Integrate with respect to x.

1. $\dfrac{1+x}{1-x}$

2. $\dfrac{x^3}{x^2+1}$

3. $\dfrac{x^3+2}{x^2-1}$

4. $\dfrac{x+2}{(x+1)(x-3)}$

5. $\dfrac{x^2-1}{4x^2-9}$

6. $\dfrac{3x+4}{(2x-1)(3x+2)}$

7. $\dfrac{1}{x^2-3}$

8. $\dfrac{1}{x^2+3}$

9. $\dfrac{1}{(x^2+1)(x+1)^2}$

10. $\dfrac{4}{x^2-2x-1}$

11. $\dfrac{1}{1+\sin x + \cos x}$

12. $\dfrac{1}{2+\cos x}$

13. $\dfrac{\sin 3x}{1 - \cos 3x}$

14. $\sec x \, \operatorname{cosec} x$

15. $\dfrac{1}{1 - \sin 2x + 2 \cos 2x}$

16. $\dfrac{3 \cos x - 7 \sin x}{2 \cos x - 5 \sin x}$

17. $\dfrac{5 \sin x + 3 \cos x}{3 \cos x + 4 \sin x}$

18. Find $\displaystyle\sum_{r=1}^{n} \dfrac{2r + 1}{r^2 (r + 1)^2}$.

19. Find $\displaystyle\sum_{r=1}^{n} \dfrac{r + 2}{2^{r+1} r(r + 1)}$

20. Find the sum of the first n terms of the series

   (i) $\dfrac{2}{1 \cdot 3 \cdot 5} + \dfrac{3}{3 \cdot 5 \cdot 7} + \dfrac{4}{5 \cdot 7 \cdot 9} + \cdots$

   (ii) $\dfrac{1}{1 \cdot 3 \cdot 5} + \dfrac{1}{3 \cdot 5 \cdot 7} + \dfrac{1}{5 \cdot 7 \cdot 9} + \cdots$

21. Expand in ascending powers of x as far as the term in x^3:

   (i) $\dfrac{x}{(1 - x)(1 - 2x)}$ (ii) $\dfrac{2x - 1}{(x - 3)(x + 2)}$.

22. Express $f(x)$, where $f(x) \equiv \dfrac{x}{(x + 2)(x + 3)(x + 4)}$, as the sum of three

   partial fractions. Hence find
   (a) $\int f(x)\,dx$

   (b) $\displaystyle\sum_{r=1}^{n} \dfrac{r}{(r + 2)(r + 3)(r + 4)}$

   (c) the nth derivative of $f(x)$ with respect to x.   [L]

23. Use partial fractions to find the nth derivative with respect to x of $(x^2 - a^2)^{-1}$.

24. Prove that $\dfrac{d^{n-1}}{dx^{n-1}} \left(\dfrac{1}{1 + x^2} \right) = (-1)^{n-1} \dfrac{(n - 1)!}{2j} \left(\dfrac{1}{(x - j)^n} - \dfrac{1}{(x + j)^n} \right)$.

   By putting $x + j = r(\cos \theta + j \sin \theta)$ and using De Moivre's Theorem

   obtain $\dfrac{d^{n-1}}{dx^{n-1}} \left(\dfrac{1}{1 + x^2} \right)$ in terms of r, θ, and n.

   Deduce that, for $0 < x < \pi/2$,

   $$\dfrac{d^n}{dx^n} (\tan^{-1} x) = (-1)^{n-1} \dfrac{(n - 1)!}{(1 + x^2)^{n/2}} \sin (n \cot^{-1} x).$$

7 Differential equations

7.1 Introduction

Equations involving derivatives are known as *differential equations*. The *order* of a differential equation is the order of the highest derivative occurring in the equation. Thus $\frac{d^2y}{dx^2} + 5\frac{dy}{dx} + 6y = 10 \sin x$ is a second-order

differential equation, while $\frac{dN}{dt} = -kN$ is a first-order differential equation.

A set of real numbers which can be represented by a finite or infinite part of the number line with no discontinuities is called an *interval*. For example $\{x : 2 \leqslant x \leqslant 3\}, \{x : x < 10\}, \{x : -100 < x \leqslant 20\}$ and $\mathbb{R}$ are intervals, but $\{x : x \neq 0\}$ is not. More formally, an interval is a set D of real numbers such that $p, r \in D$ and $p < q < r \Rightarrow q \in D$.

Suppose Y is a function whose domain is an interval. If $y = Y(x)$ satisfies a differential equation in x and y, then Y is a *solution* of the differential equation. We restrict solutions to interval domains to avoid difficulties with the constants of integration. Thus $Y_1 : x \mapsto \ln x + c_1, x > 0$ and

$Y_2 : x \mapsto \ln(-x) + c_2, x < 0$ are solutions of $\frac{dy}{dx} = \frac{1}{x}$, but

$Y : x \mapsto \ln|x| + c, x \neq 0$ is not a solution. See Vol. 1, page 393.

As usual we only specify the domain when it is not the largest possible set of real numbers. For convenience we often say that the equation $y = Y(x)$ is a solution of a differential equation when we really mean that the function Y is a solution.

Many important differential equations have solutions which can be given in the form of an explicit or implicit equation in x and y, and we deal with some of the simpler ones in §§7.3 to 7.5. However, many differential equations can be solved only approximately, by giving a table of corresponding values of x and $Y(x)$, or by drawing a graph; see §§7.2 and 7.6.

Consider the differential equation $\frac{d^3y}{dx^3} = 1$. Integrating three times we

obtain $y = \frac{1}{6}x^3 + \frac{1}{2}Ax^2 + Bx + C$, where A, B, C are arbitrary constants. Notice that this solution of a third-order differential equation contains three arbitrary constants. A *general solution* of an nth-order differential equation is a solution which involves n arbitrary constants. Care is needed when counting arbitrary constants; $y = (A + B)x + C$ is regarded as having only two arbitrary constants as $A + B$ may be replaced by the single constant D. Some differential equations have solutions which cannot be

obtained from any general solution. Such solutions are known as *singular*
solutions. For examples see Exercise 7B, Qu. 21 and Qu. 24.

A general solution may be represented graphically by a family of curves,
all of which satisfy the same differential equation. Conversely, if we define
a family of curves by an equation containing n arbitrary constants, we can,
in general, differentiate n times and eliminate the n arbitrary constants
between our $n + 1$ equations, thus forming the differential equation of the
family of curves. There may be other curves which also fit this differential
equation.

Qu. 1.   Form differential equations by eliminating the arbitrary constants
from

(i) $y = A \sin x + B \cos x$ (ii) $y = Ax\,e^x$

(iii) $y = Ae^{2x} + Be^x + \sin x$ (iv) $x^2 + y^2 + Ax = 0$.

Differential equations occur in many sciences, and the situations which
give rise to them often provide additional information or conditions which
enable us to assign values to the arbitrary constants in a general solution,
and so find a *particular integral.* These conditions are known as *boundary
conditions* (or *initial conditions* if the independent variable is t, representing
time, and the additional information is specified at $t = 0$).

The *complete* solution of a differential equation is the solution from
which, by suitable choice of the arbitrary constants, all solutions may be
obtained.

Qu. 2.   Let Y be a function such that $y = Y(x)$ is a solution of $\dfrac{dy}{dx} = ky$.

By considering $y = h(x)\,.\,e^{kx}$, show that $Y(x)$ can be expressed in
the form Ae^{kx}. This shows that $y = Ae^{kx}$ is the complete solution.

7.2 Graphical method

In a laboratory lines of force in a magnetic field are often plotted with the
aid of a small compass. We use a similar idea to sketch the family of solution
curves of a first-order differential equation.

Consider the equation $\dfrac{dy}{dx} = x^2 - y$. If there is a solution curve passing
through $P(x_1, y_1)$ its gradient at P is $x_1{}^2 - y_1$, and a short line segment
through P with this gradient is the tangent to the curve. Fig. 1 was formed
by drawing many such line segments. It illustrates what is called the *tangent
field* of the differential equation. When preparing such a diagram it is sometimes convenient to locate the points at which $\dfrac{dy}{dx}$ takes certain specific values;

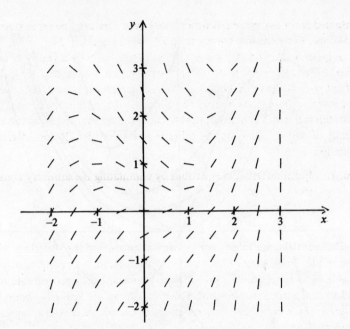

Fig. 1.

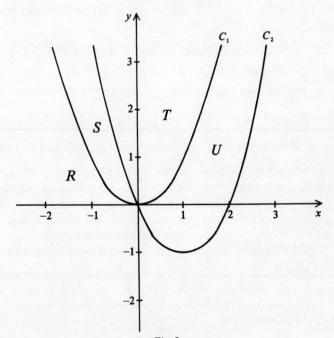

Fig. 2.

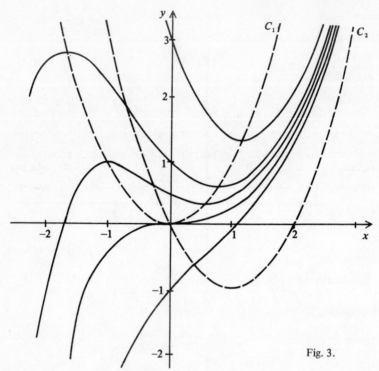

Fig. 3.

the locus of points for which $\frac{dy}{dx}$ is constant is an *isocline*. The isocline for

which $\frac{dy}{dx} = 0$ has equation $y = x^2$. It is the locus of the stationary points

of the solution curves, and is labelled C_1 in Fig. 2.

Now $\frac{dy}{dx} = x^2 - y \Rightarrow \frac{d^2y}{dx^2} = 2x - \frac{dy}{dx} = 2x - x^2 + y$. Therefore the locus

of points where $\frac{d^2y}{dx^2} = 0$ has equation $y = x^2 - 2x$. This curve is labelled C_2;

it separates the plane into two regions, one where solution curves are concave
up, and the other where they are concave down. Curves C_1, C_2 separate the
plane into four regions. The table shows how the concavity and the sign of
the gradient of solution curves vary according to the region through which
the curve is passing. We are now able to sketch various solution curves, as in
Fig. 3. We can improve the accuracy of our sketch by drawing more line
segments in Fig. 1.

	R	S	T	U
gradient	+	−	−	+
concave	down	down	up	up

Exercise 7A

1–5. Sketch the tangent field and several solution curves.

1. $dy/dx = x - y$ 2. $dy/dx = x/y$ 3. $dy/dx = xy$

4. $dy/dx = xy - 1$ 5. $dy/dx = x^2 + y^2$

6. Sketch the tangent field for the equation $\dfrac{dy}{dx} = \sqrt{y}$. Draw a separate

sketch of the tangent field for the equation $\left(\dfrac{dy}{dx}\right)^2 = |y|$. Hence sketch

sufficient curves to show fully the nature of the general solution of each
equation.   [SMP, adapted]

7. By considering various isoclines, sketch the tangent field of the differ-

ential equation $\left(\dfrac{dy}{dx}\right)^2 + x\,\dfrac{dy}{dx} = y$, and hence find a general solution.

7.3 First-order differential equations

1. Separating the variables

First-order differential equations of the form $\dfrac{dy}{dx} = \dfrac{f(x)}{g(y)}$ may be solved

by separating the variables (see Vol. 1 §8.10).

Example 1: The gravitational acceleration of a particle is inversely proportional to the square of the distance of the particle from the
centre of the earth. Take the earth to be a sphere of radius R m,
and the gravitational acceleration at the earth's surface as g m s^{-2}.
Find the minimum speed with which a particle should be
projected vertically from the earth's surface if the particle is not
to return to earth.

Solution: Let u m s^{-1} be the velocity of projection; suppose the particle is
x m from the centre of the earth, and moving with velocity
v m s^{-1} t s after launching. Then $\dfrac{d^2x}{dt^2} = -\dfrac{k}{x^2}$, where k is a

constant. $\dfrac{d^2x}{dt^2} = -g$ when $x = R \Leftrightarrow k = gR^2$.   We replace

$\dfrac{d^2x}{dt^2}$ by $v\dfrac{dv}{dx}$, to obtain $v\dfrac{dv}{dx} = -\dfrac{k}{x^2} \Leftrightarrow \int v\,dv = -\int \dfrac{k}{x^2}dx$

$$\Leftrightarrow \quad \tfrac{1}{2}v^2 = \dfrac{k}{x} + c.$$

$v = u$ when $x = R \Leftrightarrow c = \tfrac{1}{2}u^2 - \dfrac{k}{R}$.

Therefore $v^2 = u^2 - \dfrac{2k}{R} + \dfrac{2k}{x} = u^2 - 2gR + \dfrac{2gR^2}{x}$.

For $x > 0$, v^2 decreases as x increases, but if $u^2 \geqslant 2gR$, $v^2 \neq 0$. Therefore the particle will continue to move outwards from the earth if its initial speed is at least $\sqrt{(2gR)}$ m s^{-1}. This speed is known as the *escape velocity*; it is approximately 11 km s^{-1}. □

2. Integrating factor

The equation $\dfrac{dy}{dx} + y \cot x = \cos x$ appears at first sign to be intractable.

But if we multiplify by $\sin x$ we obtain $\dfrac{dy}{dx} \sin x + y \cos x = \cos x \sin x$,

in which we may (if we are lucky) recognize that the left side is the derivative of the product $y \sin x$.

Thus $\qquad \dfrac{d}{dx}(y \sin x) = \cos x \sin x$

so that $\qquad y \sin x = \displaystyle\int \cos x \sin x \, dx = \tfrac{1}{2} \sin^2 x + c$

and $\qquad y = \tfrac{1}{2} \sin x + c \operatorname{cosec} x.$

Applying the same idea to $\dfrac{dy}{dx} + y f(x) = g(x)$ we want to multiply

throughout by a suitable *integrating factor* $h(x)$ so that the left side $\dfrac{dy}{dx} h(x) + y f(x) h(x)$ is the derivative of a product.

Now $\dfrac{d}{dx}[y \, h(x)] = \dfrac{dy}{dx} h(x) + y \, h'(x),$

so we want $\quad h'(x) = f(x) h(x) \Leftrightarrow \qquad \dfrac{h'(x)}{h(x)} = f(x)$

$$\Leftrightarrow \int \frac{h'(x)}{h(x)} \, dx = \int f(x) \, dx$$

$$\Leftrightarrow \quad \ln|h(x)| = \int f(x) \, dx$$

$$\Leftrightarrow \qquad h(x) = e^{\int f(x) dx}.$$

The equation $\dfrac{dy}{dx} + y f(x) = g(x)$ can be solved by multiplying

throughout by the integrating factor $e^{\int f(x) dx}$.

In practice we use the simplest integrating factor, taking the constant of integration in $\int f(x) \, dx$ to be zero. See Ex. 7B Qu. 12 for a proof that the solution obtained in this way is the complete solution.

Example 2: Find the particular integral of $dy/dx + 2xy = 3x$ such that $y = 1$ when $x = 0$.

Solution: $\int 2x\, dx = x^2 + c_1$. Therefore the integrating factor is e^{x^2}.

$$\frac{dy}{dx} + 2xy = 3x \Leftrightarrow \frac{dy}{dx}e^{x^2} + 2xy\, e^{x^2} = 3x\, e^{x^2}$$

$$\Leftrightarrow \quad \frac{d}{dx}(y\, e^{x^2}) = 3x\, e^{x^2}$$

$$\Leftrightarrow \quad y\, e^{x^2} = \int 3x\, e^{x^2}\, dx = \tfrac{3}{2}e^{x^2} + c$$

$$\Leftrightarrow \quad y = \tfrac{3}{2} + c\, e^{-x^2}$$

$y = 1$ when $x = 0 \Leftrightarrow c = -\tfrac{1}{2}$.

Thus the required particular integral is $y = \tfrac{1}{2}(3 - e^{-x^2})$. □

Qu. 1. Solve $dy/dx + y = \sin x$.

3. Homogeneous equations

First-order differential equations which may be arranged in the form $\frac{dy}{dx} = f\left(\frac{y}{x}\right)$ are described as *homogeneous*. If we substitute v for $\frac{y}{x}$ so that $y = vx$ and $\frac{dy}{dx} = v + x\frac{dv}{dx}$ we obtain $v + x\frac{dv}{dx} = f(v)$, which is an equation that we can solve by separating the variables.

Example 3: Solve $\dfrac{dy}{dx} = \dfrac{x^2 + y^2}{xy}$, $x > 0$.

Solution: $\dfrac{dy}{dx} = \dfrac{x^2 + y^2}{xy} \Leftrightarrow v + x\dfrac{dv}{dx} = \dfrac{x^2(1 + v^2)}{x^2 v}$ where $y = vx$

$$\Leftrightarrow \quad x\frac{dv}{dx} = \frac{1}{v}$$

$$\Leftrightarrow \quad \int v\, dv = \int \frac{dx}{x}$$

$$\Leftrightarrow \quad \tfrac{1}{2}v^2 = \ln x + c_1$$

$$\Leftrightarrow \quad v^2 = \ln x^2 + c$$

$$\Leftrightarrow \quad y^2 = x^2(\ln x^2 + c).$$

Alternatively we may write $y^2 = x^2 \ln(Ax^2)$ where $A = e^c > 0$. □

Qu. 2. Solve $xy\, dy/dx = -x^2 - y^2$, $x > 0$.

Exercise 7B

1—8. Find a general solution.

1. $\dfrac{dy}{dx} = \dfrac{x^2}{y^3}$

2. $\dfrac{dy}{dx} - y = e^x$

3. $\dfrac{dy}{dx} + 2y = x + 1$

4. $\dfrac{dy}{dx} = \dfrac{y}{x^2 + 4}$

5. $\dfrac{dy}{dx} + \dfrac{2xy}{x^2 + 1} = x$

6. $\dfrac{dy}{dx} = \dfrac{2x - y}{x - 2y}$

7. $x\dfrac{dy}{dx} = x\cos x - y$

8. $\sin y \dfrac{dy}{dx} = \tan x \cos^2 y$

9. Solve the differential equation $\dfrac{dy}{dx} = xy^2 - x$, given that $y = 2$ when $x = 0$. In your answer give y in terms of x.   [O & C]

10. The gradient at any point P of a certain plane curve is proportional to the sum of the Cartesian coordinates of P. The curve passes through the point $(0, -1)$, and its gradient at this point is $-\frac{1}{2}$. Find a differential equation of the curve. Solve this equation and hence find the equation of the curve. Show that the curve has the line $y + x + 2 = 0$ as an asymptote.   [L]

11. Show that if the differential equation $6x^2 y^2 \dfrac{dy}{dx} = x^2 - 1$ is satisfied at all points on a curve, then the curve has a minimum point when $x = 1$. Find the equation of the curve when this minimum value of y is 1, and sketch the curve in this case for positive values of x.   [L]

12. (i) Show that $y = A\,e^{-\int f(x)dx}$ is the complete solution of $\dfrac{dy}{dx} + y\,f(x) = 0$. (Hint: see §7.1, Qu. 2.)

    (ii) Let p, Y be functions such that $y = Y(x)$ is a solution and $y = p(x)$ is a particular integral of $\dfrac{dy}{dx} + y\,f(x) = g(x)$. Show that $y = Y(x) - p(x)$ is a solution of $\dfrac{dy}{dx} + y\,f(x) = 0$, and deduce that the complete solution of $\dfrac{dy}{dx} + y\,f(x) = g(x)$ is $y = p(x) + A\,e^{-\int f(x)dx}$

13. Due to evaporation the mass of a spherical naphthalene moth ball is decreasing at a rate proportional to the surface area. In the first week the mass falls by 6% of its original value. Find how long the moth ball lasts.

14. When two liquids boil together in a vessel it is found that the ratio of the rates at which the liquids are evaporating is proportional to the ratio of the quantities of liquid remaining. If these quantities are x and y, prove that $y = cx^k$ where c, k are constants.

15. The rate at which a liquid runs from a container is proportional to the square root of the depth of the opening below the surface of the liquid. A cylindrical petrol storage tank is sunk in the ground with its axis vertical. There is a leak in the tank at an unknown depth. The level of the petrol in the tank, originally full, is found to drop by 20 cm in 1 hour, and by 19 cm in the next hour. Find the depth at which the leak is located.

16. When a thermometer is placed in a liquid, the rate at which the indicated temperature rises is proportional to the difference between the indicated temperature and the true temperature of the liquid. Initially the thermometer indicates $15°C$; 30 s later the indicated temperature is $20°C$, and a further 30 s later it is $21°C$. What is the true temperature of the liquid, assumed constant?

17. An appeal fund is launched with a donation of £1000; t weeks later the fund stands at £A, and is growing at the rate of £$A f(t)$ per week, where $f(t) \equiv \dfrac{1000\, t}{(t^2 + 125)^2}$ models the growth and decline of the enthusiasm of the sponsors. Find A in terms of t, and the time the appeal takes to reach its target of £50 000. To what value does the fund tend as $t \to \infty$?

18–20. Kirchhoff's loop law states that the algebraic sum of the voltage drop in a simple closed circuit is zero. The table gives the voltage drop across three electrical components and a source of electromotive force, where q coulomb is the charge on the capacitor, and $i = \dfrac{dq}{dt}$ is the current flowing in the circuit at time t seconds. The values of R, C, L are constant.

Resistor R ohm	Ri
Capacitor C farad	$\dfrac{q}{C}$
Inductor L henry	$L \dfrac{di}{dt}$
Electromotive force E volt	$-E$

18. An electromotive force, capacitor and resistor are connected in series. Show that $R \dfrac{dq}{dt} + \dfrac{q}{C} = E$.
If E is constant and $q = 0$ at $t = 0$, find q in terms of C, E, R, t and show that $q \to CE$ as $t \to \infty$.

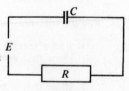

19. A capacitor with charge CE is connected in series with a resistor and a switch. At $t = 0$ the switch is closed. Show that $q = CE\, e^{-t/CR}$, and find an expression for i. If $C = 3 \times 10^{-6}, R = 1$ find the time taken for the current to fall to 1% of its original value.

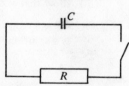

20. An electromotive force, inductor and resistor are connected in series. If $E = E_0 \sin \omega t$ and $i = 0$ at $t = 0$ show that

$$i = \dfrac{E_0}{R^2 + L^2 \omega^2} \left(R \sin \omega t - L\omega \cos \omega t + L\omega\, e^{-Rt/L} \right).$$

Also show that if t is sufficiently large $i \approx \dfrac{E_0}{\sqrt{(R^2 + L^2\omega^2)}} \sin(\omega t - \theta)$ where $\theta = \tan^{-1} \dfrac{L\omega}{R}$. This is known as the *steady state* current. Show that if $R = 0$, the maxima of the steady state current occur $\dfrac{\pi}{2\omega}$ s after the maxima of the electromotive force.

21. By differentiation, or otherwise, solve the differential equation $\left(\dfrac{dy}{dx}\right)^2 - x\dfrac{dy}{dx} + y = 0$, showing that the set of solution curves includes
    (i) precisely one parabola, and (ii) any tangent to this parabola.
    Is the curve defined by $y = \begin{cases} \frac{1}{4}x^2 & \text{if } x \leqslant 1 \\ \frac{1}{2}x - \frac{1}{4} & \text{if } x > 1 \end{cases}$ a solution curve?

    Give your reasons.   [SMP]

22. In this question mass, time, velocity, acceleration are measured in kg, s, $m\,s^{-1}$, $m\,s^{-2}$ respectively.
    (i) The mass of a raindrop is increasing at a rate proportional to its mass. If the mass is M_0 initially, show that the mass at time t is $M_0\,e^{kt}$, where k is a constant.
    (ii) The *momentum* of a body of mass M moving with velocity v is Mv newton seconds. The raindrop of part (i) starts to fall at $t = 0$, and the rate of change of momentum is proportional to the mass, the constant of proportionality being g. Find an expression for the velocity of the raindrop at time t, and show that this velocity tends to g/k as $t \to \infty$. This is known as the *terminal velocity*.

23. (M) A particle is projected vertically upwards from A with velocity $u\,m\,s^{-1}$. Air resistance is proportional to the velocity, and gravitational acceleration $g\,m\,s^{-2}$ is constant. Show that t s after projection the velocity is $(u + g/k)\,e^{-kt} - g/k\,m\,s^{-1}$, where k is constant. Also find an expression for the height of the particle above A, t s after projection.

24. For the differential equation $\dfrac{dy}{dx} = 3y^{\frac{2}{3}}$
    (i) find a general solution,
    (ii) show that $y = 0$ is a solution,
    (iii) show that $y = \begin{cases} (x - k_1)^3, & x \leqslant k_1 \\ 0 & k_1 < x < k_2 \\ (x - k_2)^3, & x \geqslant k_2 \end{cases}$

    is a solution, where k_1, k_2 are constants.

7.4 Simple harmonic motion

Consider a particle P moving along the straight line OX in such a way that its acceleration is proportional to its displacement from O, and

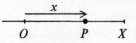

directed towards O. Let the displacement and velocity of P at time t s be x m, v m s^{-1} respectively. Then, in m s^{-2}, the acceleration is

$$\frac{d^2x}{dt^2} = v\frac{dv}{dx} = -\omega^2 x \text{ where } \omega \text{ is a positive constant. (Notice that the}$$

acceleration is in the opposite direction to the displacement.)

$$\text{Now } v\frac{dv}{dx} = -\omega^2 x \Leftrightarrow \int v\,dv = -\int \omega^2 x\,dx$$

$$\Leftrightarrow \quad \tfrac{1}{2}v^2 = -\tfrac{1}{2}\omega^2 x^2 + c_1$$

$$\Leftrightarrow \quad v^2 = c - \omega^2 x^2.$$

For v to be real, c must be positive, and it is convenient to write $c = a^2\omega^2$, where $a > 0$, so that $v^2 = a^2\omega^2 - \omega^2 x^2 \Leftrightarrow v = \omega\sqrt{(a^2-x^2)}$ (1)

$$\text{or } v = -\omega\sqrt{(a^2-x^2)}. \qquad (2)$$

Suppose $v > 0$ at some instant, so that equation (1) describes the motion.

Equation (1) $\Leftrightarrow \dfrac{dx}{dt} = \omega\sqrt{(a^2-x^2)}$

$$\Leftrightarrow \int \omega\,dt = \int \frac{dx}{\sqrt{(a^2-x^2)}}, \ x^2 \neq a^2$$

$$\Leftrightarrow \omega t + \epsilon_1 = \sin^{-1}\frac{x}{a}$$

$$\Rightarrow x = a\sin(\omega t + \epsilon_1). \qquad (3)$$

We see that x is less than a, but increasing. As $x \to a$, $v \to 0$ and $t \to \left(\dfrac{\pi}{2} - \epsilon_1\right)\Big/\omega = t_1$ say. The particle is instantaneously at rest at $t = t_1$, but as $x = a > 0$, the acceleration is negative. Therefore v becomes negative, and equation (2) describes the motion.

Equation (2) $\Leftrightarrow \dfrac{dx}{dt} = -\omega\sqrt{(a^2-x^2)}$

$$\Leftrightarrow \int \omega\,dt = -\int \frac{dx}{\sqrt{(a^2-x^2)}}, \ x^2 \neq a^2$$

$$\Leftrightarrow \omega t + \epsilon_2 = -\sin^{-1}\frac{x}{a}$$

$$\Rightarrow x = -a\sin(\omega t + \epsilon_2). \qquad (4)$$

Thus as $x \to a$, $v \to 0$ and $t \to \left(-\dfrac{\pi}{2} - \epsilon_2\right)\Big/\omega = t_1'$ say. But the particle is at rest instantaneously only, so that $t_1 = t_1' \Leftrightarrow \epsilon_2 = \epsilon_1 - \pi$.

Therefore (4) $\Leftrightarrow x = -a\sin(\omega t + \epsilon_1 - \pi) \Leftrightarrow$ (3).

Note that we have also established the validity of equations (3) and (4) at $x = a$.

Now consider the next section of the motion: $v < 0$ and the particle moves towards and past the origin, the motion being described by equations (2) and (4). As $x \to -a$, $v \to 0$, $t \to \left(\dfrac{\pi}{2} - \epsilon_2\right)\Big/\omega$. The velocity again reverses, so that equation (1) holds, but with different boundary conditions. As before, equation (1) $\Leftrightarrow \omega t + \epsilon_3 = \sin^{-1}\dfrac{x}{a}$, $x^2 \neq a^2$

$$\Rightarrow \qquad x = a \sin(\omega t + \epsilon_3). \qquad (5)$$

As previously we may show that $\epsilon_3 = \epsilon_2 - \pi$, so that equations (4) and (5) are equivalent. (It may also be shown that they are valid when $x = -a$.)

A similar argument applies to the subsequent motion, so that

> the complete solution of $\dfrac{d^2x}{dt^2} = -\omega^2 x$ is $x = a \sin(\omega t + \epsilon)$
>
> where a, ϵ are arbitrary constants.

The motion described above is an example of *simple harmonic motion*. The motion is clearly an oscillation. The time for one complete oscillation is called the *period*. The maximum displacement from the central position is known as the *amplitude*. The *frequency* is the number of complete oscillations made in unit time.

Qu. 1.   With the notation above show that the period is $\dfrac{2\pi}{\omega}$, the amplitude is a, and the frequency is $\dfrac{\omega}{2\pi}$.

Qu. 2.   Show that the complete solution $x = a \sin(\omega t + \epsilon)$ may be rewritten in the form $x = A \sin \omega t + B \cos \omega t$, where A, B are constants.

Qu. 3.   By differentiating twice, check that both forms of the complete solution satisfy the differential equation $\dfrac{d^2x}{dt^2} = -\omega^2 x$.

Qu. 4.   With the notation above, what is the particular integral
  (i) if the particle is at the origin initially,
  (ii) if the particle initially has maximum displacement?

Qu. 5.   Describe the physical significance of the constants A, B in the solution $x = A \sin \omega t + B \cos \omega t$.

Qu. 6.   By substituting $y = x - b$ show that $\dfrac{d^2x}{dt^2} = -\omega^2(x - b)$ is the defining equation for simple harmonic motion about the centre $x = b$. Write down the complete solution.

Oscillations which are simple harmonic occur in many sciences; even more common are oscillations which are approximately simple harmonic. The term simple harmonic motion should not be restricted to linear motion but may be applied to any phenomena which have variations described by the equation $\dfrac{d^2x}{dt^2} = -\omega^2 x$. It is not necessary for t to represent time.

Exercise 7C

1. A point P is moving with constant angular speed ω in a circle, radius a, centre C. The foot of the perpendicular from P to a fixed diameter is Q. Show that Q moves with simple harmonic motion with amplitude a, period $\dfrac{2\pi}{\omega}$, centre C.

2. Solve the differential equation $\dfrac{d^2y}{dx^2} + 4y = 0$

   (i)   given that $y = 0, \dfrac{dy}{dx} = -8$ at $x = 0$

   (ii) given that $y = 4, \dfrac{dy}{dx} = 0$ at $x = 0$

   (iii) given that $y = 5, \dfrac{dy}{dx} = 10\sqrt{3}$ at $x = 0$

   (iv) given that $y = 2, \dfrac{dy}{dx} = 6$ at $x = \dfrac{\pi}{4}$.

3. Solve the differential equation $\dfrac{d^2y}{dx^2} + 9y = 6$ given that $y = 2, \dfrac{dy}{dx} = 0$ at $x = 0$.

4. By using the substitution $z = \dfrac{dy}{dx}$, or otherwise, solve the differential equation $\dfrac{d^3y}{dx^3} + 4\dfrac{dy}{dx} = 0$ given that y takes the values $0, 0, 1$ for $x = 0, \dfrac{\pi}{4}, \dfrac{\pi}{2}$ respectively.   [O & C]

5. Show that the equation $x = a\cos^2 \omega t$ represents simple harmonic motion, and find the period, amplitude, and position of the centre.

6. If $\dfrac{d^2x}{dt^2} = \omega^2 x$ find $\dfrac{dx}{dt}$ in terms of x, and show that a general solution is $x = Ae^{\omega t} + Be^{-\omega t}$. (This is the complete solution.)

7. An inductor (L henry) and a charged capacitor (C farad) are connected in series. Use the data of Ex. 7B Questions 18 to 20 to show that, neglecting the resistance of the connecting wires, the current in the circuit oscillates harmonically with period $2\pi\sqrt{(LC)}$.

8. If the rod of a simple pendulum is l m long, and the rod is inclined to the vertical at an angle θ at time t s, it can be shown that
$l\dfrac{d^2\theta}{dt^2} + g\sin\theta = 0$. Assuming that θ is small show that the motion is approximately simple harmonic. If $g = 9.81$, find the length of a seconds pendulum (i.e. a pendulum which has period 2 seconds).

9. Seamen calculate the depth of tidal water from the depth at low water using the *twelfths rule*: in a 6-hour rise the rise in each hour is 1, 2, 3, 3, 2, 1 twelfths of the total rise. Show that this rule is in good agreement with the suggestion that the rise and fall of the tides obeys a simple harmonic law with period 12 hours.

10. The gravitational acceleration of a particle at the earth's surface is $9.8\ \text{m s}^{-2}$; in the interior of the earth, the gravitational acceleration is proportional to the distance from the centre of the earth. Taking the earth as a sphere of radius 6400 km, the points A, B are at opposite ends of a diameter. Show that if there was a frictionless, airless, straight tunnel connecting A and B, a particle dropped into one end of the tunnel would take about 42 minutes to reach the other end. Show that the same result applies if A and B are any two points on the earth's surface.

11. If $x = a\sin(\omega t + \delta)$, $y = b\sin(\omega t + \epsilon)$, $a, b > 0$, show that $z = x + y$ describes a simple harmonic motion with period $\dfrac{2\pi}{\omega}$ and amplitude $\sqrt{[a^2 + b^2 + 2ab\cos(\delta - \epsilon)]}$.

12. Let x, y, z be defined as in Qu. 11. Let the two-dimensional vector $\mathbf{a}$ represent the simple harmonic motion described by x, where $|\mathbf{a}| = a$ and $\mathbf{a}$ makes an angle δ with a fixed line OX. Let $\mathbf{b}$ similarly represent the simple harmonic motion described by y. Show that $\mathbf{a} + \mathbf{b}$ represents the simple harmonic motion described by z.

13. When two sources of slightly different frequencies f, f' sound together, the sound waves interfere with each other causing distinct throbs to be heard, known as *beats*. Let $x = a\sin(\omega t + \epsilon)$, $y = a\sin(\omega' t + \epsilon')$, where $a > 0$, describe the displacements produced by the two sound waves. Show that $y = a\sin(\omega t + \epsilon_1)$ where $\epsilon_1 = t\delta + \epsilon'$ and $\delta = \omega' - \omega$. Use the result of Qu. 11 to find the amplitude A of the oscillation described by $z = x + y$. Show that A varies periodically between 0 and $2a$, and find the period. The beats occur when the amplitude A is a maximum. Deduce that the number of beats per unit time is $|f' - f|$.

7.5 Linear equations

The differential equation for simple harmonic motion is an example of a *linear* differential equation. The most general second-order linear differential equation may be written as

$$f_0(x)\frac{d^2y}{dx^2} + f_1(x)\frac{dy}{dx} + f_2(x)y = g(x).$$

In this section we only consider equations in which the coefficients of y and its derivatives are constants. We use y', y'' to denote $\frac{dy}{dx}, \frac{d^2y}{dx^2}$ respectively.

Suppose Y is a solution of

$$ay'' + by' + cy = g(x), \tag{1}$$

where a, b, c are constant. Suppose also that p is a particular solution of (1).
Then $\quad a Y''(x) + b Y'(x) + c Y(x) = g(x)$
and $\quad a p''(x) + b p'(x) + c p(x) = g(x).$ Subtracting we obtain
$a [Y''(x) - p''(x)] + b [Y'(x) - p'(x)] + c [Y(x) - p(x)] = 0,$
so that $y = Y(x) - p(x)$ is a solution of

$$av'' + by' + cy = 0. \tag{2}$$

Thus every solution of (1) can be expressed in the form $y = p(x) + q(x)$ where q is a solution of (2). Conversely, by direct substitution we may show that $y = p(x) + q(x)$ satisfies (1) whenever q is a solution of (2).

The complete solution of (2) is often referred to as the *complementary function*. The argument above shows that

> the complete solution of equation (1) is the sum of
> the complementary function and a particular integral.

A similar result applies for all linear differential equations.

Example 1: Find the complete solution of $y' - 2y = x^2 - 2x + 3$.

Solution: $y' - 2y = 0 \Leftrightarrow y = A e^{2x}$. This is the complementary function.
To find a particular integral we try $y = ax^2 + bx + c$.
$$\Rightarrow y' = 2ax + b.$$
Substituting into the original differential equation:
$$2ax + b - 2a x^2 - 2bx - 2c = x^2 - 2x + 3.$$
Equating coefficients: $a = -\frac{1}{2}, b = \frac{1}{2}, c = -1$. Therefore
$y = -\frac{1}{2}x^2 + \frac{1}{2}x - 1$ is a particular integral, and the complete
solution is $y = Ae^{2x} - \frac{1}{2}x^2 + \frac{1}{2}x - 1$. □

If $g(x)$ in equation (1) is a polynomial expression, we expect a polynomial particular integral, of the same degree, because the derivatives of polynomials are polynomials of lower degree. If $g(x)$ is a multiple of $\sin kx$ or $\cos kx$, we try $y = a \sin kx + b \cos kx$. If $g(x)$ is a multiple of e^{kx}, we try $y = a e^{kx}$. But special considerations apply if g is a function which can be obtained from the complementary function, see Ex. 7D, Questions 9 and 10.

Qu. 1. Use the method above to find the complete solution of
$y' + 3y = g(x)$ where (i) $g(x) \equiv e^{2x}$ (ii) $g(x) \equiv \sin x + \cos x$.

Qu. 2. Given that c, α, β are constants, use the method above or an integrating factor to show that the complete solution of

$y' - \alpha y = c e^{\beta x}$ is $y = A e^{\alpha x} + \dfrac{c}{\beta - \alpha} e^{\beta x}$ when $\alpha \neq \beta$ and

$y = A e^{\alpha x} + cx e^{\alpha x}$ when $\alpha = \beta$.

To find the complementary function for the second-order linear equation

$$a y'' + b y' + cy = g(x), \ a \neq 0,$$

we look at the related equation

$$a y'' + b y' + cy = 0. \tag{3}$$

Now $y = Ae^{kx} \Rightarrow y' = kAe^{kx} \Rightarrow y'' = k^2 Ae^{kx}$. Substituting into (3), we obtain $Ae^{kx}(ak^2 + bk + c) = 0$.
Thus $y = Ae^{kx}$ is a solution of (3) for all A if $ak^2 + bk + c = 0$.
The equation $ak^2 + bk + c = 0$ is known as the *auxiliary equation*.

Suppose the auxiliary equation has the real roots α, β; then

$$\alpha + \beta = -\frac{b}{a} \text{ and } \alpha\beta = \frac{c}{a}.$$

Thus equation (3) $\Leftrightarrow y'' - (\alpha + \beta)y' + \alpha\beta y = 0$

$$\Leftrightarrow \frac{d}{dx}(y' - \alpha y) = \beta(y' - \alpha y)$$

$$\Leftrightarrow z' = \beta z \text{ where } z = y' - \alpha y$$

$$\Leftrightarrow z = c_1 e^{\beta x}$$

$$\Leftrightarrow y' - \alpha y = c_1 e^{\beta x}$$

$$\Leftrightarrow y = Ae^{\alpha x} + Be^{\beta x} \text{ where } B = \frac{c_1}{\beta - \alpha}, \alpha \neq \beta$$

$$\text{or } y = Ae^{\alpha x} + Bx e^{\alpha x} \text{ where } B = c_1, \alpha = \beta. \text{ (See Qu. 2)}.$$

Therefore the complete solution of $a y'' + b y' + cy = g(x)$ is

$$y = A e^{\alpha x} + B e^{\beta x} + p(x) \text{ when } \alpha \neq \beta$$

$$\text{or } y = A e^{\alpha x} + Bx e^{\alpha x} + p(x) \text{ when } \alpha = \beta$$

where α, β are the real roots of the auxiliary equation $ak^2 + bk + c = 0$, and $y = p(x)$ is a particular integral.

Example 2: Find the complete solution of $y'' + y' - 6y = \sin 3x$.

Solution: The auxiliary equation is $k^2 + k - 6 = 0 \Leftrightarrow k = 2$ or -3.
Therefore the complementary function is $y = Ae^{2x} + Be^{-3x}$.
To find a particular integral, try $y = a \sin 3x + b \cos 3x$

$$\Rightarrow y' = 3a \cos 3x - 3b \sin 3x$$
$$\Rightarrow y'' = -9a \sin 3x - 9b \cos 3x.$$

Therefore

$$y'' + y' - 6y = (-9a - 3b - 6a) \sin 3x + (-9b + 3a - 6b) \cos 3x$$
$$= \sin 3x.$$

Equating coefficients: $\left.\begin{array}{r} -15a - 3b = 1 \\ 3a - 15b = 0 \end{array}\right\} \Leftrightarrow a = -\frac{5}{78}, b = -\frac{1}{78}.$

i.e. $y = -\frac{1}{78}(5 \sin 3x + \cos 3x)$ is a particular integral. The
complete solution is $y = Ae^{2x} + Be^{-3x} - \frac{1}{78}(5 \sin 3x + \cos 3x)$.

□

Qu. 3. Find the complete solution of (i) $y'' - 3y' + 2y = 2x + 1$,
(ii) $y'' + 6y' + 9y = e^x$.

The result proved on page 188 is also true if α, β are complex
(see §13.3 for the definition of e^z, where z is complex). But we outline
below an alternative procedure for use when the roots of the auxiliary
equation are complex.

Qu. 4. Use the substitution $y = v\,e^{-px}$ to show that the complete solution
of $y'' + 2py' + (p^2 + q^2)y = 0$ is $y = Ae^{-px} \sin(qx + \epsilon)$.

Qu. 5. Use the result of Qu. 4 to find the complete solution of
$ay'' + by' + cy = 0$ where $b^2 - 4ac < 0$.

In Example 3 and in subsequent work, a dot above a character denotes
the first derivative with respect to t; two dots denote the second derivative.
Thus $\dot{x} = \dfrac{dx}{dt}$, $\ddot{x} = \dfrac{d^2x}{dt^2}$.

Example 3: The motion of a body satisfies the equation
$\ddot{x} + 2\lambda\dot{x} + \omega^2 x = 0, \lambda \geqslant 0, \omega > 0$. Describe the motion for
various values of λ.

Solution: (i) $\lambda = 0$. The particle executes simple harmonic motion about
$x = 0$, with period $\dfrac{2\pi}{\omega}$.

(ii) $0 < \lambda < \omega$. The auxiliary equation $k^2 + 2\lambda k + \omega^2 = 0$ has
complex roots. By Qu. 5 the complete solution is
$x = Ae^{-\lambda t} \sin(\omega' t + \epsilon)$ where $\omega' = \sqrt{(\omega^2 - \lambda^2)}$; $x \to 0$ as
$t \to +\infty$. The motion is a *lightly damped* oscillation, with

'period' $\frac{2\pi}{\omega}$, which is longer than the period of the

undamped simple harmonic motion in (i).

(iii) $\lambda = \omega$. The auxiliary equation has equal roots, $-\lambda$, and the complete solution is $x = (At + B)e^{-\lambda t}$; $x \to 0$ as $t \to +\infty$. The motion is not oscillatory, and is said to be *critically damped*.

(iv) $\lambda > \omega$. The roots of the auxiliary equations are the negative numbers $\alpha = -\lambda + \sqrt{(\lambda^2 - \omega^2)}$, $\beta = -\lambda - \sqrt{(\lambda^2 - \omega^2)}$, and the complete solution is $x = Ae^{\alpha t} + Be^{\beta t}$; $x \to 0$ as $t \to +\infty$ The motion is not oscillatory, and is said to be *heavily damped*.

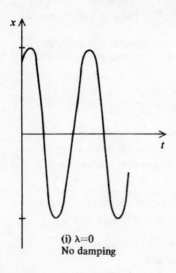

(i) $\lambda = 0$
No damping

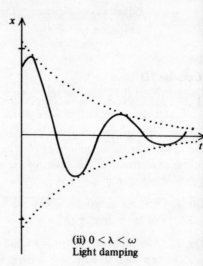

(ii) $0 < \lambda < \omega$
Light damping

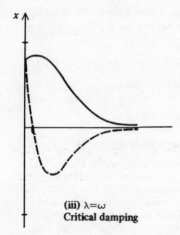

(iii) $\lambda = \omega$
Critical damping

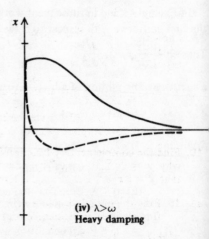

(iv) $\lambda > \omega$
Heavy damping

In the cases of critical or heavy damping the graph of x against t may cross the t-axis at most once, and this only occurs if the particle is initially projected towards the origin with a sufficiently high speed, as illustrated by the broken lines in (iii) and (iv) above.

The term $\omega^2 x$ in the differential equation is due to a force acting on the body, the force being proportional to the displacement, but in the opposite direction. The term $2\lambda\dot{x}$ is due to a force resisting motion, the resistance being proportional to the velocity. This resistance may be due to the medium in which the body is moving. We could, for example, be considering the motion of a mass supported by a spring in (i) a vacuum, (ii) air, (iii) water, (iv) treacle, respectively. Alternatively, the resistance may be due to some device attached to the body, such as shock absorbers on a car.   □

Exercise 7D

1—8. Find the complete solution. Where boundary conditions are given, find also the particular solution which satisfies them.

1. $y'' - 7y' + 10y = 4e^{3x}$.

2. $2y'' - y' - 3y = 6$; $x = 0 \Rightarrow y = y' = 0$.

3. $y'' - 2y' - y = e^x + x$.

4. $y'' - 2y' + 5y = 5x$; $x = 0 \Rightarrow y = y' = 1$.

5. $3y'' - 2y' = \sin 3x$.

6. $y'' - 2y' + 2y = 2\sin x$; $x = 0 \Rightarrow y = y' = 0$.

7. $\ddot{x} + 5\dot{x} + 6x = \sin t$.

8. $y'' + 2y' + 2y = 5\cos x$; $x = 0 \Rightarrow y = 1, y' = 0$.

9. Show that $y = Ae^{2x} + Be^{-3x}$ is the complementary function for the equation $y'' + y' - 6y = g(x)$.
   (a) If $g(x) \equiv e^{2x}$ (i) explain why $y = a\,e^{2x}$ cannot be a particular integral,
                     (ii) find a such that $y = axe^{2x}$ is a particular integral,
                     (iii) state the complete solution.
   (b) Find the complete solution if $g(x) \equiv e^{-3x}$.

10. Find the complementary function for $y'' - 6y' + 9y = e^{3x}$ and explain why $y = a\,e^{3x}$, $y = axe^{3x}$ cannot be particular integrals. Find a such that $y = ax^2e^{3x}$ is a particular integral, and give the complete solution.

11—15. Find the complete solution. When boundary conditions are given, find also the particular solution which satisfies them.

11. $y'' + 2y' - 3y = 2e^{-3x}$.

12. $y'' - y' = 3$; $x = 0 \Rightarrow y = 3$; $x = 1 \Rightarrow y = e - 1$.

13. $y'' + 4y = \cos 2x$.   14. $y'' + 4y' + 4y = e^{-2x}$.

15. $y'' - y = \cosh x$; $x = 0 \Rightarrow y = 0, y' = 1$.

16. A particle moves along a straight line so that its distance x from a fixed point O satisfies the equation $\ddot{x} + 4\dot{x} + 3x = 2ce^{-2t}$ where c is a positive constant. If the particle starts from rest at O, find the greatest value of x. Sketch the graph of x for positive values of t.   [MEI]

17. If $y = f(x)$ and $x = e^t$, show that $\dfrac{dy}{dx} = e^{-t} \dot{y}$ and find an expression for $\dfrac{d^2y}{dx^2}$ in terms of $t, \dot{y}, \ddot{y}$. By using the substitution $x = e^t$, or otherwise, find the general solution (for $x > 0$) of the differential equation
$x^2 \dfrac{d^2y}{dx^2} + 3x \dfrac{dy}{dx} - 3y = 0$.   [SMP]

18. *Resonance in an undamped oscillation.* The motion of a system satisfies the equation $\ddot{x} + \omega^2 x = a \sin kt$, $\omega > 0$, $k > 0$. Find the complete solution for (i) $k \neq \omega$; (ii) $k = \omega$. Show that in case (i) the maximum displacement is finite, but that this is not so in case (ii). Case (ii) is an example of what is known as *resonance*.

7.6 Step-by-step methods

1. Euler's method

We now consider numerical methods of tabulating x, $Y(x)$ where Y is a particular integral of a first-order differential equation. It is assumed that the equation gives y' explicitly in terms of x and y.

We use the following notation: $x_0, x_0 + h, x_0 + 2h, \ldots$ are successive values of x, where h is a small constant; x_n, y_n, y_n' are used to denote $x_0 + nh$, $Y(x_n)$, $Y'(x_n)$. We also use the notation $x = 0.0\,(0.1)\,1.0$ to mean 'values of x from 0.0 to 1.0 inclusive, at intervals of 0.1'.

The basis of *Euler's method* is the approximation $\delta y \approx \dfrac{dy}{dx} \delta x$, (see

Vol. 1 §5.15). Rewriting this in our new notation we have

$$y_{n+1} - y_n \approx h y_n'$$
$$\Leftrightarrow \qquad y_{n+1} \approx y_n + h y_n' \qquad (1)$$

If we know the value of y_0, we can assign a value to h and use equation (1) to calculate an approximation for y_1; in the same way we can find approximations for $y_2, y_3, y_4, \ldots$.

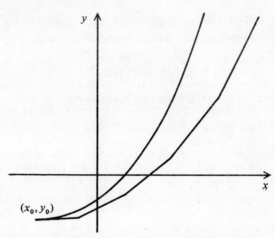

(x_0, y_0)

Graphically, this is equivalent to replacing the solution curve by a series of short straight line segments which we hope approximate sufficiently closely to the true curve. The first of these line segments is tangent to the solution curve, but the others are, in general, tangents to other solution curves.

Example 1: Solve $y' = x^2 - y$ for $x = 0.0\ (0.1)\ 0.5$ given that $y = 1$ when $x = 0$.

Solution: Let $x_0 = 0$, $h = 0.1$.
Then $y_{n+1} \approx y_n + 0.1\ y_n' = y_n + 0.1\ (x_n^2 - y_n)$
$\qquad\qquad\qquad\quad = 0.9\ y_n + 0.1\ x_n^2$.

x_n	0	0.1	0.2	0.3	0.4	0.5
y_n	1	0.9	0.81	0.73	0.67	0.62

$\qquad\qquad\qquad\qquad\qquad\qquad\qquad\qquad\qquad\qquad\qquad\qquad$ □

Qu. 1.   Use Euler's method to solve $y' = x + y$ for $x = 0.0\ (0.1)\ 0.5$ given that $y = 1$ when $x = 0$. Compare your values with the analytical solution $y = 2e^x - x - 1$.

2. Taylor series method

The rth Taylor approximation for $Y(x)$ centred at $x = x_n$ is

$$Y(x) \approx Y(x_n) + (x - x_n)\ Y'(x_n) + \frac{(x - x_n)^2}{2!}\ Y''(x_n) + \dots$$

$$+ \frac{(x - x_n)^r}{r!}\ Y^{(r)}\ (x_n).$$

(See §2.5 for the derivation of the Taylor approximations.) Putting $x = x_{n+1}$ and using the notation of this section we obtain

$$y_{n+1} \approx y_n + h\,y_n' + \frac{h^2}{2!}\,y_n'' + \ldots + \frac{h^r}{r!}\,y_n^{(r)}$$

where $y_n'', y_n^{(r)}$ denote $Y''(x_n)$, $Y^{(r)}(x_n)$ respectively.

Example 2: Use a Taylor approximation to find y when $x = 0.1$ given that $y' = x^2 - y$ and $y = 1$ when $x = 0$.

Solution: Let $x_0 = 0$, $h = 0.1$; then $x_1 = 0.1$ and $y_0 = 1$.

$$
\begin{aligned}
y' &= x^2 - y && \Rightarrow & y_0' &= -1. \\
y'' &= 2x - y' = 2x - x^2 + y && \Rightarrow & y_0'' &= 1. \\
y''' &= 2 - 2x + y' = 2 - 2x + x^2 - y && \Rightarrow & y_0''' &= 1. \\
y^{(4)} &= -2 + 2x - y' = -2 + 2x - x^2 + y && \Rightarrow & y_0^{(4)} &= -1.
\end{aligned}
$$

Using the 4th Taylor approximation centred on x_0, we obtain

$$y_1 \approx y_0 + 0.1\,y_0' + \frac{0.01}{2!}\,y_0'' + \frac{0.001}{3!}\,y_0''' + \frac{0.0001}{4!}y_0^{(4)}$$

$$\approx 1 - 0.1 + 0.005 + 0.000\,166\,7 - 0.000\,004\,17$$

$$\approx 0.905\,163.$$

Then when $x = 0.1$, $y \approx 0.905\,163$, a result which is correct to 6 D.P.   □

Theoretically the method just used provides a solution to any differential equation, but it is of limited practical value as generally the differentiations become increasingly complicated. However the Taylor approximations do provide us with alternative formulae which produce fairly accurate results without undue complications.

3. Mid-point method

Using the 2nd Taylor approximation centred on $x = x_n$, we obtain

$$y_{n+1} \approx y_n + h\,y_n' + \frac{h^2}{2!}\,y_n''$$

and

$$y_{n-1} \approx y_n - h\,y_n' + \frac{h^2}{2!}\,y_n'' \quad \text{(by replacing } h \text{ by } -h\text{).}$$

By subtraction $y_{n+1} - y_{n-1} \approx 2h\,y_n'$.

$$\boxed{\text{Therefore } y_n' \approx \frac{y_{n+1} - y_{n-1}}{2h}.}$$

We usually use this approximation in the form $y_{n+1} \approx y_{n-1} + 2h y_n'$. Graphically this is equivalent to replacing the solution curve by a chord, the gradient of the chord being the gradient of the solution curve in the middle of the interval. In the diagram, AC represents the correct value of y_{n+1}, while AB represents the value of y_{n+1} given by the approximation.

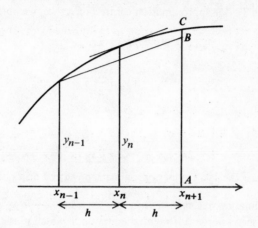

If we are given y_0, we usually find y_1 using the Taylor series method, and can then calculate y_2, y_3, etc. using the approximation above.

Example 3: Solve $y' = x^2 - y$ for $x = 0.0 \,(0.1)\, 0.5$ given that $y = 1$ when $x = 0$.

Solution: Let $x_0 = 0$, $h = 0.1$. Then $y_0 = 1$.
By the Taylor series method $y_1 \approx 0.905\,163$. (See Example 2.)

$$y_{n+1} \approx y_{n-1} + 2h y_n' = y_{n-1} + 0.2\,(x_n^2 - y_n).$$

Putting $n = 1$,
$y_2 \approx y_0 + 0.2\,(x_1^2 - y_1) \approx 1 + 0.2\,(0.01 - 0.905\,163)$
$\approx 0.820\,967$.

y_3, y_4, y_5 are calculated similarly. The results obtained by Euler's method, and from the analytical solution $y = x^2 - 2x + 2 - e^{-x}$ are given for comparison.

x_n	0	0.1	0.2	0.3	0.4	0.5
y_n (mid-point)	1	0.9052	0.8210	0.7490	0.6892	0.6431
y_n (Euler)	1	0.9	0.81	0.73	0.67	0.62
y_n (analytical)	1	0.9052	0.8213	0.7492	0.6897	0.6435

□

Qu. 2.   Use the mid-point method to solve $y' = y$ for $x = 0.0\,(0.1)\,0.5$, given that $y = 1$ when $x = 0$. Compare your values with those obtained from the exact solution.

4. Second-order differential equations

Using the 3rd Taylor approximation centred on $x = x_n$ we obtain

$$y_{n+1} \approx y_n + h\,y_n' + \frac{h^2}{2!}\,y_n'' + \frac{h^3}{3!}\,y_n'''$$

and

$$y_{n-1} \approx y_n - h\,y_n' + \frac{h^2}{2!}\,y_n'' - \frac{h^3}{3!}\,y_n'''.$$

By addition, $y_{n+1} + y_{n-1} \approx 2y_n + h^2 y_n''.$

Therefore

$$y_n'' \approx \frac{y_{n+1} - 2y_n + y_{n-1}}{h^2}.$$

This approximation, rearranged as $y_{n+1} \approx 2y_n - y_{n-1} + h^2 y_n''$, is useful when solving a second-order differential equation in which y'' is expressed in terms of x and y.

Example 4: Solve $y'' = 1 - y$ for $x = 0.0(0.1)0.5$ given that when $x = 0$, $y = y' = 1$.

Solution: Let $x_0 = 0$, $h = 0.1$. Then $y_0 = 1$, $y_0' = 1$.

$$y'' = 1 - y \Rightarrow y''' = -y' \Rightarrow y^{(4)} = -y''.$$

Therefore $y_0'' = 0, y_0''' = -1, y_0^{(4)} = 0.$

Using the 4th Taylor series approximation

$$y_1 \approx y_0 + 0.1\,y_0' + \frac{0.01}{2!}\,y_0'' + \frac{0.001}{3!}\,y_0''' + \frac{0.0001}{4!}\,y_0^{(4)}$$

$$\approx 1 + 0.1 + 0 + 0.000\,166 + 0 = 1.099\,833.$$

$$y_{n+1} \approx 2y_n - y_{n-1} + h^2 y_n'' = 2y_n - y_{n-1} + 0.01\,(1 - y_n)$$

$$= 1.99\,y_n - y_{n-1} + 0.01.$$

Substituting $n = 1(1)4$ we obtain the required results.

x_n	0	0.1	0.2	0.3	0.4	0.5
y_n	1	1.0998	1.1987	1.2955	1.3894	1.4794

These values are identical, to 4 D.P., with results obtained from the analytical solution $y = 1 + \sin x$.   □

Qu. 3. Use this method to solve $y'' = y$ for $x = 0.0(0.1)0.5$ given that $y = 1$, $y' = 0$ when $x = 0$. Compare your values with those obtainable analytically.

Example 5: Solve $y'' = 2\cos x - y$ for $x = 0.0(0.1)0.5$ given that $y = 0$ when $x = 0$, $y = 0.2397$ when $x = 0.5$.

Solution: Let $x_0 = 0$, $h = 0.1$. Then $y_0 = 0$, $y_5 = 0.2397$.

$$y_{n+1} \approx 2y_n - y_{n-1} + h^2 y_n'' = 2y_n - y_{n-1} + h^2(2\cos x_n - y_n)$$

$$\Leftrightarrow y_{n-1} - (2 - h^2)y_n + y_{n+1} = 2h^2 \cos x_n.$$

Substituting for h and setting $n = 1(1)4$ we obtain

$$
\begin{aligned}
-1.99\,y_1 + \qquad y_2 \qquad\qquad\qquad\qquad\qquad &= \quad 0.0199 \\
y_1 - 1.99\,y_2 + \qquad y_3 \qquad\qquad\qquad &= \quad 0.0196 \\
y_2 - 1.99\,y_3 + \qquad y_4 &= \quad 0.0191 \\
y_3 - 1.99\,y_4 &= \; -0.2213.
\end{aligned}
$$

Solving these equations simultaneously (the method of pivotal condensation of §9.6 is appropriate) we obtain

x_n	0	0.1	0.2	0.3	0.4	0.5
y_n	0	0.0099	0.0396	0.0886	0.1557	0.2397
y_n(analytical)	0	0.0100	0.0397	0.0887	0.1558	0.2397

The results from the numerical method are in good agreement with those from the analytical solution $y = x \sin x$. □

Exercise 7E

1. Use tables to show that when $y = \ln x$ the error in the approximation $y(a + h) \approx y(a) + h\, y'(a)$, with $a \geqslant 2$ and $h = 0.1$, is less than 1%. Calculate the values of $\ln(2.1)$, $\ln(2.2)$ and $\ln(2.3)$ by applying this approximation to the differential equation $dy/dx = 1/x$, where $y(2) = 0.6931$. Explain with the help of a diagram why the calculated values are larger than the true values. [L]

2. Given that $y' = 3x - y^2$ and $y = 1$ when $x = 0$, find an approximation to the value of y when $x = 0.1$ (i) by Euler's method; (ii) by Taylor Series, to 2 D.P.

3. The equation $y' = 1 + x y^2$ has a solution curve which passes through the origin. Show that (correct to 4 D.P.) the value of y at the point on the curve at which $x = 0.2$ is 0.2004. [L, adapted]

4. Use Euler's method with a step of 0.2 to find the value of y when $x = 0.5$ given that $y' = \frac{1}{2} + y^3$ and $y = 0$ when $x = -0.3$.

5. Write down approximations for (i) y_{n+1} in terms of h, y_n, y_n';

   (ii) y_n in terms of h, y_{n+1}, y_{n+1}'.

   Deduce that $y_{n+1} \approx y_n + \frac{1}{2}h\,(y_n' + y_{n+1}')$. This approximation is the basis of *Heun's method* of solving first-order differential equations. Explain graphically why Heun's method is in general more accurate than Euler's.

6. Given that $dy/dx = \cos x + y \sin x$, subject to $y = 0$ when $x = 0$, find the Maclaurin expansion for y in terms of x as far as the term involving x^5. Hence find the value of y when $x = 0.1$, correct to 6 S.F. Justify by means of a sketch the approximation $\left(\dfrac{dy}{dx}\right)_0 \approx \dfrac{1}{2h}\,(y_1 - y_{-1})$, and apply it to find the approximate value of y when $x = 0.2$, correct to 4 S.F. Check the accuracy of this value by using the Maclaurin expansion to evaluate y when $x = 0.2$.

   Explain briefly how to find an approximation for the value of y when

   $$x = 0.2, \quad y = e^{-\cos x} \int_0^x \cos t \,(e^{\cos t})\,dt.$$ [MEI]

7. Solve $y'' = y^2 - x^2$ for $x = 0.0(0.1)0.5$ if $y = 0$ when $x = 0$ and $y = 0.2$ when $x = 0.1$.

8. The function g satisfies the equation $g''(x) = \sin \pi x - g(x)$ and it is given that $g(-0.1) = -0.264\,66$ and $g(0) = 0$. Express $g(x + h)$ approximately in terms of $g(x)$ and $g(x - h)$ and hence evaluate the function step-by-step for $x = 0.1, 0.2, 0.3$, retaining 5 decimal places throughout the work.   [MEI, adapted]

9. The function f satisfies the differential equation $f''(x) = f(x)$, and has the values $f(1.0) = 4.629\,24$, $f(1.1) = 5.005\,56$. Evaluate approximately $f(0.9), f(1.3), f'(1.1)$. Also find explicitly the solution of the given differential equation with the given initial conditions, and hence check your previous answers.   [MEI, adapted]

10. Given that $y = F(x)$ is the solution of the equation $y'' = -y - \frac{1}{10}y^2$ for which $F(0) = 0$ and $F(\frac{1}{2}\pi) = 1$, it is required to find approximate values for $F(\frac{1}{6}\pi)$ and $F(\frac{1}{3}\pi)$. Writing $y = \sin x + G(x)$, prove that $z = G(x)$ satisfies the equation $z'' = -z - \frac{1}{10}(\sin x + z)^2$ with $G(0) = G(\frac{1}{2}\pi) = 0$. Using the approximate forms in terms of differences for z'' at $x = \frac{1}{6}\pi$ and $x = \frac{1}{3}\pi$, and neglecting the term $-\frac{1}{10}z^2$ in the above equation, form two simultaneous linear equations for $G(\frac{1}{6}\pi)$, $G(\frac{1}{3}\pi)$. Solve these and give your final values for $F(\frac{1}{6}\pi)$, $F(\frac{1}{3}\pi)$ to 3 decimal places. [Take $\pi^2/36 = 0.274\,16$.]   [MEI]

11. State an approximate formula for $f''(x)$ in terms of $f(x + h), f(x)$, $f(x - h), h$. The equation $y = f(x)$ satisfies the differential equation $y'' = y + 4x$ and $f(0) = 0, f(1) = -1$. By making three applications of the formula of the first part with $h = 0.25$, obtain approximately the values of $f(0.25), f(0.50), f(0.75)$. Hence find an approximate value for $f'(0.25)$ and check it by solving the differential equation explicitly.   [MEI, adapted]

8 Vectors II

8.1 Scalar product and projections

In mechanics the *work* done by a force on a moving body is defined as the product of the magnitude of the force and the distance moved by its point of application along the line of action of the force. The work done by force **F** producing displacement s is $Fs \cos \alpha$ where α is the angle between **F** and s. This led to the following definition.

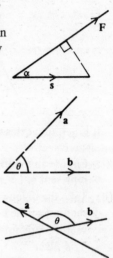

The *scalar product* of two vectors **a**, **b** is the scalar $ab \cos \theta$, where θ $(0 \leqslant \theta \leqslant \pi)$ is the smallest angle through which **b** needs to be rotated to have the same direction and sense as **a**; if **a** or **b** is zero, the scalar product is 0. The scalar product is denoted by **a** . **b** (read as 'a dot b').

The operation of forming the scalar product is not closed because the result of the operation is not itself a vector, but a scalar. Thus there can be no identity element.

Qu. 1. Prove that **a** . **b** = **b** . **a** for any **a**, **b**. This is the commutative property of scalar products.

Qu. 2. What, if anything, is meant by (i) (**a** . **b**) . **c** (ii) (**a** . **b**) **c**? Is the operation of forming scalar products associative?

Qu. 3. Prove that $m(\mathbf{a} . \mathbf{b}) = (m\mathbf{a}) . \mathbf{b}$. Notice the resemblance to an associative property: in both expressions we form a scalar product, but on the left we also multiply two scalars, while on the right we multiply a vector by a scalar.

Qu. 4. Prove that $(m\mathbf{a}) . \mathbf{b} = \mathbf{a} . (m\mathbf{b})$.

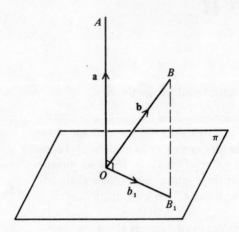

It is helpful at this stage to introduce the idea of projections. Suppose **a** is a non-zero vector.

(i)   Choose any point O. Take the unique directed line segment $\overrightarrow{OA}$ to represent **a**. Let π be the plane through O perpendicular to $\overrightarrow{OA}$.

(ii) Take the unique directed line segment $\overrightarrow{OB}$ to represent **b**. Let B_1 be the foot of the perpendicular from B to plane π. Let $\mathbf{b}_1$ be the unique vector represented by $\overrightarrow{OB_1}$. The vector $\mathbf{b}_1$ is known as the *projection of* **b** *on plane* π.

(iii) The vector $\mathbf{B_1B}$ is parallel to **a** (or possibly zero), so that we may write $\mathbf{B_1B} = \beta\hat{\mathbf{a}}$, where β is the unique scalar such that $|\beta| = B_1B$. The scalar β is known as the *projection of* **b** *on* **a**.

> Thus vector **b** can be expressed uniquely as $\mathbf{b} = \beta\hat{\mathbf{a}} + \mathbf{b}_1$
> where $\mathbf{b}_1$ is perpendicular to **a**, and β is a scalar.

Notice particularly that the projection of a vector on another vector is a scalar, while the projection of a vector on a plane is another vector.

Qu. 5.   With the notation above show that $\beta = b\cos\theta$ and $|\mathbf{b}_1| = b\sin\theta$ where $\theta\ (0 \leqslant \theta \leqslant \pi)$ is the smallest angle through which **b** needs to be rotated to have the same direction and sense as **a**. Deduce that $\mathbf{a}\cdot\mathbf{b} = a\beta$.

Qu. 6.   Let $\mathbf{v} = 3\mathbf{i} + 2\mathbf{j} - 5\mathbf{k}$. Find the projection of **v** on $7\mathbf{i}$. If **k** is vertical, find the projection of **v** on a horizontal plane.

Suppose β, γ are the projections of **b**, **c** on non-zero vector **a**; let $\mathbf{b}_1$, $\mathbf{c}_1$ be the projections of **b**, **c** on a plane π perpendicular to **a**. Then

$$b = \beta\hat{a} + b_1 \\ c = \gamma\hat{a} + c_1 \Bigg\} \Rightarrow b + c = (\beta + \gamma)\hat{a} + (b_1 + c_1).$$

The vector $b_1 + c_1$ is coplanar with b_1 and c_1.

Therefore

> the projection of $b + c$ on a is the sum of the
> projections of b and c on a, and the projection
> of $b + c$ on π is the sum of the projections of
> b and c on π.

The first of these results we use immediately, while the other result will be useful when we consider vector products in §8.5.

Using the notation above, and the result of Qu. 5

$$a.b + a.c = a\beta + a\gamma, \quad a \neq 0$$

$$= a(\beta + \gamma) \quad \text{since } a, \beta, \gamma \text{ are scalars}$$

$$= \text{product of } a \text{ with projection of } b + c \text{ on } a$$

$$= a.(b + c).$$

This result resembles a distributive property of scalar products over addition (but notice the different meanings of the two + signs), and is also clearly true if $a = 0$. The result can be extended so that $a.(b + c + d) = a.b + a.c + a.d$.

Qu. 7.   Prove that $(a + b).(c + d) = a.c + a.d + b.c + b.d$.

Qu. 8.   If a and b are non-zero, prove that

$$a \text{ and } b \text{ are perpendicular} \Leftrightarrow a.b = 0.$$

Qu. 9.   Show that $i.j = j.k = k.i = 0$ and $i.i = j.j = k.k = 1$.

Qu. 10.   Show that $a.a = a^2$.

The scalar product can be expressed conveniently in terms of the components of a and b. If $\quad a = a_1 i + a_2 j + a_3 k$

and $\qquad\qquad b = b_1 i + b_2 j + b_3 k$

then $\qquad\qquad a.b = (a_1 i + a_2 j + a_3 k).(b_1 i + b_2 j + b_3 k)$.

We can expand this product using distributivity and the results of Questions 3 and 4, getting nine terms. By Qu. 9, six of these are zero so that we are left with

$$a.b = a_1 b_1 (i.i) + a_2 b_2 (j.j) + a_3 b_3 (k.k)$$

$$= a_1 b_1 + a_2 b_2 + a_3 b_3 \quad \text{using Qu. 9 again.}$$

This form of the scalar product is often useful and should be remembered. It is a direct consequence of the fact that our set of base vectors are mutually perpendicular unit vectors. Some authors take this as the definition of scalar product.

Example 1: Prove that **a** and **b** are perpendicular where $\mathbf{a} = 3\mathbf{i} - 7\mathbf{j} + 3\mathbf{k}$, $\mathbf{b} = 2\mathbf{i} + 3\mathbf{j} + 5\mathbf{k}$.

Solution: $\mathbf{a} \cdot \mathbf{b} = (3 \times 2) + [(-7) \times 3] + (3 \times 5) = 6 - 21 + 15 = 0$.
$\mathbf{a} \cdot \mathbf{b} = 0 \Rightarrow \mathbf{a}, \mathbf{b}$ are perpendicular since neither **a** nor **b** is zero. □

Example 2: Calculate the angle between **a** and **b** where $\mathbf{a} = 2\mathbf{i} + 3\mathbf{j} + \mathbf{k}$, $\mathbf{b} = 4\mathbf{i} - 2\mathbf{j} + \mathbf{k}$.

Solution: $\mathbf{a} \cdot \mathbf{b} = 8 - 6 + 1 = 3$.

$a = \sqrt{(2^2 + 3^2 + 1^2)} = \sqrt{14}$; $b = \sqrt{[4^2 + (-2)^2 + 1^2]} = \sqrt{21}$.

$\mathbf{a} \cdot \mathbf{b} = ab \cos\theta$, where θ is the angle between **a** and **b**.

Therefore $\cos\theta = \dfrac{\mathbf{a} \cdot \mathbf{b}}{ab} = \dfrac{3}{(\sqrt{14})(\sqrt{21})} = \dfrac{3}{7\sqrt{6}} = \dfrac{\sqrt{6}}{14}$.

i.e. the angle between **a** and **b** is $\cos^{-1}(\sqrt{6}/14)$. □

Qu. 11. Find an expression for the projection of **b** on **a** in terms of **a** and **b** only.

Example 3: Use scalar products to prove that the diagonals of a rhombus are perpendicular.

Solution: Let $ABCD$ be the rhombus.
Suppose $\mathbf{AB} = \mathbf{p}$, $\mathbf{AD} = \mathbf{q}$. Then $p = q$,
and $\mathbf{AC} = \mathbf{p} + \mathbf{q}$, $\mathbf{DB} = \mathbf{p} - \mathbf{q}$.

Now
$$\begin{aligned}
\mathbf{AC} \cdot \mathbf{DB} &= (\mathbf{p} + \mathbf{q}) \cdot (\mathbf{p} - \mathbf{q}) \\
&= \mathbf{p} \cdot \mathbf{p} - \mathbf{p} \cdot \mathbf{q} + \mathbf{q} \cdot \mathbf{p} - \mathbf{q} \cdot \mathbf{q} \\
&= p^2 - q^2 = 0.
\end{aligned}$$

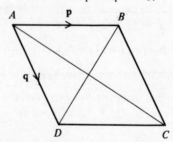

Therefore the diagonals of the rhombus are perpendicular. □

In many geometrical proofs the algebra can be simplified by a suitable choice of origin.

Summary.

---

1. $\mathbf{a}.\mathbf{b} = ab\cos\theta$ where θ is the angle between $\mathbf{a}$ and $\mathbf{b}$

$$= a_1 b_1 + a_2 b_2 + a_3 b_3 \text{ where } \mathbf{a} = \begin{pmatrix} a_1 \\ a_2 \\ a_3 \end{pmatrix}, \ \mathbf{b} = \begin{pmatrix} b_1 \\ b_2 \\ b_3 \end{pmatrix}$$

$$= a\,(\text{projection of } \mathbf{b} \text{ on } \mathbf{a}) = b\,(\text{projection of } \mathbf{a} \text{ on } \mathbf{b}).$$

2. $\mathbf{a}.\mathbf{b} = \mathbf{b}.\mathbf{a}.$

3. $k(\mathbf{a}.\mathbf{b}) = (k\mathbf{a}).\mathbf{b} = \mathbf{a}.(k\mathbf{b}).$

4. $\mathbf{a}.(\mathbf{b}+\mathbf{c}) = \mathbf{a}.\mathbf{b} + \mathbf{a}.\mathbf{c}.$

5. $\mathbf{a}.\mathbf{b} = 0 \Leftrightarrow \mathbf{a} = \mathbf{0}$ or $\mathbf{b} = \mathbf{0}$ or $\mathbf{a}$ is perpendicular to $\mathbf{b}$.

Exercise 8A

1. State which of the following are scalars, which are vectors, and which are undefined:

   (i) $(\mathbf{p}.\mathbf{q}) + \mathbf{r}$ (ii) $(\mathbf{p}.\mathbf{q})(\mathbf{r}.\mathbf{s})$ (iii) $\dfrac{\mathbf{p}.\mathbf{q}}{\mathbf{r}}$

   (iv) $(\mathbf{p}.\mathbf{q})\mathbf{r}$ (v) $(\mathbf{p}.\mathbf{q} + \mathbf{r}.\mathbf{s})\mathbf{p}$ (vi) $\dfrac{\mathbf{r}}{\mathbf{p}.\mathbf{q}}$

   (vii) $(\mathbf{p}.\mathbf{q})(\mathbf{q}.\mathbf{r})\mathbf{s}$ (viii) $[(\mathbf{p}.\mathbf{r})\mathbf{q}].\mathbf{s}$

2. Where possible evaluate the expressions in Qu. 1, given that

$$\mathbf{p} = \begin{pmatrix} 1 \\ 0 \\ 2 \end{pmatrix}, \ \mathbf{q} = \begin{pmatrix} 3 \\ -2 \\ 1 \end{pmatrix}, \ \mathbf{r} = \begin{pmatrix} 2 \\ 5 \\ 0 \end{pmatrix}, \ \mathbf{s} = \begin{pmatrix} 2 \\ 2 \\ -2 \end{pmatrix}.$$

3. Calculate the cosine of the angle between the following pairs of vectors:

   (i) $\begin{pmatrix} 2 \\ -3 \\ 6 \end{pmatrix}, \begin{pmatrix} 19 \\ 4 \\ -8 \end{pmatrix}$ (ii) $\begin{pmatrix} 8 \\ -1 \\ 4 \end{pmatrix}, \begin{pmatrix} 6 \\ 6 \\ -3 \end{pmatrix}$ (iii) $\begin{pmatrix} -2 \\ 0 \\ 5 \end{pmatrix}, \begin{pmatrix} 4 \\ -3 \\ 2 \end{pmatrix}$

4. Calculate the cosine of the acute angle between the following pairs of lines:

   (i) $\dfrac{x-3}{6} = \dfrac{y+7}{-3} = \dfrac{z-5}{6}, \qquad \dfrac{x+4}{-4} = \dfrac{y-1}{-7} = \dfrac{z+11}{4}$

   (ii) $\dfrac{x}{3} = \dfrac{y}{-9} = \dfrac{z}{\sqrt{10}}, \qquad \dfrac{x-8}{-5} = \dfrac{y+5}{1} = \dfrac{z+2}{2\sqrt{10}}$

(iii) $\mathbf{r} = \begin{pmatrix} 4 \\ 3 \\ 0 \end{pmatrix} + \lambda \begin{pmatrix} 2 \\ -1 \\ 3 \end{pmatrix}$, $\mathbf{r} = \begin{pmatrix} 0 \\ -2 \\ 7 \end{pmatrix} + \lambda \begin{pmatrix} -1 \\ 4 \\ -4 \end{pmatrix}$

(iv) $\mathbf{r} = \mathbf{a} + \lambda(\mathbf{b} - \mathbf{c})$, $\mathbf{r} = \mathbf{b} + \lambda(\mathbf{c} - \mathbf{a})$ where $\mathbf{a}, \mathbf{b}, \mathbf{c}$ are mutually perpendicular, and of equal magnitude.

5. Use scalar products to find
   (i) the acute angle between two diagonals of a cube,
   (ii) the angle between the diagonal of a cube and the base.

6. Triangle ABC is isosceles with $AB = AC$; use vector methods to prove that the median through A is perpendicular to BC.

7. Use scalar products to prove that the angle subtended at the circumference of a circle by a diameter of that same circle is a right angle.

8. Show that $|\mathbf{a} + \mathbf{b}|^2 = a^2 + b^2 + 2\mathbf{a} \cdot \mathbf{b}$. Use this result to prove the Cosine Rule for triangles.

9. If $ABCD$ is a parallelogram, prove that $AC^2 + BD^2 = 2(AB^2 + BC^2)$.

10. The altitudes through A and B of triangle ABC intersect at H. By considering $\mathbf{AH} \cdot \mathbf{BC}$ and $\mathbf{BH} \cdot \mathbf{AC}$ show that CH is perpendicular to AB. How does this prove that the altitudes of a triangle are concurrent? Use a similar method to prove that the perpendicular bisectors of the sides are concurrent.

11. The points A, B, C, D are the vertices of a tetrahedron; P, Q, R, S, T, U are the mid-points of AB, CD, BC, DA, CA, BD respectively. Prove that:

   (i) $AB^2 + BC^2 + CD^2 + DA^2 = AC^2 + BD^2 + 4TU^2$

   (ii) $AB^2 + BC^2 + CD^2 + DA^2 + AC^2 + BD^2 = 4(PQ^2 + RS^2 + TU^2)$

   (iii) $AB^2 + CD^2 = BC^2 + DA^2 \Rightarrow AC$ is perpendicular to BD

   (iv) AB, BC are perpendicular $\quad \Rightarrow \quad AC$ is perpendicular to BD and
        to CD, DA respectively $\qquad AB^2 + CD^2 = BC^2 + DA^2 = AC^2 + BD^2$

   (v) $AB = CD, BC = DA, \quad \Rightarrow \quad PQ, RS, TU$ are perpendicular to AB and
        and $AC = DB \qquad\qquad CD, BC$ and DA, AC and DB respectively

   (vi) PQ is perpendicular to $RS \Rightarrow AC = BD$.

12. Find an expression for the projection of $\mathbf{b}$ on $\mathbf{a}$ in terms of $\mathbf{a}$ and $\mathbf{b}$ and show that the projection of $\mathbf{b}$ on a plane perpendicular to $\mathbf{a}$ is $\mathbf{b} - (\mathbf{b} \cdot \hat{\mathbf{a}})\hat{\mathbf{a}}$.

13. This question involves only two dimensions. In the x-y plane a general line l has equation $ax + by + c = 0$. Write the equation of l in vector form. Write down the coordinates of a general point R on l in terms of a parameter λ. Write down a vector $\mathbf{v}$ in the x-y plane such that $\mathbf{v}$ is perpendicular to line l. P is the point (x_1, y_1); express $|\mathbf{PR} \cdot \hat{\mathbf{v}}|$ in terms of a, b, c, x_1, y_1, and explain why this gives the perpendicular distance of the point P from the line l.

14. The line l has equation $\mathbf{r} = \mathbf{a} + \lambda\mathbf{d}$; P is a point not on l. If R is an arbitrary point on l, show that

   $\mathbf{PR}.\mathbf{d} = 0 \Leftrightarrow R$ is the foot of the perpendicular from P to l.

   Use this result to calculate the perpendicular distance from $(4, -1, 1)$ to the line $\dfrac{x-3}{2} = \dfrac{y-6}{1} = \dfrac{z+10}{-2}$.

15. Calculate the perpendicular distance from these points to these lines:

   (i) $(0, 0, 0)$, $\dfrac{x-5}{-1} = \dfrac{y}{6} = \dfrac{z-13}{6}$

   (ii) $(2, 1, 3)$, $\dfrac{x-3}{3} = \dfrac{y-13}{0} = \dfrac{z+4}{4}$

   (iii) $(1, 3, -2)$, $\dfrac{x+3}{2} = \dfrac{y-8}{2} = \dfrac{z-3}{3}$.

16. In this question we assume that the Earth is a sphere, centre at O; A, B are on the equator with longitudes $0°$, $90°$E respectively, and N is the north pole. Taking OA, OB, ON as the x, y, z axes respectively, and R as the radius of the Earth, find the position vector of the point P at (latitude $\alpha°$N, longitude $\beta°$E). If Q is at $(\gamma°$N, $\delta°$E) show that

   $$\text{angle } POQ = \cos^{-1}[\cos\alpha° \cos\gamma° \cos(\beta° - \delta°) + \sin\alpha° \sin\gamma°]$$

   and hence calculate the great circle distance between London $(51°30'$N, $0°05'$W) and Rio de Janeiro $(23°00'$S, $43°12'$W) given that the radius of the earth is 6370 km.

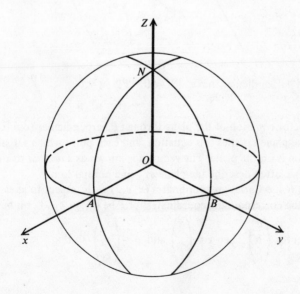

8.2 Planes

We assume that the reader is familiar with the concept of a plane. Just as we have several ways of describing a particular straight line, so a particular plane in space may be specified in a variety of ways:

(i)   by means of three points in the plane; for example many instruments are levelled by adjusting two of the three legs;

(ii)   by means of one point in the plane and two non-parallel directions; for example a billiard table is often levelled with a spirit level which is made to show when two perpendicular directions on the surface of the table are both horizontal;

(iii)   by means of one point in the plane and the direction of a line perpendicular to the plane; for example a chemical balance is sometimes checked by means of a built-in plumb line which shows that the column is vertical and thus that the base is horizontal.

We use method (iii): we want to find the position vector of any point on the plane through P given that the vector **n** is perpendicular to the plane, $\mathbf{n} \neq \mathbf{0}$. The vector **n** may be a unit vector, but this is not a necessity. Consider an arbitrary point R on the plane, with position vector **r**.

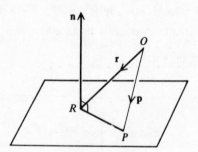

Then **PR** is perpendicular to **n** or $\mathbf{PR} = \mathbf{0} \Leftrightarrow \mathbf{PR} \cdot \mathbf{n} = 0$

$$\Leftrightarrow (\mathbf{r} - \mathbf{p}) \cdot \mathbf{n} = 0.$$

This is a vector equation of the plane through P perpendicular to n. Each point on the plane satisfies this equation, and each point which satisfies this equation is on the plane. The vector **n** is known as a *normal* to the plane, and we often describe the plane as being normal to **n**.

Suppose R has Cartesian coordinates (x, y, z) with respect to axes through O, and let the corresponding coordinates for P be (x_1, y_1, z_1). Let **n** be the non-zero vector $\begin{pmatrix} a \\ b \\ c \end{pmatrix}$. Then $\mathbf{r} = \begin{pmatrix} x \\ y \\ z \end{pmatrix}$ and $\mathbf{p} = \begin{pmatrix} x_1 \\ y_1 \\ z_1 \end{pmatrix}$ so that

$$(\mathbf{r}-\mathbf{p}).\mathbf{n}=0 \Leftrightarrow \mathbf{r}.\mathbf{n}-\mathbf{p}.\mathbf{n}=0$$
$$\Leftrightarrow ax + by + cz - (ax_1 + by_1 + cz_1) = 0$$
$$\Rightarrow ax + by + cz + d = 0,$$

where d is a constant dependent on the coordinates of P and the components of $\mathbf{n}$. Thus the coordinates of each point of a plane satisfy an equation of the form $ax + by + cz + d = 0$.

Conversely, consider the set S of points with coordinates which satisfy

$$ax + by + cz + d = 0 \qquad (1)$$

where not all a, b, c are zero. We now show that S is a plane. Equation (1) is equivalent to the equation $\mathbf{r}.\mathbf{n} + d = 0$, where R is the arbitrary point

(x, y, z) and $\mathbf{n}$ is the non-zero vector $\begin{pmatrix} a \\ b \\ c \end{pmatrix}$. The vector $\mathbf{p} = -\dfrac{d}{n^2}\mathbf{n}$ satisfies

$\mathbf{r}.\mathbf{n} + d = 0$, so the point P with position vector $\mathbf{p}$ belongs to set S, and $d = -\mathbf{p}.\mathbf{n}$. Therefore $\mathbf{r}.\mathbf{n} + d = 0 \Leftrightarrow \mathbf{r}.\mathbf{n} - \mathbf{p}.\mathbf{n} = 0$

$$\Leftrightarrow (\mathbf{r}-\mathbf{p}).\mathbf{n} = 0.$$

That is, any point R belonging to S is on the plane through P perpendicular to $\mathbf{n}$.

The *Cartesian equation* of a plane is $ax + by + cz + d = 0$,

where $\begin{pmatrix} a \\ b \\ c \end{pmatrix}$ is a normal to the plane; a, b, c may take any values

except that they cannot all be zero; d may take any value.

Example 1: Find the Cartesian equation of the plane through $(2, 3, 1)$ which is perpendicular to $5\mathbf{i} + 4\mathbf{j} - 7\mathbf{k}$.

Solution: The equation is $5x + 4y - 7z + d = 0$ where d is some constant. The plane goes through $(2, 3, 1)$; so $10 + 12 - 7 + d = 0$. That is, $d = -15$. Therefore the Cartesian equation of the plane is

$$5x + 4y - 7z - 15 = 0. \qquad \square$$

Example 2: Find the coordinates of some points on the plane $2x - 3y - 5z + 4 = 0$.

Solution: Method: choose any values for two of the coordinates, substitute, and solve to find the third coordinate.
For example: $x = 2, y = 5 \Rightarrow 4 - 15 - 5z + 4 = 0 \Rightarrow z = -\frac{7}{5}$.
$(2, 5, -\frac{7}{5})$ is on the plane.

$$y = 0, z = 0 \Rightarrow 2x + 4 = 0 \Rightarrow x = -2.$$
$(-2, 0, 0)$ is on the plane.

$x = \lambda, y = \mu$ where λ, μ are parameters
$\Rightarrow 2\lambda - 3\mu - 5z + 4 = 0 \Rightarrow z = \frac{1}{5}(2\lambda - 3\mu + 4).$
$(\lambda, \mu, \frac{1}{5}(2\lambda - 3\mu + 4))$ is a general point on
the plane.   □

Example 3: Find the Cartesian equation of the plane through $(-1, 2, 3)$
which contains the directions $8\mathbf{i} + 5\mathbf{j} + \mathbf{k}$ and $-4\mathbf{i} + 5\mathbf{j} + 7\mathbf{k}$.

Solution: We need to find a vector $\mathbf{n}$ which is normal to the plane; that is,
$\mathbf{n}$ is perpendicular to both the given directions.

Suppose $\mathbf{n} = \begin{pmatrix} a \\ b \\ c \end{pmatrix}$, then $\begin{pmatrix} a \\ b \\ c \end{pmatrix} \cdot \begin{pmatrix} 8 \\ 5 \\ 1 \end{pmatrix} = 0$ and $\begin{pmatrix} a \\ b \\ c \end{pmatrix} \cdot \begin{pmatrix} -4 \\ 5 \\ 7 \end{pmatrix} = 0.$

That is $\begin{cases} 8a + 5b + c = 0 \\ -4a + 5b + 7c = 0. \end{cases}$

Solving these simultaneously to find b, c in terms of a we obtain
$b = -2a, c = 2a$. Therefore $\mathbf{n} = \begin{pmatrix} a \\ -2a \\ 2a \end{pmatrix}$; dividing by a we get $\begin{pmatrix} 1 \\ -2 \\ 2 \end{pmatrix}$
as another normal to the plane. All non-zero scalar multiples of
$\begin{pmatrix} 1 \\ -2 \\ 2 \end{pmatrix}$ are also normals.

Therefore the equation of the plane is $x - 2y + 2z + d = 0$
where d is some constant. This equation must be satisfied by
$(-1, 2, 3)$, so $d = -1$. The required Cartesian equation is
$x - 2y + 2z - 1 = 0.$ □

Qu. 1. Use the method of Example 3 to find a vector which is perpendicular
to both $3\mathbf{i} + \mathbf{j} + 4\mathbf{k}$ and $4\mathbf{i} + 2\mathbf{j} + 3\mathbf{k}$, and hence find the equation
of the plane which contains these two directions and goes through
$(2, 1, -3)$.

Qu. 2. If $\mathbf{a} = \begin{pmatrix} a_1 \\ a_2 \\ a_3 \end{pmatrix}$ and $\mathbf{b} = \begin{pmatrix} b_1 \\ b_2 \\ b_3 \end{pmatrix}$ are non-zero vectors, prove that

$\mathbf{n} = \begin{pmatrix} a_2 b_3 - a_3 b_2 \\ a_3 b_1 - a_1 b_3 \\ a_1 b_2 - a_2 b_1 \end{pmatrix}$ is perpendicular to both $\mathbf{a}$ and $\mathbf{b}$ provided that $\mathbf{b}$

is not a scalar multiple of $\mathbf{a}$. What happens to $\mathbf{n}$ if $\mathbf{b}$ is a scalar
multiple of $\mathbf{a}$?

Qu. 3. If $\mathbf{a}$ and $\mathbf{b}$ are non-zero, non-parallel vectors and $\mathbf{n}$ is perpendicular
to both $\mathbf{a}$ and $\mathbf{b}$, prove that all vectors perpendicular to both $\mathbf{a}$ and
$\mathbf{b}$ are scalar multiples of $\mathbf{n}$. [Hint: $\mathbf{n}$ must be non-zero; $\mathbf{n}, \mathbf{a}, \mathbf{b}$ are

suitable to use as a base set; express **v** in terms of **n**, **a**, **b** and show that $\mathbf{v} \cdot \mathbf{a} = \mathbf{v} \cdot \mathbf{b} = 0 \Rightarrow \mathbf{v} = \lambda \mathbf{n}$ for some λ.]

An alternative and rather more elegant method of finding a vector perpendicular to two given vectors uses the vector product which will be dealt with in §8.5. Till then, the reader should either quote the formula of Qu. 2 or use the method of Example 3.

Example 4: Find the Cartesian equation of the plane through the points $A(2, 1, 3), B(7, 2, 3), C(5, 3, 5)$.

Solution: The plane must contain the directions **AB**, **AC**.

$$\mathbf{AB} = \mathbf{b} - \mathbf{a} = \begin{pmatrix} 5 \\ 1 \\ 0 \end{pmatrix}; \quad \mathbf{AC} = \begin{pmatrix} 3 \\ 2 \\ 2 \end{pmatrix}; \quad \begin{pmatrix} 2 \\ -10 \\ 7 \end{pmatrix} \text{ is perpendicular to}$$

both **AB** and **AC**, and is therefore a normal to the plane. Thus the equation of the plane is $2x - 10y + 7z + d = 0$ where d is some constant. The plane goes through $A(2, 1, 3)$, so $d = -15$. The Cartesian equation of plane ABC is $2x - 10y + 7z - 15 = 0$. □

Qu. 4.   Check that the equation $2x - 10y + 7z - 15 = 0$ is satisfied by each of the three points $A(2, 1, 3), B(7, 2, 3), C(5, 3, 5)$.

Qu. 5.   Find the Cartesian equation of the plane through $(1, -2, -3)$, $(4, -1, 1), (5, 0, -2)$ and check that each point satisfies the equation found.

Example 5: Find the point of intersection of the line $\dfrac{x+3}{2} = \dfrac{y-5}{-1} = \dfrac{z-2}{-3}$ with the plane $2x + 7y + 5z - 3 = 0$.

Solution: The general point on the line has coordinates $(2\lambda - 3, -\lambda + 5, -3\lambda + 2)$ where λ is a parameter. If this point is also on the plane, these coordinates must satisfy the equation of the plane: substituting we get

$$2(2\lambda - 3) + 7(-\lambda + 5) + 5(-3\lambda + 2) - 3 = 0$$
$$\Leftrightarrow \qquad\qquad -18\lambda + 36 = 0 \Leftrightarrow \lambda = 2.$$

Using this value of λ we find the coordinates of the point where the line meets the plane are $(1, 3, -4)$. □

Qu. 6.   Check that the point $(1, 3, -4)$ found in Example 5 satisfies both the equation of the plane and the equations of the line.

Exercise 8B

1. Find the Cartesian equation of the plane which is perpendicular to the stated direction, and passes through the given point:

(i) $\begin{pmatrix} 3 \\ -2 \\ 7 \end{pmatrix}$, $(3, -1, 1)$ (ii) $\begin{pmatrix} -1 \\ -2 \\ 5 \end{pmatrix}$, $(4, 0, 2)$ (iii) $\begin{pmatrix} 3 \\ 0 \\ -2 \end{pmatrix}$, $(1, 7, 4)$

(iv) the y-axis, $(2, -3, -5)$ (v) $2i - j - k$, $(4, 1, -2)$
(vi) $3i - 7j$, $(2, 0, 3)$ (vii) $-i - j + 3k$, the origin
(viii) the z-axis, the origin.

2. Find the Cartesian equation of the plane through the given point, containing the stated directions:

(i) $(1, 2, -3)$, $\begin{pmatrix} 4 \\ -2 \\ 3 \end{pmatrix}$, $\begin{pmatrix} 1 \\ 1 \\ 0 \end{pmatrix}$ (ii) $(3, 0, 0)$, $\begin{pmatrix} 1 \\ -1 \\ 0 \end{pmatrix}$, $\begin{pmatrix} 3 \\ 0 \\ 7 \end{pmatrix}$

(iii) $(0, 2, -7)$, $\begin{pmatrix} 2 \\ 6 \\ 1 \end{pmatrix}$, $\begin{pmatrix} 1 \\ 5 \\ 1 \end{pmatrix}$ (iv) $(3, -1, -5)$, the x-axis, $2i - 3j + k$

(v) the origin, $3j - k$, $2i + 5k$ (vi) $(5, -3, -7)$, the y-axis, the z-axis
(vii) $(1, -1, 1)$, $3i + 5j - 7k$, $-4i - j + 6k$
(viii) the origin, the x-axis, the y-axis.

3. Find the Cartesian equations of the planes through the following sets of three points:
   (i) $(2, 1, 3), (4, 1, 4), (2, 3, 6)$
   (ii) $(-5, -2, 7), (-3, -3, 5), (-7, -2, 5)$
   (iii) $(-11, 2, -3), (-5, 4, -5), (-16, 3, -6)$
   (iv) $(3, 0, 0), (0, 5, 0), (0, 0, 7)$
   (v) $(-3, 5, -1), (0, 0, -1), (0, 0, 0)$.

4. Find the Cartesian equation of the planes containing both the point and the line stated:

(i) $(3, -2, -7)$, $\dfrac{x - 5}{3} = \dfrac{y}{1} = \dfrac{z + 6}{4}$

(ii) $(5, 8, -4)$, $\dfrac{x}{-4} = \dfrac{y - 5}{1} = \dfrac{z + 1}{0}$

(iii) $(0, 0, 0)$, $\dfrac{x - 7}{3} = \dfrac{y + 2}{1} = \dfrac{z - 1}{-1}$

(iv) $(2, -5, 3)$, $\dfrac{x - 1}{-3} = \dfrac{y + 7}{5} = \dfrac{z - 3}{2}$

(v) the origin, the x-axis, the z-axis.

5. Find the coordinates of the points where the following lines and planes intersect:

(i) $2x - 3y + 7z - 10 = 0$, $\dfrac{x - 8}{5} = \dfrac{y - 3}{2} = \dfrac{z - 4}{3}$

(ii) $x + 2y - 3z - 17 = 0$, $\dfrac{x + 4}{-3} = \dfrac{y - 2}{1} = \dfrac{z + 1}{2}$

(iii) $x - 3y - 8z + 14 = 0$, line joining $(8, -4, -3), (-7, 6, -5)$
(v) the perpendicular from $(-6, -7, 19)$ to $5x + 2y - 9z - 5 = 0$
(vi) the perpendicular from $(8, 21, -5)$ to $5x + 9y + z - 10 = 0$.

6. Prove that the points $A(1, 3, 1), B(3, 13, 3), C(5, 2, -1), D(-2, 9, 4)$ are coplanar, but no three of them are collinear.

7. Prove that the lines $\dfrac{x+3}{2} = \dfrac{y}{3} = \dfrac{z-5}{-1}, \dfrac{x+4}{5} = \dfrac{y-2}{4} = \dfrac{z}{3}$ intersect, and find the equation of the plane containing both lines.

8. Prove that the lines $\dfrac{x+4}{3} = \dfrac{y}{-2} = \dfrac{z-1}{3}, \dfrac{x}{1} = \dfrac{y+6}{1} = \dfrac{z+3}{5}$ intersect, and find the equation of the plane containing both lines.

9. Find the equation of the plane containing the parallel lines $\dfrac{x-3}{-5} = \dfrac{y+2}{4} = \dfrac{z-1}{2}, \dfrac{x+2}{-5} = \dfrac{y-3}{4} = \dfrac{z+8}{2}$.

10. Find the equation of the plane parallel to the line $\dfrac{x-1}{2} = \dfrac{y}{3} = \dfrac{z}{1}$ containing the line $\dfrac{x+1}{3} = \dfrac{y}{1} = \dfrac{z}{2}$.

11. The point A has coordinates $(-6, 5, 7)$; l is the line $\dfrac{x-3}{3} = \dfrac{y+1}{-2} = \dfrac{z+8}{-5}$; π is the plane $2x - 4y - 3z - 5 = 0$.
    (i) Show that A is on l.
    (ii) Write down the coordinates of the general point on the line through A perpendicular to π in terms of a parameter λ.
    (iii) Find the coordinates of the foot of the perpendicular from A to the plane π.
    (iv) Find the coordinates of the image of A when it is reflected in plane π.
    (v) Find the coordinates of the point where l and π intersect.
    (vi) Find the Cartesian equations of the image of l when it is reflected in π.

12. Find the Cartesian equations of the image of line l when it is reflected in plane π in the following:
    (i) $l: \dfrac{x-8}{1} = \dfrac{y-4}{-23} = \dfrac{z-3}{2}, \quad \pi: x - 5y + 2z + 6 = 0$
    (ii) $l: \dfrac{x-1}{1} = \dfrac{y-1}{2} = \dfrac{z+7}{13}, \quad \pi: 3x + y - 2z + 24 = 0$
    (iii) $l: \dfrac{x+1}{3} = \dfrac{y+20}{2} = \dfrac{z-1}{1}, \quad \pi: 4x - 7y + 2z = 0$.

13. If l' is the image of l under the operation of reflection in plane π, find the Cartesian equation of π:
    (i) $l: \dfrac{x-2}{12} = \dfrac{y}{-1} = \dfrac{z-5}{3}, \quad l': \dfrac{x-2}{8} = \dfrac{y}{-3} = \dfrac{z-5}{9}$
    (ii) $l: \dfrac{x+2}{0} = \dfrac{y+3}{3} = \dfrac{z-4}{1}, \quad l': \dfrac{x-6}{8} = \dfrac{y+10}{-1} = \dfrac{z-7}{5}$.

14. The points A, B, C have coordinates $(a, 0, 0)$, $(0, b, 0)$, $(0, 0, c)$ respectively; P is the foot of the perpendicular from the origin O to the plane ABC.
    (a) Find the equation of the plane ABC.
    (b) If P has coordinates $(1, 2, 3)$ (i) find a, b, c
                                            (ii) find the volume of tetrahedron $OABC$
                                            (iii) prove that P is the orthocentre of triangle ABC.

8.3 Further lines and planes

At the beginning of §3.3 we mentioned that a straight line can be specified as the intersection of two planes. We are now able to point out that the Cartesian equations of a straight line

$$\frac{x - x_1}{l} = \frac{y - y_1}{m} = \frac{z - z_1}{n}$$

are equivalent to the pair of equations

$$\frac{x - x_1}{l} = \frac{y - y_1}{m}, \quad \frac{y - y_1}{m} = \frac{z - z_1}{n}$$

which are themselves equivalent to $mx - ly + d_1 = 0$, $ny - mz + d_2 = 0$ where d_1, d_2 are the constants $ly_1 - mx_1$, $mz_1 - ny_1$ respectively. These last two equations will be recognized as the equations of planes (parallel to the z- and x-axes respectively), emphasizing that a straight line is the intersection of two planes.

It is intuitively obvious that any two distinct non-parallel planes intersect in a straight line. Our next example illustrates a method of finding the equation of the line of intersection of two planes; the method (perhaps with small modifications) will be found suitable in all numerical cases.

Example 1: Find the Cartesian equations of the line of intersection of the two planes $5x - y + z - 8 = 0$, $x + 3y + z + 4 = 0$.

Solution: We need to find (a) a point on the line and (b) the direction of the line.
    (a) Choose any value for z, substitute into the two equations, and solve simultaneously for x and y. For example, if $z = 1$,
        the equations become $\begin{cases} 5x - y = 7 \\ x + 3y = -5 \end{cases} \Leftrightarrow \begin{cases} x = 1 \\ y = -2. \end{cases}$
        Thus $(1, -2, 1)$ is a point on both planes (and therefore on the intersection line).

(b) The line of intersection is in both planes and is therefore perpendicular to both normals. The vector $\begin{pmatrix} 1 \\ 1 \\ -4 \end{pmatrix}$ is perpendicular to both $\begin{pmatrix} 5 \\ -1 \\ 1 \end{pmatrix}$ and $\begin{pmatrix} 1 \\ 3 \\ 1 \end{pmatrix}$. Thus $\begin{pmatrix} 1 \\ 1 \\ -4 \end{pmatrix}$ is a direction vector for the line of intersection.

Therefore the Cartesian equations of the intersection line are
$$\frac{x-1}{1} = \frac{y+2}{1} = \frac{z-1}{-4}. \qquad \square$$

Qu. 1. Use the method of part (a) of the solution of Example 1 to find another point which is on both planes, and check that it satisfies the equations of the intersection line.

Suppose π_1, π_2 are distinct planes with common line l. Let $a_1 x + b_1 y + c_1 z + d_1 = 0, a_2 x + b_2 y + c_2 z + d_2 = 0$ be the equations of π_1, π_2 respectively. If we add multiples of these two equations we obtain the equation

$$\lambda(a_1 x + b_1 y + c_1 z + d_1) + \mu(a_2 x + b_2 y + c_2 z + d_2) = 0 \qquad \text{(A)}$$

where λ, μ are scalar constants. Provided that λ, μ are not both zero, equation (A) can be rearranged in the form $ax + by + cz + d = 0$ where not all a, b, c are zero; therefore equation (A) represents a plane. Any point which satisfies both the equation for π_1 and the equation for π_2 will clearly satisfy equation (A); thus equation (A) represents a plane containing the line common to planes π_1 and π_2.

Conversely, the equation of every plane containing l can be written in the form (A) for suitable λ, μ. To show this, we note first that (A) gives π_1 if $\mu = 0$, and π_2 if $\lambda = 0$. Next let π_3 be any plane containing l other than π_1 or π_2, and let $P(X, Y, Z)$ be a point of π_3 not on l. Then $a_1 X + b_1 Y + c_1 Z + d_1 \neq 0$ and $a_2 X + b_2 Y + c_2 Z + d_2 \neq 0$, since P is not on π_1 or π_2, and so we can choose particular values λ_1 and μ_1 such that
$$\lambda_1(a_1 X + b_1 Y + c_1 Z + d_1) + \mu_1(a_2 X + b_2 Y + c_2 Z + d_2) = 0. \qquad \text{(B)}$$
Then the plane $\lambda_1(a_1 x + b_1 y + c_1 z + d_1) + \mu_1(a_2 x + b_2 y + c_2 z + d_2) = 0$ contains l since it is of the form (A), and also contains P from (B). But π_3 is the only plane containing l and P, and so the equation of π_3 has been obtained in the form (A).

Example 2: Find the equation of the plane through $(1, 1, 1)$ which also contains the common line of the planes $x + y + z + 6 = 0$, $x - y + z + 5 = 0$.

Solution: The required equation is of the form
$$\lambda(x + y + z + 6) + \mu(x - y + z + 5) = 0.$$

This equation is satisfied by the point $(1, 1, 1)$

$\Leftrightarrow 9\lambda + 6\mu = 0 \Leftrightarrow \mu = -\frac{3}{2}\lambda.$

Therefore the required equation is

$$\lambda(x + y + z + 6) - \tfrac{3}{2}\lambda(x - y + z + 5) = 0$$
$$\Leftrightarrow 2(x + y + z + 6) - 3(x - y + z + 5) = 0$$

dividing by $\dfrac{\lambda}{2}$ which is not zero

$\Leftrightarrow -x + 5y - z - 3 = 0.$

The required plane has equation $x - 5y + z + 3 = 0.$ □

Qu. 2. Find the equation of the plane through the origin which also contains the common line of the planes $x + y + z + 6 = 0$, $x - y + z + 5 = 0$.

Qu. 3. What does equation (A) of the paragraph above represent if π_1, π_2 do not have a common line?

The distance of a point P from a plane π is the perpendicular distance; this can be calculated by finding the point Q where the line through P normal to the plane π meets π, and then using the Pythagorean distance formula to compute the length PQ. However the formula derived in the next example is useful, especially if we do not want to know the coordinates of Q.

Example 3: Prove that the distance from the point $P(x_1, y_1, z_1)$ to the plane

$$ax + by + cz + d = 0 \text{ is } \left| \frac{ax_1 + by_1 + cz_1 + d}{\sqrt{(a^2 + b^2 + c^2)}} \right|$$

Solution: Let $\mathbf{n} = a\mathbf{i} + b\mathbf{j} + c\mathbf{k}$, so that $\mathbf{n}$ is normal to the plane and the equation of the plane may be written as $\mathbf{r} \cdot \mathbf{n} + d = 0$. Suppose Q is the foot of the perpendicular from P to the plane. Then Q has position vector $\mathbf{p} - \lambda \hat{\mathbf{n}}$, where $|\lambda|$ is the length of PQ. The

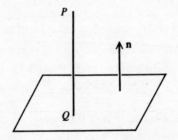

sign of λ depends on whether **PQ** and **n** have the same sense. As illustrated **PQ** and **n** have opposite senses and λ is positive.

$$Q \text{ is on the plane} \Leftrightarrow (p - \lambda\hat{n}).n + d = 0$$
$$\Leftrightarrow \mathbf{p}.\mathbf{n} + d = \lambda\hat{n}.n$$
$$\Leftrightarrow \mathbf{p}.\mathbf{n} + d = \lambda n$$
$$\Leftrightarrow \lambda = \frac{\mathbf{p}.\mathbf{n} + d}{n} = \frac{ax_1 + by_1 + cz_1 + d}{\sqrt{(a^2 + b^2 + c^2)}}.$$

i.e.
$$PQ = \left| \frac{ax_1 + by_1 + cz_1 + d}{\sqrt{(a^2 + b^2 + c^2)}} \right|. \qquad \square$$

There is a marked similarity between the formula just proved and the formula for the distance of a point from a line when we are working in two dimensions (see Vol. 1 §2.10). If $P(x_1, y_1, z_1)$ and $P'(x_2, y_2, z_2)$ are on the same side of the plane $ax + by + cz + d = 0$, then the values of the expressions $ax_1 + by_1 + cz_1 + d, ax_2 + by_2 + cz_2 + d$ have the same sign, but not otherwise.

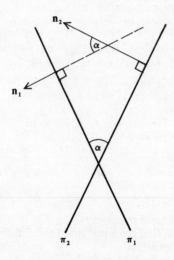

The angle between two planes is readily calculated as it is the same as the angle between their normals (or possibly the supplement of this angle, depending on the sense in which the normals are taken). An examination of the above diagram, which is an 'edge-on' view of planes π_1, π_2, should convince the reader of this fact.

Example 4: Find the end points of the common perpendicular of the lines
$$l_1: \frac{x-6}{-1} = \frac{y-1}{1} = \frac{z-1}{-1}, \quad l_2: \frac{x-5}{-1} = \frac{y+10}{4} = \frac{z+6}{2},$$
and their distance apart.

Solution: The direction vectors of l_1, l_2 are $\mathbf{d}_1 = \begin{pmatrix} -1 \\ 1 \\ -1 \end{pmatrix}$, $\mathbf{d}_2 = \begin{pmatrix} -1 \\ 4 \\ 2 \end{pmatrix}$

respectively. The arbitrary points P on l_1, Q on l_2 have coordinates $(-\lambda + 6, \lambda + 1, -\lambda + 1)$, $(-\mu + 5, 4\mu - 10, 2\mu - 6)$ respectively, where λ, μ are parameters. We must find the values of λ, μ so that $\mathbf{PQ}$ is perpendicular to both $\mathbf{d}_1, \mathbf{d}_2$.

$$\mathbf{PQ} = \begin{pmatrix} -\mu + \lambda - 1 \\ 4\mu - \lambda - 11 \\ 2\mu + \lambda - 7 \end{pmatrix}. \qquad \begin{aligned} \mathbf{PQ} \cdot \mathbf{d}_1 &= 0 \Leftrightarrow 3\mu - 3\lambda = 3. \\ \mathbf{PQ} \cdot \mathbf{d}_2 &= 0 \Leftrightarrow 21\mu - 3\lambda = 57. \end{aligned}$$

Solving simultaneously, $\lambda = 2, \mu = 3$. Therefore P is the point $(4, 3, -1)$, Q is the point $(2, 2, 0)$, and their distance apart is $\sqrt{(4 + 1 + 1)} = \sqrt{6}$ units. $\qquad\square$

Exercise 8C

1. Find the Cartesian equations of the line of intersection of the two planes:
   (i) $x + y - 6z - 4 = 0, 5x - 2y - 3z - 13 = 0$
   (ii) $14x - 4y + z + 11 = 0, 5x - 3y + 2z + 4 = 0$
   (iii) $8x - 5y - 3z - 9 = 0, 24x - 13y - 6z - 49 = 0$
   (iv) $x + y - 3z + 1 = 0, x + y + 2z - 4 = 0.$

2. Find the point of intersection of the three planes:
   (i) $2x + 3y - z - 8 = 0, 4x + 2y + 3z + 5 = 0,$
   $4x - 7y + 5z + 6 = 0$
   (ii) $5x + 3y - z + 22 = 0, x - 4y + 10z - 17 = 0,$
   $3x - 2y + z + 16 = 0$
   (iii) $x + z = 12, x + y = 4, y + z = 14.$

3. Show that these planes do not have a common point, and describe the configuration of the three planes:
   $x + 4y - z - 3 = 0, \quad 4x - 2y + 5z - 21 = 0, \quad x - 2y + 2z - 3 = 0.$

4. If l is the common line of the planes $2x + y + 3z + 1 = 0$, $4x - y + z + 1 = 0$, find the equation of the plane containing l which
   (i) also contains the origin,
   (ii) is parallel to $x + 2y + 4z = 0$,
   (iii) is perpendicular to $6\mathbf{i} - 3\mathbf{j} - \mathbf{k}$,
   (iv) is parallel to $2\mathbf{i} - 3\mathbf{k}$,
   (v) is perpendicular to $3x + 3y - z = 0$.

5. Write down the Cartesian equations of two intersecting planes which have $l: \dfrac{x - 1}{6} = \dfrac{y - 1}{1} = \dfrac{z - 1}{-4}$ as their common line. Hence find the Cartesian equation of the plane containing l
   (i) which passes through the point $(3, -1, 2)$
   (ii) which is 1 unit distant from the point $(-2, 7, -2)$.

6. Find the Cartesian equation of the plane

   (i) containing $\dfrac{x-3}{1} = \dfrac{y+1}{3} = \dfrac{z-3}{3}$ and the point $(2, 5, -9)$

   (ii) containing $\dfrac{x-2}{0} = \dfrac{y+7}{5} = \dfrac{z-3}{-2}$ and the point $(4, -1, -1)$

   (iii) containing $\dfrac{x-3}{2} = \dfrac{y+1}{-1} = \dfrac{z-5}{8}$ and perpendicular to
   $3x + 2y - 2z = 0$

   (iv) containing $\dfrac{x+7}{-3} = \dfrac{y+2}{-1} = \dfrac{z-4}{2}$ and perpendicular to
   $6x + y + 3z - 17 = 0$.

7. Find the cosine of the acute angle between these pairs of planes:
   (i) $2x - 7y + 3z - 7 = 0$, $4x + 3y + 6z - 11 = 0$
   (ii) $7x - 4y - 4z + 3 = 0$, $2x - 3y + 6z + 75 = 0$
   (iii) $5x - 14y + 2z - 13 = 0$, $6x + 7y + 6z + 23 = 0$.

8. Find the projection of the vector on the plane:
   (i) $18\mathbf{i} - 36\mathbf{j} + 45\mathbf{k}$, $8x - y - 4z + 3 = 0$
   (ii) $3\mathbf{i} + 8\mathbf{j} - 5\mathbf{k}$, $x + 2y - z + 23 = 0$
   (iii) $\mathbf{i} + 2\mathbf{j} + \mathbf{k}$, $4x - y + z - 3 = 0$.

9. Find the cosine of the acute angles between the vectors and planes of
   Qu. 8
   (a) by using the projections on the planes,
   (b) by finding the angle between the normal to the plane and the
       vector.

10. Let π be the plane with equation $\mathbf{n} \cdot \mathbf{r} + d = 0$; P' is the image of
    point P by reflection in plane π. Prove that $\mathbf{p}' = \mathbf{p} - \dfrac{2}{n^2}(\mathbf{n} \cdot \mathbf{p} + d)\mathbf{n}$.

11. Lines l, l' are skew, and P is a point on neither l nor l'. By considering
    the plane containing l and P, or otherwise, prove that if there is a line
    through P intersecting both l and l', then it is unique.

12. The Cartesian equations of l, l' are
    $\dfrac{x-2}{2} = \dfrac{y}{1} = \dfrac{z-8}{0}$, $\dfrac{x-5}{0} = \dfrac{y-7}{1} = \dfrac{z}{1}$ respectively. Find the
    Cartesian equations of the transversal of l, l' (i.e. a line meeting both l
    and l') which passes through $(0, 1, 4)$.

13. Find the distance between the lines l, l' of Qu. 12.

14. The equations $\dfrac{x-3}{2} = \dfrac{y-5}{1} = \dfrac{z-7}{-4}$, $\dfrac{x+1}{3} = \dfrac{y+4}{1} = \dfrac{z-2}{-2}$
    represent pipes A and B in a chemical plant, where length is measured in
    metres. A by-pass is to be installed connecting A and B. Find the length
    of the shortest pipe that may be fitted, and the location of its end points.

15. The point A is 4 km west of B which is 3 km south of C. Trial borings
    show that there is coal 700 m, 800 m, 950 m directly below A, B, C

respectively. Mining engineers construct a vertical shaft from the point D 1 km north-west of B. Assuming that A, B, C, D have the same elevation and that the coal seam is a plane, find the depth at which the shaft meets the coal seam. If a horizontal tunnel is cut in the plane of the coal seam find the direction of this tunnel, giving your answer as a bearing to the nearest degree.

16. Two adjacent portions of the roof of a building are parts of the planes whose equations are $z = \frac{1}{2}x + 12, z = \frac{2}{3}y + 12$; the z-axis is vertical, the origin is a point on the horizontal ground and distances are measured in metres. The line l of intersection of the planes meets the ground at P.

   (i) Find the coordinates of P.

  (ii) Express the equation of l in the form $\dfrac{x-a}{p} = \dfrac{y-b}{q} = \dfrac{z-c}{r}$.

 (iii) Write down the direction cosines $\cos\theta_1, \cos\theta_2, \cos\theta_3$ of the line l and interpret the angles $\theta_1, \theta_2, \theta_3$ in relation to the building.

 (iv) Calculate, to the nearest half degree, the angles between each of the given planes and the ground. Show that the angle inside the roof space between the two given planes is an obtuse angle (you are not required to calculate the value of the angle).   [MEI]

17. Points $A\,(3, 3, 3)$ and $B\,(5, -1, -1)$ are adjacent vertices of a cube which has its centre at the origin. Find the other vertices.

8.4 Differentiation of a vector

Suppose the function $\mathbf{f}$ maps the real (scalar) t to the vector $\mathbf{r}$; then we may write $\mathbf{r} = \mathbf{f}(t)$. If $\mathbf{r}$ is a position vector, the equation $\mathbf{r} = \mathbf{f}(t)$ defines a *curve*, and t is known as a *parameter*.

Qu. 1. Describe the curves: (i) $\mathbf{r} = \mathbf{a} + t\mathbf{d}$ (ii) $\mathbf{r} = \mathbf{c} + a\cos t\,\mathbf{i} + a\sin t\,\mathbf{k}$.

Example 1: Describe the curve $\mathbf{r} = a\cos t\,\mathbf{i} + a\sin t\,\mathbf{j} + bt\,\mathbf{k}$.

Solution: The equation $\mathbf{r} = a\cos t\,\mathbf{i} + a\sin t\,\mathbf{j}$ describes the circle which is in the plane $z = 0$, with its centre at the origin, radius a. As t varies, the point with position vector $\mathbf{r} = a\cos t\,\mathbf{i} + a\sin t\,\mathbf{j} + bt\,\mathbf{k}$ rotates about the z-axis, and moves parallel to the z-axis by an amount proportional to the increase of t. The curve is a spiral on the surface of the cylinder which has the z-axis as its axis, and radius a, the turns of the spiral being equally spaced. It is known as a circular *helix*; the distance $2\pi b$ between each turn is the *pitch*.   □

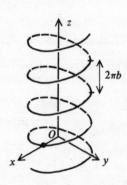

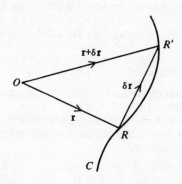

Consider the function $\mathbf{f}: t \mapsto \mathbf{r}$ and the corresponding curve C with
quation $\mathbf{r} = \mathbf{f}(t)$. Then the *derivative* of $\mathbf{r}$ with respect to t is the limit of
$\dfrac{\mathbf{f}(t + \delta t) - \mathbf{f}(t)}{\delta t}$ as δt tends to zero; this derivative is denoted by $\dfrac{d\mathbf{r}}{dt}$. Suppose
R' is the point with position vector $\mathbf{r} + \delta \mathbf{r} = \mathbf{f}(t + \delta t)$. We may write
$\dfrac{\mathbf{r}}{t} \to \dfrac{d\mathbf{r}}{dt}$ as $\delta t \to 0$. Notice that $\dfrac{\delta \mathbf{r}}{\delta t}$ is a vector parallel to $\mathbf{RR'}$ and that as δt
ends to zero, the direction of $\mathbf{RR'}$ tends to the direction of the tangent to
he curve C at the point R. Thus, if $\dfrac{d\mathbf{r}}{dt}$ exists, it is a vector which is parallel
o the tangent to the curve C at the point R.

The second derivative $\dfrac{d^2\mathbf{r}}{dt^2}$ is defined as $\dfrac{d}{dt}\left(\dfrac{d\mathbf{r}}{dt}\right)$.

Qu. 2.   Show that   (i) $\dfrac{d}{dt}(\mathbf{u} + \mathbf{v}) = \dfrac{d\mathbf{u}}{dt} + \dfrac{d\mathbf{v}}{dt}$ (ii) $\dfrac{d}{dt}(s\mathbf{v}) = \dfrac{ds}{dt}\mathbf{v} + s\dfrac{d\mathbf{v}}{dt}$

(iii) $\dfrac{d}{dt}(\mathbf{u} \cdot \mathbf{v}) = \dfrac{d\mathbf{u}}{dt} \cdot \mathbf{v} + \mathbf{u} \cdot \dfrac{d\mathbf{v}}{dt}$ (iv) $\mathbf{u}$ is constant $\Rightarrow \dfrac{d\mathbf{u}}{dt} = \mathbf{0}$.

A direct consequence of the results of Qu. 2 is that

$$\mathbf{r} = f(t)\mathbf{i} + g(t)\mathbf{j} + h(t)\mathbf{k} \Rightarrow \dfrac{d\mathbf{r}}{dt} = f'(t)\mathbf{i} + g'(t)\mathbf{j} + h'(t)\mathbf{k}.$$

Qu. 3.   Find $\dfrac{d\mathbf{r}}{dt}$ when   (i) $\mathbf{r} = 3\cos t\,\mathbf{i} + 3\sin t\,\mathbf{j} + 2t\mathbf{k}$,
(ii) $\mathbf{r} = at\,\mathbf{i} + (bt - \tfrac{1}{2}gt^2)\mathbf{j}$ where a, b, g are constants.

Qu. 4.   Find the vector equation of the tangent to the helix
$\mathbf{r} = a\cos t\,\mathbf{i} + bt\,\mathbf{j} + a\sin t\,\mathbf{k}$ at the point with parameter π.

We define *indefinite integration* as the 'reverse' of differentiation, so that
$\displaystyle\int \dfrac{d\mathbf{r}}{dt}\,dt = \mathbf{r} + \mathbf{c}$ where $\mathbf{c}$ is an arbitrary constant vector.

If $\mathbf{r}$ is the position vector of a particle at time t relative to some fixed origin O, then $\dfrac{d\mathbf{r}}{dt}$ is the rate of change of position relative to O with respect to time; that is, $\dfrac{d\mathbf{r}}{dt}$ is the velocity $\mathbf{v}$ of the particle at time t. Similarly $\dfrac{d\mathbf{v}}{dt}$ is the acceleration of the particle at time t.

Qu. 5. A particle is initially at the origin with velocity $\mathbf{u}$. It experiences a constant acceleration $\mathbf{a}$, and at time t its displacement from the origin is $\mathbf{s}$ and its velocity $\mathbf{v}$.

Show that   (i) $\mathbf{v} = \mathbf{u} + t\mathbf{a}$

(ii) $\mathbf{s} = t\mathbf{u} + \frac{1}{2}t^2\mathbf{a} = t\mathbf{v} - \frac{1}{2}t^2\mathbf{a} = \frac{1}{2}t(\mathbf{u} + \mathbf{v})$

(iii) $2\mathbf{a} \cdot \mathbf{s} = v^2 - u^2$.

For brevity we use Newton's dot notation: a dot above a character denotes the first derivative with respect to time; two dots denote the second derivative. Thus $\dot{\mathbf{r}} = \dfrac{d\mathbf{r}}{dt}$, $\ddot{\mathbf{r}} = \dfrac{d^2\mathbf{r}}{dt^2}$.

Suppose particles A, B have position vectors $\mathbf{a}, \mathbf{b}$ relative to a fixed origin O, so that $\dot{\mathbf{a}}, \dot{\mathbf{b}}$ are the respective velocities. Then $\dot{\mathbf{b}} - \dot{\mathbf{a}}$ is the velocity of B *relative to* A, often denoted by $_B\mathbf{v}_A$. Notice that the velocity of B relative to A is the derivative with respect to time of the position vector of B relative to A.

Example 2: Particles P, Q are initially at $(4, 6, 4)$, $(-11, -5, 1)$, travelling with velocities $2\mathbf{i} - \mathbf{j} - 2\mathbf{k}$, $6\mathbf{i} + 2\mathbf{j} + \mathbf{k}$ respectively, the units of distance and velocity being metres and metres per second respectively. Find the least distance between P and Q, and the time when this occurs.

Solution: The velocity of Q relative to P is $\begin{pmatrix} 6 \\ 2 \\ 1 \end{pmatrix} - \begin{pmatrix} 2 \\ -1 \\ -2 \end{pmatrix} = \begin{pmatrix} 4 \\ 3 \\ 3 \end{pmatrix} = \mathbf{v}$, say;

this is a direction vector for the path l of Q relative to P.

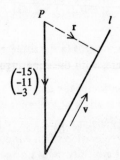

Initially the position vector of Q relative to P is
$$\begin{pmatrix} -11 \\ -5 \\ 1 \end{pmatrix} - \begin{pmatrix} 4 \\ 6 \\ 4 \end{pmatrix} = \begin{pmatrix} -15 \\ -11 \\ -3 \end{pmatrix}; \text{ at time } t \text{ the position vector of } Q$$

relative to P is $\begin{pmatrix} -15 \\ -11 \\ -3 \end{pmatrix} + t \begin{pmatrix} 4 \\ 3 \\ 3 \end{pmatrix} = \begin{pmatrix} 4t - 15 \\ 3t - 11 \\ 3t - 3 \end{pmatrix} = \mathbf{r}$, say.

P and Q are closest at time $t \Leftrightarrow \mathbf{r} \cdot \mathbf{v} = 0$
$$\Leftrightarrow 4(4t - 15) + 3(3t - 11) + 3(3t - 3) = 0$$
$$\Leftrightarrow 34t - 102 = 0 \Leftrightarrow t = 3.$$

$t = 3 \Rightarrow \mathbf{r} = \begin{pmatrix} -3 \\ -2 \\ 6 \end{pmatrix} \Rightarrow |\mathbf{r}| = 7.$

Therefore P is closest to Q after 3 s, the distance between them then being 7 m. □

Consider a particle moving in a circle of unit radius, centre O. Take O as the origin, with the x- and y-axes in the plane of the circle. Suppose that at

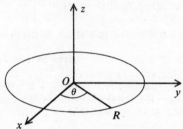

time t the particle is at R with position vector $\hat{\mathbf{r}}$; then $\hat{\mathbf{r}} = \begin{pmatrix} \cos\theta \\ \sin\theta \\ 0 \end{pmatrix}$ where θ is

the angle OR makes with the positive x-axis. Differentiating,

$$\frac{d\hat{\mathbf{r}}}{dt} = \begin{pmatrix} -\dot{\theta}\sin\theta \\ \dot{\theta}\cos\theta \\ 0 \end{pmatrix} = \dot{\theta}\hat{\mathbf{t}}, \text{ where } \hat{\mathbf{t}} \text{ denotes the unit vector } \begin{pmatrix} -\sin\theta \\ \cos\theta \\ 0 \end{pmatrix}. \text{ Vector } \hat{\mathbf{t}}$$

is known as a *transverse* vector; it is in the plane of the motion, perpendicular to the *radial* vector $\hat{\mathbf{r}}$; $\hat{\mathbf{t}}$ rotates with the same angular speed as $\hat{\mathbf{r}}$, but is $\frac{\pi}{2}$ 'in front'.

Qu. 6.   Show that if $|\mathbf{r}|$ is constant, then $\ddot{\mathbf{r}} = r\ddot{\theta}\,\hat{\mathbf{t}} - r\dot{\theta}^2\,\hat{\mathbf{r}}$. Describe the direction of the acceleration when $\dot{\theta}$ is constant.

Exercise 8D

1. If $\mathbf{a}, \mathbf{b}$ are constant show that $\mathbf{r} = \cos \omega t \, \mathbf{a} + \sin \omega t \, \mathbf{b} \Rightarrow \ddot{\mathbf{r}} + \omega^2 \mathbf{r} = \mathbf{0}$.

2. A particle moves on a helix. Its position vector at time t is
$$\mathbf{r} = a \cos \omega t \, \mathbf{i} + a \sin \omega t \, \mathbf{j} + bt \, \mathbf{k}.$$
Show that $\dot{\mathbf{r}}$ makes a constant angle with the z-axis, and that $\ddot{\mathbf{r}}$ is of constant magnitude and always directed towards the axis of the helix.

3. If $\mathbf{u} = t^2 \mathbf{i} - t^3 \mathbf{j} + 2t \, \mathbf{k}$ and $\mathbf{v} = 2t^2 \mathbf{i} + 3t \mathbf{j} + t^3 \mathbf{k}$ find $\dfrac{d}{dt}(\mathbf{u} \cdot \mathbf{v})$

   (i) by evaluating $\mathbf{u} \cdot \mathbf{v}$ and then differentiating
   (ii) by using the formula for differentiating products.

4. The position vector of a particle at time t is $\mathbf{r} = p(t)\mathbf{i} + q(t)\mathbf{j} + r(t)\mathbf{k}$ where p, q, r are polynomials of the same degree. What is this degree if
   (i) the particle moves with constant, non-zero velocity,
   (ii) the particle moves with constant, non-zero acceleration?

5. Particles A, B are projected simultaneously with different velocities, but then move freely under gravity. Prove that their relative velocity is constant.

6. By considering $\dfrac{d}{dt}(\mathbf{v} \cdot \mathbf{v})$ show that $\dfrac{d}{dt}|\mathbf{v}| = \hat{\mathbf{v}} \cdot \dfrac{d\mathbf{v}}{dt}$.

7. (M) A particle of mass m moves in a plane, and its position vector relative to an origin O at any time t is given by $\mathbf{r} = \mathbf{k} + \begin{pmatrix} a \cos nt \\ b \sin nt \end{pmatrix}$,
   where a, b and n are constants and $\mathbf{k}$ is a constant vector. Prove that the force on the particle is the resultant of forces $mn^2 \mathbf{k}$ and $-mn^2 \mathbf{r}$. Describe briefly the directions of these two forces.   [SMP]

8. Describe the curve given by $\mathbf{r} = \begin{pmatrix} 5 \cos (t^2) \\ 5 \sin (t^2) \end{pmatrix}$ where t denotes time. Find $\mathbf{a}(= \ddot{\mathbf{r}})$ and deduce the radial and transverse components of the acceleration.
   A boy slides down a bannister such that his path is given by
   $\mathbf{r} = \begin{pmatrix} 5 \cos (t^2) \\ 5 \sin (t^2) \\ kt^2 \end{pmatrix}$, where the units are metres and k is a constant. Indicate
   on a diagram the shape of the bannister and the magnitudes of the boy's acceleration components in the directions (a) towards the axis of the curve, (b) vertically downwards, (c) in a direction perpendicular to those in (a) and (b). [Note: the unit vectors for $\mathbf{r}$ are horizontal south, horizontal west, and vertically downwards.]   [SMP]

9. A destroyer sights a ship travelling with constant velocity $5\mathbf{j}$, whose position vector at the time of sighting is $2000(3\mathbf{i} + \mathbf{j})$ relative to the destroyer, distances being measured in m and speeds in m s^{-1}. The destroyer immediately begins to move with velocity $k(4\mathbf{i} + 3\mathbf{j})$, where k is a constant, in order to intercept the ship. Find k and the time to

interception. Find also the distance between the vessels when half the
time to interception has elapsed.   [O & C]

10. In this question the units of time and distance are the second and metre
respectively. Two particles A and B start simultaneously from the points
which have position vectors $3\mathbf{i} + 4\mathbf{j}$ and $8\mathbf{i} + 4\mathbf{j}$ respectively. Each moves
with constant velocity, the velocity of B being $\mathbf{i} + 4\mathbf{j}$.
  (i) If the velocity of A is $5\mathbf{i} + 2\mathbf{j}$, show that the least distance between
      A and B subsequently is $\sqrt{5}$ metres.
  (ii) If the speed of A is 5 m s^{-1} and the particles collide, find the velocity
      of A.   [L]

11. An aircraft takes off from the end of a runway in a southerly direction
and climbs at an angle of $\tan^{-1}(\frac{1}{2})$ to the horizontal at a speed of
$225\sqrt{5} \text{ km h}^{-1}$. Show that t seconds after take-off the position vector $\mathbf{r}$
of the aircraft with respect to the end of the runway is given by
$\mathbf{r} = (t/16)(2\mathbf{i} + \mathbf{k})$ where $\mathbf{i}, \mathbf{j}, \mathbf{k}$ represent vectors of length 1 km in
directions south, east and vertically upwards. At time $t = 0$ a second
aircraft flying horizontally south-west at $720\sqrt{2} \text{ km h}^{-1}$ has position
vector $-1.2\mathbf{i} + 3.2\mathbf{j} + \mathbf{k}$. Find its position vector at time t in terms of
$\mathbf{i}, \mathbf{j}, \mathbf{k}$ and t. Show that there will be a collision unless courses are
changed and state at what time it will occur.   [SMP]

12. An axle OA of length a is rotating in a horizontal plane about O with
constant angular speed $\dot{\theta}$. A wheel of radius b is mounted on the axle at
A, the plane of the wheel being perpendicular to OA. Let $\mathbf{u}$ be the unit

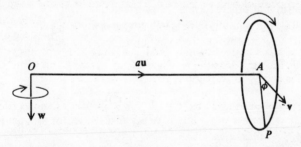

vector in the direction OA, $\mathbf{v}$ the unit vector in the direction in which A
is moving, and $\mathbf{w}$ the unit vector vertically downwards. The point P is on
the rim of the wheel, such that AP makes an angle ϕ with $\mathbf{v}$, and the
wheel rotates about the axle with constant angular speed $\dot{\phi}$. Show that
the position vector of P is

$$\mathbf{p} = a\mathbf{u} + b\cos\phi\,\mathbf{v} + b\sin\phi\,\mathbf{w}.$$

Write down $\dot{\mathbf{u}}$ in terms of $\mathbf{v}$, and $\dot{\mathbf{v}}$ in terms of $\mathbf{u}$. Find $\dot{\mathbf{p}}$, and rearrange
your expression in the form

  (velocity of P relative to A) + (velocity of A relative to O).

13. Two ships are moving with constant, but different, velocities $\mathbf{v}_1$ and $\mathbf{v}_2$.
At noon the ships have position vectors $\mathbf{a}_1$ and $\mathbf{a}_2$ respectively with
respect to a fixed origin. The distance between the ships is decreasing.

Show that $\qquad (\mathbf{v}_1 - \mathbf{v}_2) \cdot (\mathbf{a}_1 - \mathbf{a}_2) < 0.$

The ship with velocity $\mathbf{v}_1$ will sight the second ship if the distance between the ships is, at any time, less than or equal to d, where $d < |\mathbf{a}_1 - \mathbf{a}_2|$. Show that sighting will occur provided

$$d^2 \geqslant |\mathbf{a}_1 - \mathbf{a}_2|^2 - \frac{[(\mathbf{v}_1 - \mathbf{v}_2) \cdot (\mathbf{a}_1 - \mathbf{a}_2)]^2}{|\mathbf{v}_1 - \mathbf{v}_2|^2}.$$ If this condition is just

satisfied, find the time after noon when sighting occurs.   [MEI]

8.5 Vector product

In 1820 Ampère discovered that when a wire carrying a current is placed in a magnetic field, the wire is subject to a force which is perpendicular and proportional to both the magnetic field and the current. In many other applications of vectors we need to find a vector which is perpendicular to two given vectors.

The *vector product* of $\mathbf{a}$ and $\mathbf{b}$ is written $\mathbf{a} \times \mathbf{b}$ and is defined by $\mathbf{a} \times \mathbf{b} = a\,b \sin\theta\,\hat{\mathbf{n}}$ where θ is the angle between $\mathbf{a}$ and $\mathbf{b}$ (see §8.1) and $\hat{\mathbf{n}}$ is a unit vector perpendicular to both $\mathbf{a}$ and $\mathbf{b}$, and such that $\mathbf{a}, \mathbf{b}, \hat{\mathbf{n}}$ (in that order) form a right-handed triad (see §3.2). If $\mathbf{a}$ or $\mathbf{b}$ is $\mathbf{0}$, or if $\mathbf{a}, \mathbf{b}$ are parallel, $\mathbf{a} \times \mathbf{b} = \mathbf{0}$. The vector product is sometimes written as $\mathbf{a} \wedge \mathbf{b}$.

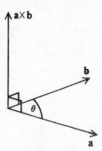

Qu. 1.   Show that $\mathbf{a} \times \mathbf{b} = -\mathbf{b} \times \mathbf{a}$; we say that the vector product is *anti-commutative*.

Qu. 2.   What is $\mathbf{a} \times \mathbf{a}$?

Qu. 3.   Is the vector product associative? Justify your answer.

Qu. 4.   Show that $(m\mathbf{a}) \times (n\mathbf{b}) = mn(\mathbf{a} \times \mathbf{b})$.

Qu. 5.   Show that the parallelogram $OACB$ has area $|\mathbf{a} \times \mathbf{b}|$.

Qu. 6.   Find an expression for the area of triangle PQR in terms of $\mathbf{p}, \mathbf{q}, \mathbf{r}$.

If $\mathbf{a} \neq \mathbf{0}$, the following steps describe a geometrical method of finding $\mathbf{a} \times \mathbf{b}$.

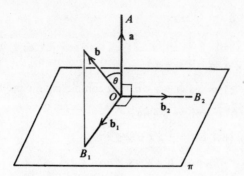

(i)   Choose any point O, and take $\overrightarrow{OA}$ to represent **a**; let π be the plane
      through O perpendicular to $\overrightarrow{OA}$.

(ii) Let $\mathbf{b}_1$ be the projection of **b** on plane π (see §8.1). Then
      $|\mathbf{b}_1| = b \sin\theta$ where θ is the angle between **a** and **b**.

(iii) Take $\overrightarrow{OB_1}$ to represent $\mathbf{b}_1$; B_1 is in plane π. Let $\overrightarrow{OB_2}$ be the image of $\overrightarrow{OB_1}$
      under a rotation of plane π about O, $90°$ anti-clockwise as viewed from
      A. Let $\mathbf{b}_2$ be the unique vector represented by $\overrightarrow{OB_2}$. Then $|\mathbf{b}_2| = b \sin\theta$.

(iv) Vector $\mathbf{b}_2$ is perpendicular to both **a** and **b**, and **a**, **b**, $\mathbf{b}_2$ in that order
      form a right-handed triad, so that $\mathbf{a} \times \mathbf{b} = a\mathbf{b}_2$.

We now prove the distributive property

$$\mathbf{a} \times (\mathbf{b} + \mathbf{c}) = \mathbf{a} \times \mathbf{b} + \mathbf{a} \times \mathbf{c}. \tag{A}$$

If $\mathbf{a} = \mathbf{0}$, the result is clearly true as both sides reduce to **0**. If $\mathbf{a} \neq \mathbf{0}$ we define
plane π as in (i) above. In §8.1 we showed that the projection of the sum of
two vectors on plane π is equal to the sum of the projections of the two
vectors on π. Thus if $\mathbf{b} + \mathbf{c} = \mathbf{d}$, and $\mathbf{b}_1, \mathbf{c}_1, \mathbf{d}_1$ are the projections of **b**, **c**, **d**
respectively on plane π, then $\mathbf{b}_1 + \mathbf{c}_1 = \mathbf{d}_1$. Rotate plane π about O, $90°$ anticlockwise as viewed from A, to obtain $\mathbf{b}_2, \mathbf{c}_2, \mathbf{d}_2$ as in (iii) above.

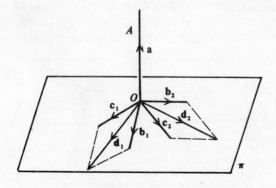

Then $d_2 = b_2 + c_2$

$\Rightarrow$ $a\,d_2 = a\,b_2 + a\,c_2$

$\Rightarrow$ $a \times d = a \times b + a \times c$

$\Rightarrow a \times (b + c) = a \times b + a \times c$, thus completing the proof.

Multiplying (A) by -1 we obtain

$$-a \times (b + c) = -a \times b - a \times c$$

$\Rightarrow$ $(b + c) \times a = b \times a + c \times a,$

an alternative form of the distributive property.

Qu. 7.   Find $(a - b) \times (a + b)$.

Qu. 8.   Verify that $i \times i = j \times j = k \times k = 0$, and that $i \times j = k$,
$j \times k = i, k \times i = j, j \times i = -k, k \times j = -i, i \times k = -j$.

Qu. 9.   Use the distributive properties and the result of Qus. 4 and 8 to
show that

$$a = \begin{pmatrix} a_1 \\ a_2 \\ a_3 \end{pmatrix}, b = \begin{pmatrix} b_1 \\ b_2 \\ b_3 \end{pmatrix} \Rightarrow a \times b = \begin{pmatrix} a_2 b_3 - a_3 b_2 \\ a_3 b_1 - a_1 b_3 \\ a_1 b_2 - a_2 b_1 \end{pmatrix}.$$

The result of Qu. 9 which enables us to express a vector product in
terms of its components may be rewritten using 2×2 determinants:

$$a \times b = \begin{vmatrix} a_2 & a_3 \\ b_2 & b_3 \end{vmatrix} i - \begin{vmatrix} a_1 & a_3 \\ b_1 & b_3 \end{vmatrix} j + \begin{vmatrix} a_1 & a_2 \\ b_1 & b_2 \end{vmatrix} k.$$

However, the result is best remembered as a 3×3 determinant (see §9.3).

Consider the parallelepiped *OBDCAEFG*. The volume V of the parallelepiped is the product of the base area and the height. If we take parallelogram
OBDC (with area $|b \times c|$) as the base, the height is OP, where P is the foot of
the perpendicular from O to plane *AEFG*. If angle AOP is θ, $AP = |a| \cos \theta$
and the volume V is $|a| |b \times c| \cos \theta$. If a, b, c is a right-handed triad (as
illustrated), $b \times c$ has the same direction and sense as **OP**, and $V = a \cdot (b \times c)$.

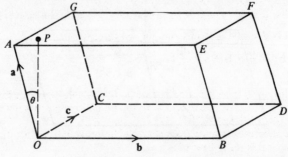

But if $\mathbf{a}, \mathbf{b}, \mathbf{c}$ is left-handed, $\mathbf{b} \times \mathbf{c}$ and $\mathbf{OP}$ are parallel with opposite sense, so that $V = -\mathbf{a} . (\mathbf{b} \times \mathbf{c})$.

As the volume of the parallelepiped is independent of which parallelogram is taken as the base,

$$\mathbf{a} . (\mathbf{b} \times \mathbf{c}) = \mathbf{b} . (\mathbf{c} \times \mathbf{a}) = \mathbf{c} . (\mathbf{a} \times \mathbf{b})$$
$$= -\mathbf{a} . (\mathbf{c} \times \mathbf{b}) = -\mathbf{b} . (\mathbf{a} \times \mathbf{c}) = -\mathbf{c} . (\mathbf{b} \times \mathbf{a}).$$

The scalar $\mathbf{a} . (\mathbf{b} \times \mathbf{c})$ is called the *scalar triple product* of $\mathbf{a}, \mathbf{b}, \mathbf{c}$. Cyclic interchange of the letters does not affect the value of the scalar triple product, though non-cyclic interchange of the letters multiplies the product by -1.

Qu. 10. Show that the volume of the tetrahedron $OABC$ is $\frac{1}{6}|\mathbf{a} . (\mathbf{b} \times \mathbf{c})|$.

Qu. 11. Show that $\mathbf{a} . (\mathbf{b} \times \mathbf{c}) = 0 \Leftrightarrow O, A, B, C$ are coplanar.

Qu. 12. Simplify $\mathbf{p} . (\mathbf{p} \times \mathbf{q})$.

Exercise 8E

1. Let $\mathbf{a} = 3\mathbf{i} + 2\mathbf{j} + 5\mathbf{k}$, $\mathbf{b} = 4\mathbf{i} + 3\mathbf{j} + 2\mathbf{k}$, $\mathbf{c} = 2\mathbf{i} + \mathbf{j} + 10\mathbf{k}$. Calculate
   (i) $\mathbf{a} \times \mathbf{b}$ (ii) $(\mathbf{a} \times \mathbf{b}) . \mathbf{c}$.
   Without doing further calculations find $\mathbf{b} . (\mathbf{a} \times \mathbf{c})$.

2. Let θ be the angle between $\mathbf{a}$ and $\mathbf{b}$ where $\mathbf{a} = 2\mathbf{i} + \mathbf{j} + 2\mathbf{k}$,
   $\mathbf{b} = 3\mathbf{i} + 2\mathbf{j} + 6\mathbf{k}$. Use $\mathbf{a} \times \mathbf{b}$ to find $\sin \theta$; use $\mathbf{a} . \mathbf{b}$ to find $\cos \theta$. Check that $\sin^2 \theta + \cos^2 \theta = 1$.

3. Find a unit vector perpendicular to both $\mathbf{a} = 4\mathbf{i} - \mathbf{k}$ and $\mathbf{b} = 4\mathbf{i} + 3\mathbf{j} - 2\mathbf{k}$.

4. Find the area of triangle ABC where $A \equiv (4, -8, -13), B \equiv (5, -2, -3)$, $C \equiv (5, 4, 10)$.

5. Find the volume of tetrahedron $PQRS$ given that P, Q, R, S have coordinates $(4, 1, -2), (3, -2, -1), (1, -6, -4), (2, -3, 1)$ respectively.

6. Show that $W(4, 1, 1), X(3, -3, -2), Y(2, 0, -3), Z(5, -2, 2)$ are coplanar.

7. Prove that $\mathbf{a} + \mathbf{b} + \mathbf{c} = 0 \Rightarrow \mathbf{a} \times \mathbf{b} = \mathbf{b} \times \mathbf{c} = \mathbf{c} \times \mathbf{a}$. Is the converse true?

8. Expand $(\mathbf{a} + \delta\mathbf{a}) \times (\mathbf{b} + \delta\mathbf{b})$ and deduce that
   $$\frac{d}{dt}(\mathbf{a} \times \mathbf{b}) = \frac{d\mathbf{a}}{dt} \times \mathbf{b} + \mathbf{a} \times \frac{d\mathbf{b}}{dt}.$$

9. The points A, B have position vectors $\mathbf{a}, \mathbf{b}$; $\hat{\mathbf{n}}$ is a unit vector. Prove that
   (i) the shortest distance from A to the plane through B with normal vector $\hat{\mathbf{n}}$ is $|(\mathbf{a} - \mathbf{b}) . \hat{\mathbf{n}}|$
   (ii) the shortest distance from A to the line through B with direction vector $\hat{\mathbf{n}}$ is $|(\mathbf{a} - \mathbf{b}) \times \hat{\mathbf{n}}|$.

10. Find an expression in terms of **a**, **b**, **d**, **e** for the shortest distance between the skew lines $\mathbf{r} = \mathbf{a} + \lambda\mathbf{d}$ and $\mathbf{r} = \mathbf{b} + \lambda\mathbf{e}$.

11. A rigid body is rotating with angular speed $\dot{\theta}$ about a fixed axis l. Let O be any point on l, and π the plane through O perpendicular to l. Let A be a point on l such that points of plane π rotate anticlockwise about O when π is viewed from A. The *angular velocity* of the rigid body is the vector $\boldsymbol{\omega} = \dot{\theta}\,\hat{\mathbf{a}}$. (It may be proved that angular velocity is a vector quantity.) Show that the velocity of the point R is $\boldsymbol{\omega} \times \mathbf{r}$.

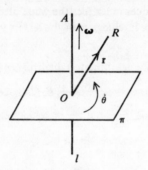

The drum of a cement mixer has angular velocity $2\mathbf{i} - \mathbf{j} + 2\mathbf{k}$ rad s^{-1} about an axis through $(4, 3, 2)$. Find how many revolutions the drum makes per second, and the speed of the point on the drum with coordinates $(5, 2, 3)$, where distances are measured in metres.

12. The position vectors of the non-collinear points A, B, C relative to a given origin O are denoted by **a**, **b**, **c** respectively. Prove that $(\mathbf{a} \times \mathbf{b}) \cdot \mathbf{a} = 0$, and show that the vector $\mathbf{b} \times \mathbf{c} + \mathbf{c} \times \mathbf{a} + \mathbf{a} \times \mathbf{b}$ is perpendicular to the plane ABC. Obtain the Cartesian equation of the plane containing the points $A(1, 0, -1)$, $B(2, 1, 0)$, $C(2, -2, -1)$ and find the coordinates of the point obtained by reflecting the point $P(1, 1, 4)$ in this plane.   [O & C]

13. A triangle ABC is moving in its own plane such that the velocity of its centroid is constant and represented by vector **v**, and is also rotating with constant angular velocity represented by vector $\boldsymbol{\omega}$ normal to the plane of the triangle. Show that if the velocity vectors of A, B, C are $\mathbf{v}_a, \mathbf{v}_b, \mathbf{v}_c$, then $\mathbf{v}_a + \mathbf{v}_b + \mathbf{v}_c = 3\mathbf{v}$ and $\mathbf{v}_a = \mathbf{v} + \frac{1}{3}\boldsymbol{\omega} \times (2\mathbf{a} - \mathbf{b} - \mathbf{c})$, where **a**, **b** and **c** are the position vectors of A, B and C referred to a point O fixed in the plane.   [O & C]

14. The *moment of force* **F** *about the point* A is $\mathbf{p} \times \mathbf{F}$ where **p** is the position vector (relative to A) of a point P on the line of action of **F**.
    (i) Prove that the moment of **F** about A is independent of the position of P on the line of action of **F**.
    (ii) The vector sum of a set of n forces is **0**, and the vector sum of the moments of these n forces about the point O is also **0**. Prove that the vector sum of the moments of the n forces about any other point is **0**.

(iii) The sums of the moments of n coplanar forces about A, B and C are zero. If A, B, C are not collinear, prove that the sum of the n forces is zero.

15. The *moment of force* **F** *about the line* l is $(\mathbf{p} \times \mathbf{F}) . \hat{\mathbf{d}}$ where **p** is the position vector of a point P on the line of action of **F** relative to a point A on l, and $\hat{\mathbf{d}}$ is a unit vector in the direction of l.
   (i) Prove that the moment of **F** about l does not depend on the particular choice of the points P and A.
   (ii) A force of magnitude 50 units acts along a diagonal of a face of a cube whose edges have length 10 units. Calculate the moment of the force about one of the diagonals of the cube which it does not intersect.

16. Prove that $\mathbf{a} \times (\mathbf{b} \times \mathbf{c}) = (\mathbf{a} . \mathbf{c})\mathbf{b} - (\mathbf{a} . \mathbf{b})\mathbf{c}$.
    Deduce that $\mathbf{a} \times (\mathbf{b} \times \mathbf{c}) + \mathbf{b} \times (\mathbf{c} \times \mathbf{a}) + \mathbf{c} \times (\mathbf{a} \times \mathbf{b}) = \mathbf{0}$.

17. (i) Points L, M, N are taken on the sides BC, CA, AB of triangle ABC such that $\dfrac{BL}{LC} = \lambda, \dfrac{CM}{MA} = \mu, \dfrac{AN}{NB} = \nu$, the sense of each line segment being taken into account. Prove that

$$\triangle LMN = \frac{1 + \lambda\mu\nu}{(1 + \lambda)(1 + \mu)(1 + \nu)} . \triangle ABC$$

   (ii) Deduce the *theorem of Menelaus* (c. 100 A.D.):

$$LMN \text{ is a straight line } \Leftrightarrow \lambda\mu\nu = -1.$$

   (iii) With the same notation, prove the *theorem of Ceva* (1678):

$$AL, BM, CN \text{ are concurrent } \Leftrightarrow \lambda\mu\nu = +1.$$

[Hint for $\Rightarrow$: suppose that AL, BM, CN concur at P, and apply the theorem of Menelaus to $\triangle ABL$ with transversal NPC and to $\triangle ALC$ with transversal MPB.]

9 Matrices

9.1 Linear transformations

In Vol. 1, §§1.5 and 1.6 we saw how the algebra of 2 × 2 matrices arises naturally from certain plane transformations; we now consider extensions of this. It is convenient to use $\mathbf{R}$ for the set of real numbers, $\mathbf{R}^2$ for the set of ordered pairs of real numbers, i.e. $\mathbf{R}^2 = \{(x,y) : x,y \in \mathbf{R}\}$, and $\mathbf{R}^3$ for the set of ordered triples (x,y,z) of real numbers. These sets $\mathbf{R}$, $\mathbf{R}^2$, $\mathbf{R}^3$ will be interpreted as the sets of points of a line, a plane, or three-dimensional space respectively, referred to fixed coordinate axes. Thus transformations of a plane map $\mathbf{R}^2$ into $\mathbf{R}^2$.

The following example deals with two transformations of three-dimensional space.

Example: Find the matrices of the following transformations of $\mathbf{R}^3$ into $\mathbf{R}^3$.
(i) Orthogonal projection onto the plane π with equation
$x + 2y - 3z = 0$.
(ii) Reflection in π.

Solution (i): Let $P(X, Y, Z)$ be an arbitrary point of $\mathbf{R}^3$. The image of P under orthogonal projection onto π is the foot of the perpendicular from P to π; let this be $P'(X', Y', Z')$.

Then $\mathbf{PP'}$ is parallel to the normal vector $\begin{pmatrix} 1 \\ 2 \\ -3 \end{pmatrix}$, so that

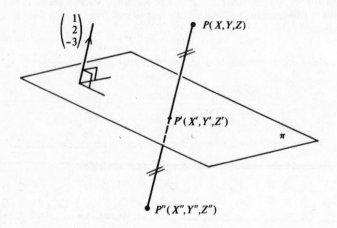

$$\begin{pmatrix} X' \\ Y' \\ Z' \end{pmatrix} = \begin{pmatrix} X \\ Y \\ Z \end{pmatrix} + \lambda \begin{pmatrix} 1 \\ 2 \\ -3 \end{pmatrix} = \begin{pmatrix} X + \lambda \\ Y + 2\lambda \\ Z - 3\lambda \end{pmatrix}.$$

Also P' lies on π, so that $X' + 2Y' - 3Z' = 0$,
i.e. $X + \lambda + 2(Y + 2\lambda) - 3(Z - 3\lambda) = 0$.

Therefore $\lambda = -\dfrac{X + 2Y - 3Z}{14}$,

and $\quad X' = X + \lambda \quad = \tfrac{1}{14}(13X - 2Y + 3Z)$

$\quad\quad\quad Y' = Y + 2\lambda = \tfrac{1}{7}(-X + 5Y + 3Z)$

$\quad\quad\quad Z' = Z - 3\lambda = \tfrac{1}{14}(3X + 6Y + 5Z).$

The matrix of this transformation is the array of coefficients
in the simultaneous equations giving X', Y', Z' in terms of
X, Y, Z:

$$\begin{pmatrix} \tfrac{13}{14} & -\tfrac{1}{7} & \tfrac{3}{14} \\ -\tfrac{1}{7} & \tfrac{5}{7} & \tfrac{3}{7} \\ \tfrac{3}{14} & \tfrac{3}{7} & \tfrac{5}{14} \end{pmatrix}$$

(*ii*):   To reach the reflection $P''(X'', Y'', Z'')$ of P in π we move
from P to P' as in (i) and continue along PP' to P'' so that

$\mathbf{P'P''} = \mathbf{PP'}$. Thus $\begin{pmatrix} X'' \\ Y'' \\ Z'' \end{pmatrix} = \begin{pmatrix} X \\ Y \\ Z \end{pmatrix} + 2\lambda \begin{pmatrix} 1 \\ 2 \\ -3 \end{pmatrix}$, where λ is as in (i).

This gives $\quad X'' = \tfrac{1}{7}(6X - 2Y + 3Z)$

$\quad\quad\quad\quad Y'' = \tfrac{1}{7}(-2X + 3Y + 6Z)$

$\quad\quad\quad\quad Z'' = \tfrac{1}{7}(3X + 6Y + 10Z).$

The matrix is $\quad\quad \begin{pmatrix} \tfrac{6}{7} & -\tfrac{2}{7} & \tfrac{3}{7} \\ -\tfrac{2}{7} & \tfrac{3}{7} & \tfrac{6}{7} \\ \tfrac{3}{7} & \tfrac{6}{7} & \tfrac{10}{7} \end{pmatrix}.$ $\square$

Qu. 1.   Write down the matrices of the following transformations of $\mathbb{R}^3$
into $\mathbb{R}^3$:

   (i) rotation $180°$ about the z-axis

   (ii) reflection in the plane $y = z$

   (iii) englargement, centre O, scale factor $-\tfrac{1}{3}$

   (iv) one-way stretch from O in the z-direction, scale factor 2.

Qu. 2.   Describe the transformations with these matrices:

(i) $\begin{pmatrix} 2 & 0 & 0 \\ 0 & 2 & 0 \\ 0 & 0 & 2 \end{pmatrix}$ (ii) $\begin{pmatrix} 1 & 0 & 0 \\ 0 & -1 & 0 \\ 0 & 0 & -1 \end{pmatrix}$ (iii) $\begin{pmatrix} 0 & 1 & 0 \\ 1 & 0 & 0 \\ 0 & 0 & 1 \end{pmatrix}$ (iv) $\begin{pmatrix} 1 & 0 & 3 \\ 0 & 1 & 0 \\ 0 & 0 & 1 \end{pmatrix}$.

The domain and codomain of a transformation need not be the same set.
For example, when we draw the plane of a building with scale 1 : 100, we

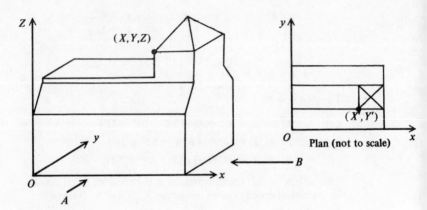

start with an object in three dimensional space and finish with a plane drawing, so this is a mapping from $\mathbb{R}^3$ to $\mathbb{R}^2$.

If the $z = 0$ plane in $\mathbb{R}^3$ is horizontal, then under this transformation $(X, Y, Z) \mapsto (X', Y')$ where $X' = \frac{1}{100} X$, $Y' = \frac{1}{100} Y$. The matrix of this transformation is $\begin{pmatrix} \frac{1}{100} & 0 & 0 \\ 0 & \frac{1}{100} & 0 \end{pmatrix}$. A similar transformation is the basis of the geographer's orthographic projection, sometimes used for mapping the polar regions of the world.

Qu. 3. Sketch the elevations of the building as seen from A and B. Give the matrices of the transformations which produce these elevations with scale $1 : 100$.

Notice how the shape of the matrix depends on the dimensions of the domain and codomain. Thus when mapping from $\mathbb{R}^3$ to $\mathbb{R}^2$, we use a 2×3 matrix (i.e. 2 rows, 3 columns), and similarly a $m \times n$ matrix defines a transformation from $\mathbb{R}^n$ to $\mathbb{R}^m$, where for the moment the values of m and n are restricted to 1, 2 or 3.

Qu. 4. Let T be the transformation from $\mathbb{R}^2$ to $\mathbb{R}^3$ with matrix $\begin{pmatrix} 2 & 3 \\ 1 & -1 \\ 4 & 2 \end{pmatrix}$.

Let $\mathbf{p} = \begin{pmatrix} x_1 \\ y_1 \end{pmatrix}$, $\mathbf{q} = \begin{pmatrix} x_2 \\ y_2 \end{pmatrix}$ and $\mathbf{r} = 3\mathbf{p} + 2\mathbf{q}$.

Find in terms of x_1, y_1, x_2, y_2 (i) $\mathbf{r}$ (ii) $T(\mathbf{p})$ (iii) $T(\mathbf{q})$ (iv) $T(\mathbf{r})$.

Show that $T(3\mathbf{p} + 2\mathbf{q}) = 3T(\mathbf{p}) + 2T(\mathbf{q})$.

Qu. 5. Repeat Qu. 4 (i)–(iv) when the matrix is $\begin{pmatrix} a & d \\ b & e \\ c & f \end{pmatrix}$ and $\mathbf{r} = \lambda\mathbf{p} + \mu\mathbf{q}$,

where λ, μ are arbitrary numbers.

Show that $T(\lambda\mathbf{p} + \mu\mathbf{q}) = \lambda T(\mathbf{p}) + \mu T(\mathbf{q})$.

The property proved in Qu. 5 is common to many transformations, and leads to the following definition.

> A *linear transformation* T is a transformation such that
> $T(\lambda \mathbf{p} + \mu \mathbf{q}) = \lambda T(\mathbf{p}) + \mu T(\mathbf{q})$ for all $\mathbf{p}$, $\mathbf{q}$ in the domain
> and all numbers λ, μ.

Qu. 6. By writing $\mathbf{p}$ as $\mathbf{p} + \mathbf{0}$, prove that for every linear transformation $T(\mathbf{0}) = \mathbf{0}$. Deduce that non-zero translations are not linear transformations.

Qu. 7. Which of the following are linear transformations from $\mathbb{R}^2$ to $\mathbb{R}^2$?
   (i) reflection in $y = x$ (ii) reflection in $y = x + 1$
  (iii) rotation $-90°$ about $(0, 4)$ (iv) half turn about $(0, 0)$
   (v) enlargement, scale factor -2, centre $(3, 5)$.

Working similar to that of Qu. 5, which is typical of the general case, shows that all transformations of the form $\mathbf{p} \mapsto \mathbf{Mp}$, where $\mathbf{M}$ is a matrix, are linear transformations. Surprisingly, the converse is also true. For suppose that T is a linear transformation of $\mathbb{R}^3$ such that $T(\mathbf{i}) = \mathbf{a}$, $T(\mathbf{j}) = \mathbf{b}$, $T(\mathbf{k}) = \mathbf{c}$. Then

$$
\begin{aligned}
T([x\mathbf{i} + y\mathbf{j}] + z\mathbf{k}) &= T(x\mathbf{i} + y\mathbf{j}) + zT(\mathbf{k}) \quad \text{by definition of linear} \\
&\qquad\qquad\qquad\qquad\qquad\qquad\text{transformation} \\
&= xT(\mathbf{i}) + yT(\mathbf{j}) + zT(\mathbf{k}) \\
&= x\mathbf{a} + y\mathbf{b} + z\mathbf{c} \\
&= \mathbf{M}\begin{pmatrix} x \\ y \\ z \end{pmatrix},
\end{aligned}
$$

where $\mathbf{M} = (\mathbf{a}\ \mathbf{b}\ \mathbf{c})$ is the matrix which has the components of column vectors $\mathbf{a}, \mathbf{b}, \mathbf{c}$ as its first, second, third columns respectively, the components remaining in their correct order. A similar proof applies if the domain of T is $\mathbb{R}$ or $\mathbb{R}^2$. Therefore

> T is a linear transformation $\Leftrightarrow$ T is a transformation such that
> $\mathbf{p} \mapsto \mathbf{Mp}$, where $\mathbf{M}$ is a matrix.

Two transformations T and U are said to be equal if they have the same domain and $T(\mathbf{p}) = U(\mathbf{p})$ for all $\mathbf{p}$ in this domain. If $\mathbf{A}$ and $\mathbf{B}$ are the matrices of linear transformations T and U, then we say that $\mathbf{A} = \mathbf{B}$ if and only if $T = U$. In this case T and U have the same domain and the same range, so that $\mathbf{A}$ and $\mathbf{B}$ must have the same order.

Also $T = U \Rightarrow T(\mathbf{i}) = U(\mathbf{i}) \Rightarrow \mathbf{Ai} = \mathbf{Bi}$

                                   $\Rightarrow$ **A** and **B** have the same first column.

A similar argument applies to the remaining columns. Thus if a_{ij} is the element in the ith row and jth column of **A**, with b_{ij} defined similarly for **B**, then

A = **B** $\Leftrightarrow$ **A** and **B** have the same order and $a_{ij} = b_{ij}$ for all possible i,j.

The sum $T + U$ of two transformations T and U with the same domain is defined by $(T + U)(\mathbf{p}) = T(\mathbf{p}) + U(\mathbf{p})$ for all $\mathbf{p}$ in this domain.

Qu. 8. Prove that if T and U are linear transformations, then so is $T + U$.

Qu. 9. If **A**, **B** are the matrices of linear transformations T, U, then **A** + **B** is defined to be the matrix of $T + U$.
   (i) Explain why **A** + **B** cannot exist unless **A** and **B** have the same order.
   (ii) Prove that the element in the ith row, jth column of **A** + **B** is $a_{ij} + b_{ij}$.

Qu. 10. Suggest suitable definitions for the transformation λT and the matrix $\lambda\mathbf{A}$, where λ is a number.

Qu. 11. Prove the following matrix properties:
   (i) **A** + **B** = **B** + **A**   (ii) (**A** + **B**) + **C** = **A** + (**B** + **C**)
   (iii) λ(**A** + **B**) = λ**A** + λ**B**   (iv) $(\lambda + \mu)$**A** = λ**A** + μ**A**.
   Comment on the use of + in (iv).

Multiplication of matrices seems more complicated, but in fact it comes immediately from successive linear transformations. For example, let S be the transformation from $\mathbb{R}^2$ to $\mathbb{R}^2$ with 2×2 matrix $\begin{pmatrix} a & c \\ b & d \end{pmatrix}$, and let T be the transformation from $\mathbb{R}^2$ to $\mathbb{R}^3$ with 3×2 matrix $\begin{pmatrix} e & h \\ f & i \\ g & j \end{pmatrix}$. Suppose that

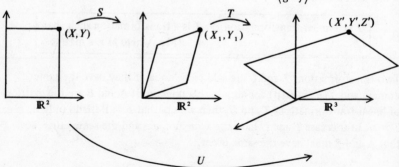

S maps (X, Y) to (X_1, Y_1) and that T maps (X_1, Y_1) to (X', Y', Z'). Since

$$\begin{pmatrix} X' \\ Y' \\ Z' \end{pmatrix} = \begin{pmatrix} e & h \\ f & i \\ g & j \end{pmatrix} \begin{pmatrix} X_1 \\ Y_1 \end{pmatrix} \text{ and } \begin{pmatrix} X_1 \\ Y_1 \end{pmatrix} = \begin{pmatrix} a & c \\ b & d \end{pmatrix} \begin{pmatrix} X \\ Y \end{pmatrix} \text{ we naturally write}$$

$$\begin{pmatrix} X' \\ Y' \\ Z' \end{pmatrix} = \begin{pmatrix} e & h \\ f & i \\ g & j \end{pmatrix} \begin{pmatrix} a & c \\ b & d \end{pmatrix} \begin{pmatrix} X \\ Y \end{pmatrix}. \quad \text{But by substitution}$$

$$X' = e(aX + cY) + h(bX + dY) = (ea + hb)X + (ec + hd)Y$$
$$Y' = f(aX + cY) + i(bX + dY) = (fa + ib)X + (fc + id)Y$$
$$Z' = g(aX + cY) + j(bX + dY) = (ga + jb)X + (gc + jd)Y$$

i.e.
$$\begin{pmatrix} X' \\ Y' \\ Z' \end{pmatrix} = \begin{pmatrix} ea + hb & ec + hd \\ fa + ib & fc + id \\ ga + jb & gc + jd \end{pmatrix} \begin{pmatrix} X \\ Y \end{pmatrix}.$$

So the transformation U mapping (X, Y) directly to (X', Y', Z') is a linear transformation, whose matrix is said to be the *product* of the matrices

$\begin{pmatrix} e & h \\ f & i \\ g & j \end{pmatrix}$ and $\begin{pmatrix} a & c \\ b & d \end{pmatrix}$. Notice that the order of combination is significant; it is

impossible to have T followed by S, since the codomain of T is $\mathbb{R}^3$ but the domain of S is $\mathbb{R}^2$.

The corresponding general definition was given by Arthur Cayley in 1858. If **A** and **B** are matrices of orders $p \times q$ and $q \times r$ respectively, then

product **AB** is the $p \times r$ matrix which has $\sum\limits_{u=1}^{q} a_{iu} b_{uj}$ as the element in the

ith row and jth column. The product is not defined unless the orders of the matrices fit according to the 'domino pattern': $p \times q \; q \times r$.

If **C** is a $r \times s$ matrix, then we can form a linear transformation from $\mathbb{R}^s$ to $\mathbb{R}^p$ by applying successively the transformations with matrices **C**, **B**, **A**.

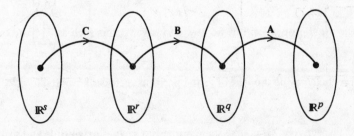

The same transformation can be performed by-passing $\mathbb{R}^q$; this corresponds to the matrix product $(\mathbf{AB})\mathbf{C}$.

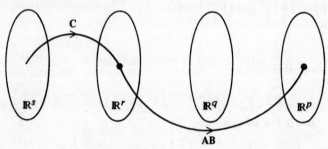

Alternatively we can by-pass $\mathbb{R}^r$, which means using the matrix product $\mathbf{A}(\mathbf{BC})$.

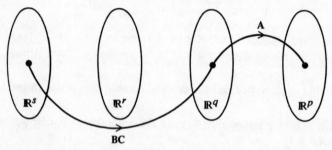

Since the overall transformation is the same in each case, it follows that $(\mathbf{AB})\mathbf{C} = \mathbf{A}(\mathbf{BC})$. Therefore matrix multiplication is associative.

Qu. 12. Prove that $\mathbf{L}(\mathbf{M} + \mathbf{N}) = \mathbf{LM} + \mathbf{LN}$ and $(\mathbf{P} + \mathbf{Q})\mathbf{R} = \mathbf{PR} + \mathbf{QR}$, provided that all the sums and products are defined.

We originally defined vectors as geometric entities, having magnitude, direction and sense, and then found that a vector can be represented by a $n \times 1$ matrix (or column vector), where $n = 2$ or 3. We have now reached the stage where it is useful to redefine vectors as $n \times 1$ matrices. Then when $n = 1, 2$ or 3 these new vectors may be represented by directed line segments, with the usual properties. But we now also have vectors such as $\begin{pmatrix} 3 \\ 1 \\ 2 \\ 5 \end{pmatrix}$ and $\begin{pmatrix} 4 \\ -2 \\ 0 \\ 6 \\ 1 \end{pmatrix}$, which may be thought of as the position vectors of 'points' in $\mathbb{R}^4$ and $\mathbb{R}^5$ respectively. Of course some properties such as magnitude and scalar product have to be redefined, and lose some of their obvious physical significance. However, we can now say that a 5 × 4 matrix represents a transformation from $\mathbb{R}^4$ to $\mathbb{R}^5$. With this interpretation, the transformation arguments of this section apply to matrices of all orders.

Exercise 9A

1. (i) By considering the images of **i** and **j** show that the matrix $\mathbf{P}(\theta)$ of the linear transformation of $\mathbb{R}^2$ which reflects each point in the line $y = x \tan \theta$ is
$$\mathbf{P}(\theta) = \begin{pmatrix} \cos 2\theta & \sin 2\theta \\ \sin 2\theta & -\cos 2\theta \end{pmatrix}.$$

   (ii) What linear transformation of $\mathbb{R}^2$ has the matrix
$$\mathbf{Q}(\phi) = \begin{pmatrix} \cos \phi & -\sin \phi \\ \sin \phi & \cos \phi \end{pmatrix}?$$

   (iii) Prove that $\mathbf{P}(\alpha)\,\mathbf{P}(\beta) = \mathbf{Q}(2\,[\alpha - \beta])$, and give a geometrical interpretation.

2. Find the equation of the axis of the rotation of $\mathbb{R}^3$ given by the matrix
$$\mathbf{M} = \begin{pmatrix} 0 & 0 & 1 \\ 1 & 0 & 0 \\ 0 & 1 & 0 \end{pmatrix}.$$ Find $\mathbf{M}^2$ and $\mathbf{M}^3$. What is the angle of rotation?

3. Prove from the definition that the following are not linear transformations:

   (i) $T : \begin{pmatrix} x \\ y \end{pmatrix} \mapsto \begin{pmatrix} x+1 \\ y \end{pmatrix}$ (ii) $T : \begin{pmatrix} x \\ y \end{pmatrix} \mapsto \begin{pmatrix} x \\ xy \end{pmatrix}$ (iii) $T : \begin{pmatrix} x \\ y \end{pmatrix} \mapsto \begin{pmatrix} |x| \\ y \end{pmatrix}$.

4. The function $f : x \mapsto ax + b$ is a linear function. The transformation $T : x \mapsto ax + b$ maps $\mathbb{R}$ into $\mathbb{R}$. Prove that T is linear if and only if $b = 0$.

5. Transformation T maps $\mathbb{R}^2$ into $\mathbb{R}^2$ as follows: $T(P)$ is the foot of the perpendicular from P to the line $y = 2x$. Express T in the form $\begin{pmatrix} X \\ Y \end{pmatrix} \mapsto \begin{pmatrix} X' \\ Y' \end{pmatrix}$. Is T linear?

6. A *shear* of $\mathbb{R}^3$ has a plane π of invariant points; each point P is mapped to P' such that $\mathbf{PP'} = k\mathbf{v}$ where $\mathbf{v}$ is a constant vector parallel to π, and $|k|$ is proportional to the distance from P to π; k is positive for points on one side of π and negative for points on the other side of π. Shear S maps $(0, 0, 1)$ to $(3, -1, 1)$ and has the plane $z = 0$ invariant.
   (a) Find the images of the following points: (i) $(4, -3, 0)$ (ii) $(7, 2, 1)$ (iii) $(-5, 2, 3)$ (iv) $(2, 3, -2)$ (v) $(0, 0, 20)$ (vi) (X, Y, Z).
   (b) Find the matrix representing the transformation and deduce that S is linear.

7. (i) Write down the vector equation of a straight line l. If T is a linear transformation, find the vector equation of $T(l)$ and deduce that $T(l)$ is either a straight line or a single point.

(ii) Let T be a transformation, and l a straight line. Show that
$T(l)$ is straight does not imply that T is linear.

8. Let T be a linear transformation, and A, B, C three points such that C
divides AB in the ratio $m : n$. Prove that $T(C)$ divides $T(AB)$ in the ratio
$m : n$.

9. If linear transformation T maps A, B, C to A', B', C' respectively, and G
is the centroid of triangle ABC, prove that $T(G)$ is the centroid of
triangle $A'B'C'$.

10. Let T be a linear transformation and π be a plane containing the point
P. Let $\mathbf{a}, \mathbf{b}$ be non-zero, non-parallel vectors such that $\mathbf{a}, \mathbf{b}$ are both
parallel to π. Show that an arbitrary point R on π has position vector
$\mathbf{r} = \mathbf{p} + \lambda\mathbf{a} + \mu\mathbf{b}$ where λ, μ are parameters. Show that $T(\pi)$ is a plane
or a line or a single point. Under what conditions is $T(\pi)$ a single point?

11. Let T be a transformation. Show that, for all vectors, $\mathbf{a}, \mathbf{b}$, and all
scalars λ:
$$\left.\begin{aligned} T(\mathbf{a} + \mathbf{b}) &= T(\mathbf{a}) + T(\mathbf{b}) \\ T(\lambda\mathbf{a}) &= \lambda\, T(\mathbf{a}) \end{aligned}\right\} \Leftrightarrow T \text{ is a linear transformation.}$$

12. The origin O and the points A, B, C, are distinct, non-coplanar points in
$\mathbb{R}^3$. Show that the position vector of an arbitrary point R may be
expressed uniquely as $\alpha\mathbf{a} + \beta\mathbf{b} + \gamma\mathbf{c}$ where α, β, γ, are scalars. If the linear
transformation T maps A, B, C, R, to A', B', C', R', find $\mathbf{r}'$ in terms of
$\mathbf{a}', \mathbf{b}', \mathbf{c}'$. Deduce that there is only one linear transformation of $\mathbb{R}^3$
which has three invariant points (distinct and not coplanar with the
origin), and describe it.

13. Let T be a linear transformation of the plane. It is found that n points
are invariant under T. State the minimum value of n and a condition on
these n invariant points which together imply that T is the identity
transformation. Justify your answer.

14. A is a 3×3 matrix whose elements are 0, 1 or -1 and each row and
each column of A contains exactly one non-zero element. Prove that
$\mathbf{A}^2, \mathbf{A}^3, \ldots, \mathbf{A}^n$ are all of the same form and deduce that $\mathbf{A}^h = \mathbf{I}$ for
some positive integer $h \leqslant 48$. Interpret the action of A on a vector
(x, y, z) geometrically.   [MEI]

15. Multiply up the equation $\begin{pmatrix} x' \\ y' \\ 1 \end{pmatrix} = \begin{pmatrix} a & b & c \\ d & e & f \\ 0 & 0 & 1 \end{pmatrix} \begin{pmatrix} x \\ y \\ 1 \end{pmatrix}$ to obtain two

equations which define a transformation of the plane in which the point
(x, y) is mapped to (x', y'). Find the appropriate values of a, b, c, d, e, f
if this transformation is
(i) a translation T mapping (x, y) to $(x + 3, y + 5)$
(ii) a quarter turn Q in the positive sense about the origin.
Hence, from the corresponding matrix equations, find the values of
a, b, c, d, e, f which give the transformation 'T followed by Q'. Use your
answer to verify that the point $(-4, -1)$ stays fixed under this compound
transformation.   [SMP]

16. The scalar product $\mathbf{x} \cdot \mathbf{y}$ of two vectors $\mathbf{x} = \begin{pmatrix} x_1 \\ x_2 \\ \vdots \\ x_n \end{pmatrix}$ and $\mathbf{y} = \begin{pmatrix} y_1 \\ y_2 \\ \vdots \\ y_n \end{pmatrix}$ in $\mathbb{R}^n$ is

defined by $\mathbf{x} \cdot \mathbf{y} = x_1 y_1 + x_2 y_2 + \ldots + x_n y_n$. Prove that
$\mathbf{x} \cdot (\mathbf{y} + \mathbf{z}) = \mathbf{x} \cdot \mathbf{y} + \mathbf{x} \cdot \mathbf{z}$.
Suggest suitable definitions for
 (i) the magnitude $|\mathbf{x}|$ of the vector $\mathbf{x}$
 (ii) the angle between $\mathbf{x}$ and $\mathbf{y}$.

Find the angle between $\begin{pmatrix} 6 \\ 9 \\ 5 \\ 1 \end{pmatrix}$ and $\begin{pmatrix} 3 \\ 5 \\ -7 \\ 2 \end{pmatrix}$, and find a vector perpendicular to both of these.

17. (P) The two variables x and y are observed to take n pairs of values (x_i, y_i), $i = 1, 2, \ldots, n$. If $\bar{x} = \Sigma x_i/n$, $\bar{y} = \Sigma y_i/n$, and if $u_i = x_i - \bar{x}$, $v_i = y_i - \bar{y}$ are the components of vectors $\mathbf{u}$, $\mathbf{v}$ in $\mathbb{R}^n$, prove that the product-moment correlation coefficient of x and y equals the cosine of the angle between $\mathbf{u}$ and $\mathbf{v}$.

9.2 Transpose

If the elements of a matrix $\mathbf{M}$ are rearranged to form a new matrix $\mathbf{N}$ in such a way that the first row of $\mathbf{M}$ becomes the first column of $\mathbf{N}$, the second row of $\mathbf{M}$ becomes the second column of $\mathbf{N}$, and so on, then the matrix $\mathbf{N}$ is known as the *transpose* of $\mathbf{M}$, and is denoted by $\mathbf{M}^\mathrm{T}$. Any matrix

can be transposed: the transposes of $\begin{pmatrix} 1 & 2 & 3 \\ 4 & 5 & 6 \end{pmatrix}$, $(7 \ \ 8)$, $\begin{pmatrix} 9 \\ 10 \\ 11 \\ 12 \end{pmatrix}$, $\begin{pmatrix} 13 & 15 \\ 14 & 16 \end{pmatrix}$ are

$\begin{pmatrix} 1 & 4 \\ 2 & 5 \\ 3 & 6 \end{pmatrix}$, $\begin{pmatrix} 7 \\ 8 \end{pmatrix}$, $(9 \ \ 10 \ \ 11 \ \ 12)$, $\begin{pmatrix} 13 & 14 \\ 15 & 16 \end{pmatrix}$ respectively.

The transpose of a matrix of order $p \times q$ is a matrix of order $q \times p$; in particular the transpose of a column matrix is a row matrix, and *vice versa*. Transposing the transpose restores the matrix to its original form, so that

$$(\mathbf{M}^\mathrm{T})^\mathrm{T} = \mathbf{M} \quad \text{for all matrices } \mathbf{M}.$$

A matrix $\mathbf{M}$ which is unchanged by transposition (that is $\mathbf{M}^\mathrm{T} = \mathbf{M}$) is called *symmetric*. A matrix $\mathbf{M}$ which is made its negative by transposition (that is $\mathbf{M}^\mathrm{T} = -\mathbf{M}$) is called *skew-symmetric*.

Qu. 1. Explain why all symmetric and skew-symmetric matrices are square.

Qu. 2. If matrix $\mathbf{A}$ is both symmetric and skew-symmetric, what can be said about $\mathbf{A}$?

Qu. 3. Prove that $(\lambda\mathbf{A} + \mu\mathbf{B})^T = \lambda\mathbf{A}^T + \mu\mathbf{B}^T$ where λ, μ are any scalars, and $\mathbf{A}$, $\mathbf{B}$ are any two matrices of the same order.

Qu. 4. If $\mathbf{A}$, $\mathbf{B}$ are symmetric matrices of the same order, prove that $\mathbf{A} + \mathbf{B}$ is symmetric. State and prove the corresponding result for skew-symmetric matrices.

Qu. 5. If $\mathbf{S}$ is a square matrix, prove that $\mathbf{S} + \mathbf{S}^T$ is symmetric and that $\mathbf{S} - \mathbf{S}^T$ is skew-symmetric. Hence show that every square matrix of real elements can be expressed uniquely as the sum of a symmetric and a skew-symmetric matrix.

If $\mathbf{y} = \begin{pmatrix} y_1 \\ y_2 \\ \vdots \\ y_n \end{pmatrix}$ and $\mathbf{z} = \begin{pmatrix} z_1 \\ z_2 \\ \vdots \\ z_n \end{pmatrix}$

then $\quad \mathbf{y}^T\mathbf{z} = (y_1 \ y_2 \ \dots \ y_n)\begin{pmatrix} z_1 \\ z_2 \\ \vdots \\ z_n \end{pmatrix}$

$$= (y_1 z_1 + y_2 z_2 + \dots + y_n z_n)$$

$$= (\mathbf{y} \cdot \mathbf{z}),$$

the 1×1 matrix whose single element is the number $\mathbf{y} \cdot \mathbf{z}$.

We can use this to find the transpose of the product $\mathbf{UV}$, where $\mathbf{U}$ is of order $p \times q$ and $\mathbf{V}$ is of order $q \times r$. Let the ith row of $\mathbf{U}$ be $\mathbf{u}_i^T$ $(1 \leqslant i \leqslant p)$, and let the jth column of $\mathbf{V}$ be $\mathbf{v}_j (1 \leqslant j \leqslant r)$, where $\mathbf{u}_i$ and $\mathbf{v}_j$ each have q elements. Then the element in the ith row, jth column of $\mathbf{UV}$ is $\mathbf{u}_i \cdot \mathbf{v}_j$. On transposition this becomes the element in the jth row, ith column of $(\mathbf{UV})^T$. But since $(\mathbf{u}_i \cdot \mathbf{v}_j) = (\mathbf{v}_j \cdot \mathbf{u}_i) = \mathbf{v}_j^T \mathbf{u}_i$, so $\mathbf{u}_i \cdot \mathbf{v}_j$ is also the element in the jth row, ith column of the product $\mathbf{V}^T \mathbf{U}^T$. Thus the elements of $(\mathbf{UV})^T$ and $\mathbf{V}^T \mathbf{U}^T$ in corresponding positions are equal, and so

$$\boxed{(\mathbf{UV})^T = \mathbf{V}^T\mathbf{U}^T.}$$

Qu. 6. Prove that $(\mathbf{ABC})^T = \mathbf{C}^T\mathbf{B}^T\mathbf{A}^T$.

Exercise 9B

1. Simplify $(\mathbf{A}\,\mathbf{B}^{\mathrm{T}}\mathbf{C}^{\mathrm{T}})^{\mathrm{T}}$.

2. Suppose $\mathbf{A}, \mathbf{B}$ are symmetric matrices of the same order. Prove that $\mathbf{AB}$ is symmetric if and only if $\mathbf{AB} = \mathbf{BA}$. Deduce a property of $\mathbf{A}^k$, where k is a positive integer.

3. Suppose $\mathbf{A}$ is a symmetric matrix, and that $\mathbf{B}$ is a skew-symmetric matrix of the same order.
   (a) Prove that $\mathbf{AB}$ is skew-symmetric if and only if $\mathbf{AB} = \mathbf{BA}$.
   (b) State and prove a similar property concerning $\mathbf{BC}$, where $\mathbf{C}$ is a skew-symmetric matrix of the same order as $\mathbf{B}$.
   (c) What can be said of $\mathbf{B}^k$, where k is a positive integer?

4. Prove that $\mathbf{MM}^{\mathrm{T}}$, $\mathbf{M}^{\mathrm{T}}\mathbf{M}$ exist and are symmetric whatever the order of matrix $\mathbf{M}$.

5. Let $\mathbf{A}$ be a square matrix of order $m \times m$, and $\mathbf{P}$ a $m \times n$ matrix. Prove that $\mathbf{P}^{\mathrm{T}}\mathbf{AP}$ is symmetric if $\mathbf{A}$ is symmetric, but skew-symmetric if $\mathbf{A}$ is skew-symmetric.

6. A *diagonal* matrix is a square matrix such that the only non-zero elements are on the leading diagonal (that is, the element in the ith row, jth column is zero whenever $i \neq j$). If $\mathbf{D}$ is a diagonal matrix and $\mathbf{P}$ is a square matrix, prove that $\mathbf{M} = \mathbf{P}^{\mathrm{T}}\mathbf{DP} \Rightarrow \mathbf{M}$ is symmetric.

7. If $\mathbf{r} = \begin{pmatrix} x \\ y \\ 1 \end{pmatrix}$ and $\mathbf{A} = \begin{pmatrix} 1 & 0 & g \\ 0 & 1 & f \\ g & f & c \end{pmatrix}$ show that $\mathbf{r}^{\mathrm{T}}\mathbf{A}\mathbf{r} = \mathbf{0}$ is the equation of the circle in $\mathbb{R}^2$ with centre $(-g, -f)$ and radius $\sqrt{(g^2 + f^2 - c)}$.

8. In the skew-symmetric matrix $\mathbf{S} = \begin{pmatrix} 0 & -n & m \\ n & 0 & -l \\ -m & l & 0 \end{pmatrix}$ the numbers l, m, n
   are such that $l^2 + m^2 + n^2 = 1$. Show that $\mathbf{S}^2$ is symmetric and that $\mathbf{S}^3 = -\mathbf{S}$.
   The matrix $\mathbf{A}(\theta)$ is defined by $\mathbf{A}(\theta) = \mathbf{I} + \mathbf{S}\sin\theta + \mathbf{S}^2(1 - \cos\theta)$, where $\mathbf{I}$ is the unit matrix of order 3. Show that $\mathbf{A}(\theta)\,\mathbf{A}(\phi) = \mathbf{A}(\theta + \phi)$, and deduce that $\{\mathbf{A}(\theta)\}^k = \mathbf{A}(k\theta)$ for any positive integer k.   [L]

9. (P) The two variables x and y are observed to take n pairs of values $(x_i, y_i), i = 1, 2, \ldots, n$.
   Let $\mathbf{X} = \begin{pmatrix} x_1 & 1 \\ x_2 & 1 \\ \vdots & \vdots \\ x_n & 1 \end{pmatrix}$ and $\mathbf{Y} = \begin{pmatrix} y_1 \\ y_2 \\ \vdots \\ y_n \end{pmatrix}$.
   Show that the equation of the regression line of y on x is $y = mx + c$, where $\mathbf{X}^{\mathrm{T}}\mathbf{X}\begin{pmatrix} m \\ c \end{pmatrix} = \mathbf{X}^{\mathrm{T}}\mathbf{Y}$.

9.3 Determinants

We now turn to the problem of 'undoing' a linear transformation. For the transformation $\mathbf{p} \mapsto \mathbf{Mp}$ with a given 3×3 matrix $\mathbf{M}$, it is easy to find the image $\mathbf{Mp}$ of any known object $\mathbf{p}$ by matrix multiplication. It is not so easy to find the object (or objects) $\mathbf{p}$ which will produce a given image.

Suppose that $\mathbf{M} = \begin{pmatrix} a_1 & b_1 & c_1 \\ a_2 & b_2 & c_2 \\ a_3 & b_3 & c_3 \end{pmatrix}$, $\mathbf{p} = \begin{pmatrix} x \\ y \\ z \end{pmatrix}$, and that the given image is

$\mathbf{d} = \begin{pmatrix} d_1 \\ d_2 \\ d_3 \end{pmatrix}$. Then we want to solve for $\mathbf{p}$ the matrix equation $\mathbf{Mp} = \mathbf{d}$, which

is equivalent to the set of three simultaneous equations

$$a_1 x + b_1 y + c_1 z = d_1$$
$$a_2 x + b_2 y + c_2 z = d_2$$
$$a_3 x + b_3 y + c_3 z = d_3$$

in the three unknowns x, y, z.

We can write these equations in terms of column vectors:

$$x \begin{pmatrix} a_1 \\ a_2 \\ a_3 \end{pmatrix} + y \begin{pmatrix} b_1 \\ b_2 \\ b_3 \end{pmatrix} + z \begin{pmatrix} c_1 \\ c_2 \\ c_3 \end{pmatrix} = \begin{pmatrix} d_1 \\ d_2 \\ d_3 \end{pmatrix}$$

or $$x\mathbf{a} + y\mathbf{b} + z\mathbf{c} = \mathbf{d} \qquad\qquad \text{(A)}$$

where $$\mathbf{a} = \begin{pmatrix} a_1 \\ a_2 \\ a_3 \end{pmatrix}, \ \mathbf{b} = \begin{pmatrix} b_1 \\ b_2 \\ b_3 \end{pmatrix}, \ \mathbf{c} = \begin{pmatrix} c_1 \\ c_2 \\ c_3 \end{pmatrix}.$$

The problem then is to find the components x, y, z of $\mathbf{d}$ relative to the base vectors $\mathbf{a}, \mathbf{b}, \mathbf{c}$ (see §3.2).

Equation (A) can be interpreted in terms of forces: we have to find forces $x\mathbf{a}, y\mathbf{b}, z\mathbf{c}$ having the directions of given vectors $\mathbf{a}, \mathbf{b}, \mathbf{c}$ and having resultant $\mathbf{d}$. One useful method in mechanics is to resolve perpendicular to an unknown force. Since resolving can be expressed in terms of scalar products, this suggests taking the scalar products of both sides of (A) with a vector perpendicular to both $\mathbf{b}$ and $\mathbf{c}$, thus eliminating y and z. The obvious vector to use for this purpose is $\mathbf{b} \times \mathbf{c}$; we then have

$$x\mathbf{a} + y\mathbf{b} + z\mathbf{c} = \mathbf{d}$$

$\Rightarrow \qquad x\mathbf{a}.(\mathbf{b} \times \mathbf{c}) + y\mathbf{b}.(\mathbf{b} \times \mathbf{c}) + z\mathbf{c}.(\mathbf{b} \times \mathbf{c}) = \mathbf{d}.(\mathbf{b} \times \mathbf{c})$

$\Rightarrow \qquad\qquad\qquad x\mathbf{a}.(\mathbf{b} \times \mathbf{c}) = \mathbf{d}.(\mathbf{b} \times \mathbf{c})$

since $\mathbf{b}.(\mathbf{b} \times \mathbf{c}) = \mathbf{c}.(\mathbf{b} \times \mathbf{c}) = 0$

$\Rightarrow$
$$x = \frac{\mathbf{d}.(\mathbf{b} \times \mathbf{c})}{\mathbf{a}.(\mathbf{b} \times \mathbf{c})}, \text{ provided that } \mathbf{a}.(\mathbf{b} \times \mathbf{c}) \neq 0.$$

Similarly   (A) $\Rightarrow$ $y = \dfrac{\mathbf{d}.(\mathbf{c} \times \mathbf{a})}{\mathbf{b}.(\mathbf{c} \times \mathbf{a})},$ provided that $\mathbf{b}.(\mathbf{c} \times \mathbf{a}) \neq 0$

and   (A) $\Rightarrow$ $z = \dfrac{\mathbf{d}.(\mathbf{a} \times \mathbf{b})}{\mathbf{c}.(\mathbf{a} \times \mathbf{b})},$ provided that $\mathbf{c}.(\mathbf{a} \times \mathbf{b}) \neq 0.$

Subject to these provisos these values of x, y, z give the only possible solution of (A).

Now we saw in §8.5 that $\mathbf{a}.(\mathbf{b} \times \mathbf{c})$ is the signed volume of the parallelepiped with edges $\mathbf{a}, \mathbf{b}, \mathbf{c}$, and that $\mathbf{a}.(\mathbf{b} \times \mathbf{c}) = \mathbf{b}.(\mathbf{c} \times \mathbf{a}) = \mathbf{c}.(\mathbf{a} \times \mathbf{b})$.
So these three provisos are equivalent; they are stating that the volume of the parallelepiped should not be zero, i.e. that the vectors $\mathbf{a}, \mathbf{b}, \mathbf{c}$ should not be coplanar. If $\mathbf{a}, \mathbf{b}, \mathbf{c}$ are not coplanar we know by the Unique Components Theorem of §3.2 that (A) has a unique solution, which can only be the one we have just found.

We introduce some new notation which emphasizes the symmetry of this unique solution condition. The expression $\Delta = \mathbf{a}.(\mathbf{b} \times \mathbf{c})$ is called the *determinant* of $\mathbf{M}$, and we write

$$\Delta = \det \mathbf{M} = \begin{vmatrix} a_1 & b_1 & c_1 \\ a_2 & b_2 & c_2 \\ a_3 & b_3 & c_3 \end{vmatrix} = |\mathbf{a} \quad \mathbf{b} \quad \mathbf{c}|.$$

Since $\mathbf{a} = \begin{pmatrix} a_1 \\ a_2 \\ a_3 \end{pmatrix}$ and $\mathbf{b} \times \mathbf{c} = \begin{pmatrix} b_2 c_3 - b_3 c_2 \\ b_3 c_1 - b_1 c_3 \\ b_1 c_2 - b_2 c_1 \end{pmatrix}$,

$$\Delta = a_1(b_2 c_3 - b_3 c_2) + a_2(b_3 c_1 - b_1 c_3) + a_3(b_1 c_2 - b_2 c_1).$$

This expression can be written in terms of 2×2 determinants:

$$\Delta = a_1 \begin{vmatrix} b_2 & c_2 \\ b_3 & c_3 \end{vmatrix} + a_2 \begin{vmatrix} b_3 & c_3 \\ b_1 & c_1 \end{vmatrix} + a_3 \begin{vmatrix} b_1 & c_1 \\ b_2 & c_2 \end{vmatrix}$$

$$= a_1 \begin{vmatrix} b_2 & c_2 \\ b_3 & c_3 \end{vmatrix} - a_2 \begin{vmatrix} b_1 & c_1 \\ b_3 & c_3 \end{vmatrix} + a_3 \begin{vmatrix} b_1 & c_1 \\ b_2 & c_2 \end{vmatrix}.$$

The reason for the change of sign in the second line is that each 2×2 determinant can then be obtained from $\det \mathbf{M}$ by deleting the row and column which contain the corresponding a. Thus $\begin{vmatrix} b_1 & c_1 \\ b_2 & c_2 \end{vmatrix}$ is the determinant of what is left when the row and column containing a_3 are deleted:
$\begin{vmatrix} \cancel{a_1} & b_1 & c_1 \\ \cancel{a_2} & b_2 & c_2 \\ \cancel{a_3} & \cancel{b_3} & \cancel{c_3} \end{vmatrix}$. For this reason $\begin{vmatrix} b_1 & c_1 \\ b_2 & c_2 \end{vmatrix}$ is called the *minor* of a_3. Similarly the

minor of b_1 is $\begin{vmatrix} a_2 & c_2 \\ a_3 & c_3 \end{vmatrix}$, obtained thus: $\begin{vmatrix} a_1 & b_1 & c_1 \\ a_2 & b_2 & c_2 \\ a_3 & b_3 & c_3 \end{vmatrix}$.

The $\mathbf{b} \cdot (\mathbf{c} \times \mathbf{a})$ version of Δ is $-b_1 \begin{vmatrix} a_2 & c_2 \\ a_3 & c_3 \end{vmatrix} + b_2 \begin{vmatrix} a_1 & c_1 \\ a_3 & c_3 \end{vmatrix} - b_3 \begin{vmatrix} a_1 & c_1 \\ a_2 & c_2 \end{vmatrix}$

and the $\mathbf{c} \cdot (\mathbf{a} \times \mathbf{b})$ version is $c_1 \begin{vmatrix} a_2 & b_2 \\ a_3 & b_3 \end{vmatrix} - c_2 \begin{vmatrix} a_1 & b_1 \\ a_3 & b_3 \end{vmatrix} + c_3 \begin{vmatrix} a_1 & b_1 \\ a_2 & b_2 \end{vmatrix}$.

The signs attached to the minors occur in the chessboard pattern

$$\begin{vmatrix} + & - & + \\ - & + & - \\ + & - & + \end{vmatrix}.$$

It is convenient to use the term *cofactor* for a minor with the correct sign attached in this way, and to denote the cofactor of each element by the corresponding capital letter. Thus

$$A_3 = + \begin{vmatrix} b_1 & c_1 \\ b_2 & c_2 \end{vmatrix}, \quad B_1 = - \begin{vmatrix} a_2 & c_2 \\ a_3 & c_3 \end{vmatrix}, \text{ and so on.}$$

Qu. 1. Write down A_2, B_3, C_3.

In this cofactor notation $\mathbf{b} \times \mathbf{c} = \begin{pmatrix} A_1 \\ A_2 \\ A_3 \end{pmatrix}$, $\mathbf{c} \times \mathbf{a} = \begin{pmatrix} B_1 \\ B_2 \\ B_3 \end{pmatrix}$, $\mathbf{a} \times \mathbf{b} = \begin{pmatrix} C_1 \\ C_2 \\ C_3 \end{pmatrix}$,

and
$$\Delta = a_1 A_1 + a_2 A_2 + a_3 A_3 \tag{1}$$

$$= b_1 B_1 + b_2 B_3 + b_3 B_3 \tag{2}$$

$$= c_1 C_1 + c_2 C_2 + c_3 C_3. \tag{3}$$

We call (1), (2), (3) the expansions of det $\mathbf{M}$ by the first, second, and third columns respectively.

Example 1: Prove that interchanging any two columns of a 3 × 3 determinant reverses the sign of the determinant.

Solution (i):   First note that the result is also true for 2 × 2 determinants,

since $\begin{vmatrix} p_1 & q_1 \\ p_2 & q_2 \end{vmatrix} = p_1 q_2 - p_2 q_1 = -(q_1 p_2 - q_2 p_1) = - \begin{vmatrix} q_1 & p_1 \\ q_2 & p_2 \end{vmatrix}$

In the 3 × 3 case, consider the expansion by the column which is unaltered when the other two are interchanged. Each of the cofactors in this expansion is a 2 × 2 determinant in which the columns are interchanged. The sign of each cofactor is reversed and therefore so is the sign of the whole expression.   □

Solution (ii):   The result follows immediately from the facts that
  (i) |a b c| is the signed volume of the parallelepiped with
      edges a, b, c;
  (ii) interchanging two columns reverses the handedness of
       a, b, c, which is what determines the sign of this volume.
                                                                                    □

Qu. 2.   Deduce from Example 1 that a determinant with two identical
columns is zero. Give a geometrical interpretation.

Qu. 3.   Prove that a cyclic interchange of the columns of a 3 × 3 determinant leaves its value unchanged. What happens when there is a
non-cyclic interchange?

Other general properties of determinants are developed in Ex. 9C. The
early work on determinants as such was done by Vandermonde (1776). The
word 'determinant' is due to Cauchy (1815); the vertical lines were introduced in 1841 by Arthur Cayley, a pioneer in the theory of matrices.

Example 2: Factorize the determinant $\Delta \equiv \begin{vmatrix} a^3 & b^3 & c^3 \\ a & b & c \\ 1 & 1 & 1 \end{vmatrix}$.

Solution: First regard Δ as a polynomial in a, with coefficients involving b
and c. When $a = b$, the first and second columns are identical,
and so by Qu. 2, $\Delta = 0$. Therefore by the Remainder Theorem,
$(a - b)$ is a factor of Δ.
Similarly $(b - c)$ and $(c - a)$ are factors of Δ.
A cyclic interchange of a, b, c leaves both Δ and
$(a - b)(b - c)(c - a)$ unchanged, while a non-cyclic interchange
reverses the signs of both expressions. Therefore any remaining
factor of Δ must be symmetrical in a, b, c. Since Δ is of degree 4
in a, b, c, the remaining factor is of degree 1, and so must be
$k(a + b + c)$, where k is a number. By considering the coefficient
of a^3b in Δ and in $(\underline{a} - b)(\underline{b} - c)(c - \underline{a})k(\underline{a} + b + c)$ we see that
$k = -1$. Therefore

$$\Delta \equiv -(a - b)(b - c)(c - a)(a + b + c).$$ □

The solution of equation (A) which we obtained earlier can be written in
terms of determinants.

If |a   b   c| $\neq 0$ then the equation $x\mathbf{a} + y\mathbf{b} + z\mathbf{c} = \mathbf{d}$ has the

unique solution $x = \dfrac{|\mathbf{d} \ \ \mathbf{b} \ \ \mathbf{c}|}{|\mathbf{a} \ \ \mathbf{b} \ \ \mathbf{c}|}$, $y = \dfrac{|\mathbf{a} \ \ \mathbf{d} \ \ \mathbf{c}|}{|\mathbf{a} \ \ \mathbf{b} \ \ \mathbf{c}|}$, $z = \dfrac{|\mathbf{a} \ \ \mathbf{b} \ \ \mathbf{d}|}{|\mathbf{a} \ \ \mathbf{b} \ \ \mathbf{c}|}$.

This is known as *Cramer's rule* after Gabriel Cramer who published it in
1750, though Maclaurin seems to have known it as early as 1729.

Exercise 9C [The notation in this Exercise is the same as in §9.3.]

1. Evaluate these determinants

   (i) $\begin{vmatrix} 4 & 6 & 7 \\ 2 & 3 & 8 \\ 1 & 9 & 0 \end{vmatrix}$ (ii) $\begin{vmatrix} 2 & -3 & 0 \\ -4 & 5 & 1 \\ 0 & 6 & 8 \end{vmatrix}$ (iii) $\begin{vmatrix} 1 & 2 & 3 \\ 4 & 5 & 6 \\ 7 & 8 & 9 \end{vmatrix}$

2. Expand and simplify

   (i) $\begin{vmatrix} 1 & h & 2 \\ 5 & 10 & 10 \\ 2 & 4 & k \end{vmatrix}$ (ii) $\begin{vmatrix} p & q & r \\ q & r & p \\ p-q & q-r & r-p \end{vmatrix}$ (iii) $\begin{vmatrix} a & h & g \\ h & b & f \\ g & f & c \end{vmatrix}$.

3. Check that $\det \mathbf{M}$ can be expanded by this method, which was devised by P. F. Sarrus (1798–1861).

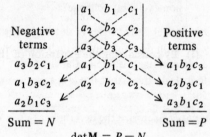

   Evaluate the determinants of Qu. 1 (i), (iii) again by this method.

4. Use Cramer's rule to solve the equations

   (i) $\begin{aligned} 3x + 4y + 2z &= 4 \\ x - 2y - 4z &= 2 \\ 2x + 5y + 3z &= -1 \end{aligned}$ (ii) $\begin{aligned} 4x + 4y + 2z &= 3 \\ 2x - 2y - 4z &= 1 \\ -x + 5y + 3z &= 2. \end{aligned}$

   Write down another set of three equations in x, y, z which can be solved by Cramer's rule without evaluating any more determinants, and solve these equations.

5. Prove that $\mathbf{a} \times \mathbf{b} = \begin{vmatrix} \mathbf{i} & a_1 & b_1 \\ \mathbf{j} & a_2 & b_2 \\ \mathbf{k} & a_3 & b_3 \end{vmatrix}$

6. Prove that if all the elements of one column of a determinant are multiplied by a constant λ, then the value of the determinant is multiplied by λ. Give a geometrical interpretation in terms of volumes.

7. Given that $\begin{vmatrix} 5 & 4 & 7 \\ 7 & 2 & 2 \\ 6 & 3 & 1 \end{vmatrix} = 63$ use Qu. 6 to find without expansion

   (i) $\begin{vmatrix} 50 & 4 & 7 \\ 70 & 2 & 2 \\ 60 & 3 & 1 \end{vmatrix}$ (ii) $\begin{vmatrix} 15 & 20 & -56 \\ 21 & 10 & -16 \\ 18 & 15 & -8 \end{vmatrix}$ (iii) $\begin{vmatrix} 5x^3 & 2/x & 21y^2 \\ 7x^3 & 1/x & 6y^2 \\ 6x^3 & 3/2x & 3y^2 \end{vmatrix}$.

8. Express $\begin{vmatrix} 42 & 90 & 39 \\ 126 & 60 & 26 \\ 84 & 45 & 52 \end{vmatrix}$ as the product of prime factors.

9. Prove that $|p + q \quad r \quad s| = |p \quad r \quad s| + |q \quad r \quad s|$.
   Express $|a + b \quad c + d \quad e + f|$ as the sum of eight determinants.

10. Show that $\begin{vmatrix} b_1 + c_1 & c_1 + a_1 & a_1 + b_1 \\ b_2 + c_2 & c_2 + a_2 & a_2 + b_2 \\ b_3 + c_3 & c_3 + a_3 & a_3 + b_3 \end{vmatrix} = 2 \begin{vmatrix} a_1 & b_1 & c_1 \\ a_2 & b_2 & c_2 \\ a_3 & b_3 & c_3 \end{vmatrix}$.   [MEI]

11. Prove that $|a + \lambda b \quad b \quad c| = |a \quad b \quad c|$. Give a geometrical interpretation in terms of volumes. Use this and other properties previously established to show that the value of a determinant is unaltered if a multiple of any column is added to any other column.

12. Using Qu. 11, evaluate

   (i) $\begin{vmatrix} 73 & 7 & 1 \\ 82 & 8 & 2 \\ 65 & 6 & 1 \end{vmatrix}$ (ii) $\begin{vmatrix} 21 & 43 & 63 \\ 35 & 70 & 106 \\ 43 & 85 & 135 \end{vmatrix}$ (iii) $\begin{vmatrix} 5 & 14 & 23 \\ 2 & 16 & 30 \\ -3 & 23 & 49 \end{vmatrix}$.

13. (i) By considering the expansion of $|b \quad b \quad c|$ by the first column show that $b_1 A_1 + b_2 A_2 + b_3 A_3 = 0$.
    (ii) Prove that $a_1 C_1 + a_2 C_2 + a_3 C_3 = 0$.
    (iii) Write down four other similar expressions, all of which equal zero.
    [These six expressions equalling zero are sometimes called the expansions of $\det M$ by *alien cofactors*.]

14. Prove that $\begin{vmatrix} x_1 & y_1 & 1 \\ x_2 & y_2 & 1 \\ x_3 & y_3 & 1 \end{vmatrix} = \begin{vmatrix} x_2 - x_1 & y_2 - y_1 \\ x_3 - x_1 & y_3 - y_1 \end{vmatrix}$.

   Deduce that the area of the triangle formed by the points in $\mathbb{R}^2$ with

   coordinates $(x_1, y_1), (x_2, y_2), (x_3, y_3)$ is $\pm \frac{1}{2} \begin{vmatrix} x_1 & y_1 & 1 \\ x_2 & y_2 & 1 \\ x_3 & y_3 & 1 \end{vmatrix}$.

15. The points P_1, P_2, P_3, P_4 lie on the parabola $y = x^2$ and have x-coordinates $a, a + 1, a + 2, a + 3$ respectively. The chords $P_1 P_3, P_2 P_4$ meet at P. Find the area of triangle $P_1 P_2 P_3$, and show that the areas of $P_1 P_2 P$ and $P_3 P_4 P$ are equal.

16. Prove that the equation of the straight line in $\mathbb{R}^2$ through (a, b) and

   (c, d) is $\begin{vmatrix} x & y & 1 \\ a & b & 1 \\ c & d & 1 \end{vmatrix} = 0$.

17. Show that $x = 0$ is a solution of the equation $\begin{vmatrix} x - 1 & 4 & -1 \\ 1 & x + 2 & 1 \\ 2x - 4 & 4 & x - 4 \end{vmatrix} = 0$

   and find the other two roots.   [L]

18. Prove that $\begin{vmatrix} a & b & c \\ c & a & b \\ b & c & a \end{vmatrix} \equiv (a + b + c) \begin{vmatrix} 1 & b & c \\ 1 & a & b \\ 1 & c & a \end{vmatrix}$ and deduce that

   $a^3 + b^3 + c^3 - 3abc \equiv (a + b + c)(a^2 + b^2 + c^2 - bc - ca - ab)$.

19. Factorize these determinants

(i) $\begin{vmatrix} 1 & 1 & 1 \\ x & y & z \\ x^2 & y^2 & z^2 \end{vmatrix}$ (ii) $\begin{vmatrix} x & y & z \\ x^2 & y^2 & z^2 \\ y+z & z+x & x+y \end{vmatrix}$ (iii) $\begin{vmatrix} 1 & 1 & 1 \\ 1 & x & x^2 \\ 1 & x^2 & x^4 \end{vmatrix}$.

20. Prove that $\begin{vmatrix} (b+c)^2 & a^2 & a^2 \\ b^2 & (c+a)^2 & b^2 \\ c^2 & c^2 & (a+b)^2 \end{vmatrix} \equiv 2abc(a+b+c)^3$.

21. Verify the determinantal identity

$$\begin{vmatrix} t_1^3 & t_1^2 & 1 \\ t_2^3 & t_2^2 & 1 \\ t_3^3 & t_3^2 & 1 \end{vmatrix} = (t_1 t_2 + t_2 t_3 + t_3 t_1) \begin{vmatrix} t_1^2 & t_1 & 1 \\ t_2^2 & t_2 & 1 \\ t_3^2 & t_3 & 1 \end{vmatrix}$$

and deduce that the three distinct points t_1, t_2, t_3 of the curve given parametrically by $x = t(t^2 + 1)$, $y = t^2$ are collinear if and only if $t_1 t_2 + t_2 t_3 + t_3 t_1 = 1$.

Assuming that the result holds, with the obvious interpretation, when the points are no longer distinct, show that the three points where the tangents at the points $t_1 = -\frac{1}{3}$, $t_2 = 1$, $t_3 = 2$ meet the curve again are collinear. [MEI]

9.4 The inverse of a 3 × 3 matrix

When seeking to 'undo' a linear transformation with 3 × 3 matrix **M**, we want to find a matrix **L** such that **LM** = **I**, the identity matrix. For then

$$\mathbf{Mp} = \mathbf{d} \Rightarrow \mathbf{LMp} = \mathbf{Ld} \Rightarrow \mathbf{p} = \mathbf{Ld}.$$

We can use the ideas and notation of the preceding section to find **L** by taking three special cases. We suppose that $\Delta = \det \mathbf{M} \neq 0$. If **d** is the unit

vector $\mathbf{i} = \begin{pmatrix} 1 \\ 0 \\ 0 \end{pmatrix}$ then $\mathbf{Ld} = \mathbf{Li}$, which is simply the first column of **L**.

But when $\mathbf{d} = \mathbf{i}$ the solution of $\mathbf{Mp} = \mathbf{d}$ is

$$x = \frac{\mathbf{i} \cdot (\mathbf{b} \times \mathbf{c})}{\Delta} = \frac{A_1}{\Delta}, y = \frac{\mathbf{i} \cdot (\mathbf{c} \times \mathbf{a})}{\Delta} = \frac{B_1}{\Delta}, z = \frac{\mathbf{i} \cdot (\mathbf{a} \times \mathbf{b})}{\Delta} = \frac{C_1}{\Delta}.$$

Therefore the first column of **L** is $\begin{pmatrix} A_1/\Delta \\ B_1/\Delta \\ C_1/\Delta \end{pmatrix}$. Similarly, by taking $\mathbf{d} = \mathbf{j}$ and

then $\mathbf{d} = \mathbf{k}$, the second and third columns of **L** are $\begin{pmatrix} A_2/\Delta \\ B_2/\Delta \\ C_2/\Delta \end{pmatrix}$ and $\begin{pmatrix} A_3/\Delta \\ B_3/\Delta \\ C_3/\Delta \end{pmatrix}$

respectively. Therefore $\mathbf{L} = \dfrac{1}{\Delta} \begin{pmatrix} A_1 & A_2 & A_3 \\ B_1 & B_2 & B_3 \\ C_1 & C_2 & C_3 \end{pmatrix}$.

The matrix $\begin{pmatrix} A_1 & A_2 & A_3 \\ B_1 & B_2 & B_3 \\ C_1 & C_2 & C_3 \end{pmatrix}$ is called the *adjugate* (or adjoint) of M, denoted by

adj M.

> The adjugate is formed by replacing each element of M by its cofactor and then transposing.

Qu. 1. Find adj M and L when $M = \begin{pmatrix} 2 & 1 & 4 \\ 7 & -2 & 6 \\ 5 & -2 & 4 \end{pmatrix}$. Check that $LM = I$.

The matrix L is a *left-inverse* of M, since $LM = I$. We can easily show that L is a right-inverse of M too. For since $d = Mp \Rightarrow p = Ld$ we have $d = M(Ld) = (ML)d$ for all d. Therefore the transformation with matrix ML is the identity transformation, and so $ML = I$.

There can be no other right- or left-inverse of M. For if N is a left-inverse,

then
$$\begin{aligned} N &= N(ML) && \text{since } L \text{ is a right-inverse} \\ &= (NM)L && \\ &= IL && \text{since } N \text{ is a left-inverse} \\ &= L. \end{aligned}$$

Qu. 2. Prove similarly that N is a right-inverse $\Rightarrow N = L$.

Therefore L is the unique left- and right-inverse of M, and we are justified in calling L *the* inverse of M, which is denoted by M^{-1}. Thus $M^{-1} = \dfrac{\text{adj } M}{\det M}$.

Since M^{-1} is a right-inverse, M adj $M = (\det M)I$,

i.e.
$$\begin{pmatrix} a_1 & b_1 & c_1 \\ a_2 & b_2 & c_2 \\ a_3 & b_3 & c_3 \end{pmatrix} \begin{pmatrix} A_1 & A_2 & A_3 \\ B_1 & B_2 & B_3 \\ C_1 & C_2 & C_3 \end{pmatrix} = \begin{pmatrix} \Delta & 0 & 0 \\ 0 & \Delta & 0 \\ 0 & 0 & \Delta \end{pmatrix}.$$

Taking the top left-hand element of the product, we have $a_1 A_1 + b_1 B_1 + c_1 C_1 = \Delta$. But $a_1 A_1 + b_1 B_1 + c_1 C_1$ is the expansion by the first column of $\det M^T$, where M^T is the transpose of M. Therefore

> $$\det M^T = \Delta = \det M;$$
> the determinant of a 3 × 3 matrix equals the determinant of its transpose.

It follows that a determinant can be expanded by rows as well as by columns, and that there are row properties corresponding to all the column properties established in the preceding section and exercise.

Qu. 3.   Prove that det $\mathbf{X}^T$ = det $\mathbf{X}$ if $\mathbf{X}$ is a 2 × 2 matrix.

Qu. 4.   Prove that $\mathbf{M}(\text{adj } \mathbf{M}) = (\text{adj } \mathbf{M})\mathbf{M} = (\det \mathbf{M})\mathbf{I}$ whether or not $\det \mathbf{M} = 0$.

Finally we must see whether $\mathbf{M}$ can have an inverse if $\det \mathbf{M} = 0$. To do this we shall use a result which is difficult to prove algebraically, but which is very straightforward when considered geometrically. The transformation with matrix $\mathbf{M}$ maps the unit cube with edges $\mathbf{i}, \mathbf{j}, \mathbf{k}$ to the parallelepiped with edges $\mathbf{a}, \mathbf{b}, \mathbf{c}$, which has signed volume $\det \mathbf{M}$. The volume scale factor of the transformation is therefore $\det \mathbf{M}$, with a negative scale factor showing change of handedness. Successive transformations with matrices $\mathbf{M}$ and $\mathbf{N}$ are equivalent to the single transformation with matrix $\mathbf{NM}$.

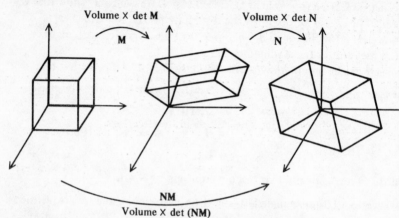

Volume × det $\mathbf{M}$ Volume × det $\mathbf{N}$

$\mathbf{M}$ $\mathbf{N}$

$\mathbf{NM}$

Volume × det $(\mathbf{NM})$

By considering the volume scale factors we see immediately that

> $\det(\mathbf{NM}) = \det \mathbf{N} \times \det \mathbf{M}$;
> the determinant of a product is the product of the determinants.

Qu. 5.   For 3 × 3 matrices $\mathbf{M}, \mathbf{N}$
   (i) is it always true that $\mathbf{MN} = \mathbf{NM}$?
   (ii) is it always true that $\det(\mathbf{MN}) = \det(\mathbf{NM})$?

If $\det \mathbf{M} = 0$ then $\det(\mathbf{NM}) = \det \mathbf{N} \times 0 = 0$ for all $\mathbf{N}$. So $\mathbf{NM}$ can never equal $\mathbf{I}$, since $\det \mathbf{I} = 1 \neq 0$. Therefore $\mathbf{M}$ can have no inverse if $\det \mathbf{M} = 0$. A matrix which has no inverse is said to be *singular*, and we can summarize our results as follows.

> $\mathbf{M}$ is non-singular $\Leftrightarrow$ det $\mathbf{M} \neq 0$ $\Leftrightarrow$ $\mathbf{M}$ has the unique inverse $\dfrac{\text{adj } \mathbf{M}}{\det \mathbf{M}}$.

Exercise 9D

1. If $M = \begin{pmatrix} 2 & 4 & 5 \\ 1 & 3 & 2 \\ 1 & 6 & 4 \end{pmatrix}$ find adj M, and evaluate $M(\text{adj } M)$ and $(\text{adj } M)M$.

   What is det M?

2. With M as in Qu. 1, find $\det(\text{adj } M)$ and $\text{adj}(\text{adj } M)$. Comment on your results.

3. Find, if possible, the inverses of these matrices.

   (i) $\begin{pmatrix} 1 & -1 & 3 \\ 2 & 1 & 4 \\ 0 & 1 & 1 \end{pmatrix}$ (ii) $\begin{pmatrix} 1 & 3 & -2 \\ -2 & -9 & 5 \\ 1 & 10 & 4 \end{pmatrix}$

   (iii) $\begin{pmatrix} 3 & 2 & 6 \\ 5 & 3 & 11 \\ 7 & 4 & 16 \end{pmatrix}$ (iv) $\begin{pmatrix} -4 & 3 & 5 \\ 8 & 9 & -6 \\ 2 & 6 & -1 \end{pmatrix}$

4. If $M = \begin{pmatrix} 2 & 3 & 6 \\ 6 & 2 & -3 \\ 3 & -6 & 2 \end{pmatrix}$ form the product MM^T and show that

   $M^{-1} = \frac{1}{49} M^T$. Without multiplying out, state the product $M^T M$, giving reasons for your answer.

   Hence, or otherwise, find the solution of the equations.

   $$2x + 3y + 6z = 1$$
   $$6x + 2y - 3z = 1$$
   $$3x - 6y + 2z = 2. \qquad \text{[MEI, adapted]}$$

5. Find the inverse of the matrix $M = \begin{pmatrix} 1 & -1 & 1 \\ 3 & -9 & 5 \\ 1 & -3 & 3 \end{pmatrix}$.

   Rewrite the set of equations $\qquad x - y + z = 3$
   $$3x - 9y + 5z = 6$$
   $$x - 3y + 3z = 13$$

   in a form using the matrix M, and hence solve the equations.   [MEI]

6. Find the adjugate and the inverse of $\begin{pmatrix} a & b \\ c & d \end{pmatrix}$, where $ad - bc \neq 0$.

7. Find the values of λ for which the following are singular.

   (i) $\begin{pmatrix} 4 & 6 & -1 \\ -1 & 2 & -3 \\ 5 & \lambda & 15 \end{pmatrix}$ (ii) $\begin{pmatrix} 6 & 7 & -1 \\ 3 & \lambda & 5 \\ 9 & 11 & \lambda \end{pmatrix}$ (iii) $\begin{pmatrix} 1-\lambda & 2 \\ -1 & 4-\lambda \end{pmatrix}$

   (iv) $\begin{pmatrix} 1-\lambda & 1 & -2 \\ -1 & 2-\lambda & 1 \\ 0 & 1 & -1-\lambda \end{pmatrix}$.

8. If $\mathbf{M} = \begin{pmatrix} 1 & 3 & 2 \\ -1 & 2 & 0 \\ 4 & 1 & -3 \end{pmatrix}$, $\mathbf{N} = \begin{pmatrix} 2 & -1 & 5 \\ -4 & 0 & 2 \\ 3 & 1 & 0 \end{pmatrix}$ find

   (i) $\mathbf{MN}$ (ii) $\det(\mathbf{MN})$ (iii) $\mathbf{NM}$ (iv) $\det(\mathbf{NM})$ (v) $\det \mathbf{M}$
   (vi) $\det \mathbf{N}$.
   Verify that $\det(\mathbf{MN}) = \det(\mathbf{NM}) = \det \mathbf{M} \times \det \mathbf{N}$.

9. Evaluate the determinant $\begin{vmatrix} 2 & 1+j & 2 \\ 1-j & -1 & 4-j \\ 2 & 4+j & 3 \end{vmatrix}$, where $j^2 = -1$.

   Explain how, without evaluation, it could have been concluded that the
   value of the determinant was real.   [MEI]

10. By considering $\det(\mathbf{PQ})$, where $\mathbf{P} = \begin{pmatrix} a^2 & a & 1 \\ b^2 & b & 1 \\ c^2 & c & 1 \end{pmatrix}$, $\mathbf{Q} = \begin{pmatrix} 1 & 1 & \\ -2x & -2y & - \\ x^2 & y^2 & \end{pmatrix}$

   show that
   $$\Delta \equiv \begin{vmatrix} (a-x)^2 & (a-y)^2 & (a-z)^2 \\ (b-x)^2 & (b-y)^2 & (b-z)^2 \\ (c-x)^2 & (c-y)^2 & (c-z)^2 \end{vmatrix}$$

   $$\equiv 2(b-c)(c-a)(a-b)(y-z)(z-x)(x-y).$$

   Give an alternative proof, working directly from Δ.

11. (i) Prove that: $\mathbf{P}$ and $\mathbf{Q}$ are non-singular $\Leftrightarrow \mathbf{PQ}$ is non-singular.
    (ii) If $\mathbf{PQ}$ is non-singular, prove that
         (a) $(\mathbf{PQ})^{-1} = \mathbf{Q}^{-1}\mathbf{P}^{-1}$ (b) $\mathrm{adj}(\mathbf{PQ}) = \mathrm{adj}\,\mathbf{Q} \cdot \mathrm{adj}\,\mathbf{P}$.

12. Prove that if $\mathbf{M}$ is a non-singular 3×3 matrix then
    (i) $\det(\mathrm{adj}\,\mathbf{M}) = (\det \mathbf{M})^2$ (ii) $\mathrm{adj}\,(\mathrm{adj}\,\mathbf{M}) = (\det \mathbf{M})\mathbf{M}$.
    Do these results still hold if $\mathbf{M}$ is singular?

13. A square matrix $\mathbf{M}$ is said to be *orthogonal* if $\mathbf{M}^T\mathbf{M} = \mathbf{I}$.
    (i) What are the possible values of $\det \mathbf{M}$ if $\mathbf{M}$ is orthogonal?
    (ii) Prove that the only possible 2×2 orthogonal matrices are of the
         form $\begin{pmatrix} \cos\theta & -\sin\theta \\ \sin\theta & \cos\theta \end{pmatrix}$ or $\begin{pmatrix} \cos\theta & \sin\theta \\ \sin\theta & -\cos\theta \end{pmatrix}$. Describe geometrically
         the transformations which have these matrices.
    (iii) Prove that the 3×3 matrix $\mathbf{M} = (\mathbf{a}\quad \mathbf{b}\quad \mathbf{c})$ is orthogonal if and only
         if $\mathbf{a}, \mathbf{b}, \mathbf{c}$ are mutually orthogonal (i.e. perpendicular) unit vectors.
    (iv) Prove that a linear transformation preserves size and shape if and
         only if its matrix is orthogonal.

14. Find the inverse of $\begin{pmatrix} \cos\theta\cos\phi & \sin\theta\cos\phi & -\sin\phi \\ \cos\theta\sin\phi & \sin\theta\sin\phi & \cos\phi \\ \sin\theta & -\cos\theta & 0 \end{pmatrix}$.

    Give a geometrical interpretation to explain the result.

15. The non-singular matrix $\mathbf{B}$ has the property $\mathbf{B}\mathbf{B}^T = \mathbf{B}^T\mathbf{B}$. Prove that $\mathbf{B}^T\mathbf{B}^{-1} = \mathbf{B}^{-1}\mathbf{B}^T$. Prove also that if $\mathbf{C} = \mathbf{B}^{-1}\mathbf{B}^T$, then $\mathbf{C}\mathbf{C}^T$ is the identity

matrix. Find $\mathbf{B}^T$, $\mathbf{B}\mathbf{B}^T$, $\mathbf{B}^{-1}$ and $\mathbf{C}$ when $\mathbf{B} = \begin{pmatrix} 2 & 2 & 1 \\ -2 & 1 & 2 \\ 1 & -2 & 2 \end{pmatrix}$. [MEI]

16. A matrix $\mathbf{A}$ satisfies the equation $(\mathbf{I} + \mathbf{S}) = (\mathbf{I} - \mathbf{S})\mathbf{A}$ where $\mathbf{I}$ is the

3×3 identity matrix and $\mathbf{S} = \begin{pmatrix} 0 & 1 & 2 \\ -1 & 0 & 3 \\ -2 & -3 & 0 \end{pmatrix}$.

Without computing $\mathbf{A}$, show that $\mathbf{A}^T\mathbf{A} = \mathbf{I}$ and find the value of the determinant of $\mathbf{A}$. Check your answers by computing $\mathbf{A}$. [MEI]

17. (i) Let $S_n = \alpha^n + \beta^n + \gamma^n$. By considering $\det(\mathbf{A}\mathbf{A}^T)$, where

$\mathbf{A} = \begin{pmatrix} 1 & 1 & 1 \\ \alpha & \beta & \gamma \\ \alpha^2 & \beta^2 & \gamma^2 \end{pmatrix}$, show that

$$\Delta = \begin{vmatrix} S_0 & S_1 & S_2 \\ S_1 & S_2 & S_3 \\ S_2 & S_3 & S_4 \end{vmatrix} = \{(\beta - \gamma)(\gamma - \alpha)(\alpha - \beta)\}^2.$$

(ii) If α, β, γ are the roots of the equation $x^3 + px + q = 0$ show that $\Delta = -4p^3 - 27q^2$.

(iii) Deduce that this equation has 1, 2 or 3 distinct real roots according as $4p^3 + 27q^2$ is positive, zero or negative.

9.5 Linear equations: the singular case

We have seen that if $\det \mathbf{M} \neq 0$, then the simultaneous equations

$$\left.\begin{array}{l} a_1 x + b_1 y + c_1 z = d_1 \\ a_2 x + b_2 y + c_2 z = d_2 \\ a_3 x + b_3 y + c_3 z = d_3 \end{array}\right\} \text{(A)}$$

have a unique solution given by Cramer's rule. Now we look at what happens in the singular case, when $\det \mathbf{M} = 0$.

A different geometrical interpretation is helpful here. The equations (A) are the Cartesian equations of three planes in $\mathbb{R}^3$. The elements of the solution set are the coordinates (x, y, z) of points which lie on all three planes. So we first consider the various ways in which three planes may intersect. There are eight distinct configurations, which are described in the following table (where configuration **6** occurs twice, and **7** can arise as a special case of **3**).

If two planes meet in a line			
then the third plane	**1** meets line in a point	**2** is parallel to line	**3** contains line
The solution set is	unique solution	ϕ	points of line (one parameter)

If two planes are parallel			
then the third plane	**4** meets in parallel lines	**5** is parallel	**6** coincides with one plane
The solution set is	ϕ	ϕ	ϕ

If two planes coincide			
then the third plane	**7** meets in a line	**6** is parallel	**8** coincides
The solution set is	points of line (one parameter)	ϕ	points of plane (two parameters)

This indicates that three equations in three unknowns may have no solution (in which case they are said to be *inconsistent*), one solution, a 'single infinity' of solutions (when all solutions can be given in terms of a single parameter), or a 'double infinity' of solutions (when two parameters are needed).

Example 1: Solve the equations
$$x + 3y - 2z = 7$$
$$2x - 2y + z = 3$$
$$3x + y - z = k$$

      (i) when $k = 10$ (ii) when $k = 12$.

Solution: (i) When $k = 10$, we notice that the third equation is the sum of the other two, so that the third plane contains the line of intersection of the other two (§8.3), and we have configuration 3. Finding the equations of the line of intersection of these planes in the usual way, we obtain the single parameter family of solutions

$$x = 2 + \lambda, \ y = -1 + 5\lambda, \ z = -3 + 8\lambda.$$

    (ii) When $k = 12$ the third plane is parallel to the line of intersection of the other two, as in configuration 2. The equations are inconsistent.   □

Example 2: Investigate for various values of a and b the type of solution set of the equations
$$x + 3y - 2z = 7 \tag{1}$$
$$ax + 6y - 4z = 2 - 3b \tag{2}$$
$$2x + 6y + bz = 14 \tag{3}$$

Solution:
$$\begin{vmatrix} 1 & 3 & -2 \\ a & 6 & -4 \\ 2 & 6 & b \end{vmatrix} = 24 - 12a + 6b - 3ab = 3(2-a)(4+b).$$

    (i)   If $a \neq 2$ and $b \neq -4$ there is a unique solution.

    (ii) If $a = 2$ but $b \neq -4$ (1) and (2) represent parallel planes, which the plane with equation (3) intersects, as in configuration 4. The equations are inconsistent.

    (iii) If $b = -4$ but $a \neq 2$, (3) is twice (1) but (2) is not a multiple of these. We have two coincident planes with an intersecting plane, configuration 7, and therefore a single parameter set of solutions.

    (iv) If $a = 2$ and $b = -4$, the planes with equations (1), (2), (3) all coincide, configuration 8. There is a two-parameter set of solutions, for example
$$x = \lambda, y = \mu, z = \tfrac{1}{2}(\lambda + 3\mu - 7).$$
                                                  □

The equations $\mathbf{Mp} = \mathbf{0}$ or
$$a_1 x + b_1 y + c_1 z = 0$$
$$a_2 x + b_2 y + c_2 z = 0$$
$$a_3 x + b_3 y + c_3 z = 0$$

are called *homogeneous*. Clearly homogeneous equations are never inconsistent since they always have the solution $x = y = z = 0$; this is called the *trivial* solution. If $\det M \neq 0$ this solution is unique, so there can be no non-trivial solution. If $\det M = 0$ there is a one or two-parameter set of solutions which contains both trivial and non-trivial solutions. Therefore

> the homogeneous equations $Mp = 0$ have non-trivial solutions
> $\Leftrightarrow \det M = 0$.

Qu. 1. Show that, with the usual cofactor notation, $x = A_i, y = B_i, z = C_i$ is a solution of $Mp = 0$ for $i = 1, 2, 3$. Unless all the cofactors are zero, this gives at least one non-trivial solution explicitly.

Qu. 2. Describe geometrically what is happening when all the cofactors are zero. Give an explicit non-trivial solution of $Mp = 0$ in this case.

Qu. 3. Show that if $p = p_1$ and $p = p_2$ are solutions of $Mp = 0$ then so is $p = \lambda p_1 + \mu p_2$.

Exercise 9E

1. Show that the equations $\begin{aligned} x + y + z &= 4 \\ 2x + 3y - 4z &= 3 \\ 5x + 8y - 13z &= 6. \end{aligned}$ are inconsistent.

   Describe the configuration formed by the planes with these equations.

2. Show that the matrix $A = \begin{pmatrix} 3 & 1 & -5 \\ 5 & 0 & -4 \\ 2 & -1 & 1 \end{pmatrix}$ is singular. Discuss the solution

   of the equations (i) $A \begin{pmatrix} x \\ y \\ z \end{pmatrix} = \begin{pmatrix} 2 \\ 5 \\ 4 \end{pmatrix}$ (ii) $A \begin{pmatrix} x \\ y \\ z \end{pmatrix} = \begin{pmatrix} 2 \\ 6 \\ 4 \end{pmatrix}$.

   Interpret your results geometrically.

3. Find the two values of k for which the equations
   $$\begin{aligned} x + 3y + 2z &= k \\ 2x + ky + 4z &= 12 \\ kx + 2y - 4z &= 20 \end{aligned}$$
   do not have a unique solution for x, y, z. Find the solution set for each of these values of k.

4. Evaluate the determinant of the matrix $A = \begin{pmatrix} 2 & 4 & 1 \\ 3 & 5 & 2 \\ 1 & 5 & -1 \end{pmatrix}$.
   Solve the equation $A \begin{pmatrix} x \\ y \\ z \end{pmatrix} = \begin{pmatrix} 3 \\ 7 \\ -6 \end{pmatrix}$.

5. Solve the equations
$$4x + 2y + z = 7$$
$$8x + ay + 2z = 10 + a$$
$$bx + 6y + 3z = 1 + 5a$$

in each of the particular cases (i) $a = 3, b = 11$ (ii) $a = 3, b = 12$
(iii) $a = 4, b = 11$ (iv) $a = 4, b = 12$.

6. Discuss the solution of the equations $ax + by = c, dx + ey = f$ for x, y, considering all the special cases that can arise and giving geometrical interpretations.

7. Find the value of k for which the equations

$$x + 2y = 0$$
$$3x + ky - z = 0$$
$$2x + 5y - 2z = 0$$

have a solution other than $x = y = z = 0$. Find the solution set for this value of k.   [NI]

8. Show that there is just one real value of the constant λ for which the equations
$$\lambda x + y + z = 0$$
$$2x + \lambda y + 3z = 0$$
$$4x - 2y + \lambda z = 0$$

have common solutions other than $x = y = z = 0$. Find, for this value of λ, the solution of these equations which satisfies also the equation

$$x + y + z = 3.$$ [MEI]

9. Show that, for a triangle ABC with the usual notation, $a = b \cos C + c \cos B$, and write down similar expressions for b and c. Treating these as homogeneous equations for a, b, c, write down the condition for non-trivial solutions, and deduce that

$$\cos^2 A + \cos^2 B + \cos^2 C = 1 - 2 \cos A \cos B \cos C.$$

10. Let A be a 3×3 real matrix and b a 3-rowed column vector. It is proposed to solve the equation $Ax = b$, and a particular solution $x = x_0$ is noted. Prove that any other solution may be written in the form $x = x_0 + u$ where u is a solution of the equation $Ax = 0$. Prove conversely that any vector $x_0 + u$ is a solution of the equation, where x_0 is a fixed particular solution of the equation and u is any vector such that $Au = 0$.
Interpret geometrically the equation and its solutions if A is singular
(i) when an x_0 exists,
(ii) when there is no such particular solution x_0.   [MEI]

11. Show that the equations $2x + 3y = 8$ are inconsistent unless k takes
$$5x - 4y = -3$$
$$9x + 2y = k$$

one particular value. Solve the equations when k has this value.

12. Discuss for various values of a the solution of
$$x + 2y = 3$$
$$3x + ay = 11$$
$$ax - 2y = 22.$$

13. Consider the three equations $a_1 x + b_1 y + c_1 = 0$
$$a_2 x + b_2 y + c_2 = 0$$
$$a_3 x + b_3 y + c_3 = 0$$
as the equations of lines in $\mathbb{R}^2$.

   (i) Draw the seven possible configurations formed by these lines, and state in each case the type of solution set.

   (ii) Show that $\Delta = \begin{vmatrix} a_1 & b_1 & c_1 \\ a_2 & b_2 & c_2 \\ a_3 & b_3 & c_3 \end{vmatrix} = 0$ is a necessary condition for the

   consistency of these equations. Give a counter-example which shows that $\Delta = 0$ is not a sufficient condition for consistency.

   (iii) Show that, with the usual notation,
   the equations have a unique solution $\Leftrightarrow \Delta = 0$ and C_1, C_2, C_3 are not all zero.

   (iv) Find a necessary and sufficient condition for the equations to have infinitely many solutions.

14. (i) The equations $2x + 4y = 1$ are written in the matrix form
   $$-3x + \ y = 1$$
   $$x + 2y = -1$$

   $\mathbf{Pr} = \mathbf{s}$, where $\mathbf{P} = \begin{pmatrix} 2 & 4 \\ -3 & 1 \\ 1 & 2 \end{pmatrix}$, $\mathbf{r} = \begin{pmatrix} x \\ y \end{pmatrix}$, $\mathbf{s} = \begin{pmatrix} 1 \\ 1 \\ -1 \end{pmatrix}$. Carry out

   numerically the stages of the following argument:

   $$\mathbf{Pr} = \mathbf{s}$$
   $$\Rightarrow \qquad \mathbf{P}^T\mathbf{Pr} = \mathbf{P}^T\mathbf{s}$$
   $$\Rightarrow (\mathbf{P}^T\mathbf{P})^{-1}\,\mathbf{P}^T\mathbf{Pr} = (\mathbf{P}^T\mathbf{P})^{-1}\,\mathbf{P}^T\mathbf{s}$$
   $$\Rightarrow \qquad \mathbf{r} = (\mathbf{P}^T\mathbf{P})^{-1}\,\mathbf{P}^T\mathbf{s}.$$

   Check that the values of x and y found in this way do not satisfy all of the original equations. What conclusion do you draw about these equations?

   (ii) Consider the linear transformation T from $\mathbb{R}^2$ to $\mathbb{R}^3$ with matrix $\mathbf{P}$. Show that the range of T is the plane π through the origin

   containing the vectors $\begin{pmatrix} 2 \\ -3 \\ 1 \end{pmatrix}$ and $\begin{pmatrix} 4 \\ 1 \\ 2 \end{pmatrix}$. Find the Cartesian equation

   of π, and check that the point S with coordinates $(1, 1, -1)$ is not in π.

   (iii) Find the image under T of the point of $\mathbb{R}^2$ whose coordinates were found in (i). Let this image be S'. Prove that SS' is perpendicular to π. This shows that, although the method in (i) fails to produce a solution (since there is no solution), it does give the point of $\mathbb{R}^2$ which maps to the point in π nearest to $(1, 1, -1)$.

9.6 Systematic elimination

In Ex. 6A, Qu. 12, we found $\sum\limits_{r=1}^{n} r^4$ by expressing r^4 in the form

$Ar + Br(r + 1) + Cr(r + 1)(r + 2) + Dr(r + 1)(r + 2)(r + 3)$ and then summing. To find the constants A, B, C, D, we can equate the coefficients of r, r^2, r^3, r^4:

$$A + B + 2C + 6D = 0$$
$$B + 3C + 11D = 0$$
$$C + 6D = 0$$
$$D = 1.$$

We have here four simultaneous equations in A, B, C, D. It would be a waste of time to use a sophisticated matrix or determinant method in this case, since we can solve these equations very easily by *back-substitution*, working from the last equation:

$$D = 1$$
$$C = -6D = -6$$
$$B = -3C - 11D = 18 - 11 = 7$$
$$A = -B - 2C - 6D = -7 + 12 - 6 = -1.$$

The matrix of coefficients in this example is $\begin{pmatrix} 1 & 1 & 2 & 6 \\ 0 & 1 & 3 & 11 \\ 0 & 0 & 1 & 6 \\ 0 & 0 & 0 & 1 \end{pmatrix}$. Because all the

non-zero elements are on or above the leading diagonal, this is called an

upper triangular matrix. Similarly a matrix of the form $\begin{pmatrix} * & 0 & 0 & 0 \\ * & * & 0 & 0 \\ * & * & * & 0 \\ * & * & * & * \end{pmatrix}$, where

each * is a number, is called a *lower triangular* matrix. A set of linear equations with a triangular matrix is said to be in *échelon form*; the original meaning of échelon is ladder, but it is now the military term for an arrangement of troops in parallel groups each with its end clear of those ahead and behind (like the *'s in the matrix above).

If we can reduce a set of equations to échelon form, then we can easily complete the solution by substitution. The process used in this reduction is adding to one equation a multiple of another equation in order to produce a more convenient set of equations with the same solution set.

For example, if $x + 2y + 3z = 26$ (1)

$$3x + 2y + z = 39 \qquad (2)$$

$$2x + 3y + z = 34 \qquad (3)$$

then taking $3 \times (1)$ from (2) gives $-4y - 8z = -39$ (4)

taking $2 \times (1)$ from (3) gives $-y - 5z = -18$ (5)

and taking $4 \times (5)$ from (4) gives $12z = 33$ (6)

Equations (1), (5), (6) are now in échelon form:

$$x + 2y + 3z = 26$$

$$-y - 5z = -18$$

$$12z = 33,$$

from which $z = \frac{11}{4}$, $y = 18 - \frac{55}{4} = \frac{17}{4}$, $x = 26 - \frac{34}{4} - \frac{33}{4} = \frac{37}{4}$.

To ensure accuracy, it is essential to set this work out systematically. There is no need to keep writing x, y, z, since all the calculations are done on the coefficients. We therefore write the original equations as

x	y	z	
1	2	3	26
3	2	1	39
2	3	1	34

At each stage in the elimination we write down a complete set of three equations, indicating how any new rows are obtained from the previous equations. For example, to show that the new row 2 is the old row 2 minus $3 \times$ the old row 1 we write $r_2' := r_2 - 3r_1$. The complete working is then

	x	y	z	
	1	2	3	26
	3	2	1	39
	2	3	1	34
	1	2	3	26
$r_2' := r_2 - 3r_1$	0	-4	-8	-39
$r_3' := r_3 - 2r_1$	0	-1	-5	-18
	1	2	3	26
$r_2' := -r_3$	0	1	5	18
$r_3' := r_2 - 4r_3$	0	0	12	33

$\Leftrightarrow z = \frac{11}{4}$, $y = \frac{17}{4}$, $x = \frac{37}{4}$ as before.

This method of systematic elimination is very ancient. In fact these same equations are solved in this way in a Chinese book called *Nine Chapters on the Mathematical Art* written about 250 B.C.

Qu. 1.   Show that the equations $x + 2y + 3z = 5$, $2x + 3y + z = 4$, $3x + 5y + 4z = 10$ are inconsistent. Try to solve them by systematic elimination, and find where the method breaks down.

Qu. 2.   In the equations of Qu. 1, replace the 10 by 9. Investigate the solution of these new equations by systematic elimination.

When linear equations arise from a practical problem, the coefficients are unlikely to be small integers and the arithmetic can get heavy. It is worth introducing two refinements of the method.

1. In order to guard against mistakes, we use a row-sum check. At the end of each row we write the sum of all the entries. When a new row is formed the same row operation is carried out on these sums, and we then check that the new value that has been calculated is indeed the sum of the entries in the new row.

2. When eliminating x, we look first for the numerically largest coefficient of x; this is called the *pivot*, and the equation in which it occurs is the *pivotal equation*. We put the pivotal equation at the top, and then eliminate x from all the other equations by subtracting multiples of the pivotal equation from them. Leaving the top equation unchanged, we next choose a new pivotal equation from the others by finding the numerically greatest coefficient of y, and then carry on as before. The advantage of using the pivotal equation is that the multipliers used are then numerically less than 1; this keeps the rounding errors under control, and means that we can work to a constant number of decimal places.

Systematic elimination using pivots and a row-sum check is called *pivotal condensation* or *Gaussian elimination*.

Example 1: Solve these equations by pivotal condensation, working to 3 D.P.

$$0.61x + 0.12y + 0.37z = 0.40$$
$$0.22x + 0.39y - 0.11z = 0.16$$
$$0.18x - 0.54y + 0.28z = 0.42.$$

Solution: The pivots are shown in heavy type (overleaf).

	x	y	z		Σ	
	0.61	0.12	0.37	0.40	1.50	
	0.22	0.39	−0.11	0.16	0.66	
	0.18	−0.54	0.28	0.42	0.34	
	0.61	0.12	0.37	0.40	1.50	R
$r_2' := r_2 - \dfrac{0.22}{0.61}r_1$	0	0.347	−0.243	0.016	0.119 ✓ *	
$r_3' := r_3 - \dfrac{0.18}{0.61}r_1$	0	**−0.575**	0.171	0.302	−0.103 ✓	
	0.61	0.12	0.37	0.40	1.50	R
$r_2' := r_3$	0	**−0.575**	0.171	0.302	−0.103	R
$r_3' := r_2$	0	0.347	−0.243	0.016	0.119	R
	0.61	0.12	0.37	0.40	1.50	R
	0	−0.575	0.171	0.302	−0.103	R
$r_3' := r_3 + \dfrac{0.347}{0.575}r_2$	0	0	−0.140	0.198	0.057 ✓	

$$z = \frac{0.198}{-0.140} = -1.414, \quad y = \frac{0.302 + 0.171 \times 1.414}{-0.575} = -0.946$$

$$x = \frac{0.40 + 0.12 \times 0.946 + 0.37 \times 1.414}{0.61} = 1.700.$$

As a final check we substitute into the sum of the original equations, $1.01x - 0.03y + 0.54z = 0.98$. With these values of x, y, z the left-hand side equals
$1.01 \times 1.700 + 0.03 \times 0.946 - 0.54 \times 1.414 = 0.982$.
The discrepancy between this and the right-hand side is called the *residual*, 0.002. The residual is not zero because of rounding errors caused by working to 3 D.P.

Note (i) The entry * is calculated as $0.66 - \dfrac{0.22}{0.61} \times 1.50 = 0.119$
and checked as $0.347 - 0.243 + 0.016 = 0.120$. The discrepancy is due to rounding.

(ii) It is possible to shorten the working by omitting the rows marked R, which are repetitions of previous rows.

□

The row operations used in elimination are
(i) multiplying a row by a constant
(ii) interchanging two rows
(iii) adding to one row a constant multiple of another row.
The rows of a matrix **M** can be manipulated in the same ways by multiplying on the left by suitable matrices. For example

(i) $\begin{pmatrix} k & 0 & 0 \\ 0 & 1 & 0 \\ 0 & 0 & 1 \end{pmatrix} \begin{pmatrix} a_1 & b_1 & c_1 \\ a_2 & b_2 & c_2 \\ a_3 & b_3 & c_3 \end{pmatrix} = \begin{pmatrix} ka_1 & kb_1 & kc_1 \\ a_2 & b_2 & c_2 \\ a_3 & b_3 & c_3 \end{pmatrix}$

(ii) $\begin{pmatrix} 0 & 1 & 0 \\ 1 & 0 & 0 \\ 0 & 0 & 1 \end{pmatrix} \begin{pmatrix} a_1 & b_1 & c_1 \\ a_2 & b_2 & c_2 \\ a_3 & b_3 & c_3 \end{pmatrix} = \begin{pmatrix} a_2 & b_2 & c_2 \\ a_1 & b_1 & c_1 \\ a_3 & b_3 & c_3 \end{pmatrix}$

(iii) $\begin{pmatrix} 1 & k & 0 \\ 0 & 1 & 0 \\ 0 & 0 & 1 \end{pmatrix} \begin{pmatrix} a_1 & b_1 & c_1 \\ a_2 & b_2 & c_2 \\ a_3 & b_3 & c_3 \end{pmatrix} = \begin{pmatrix} a_1 + ka_2 & b_1 + kb_2 & c_1 + kc_2 \\ a_2 & b_2 & c_2 \\ a_3 & b_3 & c_3 \end{pmatrix}.$

The matrices used on the left here are called *elementary* matrices. Each one can be found by applying the required row operation to the identity matrix.

Qu. 3. Write down the elementary matrix for
(i) interchanging rows 1 and 3   (ii) changing the signs of row 2
(iii) subtracting 4 times row 2 from row 3

Qu. 4. What is the effect of multiplying **M** on the right by an elementary matrix?

Elementary matrices provide an important alternative way of finding the inverse of a matrix. If by multiplying by a sequence of elementary matrices $E_1, E_2, \ldots, E_n$, we can reduce **M** to the identity matrix **I**, we then have

$$E_n \ldots E_2 E_1 M = I$$

so that $\qquad\qquad\qquad M^{-1} = E_n \ldots E_2 E_1.$

It is not necessary to write down the individual elementary matrices. All we need is their product, which we can find by applying the same row operations to the identity matrix, since $E_n \ldots E_2 E_1 I = E_n \ldots E_2 E_1$. The work is set out in parallel columns as shown below. The first stages are the same as the reduction to échelon form we have used before, but having reached échelon form we continue so as to eliminate the entries above the leading diagonal too.

$$\begin{pmatrix} 1 & 2 & 3 \\ 2 & 3 & 1 \\ 3 & 2 & 1 \end{pmatrix} \quad \vdots \quad \begin{pmatrix} 1 & 0 & 0 \\ 0 & 1 & 0 \\ 0 & 0 & 1 \end{pmatrix}$$

$r_2' := r_2 - 2r_1$
$r_3' := r_3 - 3r_1$
$$\begin{pmatrix} 1 & 2 & 3 \\ 0 & -1 & -5 \\ 0 & -4 & -8 \end{pmatrix} \quad \vdots \quad \begin{pmatrix} 1 & 0 & 0 \\ -2 & 1 & 0 \\ -3 & 0 & 1 \end{pmatrix}$$

$r_2' := -r_2$
$r_3' := r_3 - 4r_2$
$$\begin{pmatrix} 1 & 2 & 3 \\ 0 & 1 & 5 \\ 0 & 0 & 12 \end{pmatrix} \quad \vdots \quad \begin{pmatrix} 1 & 0 & 0 \\ 2 & -1 & 0 \\ 5 & -4 & 1 \end{pmatrix}$$

$$r_3' := r_3 \div 12 \quad \begin{pmatrix} 1 & 2 & 3 \\ 0 & 1 & 5 \\ 0 & 0 & 1 \end{pmatrix} \;\vdots\; \begin{pmatrix} 1 & 0 & 0 \\ 2 & -1 & 0 \\ \frac{5}{12} & -\frac{1}{3} & \frac{1}{12} \end{pmatrix}$$

$$\begin{aligned} r_1' &:= r_1 - 3r_3 \\ r_2' &:= r_2 - 5r_3 \end{aligned} \quad \begin{pmatrix} 1 & 2 & 0 \\ 0 & 1 & 0 \\ 0 & 0 & 1 \end{pmatrix} \;\vdots\; \begin{pmatrix} -\frac{1}{4} & 1 & -\frac{1}{4} \\ -\frac{1}{12} & \frac{2}{3} & -\frac{5}{12} \\ \frac{5}{12} & -\frac{1}{3} & \frac{1}{12} \end{pmatrix}$$

$$r_1' := r_1 - 2r_2 \quad \begin{pmatrix} 1 & 0 & 0 \\ 0 & 1 & 0 \\ 0 & 0 & 1 \end{pmatrix} \;\vdots\; \begin{pmatrix} -\frac{1}{12} & -\frac{1}{3} & \frac{7}{12} \\ -\frac{1}{12} & \frac{2}{3} & -\frac{5}{12} \\ \frac{5}{12} & -\frac{1}{3} & \frac{1}{12} \end{pmatrix}$$

The final matrix on the right is the required inverse.

For a 3×3 matrix with simple numbers, it is probably quicker to find the inverse by the adjoint method. The row operation method is definitely better for larger matrices, since it requires many fewer arithmetic steps. For example, to invert a 10×10 matrix takes about 10^8 steps by the adjoint method, but only about 3000 steps by row operations. The row operation method is easy to programme for a computer, which is another great advantage.

Exercise 9F

1. Solve the equations
$$\begin{aligned} 2x + 4y + 5z &= -3 \\ 4x - y - 7z &= 6 \\ 6x + 3y - z &= 3 \end{aligned}$$
by systematic elimination.

2. Solve the equations
$$\begin{aligned} p + 2q + 3r - s &= -4 \\ 4p - q - 7r - 3s &= 0 \\ 2p + 3q + r - 2s &= -1 \\ -3p - 4q + 5r + 6s &= 5 \end{aligned}$$
by systematic elimination, using a row-sum check.

3. The chemical equation for the explosion of nitroglycerine is
$$x\,C_3H_5N_3O_9 \rightarrow y\,CO_2 + z\,H_2O + t\,N_2 + u\,O_2.$$
Find the smallest integer values of x, y, z, t, u which make the equation balance.

4. Solve the equations
$$\begin{aligned} 4.3x + 2.1y &= 8.4 \\ 2.7x + 3.9y &= 5.2 \end{aligned}$$
by pivotal condensation, working to 2 D.P.

5. Illustrate the method of pivotal condensation by solving the system of equations
$$\begin{aligned} 4x_1 + x_2 + x_3 &= 1 \\ x_1 + 3x_2 + x_3 &= 1 \\ 2x_1 + x_2 + 3x_3 &= 2. \end{aligned}$$

Give your solution to three places of decimals. Show what checking procedure can be applied to your results.

Calculate the residuals and state whether your results imply anything about the accuracy of your solution.   [MEI]

6. Use the method of pivotal condensation, with a sum check, or any other suitable method to find the values of x_1, x_2, x_3 from the following simultaneous equations. Tabulate the details of your working, and give your results to 3 S.F.

$$9.37x_1 + 3.04x_2 - 2.44x_3 = 9.23$$
$$3.04x_1 + 6.18x_2 + 1.22x_3 = 8.20$$
$$-2.44x_1 + 1.22x_2 + 8.44x_3 = 3.93.$$

Find the residual when your results are substituted in the second equation. Comment on the value of this residual as a guide to the accuracy of your results.   [MEI]

7. It can be shown that the number of arithmetic steps needed to solve n simultaneous equations in n unknowns is $\dfrac{n}{3}(n^2 + 3n - 1)$ by systematic elimination and approximately $1.72n \times n!$ by Cramer's rule. Allowing $1 \, \mu s$ (10^{-6} second) per step, estimate the time taken by a computer to solve 100 equations in 100 unknowns by each method. ($100! \approx 10^{158}$).

8. Use row operations to find the inverses of

(i) $\begin{pmatrix} 2 & 8 \\ 3 & 5 \end{pmatrix}$ (ii) $\begin{pmatrix} 4 & 2 & 1 \\ 3 & 1 & 2 \\ 3 & 5 & 1 \end{pmatrix}$ (iii) $\begin{pmatrix} 1 & 8 & 5 \\ 2 & 10 & 7 \\ 9 & 7 & 3 \end{pmatrix}$.

9. Try to find the inverse of $\begin{pmatrix} 4 & 1 & 12 \\ 3 & 5 & -25 \\ 7 & 4 & 3 \end{pmatrix}$ by using row operations.

Comment on what happens.

10. (i) Find the lower triangular matrix $\mathbf{L} = \begin{pmatrix} * & 0 & 0 \\ * & * & 0 \\ * & * & * \end{pmatrix}$ and the upper triangular matrix $\mathbf{U} = \begin{pmatrix} 1 & * & * \\ 0 & 1 & * \\ 0 & 0 & 1 \end{pmatrix}$ such that $\mathbf{LU} = \mathbf{M}$, where

$$\mathbf{M} = \begin{pmatrix} 2 & 10 & -2 \\ -1 & -4 & 3 \\ 4 & 22 & 3 \end{pmatrix}.$$

(ii) Show that the equation $\mathbf{Mp} = \mathbf{d}$ can be solved for $\mathbf{p}$ by first solving $\mathbf{Lr} = \mathbf{d}$ for $\mathbf{r}$ and then solving $\mathbf{Up} = \mathbf{r}$ for $\mathbf{p}$.

(iii) Use this method to solve the equations

$$2x + 10y - 2z = 26$$
$$-x - 4y + 3z = -15$$
$$4x + 22y + 3z = 45.$$

[This is *Choleski's method* for solving linear equations.]

11. Solve these equations by Choleski's method.

    (i) $3x + 7y + 8z = 6$
       $x + 6y + 2z = 2$
     $2x + 10y + 5z = 12$

    (ii) $3x + 7y + 5z = 0$
        $x + 6y - 2z = -11$
        $x + 5y - z = -8.$

12. Express $\mathbf{M} = \begin{pmatrix} 3 & 4 & 1 \\ 2 & -2 & 3 \\ 5 & 1 & 6 \end{pmatrix}$ as the product of lower and upper triangular

matrices $\mathbf{L}$ and $\mathbf{U}$ as in Qu. 10(i). Use the fact that $(\mathbf{LU})^{-1} = \mathbf{U}^{-1}\mathbf{L}^{-1}$ to find $\mathbf{M}^{-1}$.

13. Find the exact solutions of (i) $x + 2y = 2$ (ii) $x + 2y = 2.$
                           $2x + 4.01y = 5$ $2x + 3.99y = 5$

Equations such as these, in which a small change in a coefficient can produce large changes in the solutions, are called *ill-conditioned*. Sketch these two pairs of lines in $\mathbb{R}^2$ to illustrate what is happening.

14. (i) Prove that $\mathbf{M} = \begin{pmatrix} 1 & k & 3 \\ 1 & 3 & -1 \\ 2 & 5 & k \end{pmatrix}$ is singular when $k = 2$ and when $k = -1$.

    (ii) Find the general solution of the equations $\mathbf{M}\begin{pmatrix} x \\ y \\ z \end{pmatrix} = \begin{pmatrix} -3 \\ 5 \\ 2 \end{pmatrix}$ when

    $k = 2$. Find also the solutions of these equations
    (a) when $k = 2.1$ (b) when $k = 1.9$.

    (iii) Show that the equations $\mathbf{M}\begin{pmatrix} x \\ y \\ z \end{pmatrix} = \begin{pmatrix} -3 \\ 5 \\ 2 \end{pmatrix}$ are inconsistent when
    $k = -1$.
    Solve these equations (a) when $k = -1.1$ (b) when $k = -0.9$.

    (iv) What does this example suggest about the circumstances in which equations are ill-conditioned?

9.7 Eigenvectors and eigenvalues

Instead of concentrating on the effect of a linear transformation on points, we now consider how position vectors are transformed. For example, if the matrix of the transformation is $\begin{pmatrix} 4 & 2 \\ 1 & 3 \end{pmatrix}$ then the image of $\begin{pmatrix} 1 \\ 1 \end{pmatrix}$ is

$\begin{pmatrix} 4 & 2 \\ 1 & 3 \end{pmatrix}\begin{pmatrix} 1 \\ 1 \end{pmatrix} = \begin{pmatrix} 6 \\ 4 \end{pmatrix}$, and similarly the image of $\begin{pmatrix} k \\ k \end{pmatrix}$ is $\begin{pmatrix} 6k \\ 4k \end{pmatrix}$ for all k. The points with position vectors $\begin{pmatrix} k \\ k \end{pmatrix}$ form the line $y = x$, and the points with position vectors $\begin{pmatrix} 6k \\ 4k \end{pmatrix}$ form the image of this, the line $y = \frac{2}{3}x$. In the same way, since $\begin{pmatrix} 1 \\ 2 \end{pmatrix}$ maps to $\begin{pmatrix} 8 \\ 7 \end{pmatrix}$, the image of the line $y = 2x$ is the line $y = \frac{7}{8}x$.

Qu. 1. Find the images under this transformation of

(i) $\begin{pmatrix} 1 \\ 0 \end{pmatrix}$ (ii) $\begin{pmatrix} 2 \\ 1 \end{pmatrix}$ (iii) $\begin{pmatrix} 0 \\ 1 \end{pmatrix}$ (iv) $\begin{pmatrix} -1 \\ 2 \end{pmatrix}$ (v) $\begin{pmatrix} -1 \\ 1 \end{pmatrix}$ (vi) $\begin{pmatrix} -2 \\ 1 \end{pmatrix}$.

Qu. 2. Write down the equations of the images under this transformation of the lines   (i) $y = 0$ (ii) $y = \frac{1}{2}x$ (iii) $x = 0$

            (iv) $y = -2x$ (v) $y = -x$ (vi) $y = -\frac{1}{2}x$.

The information we have gathered about this transformation is represented on this diagram. Parts of the object lines are shown inside the ring, and parts of their images are shown outside. The arrows indicate the directions in which the lines move when they are transformed.

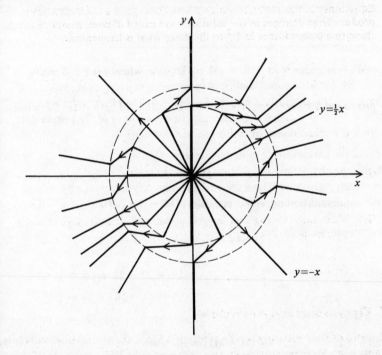

We notice that there are two *invariant* lines, $y = \frac{1}{2}x$ and $y = -x$, which map into themselves under this transformation. The other lines bunch towards $y = \frac{1}{2}x$, moving away from $y = -x$. This diagram prompts several questions. Might there be other invariant lines which we have not spotted? Why does $y = \frac{1}{2}x$ attract and $y = -x$ repel? Do all transformations have invariant lines? How can we find such lines efficiently?

Before answering these questions we introduce some terminology. A non-zero vector whose direction is unchanged by a transformation with matrix **M** is called an *eigenvector* of **M** (or characteristic vector or latent vector; 'eigen' is German for 'characteristic'). Thus $\begin{pmatrix} 2 \\ 1 \end{pmatrix}$, $\begin{pmatrix} -1 \\ 1 \end{pmatrix}$ and any scalar multiples of

these are eigenvectors of $\begin{pmatrix} 4 & 2 \\ 1 & 3 \end{pmatrix}$. Although the direction of an eigenvector is

unchanged by the transformation, its length may be altered. Thus

$\begin{pmatrix} 4 & 2 \\ 1 & 3 \end{pmatrix} \begin{pmatrix} 2k \\ k \end{pmatrix} = \begin{pmatrix} 10k \\ 5k \end{pmatrix} = 5\begin{pmatrix} 2k \\ k \end{pmatrix}$; the eigenvector is enlarged with scale

factor 5. This scale factor is called the *eigenvalue* (or characteristic or latent

value) corresponding to this eigenvector. Similarly, since

$\begin{pmatrix} 4 & 2 \\ 1 & 3 \end{pmatrix} \begin{pmatrix} -k \\ k \end{pmatrix} = 2\begin{pmatrix} -k \\ k \end{pmatrix}$ the eigenvector $\begin{pmatrix} -k \\ k \end{pmatrix}$ has eigenvalue 2.

> The non-zero vector **s** is an eigenvector of the square matrix **M**
> with eigenvalue λ if and only if $\mathbf{Ms} = \lambda \mathbf{s}$.

Now $\mathbf{Ms} = \lambda \mathbf{s} \Leftrightarrow \mathbf{Ms} - \lambda \mathbf{s} = \mathbf{0} \Leftrightarrow (\mathbf{M} - \lambda \mathbf{I})\mathbf{s} = \mathbf{0}$, where **I** is the identity
matrix. By §9.5 the condition for homogeneous equations $(\mathbf{M} - \lambda \mathbf{I})\mathbf{s} = \mathbf{0}$
to have non-trivial solutions is $\det(\mathbf{M} - \lambda \mathbf{I}) = 0$. The polynomial equation
$\det(\mathbf{M} - \lambda \mathbf{I}) = 0$ is called the *characteristic equation* of **M**. The procedure
for finding eigenvectors is as follows.

1. Form the characteristic equation $\det(\mathbf{M} - \lambda \mathbf{I}) = 0$.
2. Solve the characteristic equation to find the eigenvalues.
3. For each eigenvalue λ find the corresponding eigenvectors **s** by solving
   the consistent homogeneous equations $(\mathbf{M} - \lambda \mathbf{I})\mathbf{s} = \mathbf{0}$.

For example, when $\mathbf{M} = \begin{pmatrix} 4 & 2 \\ 1 & 3 \end{pmatrix}$,

$$\det(\mathbf{M} - \lambda \mathbf{I}) = 0 \Leftrightarrow \begin{vmatrix} 4-\lambda & 2 \\ 1 & 3-\lambda \end{vmatrix} = 0 \Leftrightarrow \lambda^2 - 7\lambda + 10 = 0$$
$$\Leftrightarrow (\lambda - 2)(\lambda - 5) = 0$$
$$\Leftrightarrow \lambda = 2 \text{ or } 5.$$

When $\lambda = 2$, $(\mathbf{M} - \lambda \mathbf{I})\mathbf{s} = \mathbf{0}$

$$\Leftrightarrow \begin{pmatrix} 2 & 2 \\ 2 & 1 \end{pmatrix} \begin{pmatrix} x \\ y \end{pmatrix} = 0, \text{ where } \mathbf{s} = \begin{pmatrix} x \\ y \end{pmatrix},$$

$$\Leftrightarrow x + y = 0 \Leftrightarrow \mathbf{s} = \begin{pmatrix} -k \\ k \end{pmatrix}.$$

When $\lambda = 5$, $(\mathbf{M} - \lambda \mathbf{I})\mathbf{s} = \mathbf{0}$

$$\Leftrightarrow \begin{pmatrix} -1 & 2 \\ 1 & -2 \end{pmatrix} \begin{pmatrix} x \\ y \end{pmatrix} = \mathbf{0}$$

$$\Leftrightarrow -x + 2y = 0 \Leftrightarrow \mathbf{s} = \begin{pmatrix} 2k \\ k \end{pmatrix}.$$

Thus we obtain the same eigenvalues and eigenvectors as before; we have now also shown that there can be no others for this **M**.

Qu. 3. Use this method to find the eigenvalues and eigenvectors of

(i) $\begin{pmatrix} 3 & 0 \\ 0 & -4 \end{pmatrix}$ (ii) $\begin{pmatrix} 1 & 2 \\ 0 & 1 \end{pmatrix}$ (iii) $\begin{pmatrix} 0 & -1 \\ 1 & 0 \end{pmatrix}$

(iv) $\begin{pmatrix} 3 & 0 \\ 0 & 3 \end{pmatrix}$ (v) $\begin{pmatrix} 2 & 4 \\ 3 & 6 \end{pmatrix}$.

Give a geometrical interpretation in each case.

We gain new insight into a transformation if we use its eigenvectors as base vectors, instead of **i** and **j**. Continuing with our example, the eigenvectors

$s_1 = \begin{pmatrix} -1 \\ 1 \end{pmatrix}$ and $s_2 = \begin{pmatrix} 2 \\ 1 \end{pmatrix}$ are non-parallel and non-zero, and so the position

vector **p** of any point in $\mathbb{R}^2$ can be expressed uniquely in the form $\mathbf{p} = \alpha s_1 + \beta s_2$. Then $\mathbf{Mp} = \alpha \mathbf{M} s_1 + \beta \mathbf{M} s_2 = 2\alpha s_1 + 5\beta s_2$.

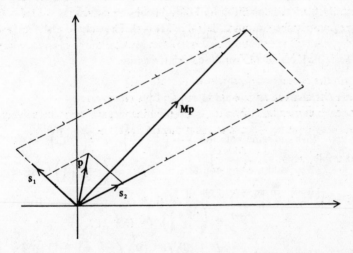

The s_1 and s_2 components have been multiplied by the eigenvalues 2 and 5 respectively. This explains why the lines in our original diagram swing towards s_2 and away from s_1: the eigenvector which attracts is the one with the numerically larger eigenvalue.

The two statements $\begin{pmatrix} 4 & 2 \\ 1 & 3 \end{pmatrix}\begin{pmatrix} -1 \\ 1 \end{pmatrix} = 2\begin{pmatrix} -1 \\ 1 \end{pmatrix}$ and $\begin{pmatrix} 4 & 2 \\ 1 & 3 \end{pmatrix}\begin{pmatrix} 2 \\ 1 \end{pmatrix} = 5\begin{pmatrix} 2 \\ 1 \end{pmatrix}$ can

be combined in the single statement $\begin{pmatrix} 4 & 2 \\ 1 & 3 \end{pmatrix}\begin{pmatrix} -1 & 2 \\ 1 & 1 \end{pmatrix} = \begin{pmatrix} -1 & 2 \\ 1 & 1 \end{pmatrix}\begin{pmatrix} 2 & 0 \\ 0 & 5 \end{pmatrix}$, or

$\mathbf{MS} = \mathbf{S}\Lambda$, where $\mathbf{S} = \begin{pmatrix} -1 & 2 \\ 1 & 1 \end{pmatrix}$, $\Lambda = \begin{pmatrix} 2 & 0 \\ 0 & 5 \end{pmatrix}$.

Similarly if any 2×2 matrix $\mathbf{M}$ has eigenvalues λ_1, λ_2 with corresponding eigenvectors $\mathbf{s}_1, \mathbf{s}_2$ then $\mathbf{M}\mathbf{s}_1 = \lambda_1 \mathbf{s}_1$ and $\mathbf{M}\mathbf{s}_2 = \lambda_2 \mathbf{s}_2$ so that $\mathbf{MS} = \mathbf{S}\Lambda$, where $\mathbf{S} = (\mathbf{s}_1 \ \mathbf{s}_2)$, the matrix whose columns are eigenvectors, and $\Lambda = \begin{pmatrix} \lambda_1 & 0 \\ 0 & \lambda_2 \end{pmatrix}$, the diagonal matrix of corresponding eigenvalues. If $\mathbf{S}$ is non-singular then multiplying $\mathbf{MS} = \mathbf{S}\Lambda$ on the left by $\mathbf{S}^{-1}$ gives $\mathbf{S}^{-1}\mathbf{MS} = \Lambda$; we say that $\mathbf{M}$ has been *reduced to diagonal form*.

Qu. 4. Reduce $\begin{pmatrix} 6 & -3 \\ 7 & -4 \end{pmatrix}$ to diagonal form by this method.

Qu. 5. Show that this method fails to reduce $\begin{pmatrix} 6 & -5 \\ 5 & -4 \end{pmatrix}$ to diagonal form.

Multiplying $\mathbf{MS} = \mathbf{S}\Lambda$ on the right by $\mathbf{S}^{-1}$ gives $\mathbf{M} = \mathbf{S}\Lambda\mathbf{S}^{-1}$. We can use this to calculate powers of $\mathbf{M}$. For example

$$\mathbf{M}^3 = (\mathbf{S}\Lambda\mathbf{S}^{-1})(\mathbf{S}\Lambda\mathbf{S}^{-1})(\mathbf{S}\Lambda\mathbf{S}^{-1})$$
$$= \mathbf{S}\Lambda(\mathbf{S}\mathbf{S}^{-1})\Lambda(\mathbf{S}\mathbf{S}^{-1})\Lambda\mathbf{S}^{-1} = \mathbf{S}\Lambda^3\mathbf{S}^{-1}.$$

and similarly (or formally by induction) $\mathbf{M}^n = \mathbf{S}\Lambda^n\mathbf{S}^{-1}$. Since Λ is diagonal, $\Lambda^n = \begin{pmatrix} \lambda_1^n & 0 \\ 0 & \lambda_2^n \end{pmatrix}$, and so $\mathbf{M}^n$ can be calculated easily.

Example 1: Calculate $\mathbf{M}^n$, where $\mathbf{M} = \begin{pmatrix} 4 & 2 \\ 1 & 3 \end{pmatrix}$.

Solution: With the same notation, $\mathbf{S} = \begin{pmatrix} -1 & 2 \\ 1 & 1 \end{pmatrix}$, $\Lambda = \begin{pmatrix} 2 & 0 \\ 0 & 5 \end{pmatrix}$.

(We could use any non-zero multiples of $\begin{pmatrix} -1 \\ 1 \end{pmatrix}, \begin{pmatrix} 2 \\ 1 \end{pmatrix}$ in $\mathbf{S}$; we choose these for simplicity.)

$$\mathbf{S}^{-1} = \begin{pmatrix} -\frac{1}{3} & \frac{2}{3} \\ \frac{1}{3} & \frac{1}{3} \end{pmatrix}, \ \Lambda^n = \begin{pmatrix} 2^n & 0 \\ 0 & 5^n \end{pmatrix}.$$

Therefore $\mathbf{M}^n = \begin{pmatrix} -1 & 2 \\ 1 & 1 \end{pmatrix}\begin{pmatrix} 2^n & 0 \\ 0 & 5^n \end{pmatrix}\begin{pmatrix} -\frac{1}{3} & \frac{2}{3} \\ \frac{1}{3} & \frac{1}{3} \end{pmatrix}$

$$= \tfrac{1}{3}\begin{pmatrix} 2.5^n + 2^n & 2.5^n - 2^{n+1} \\ 5^n - 2^n & 5^n + 2^{n+1} \end{pmatrix}.$$

As a check, when $n = 1$ this gives $\tfrac{1}{3}\begin{pmatrix} 10+2 & 10-4 \\ 5-2 & 5+4 \end{pmatrix} = \mathbf{M}$. $\square$

Qu. 6. Does this result give $\mathbf{M}^n$ (i) when $n = 0$ (ii) when $n = -1$?

Qu. 7. Prove that $\lambda_1 \neq \lambda_2 \Rightarrow \mathbf{S}$ is non-singular.
[Hint: one proof takes the form
$\mathbf{S}$ singular $\Rightarrow \mathbf{s}_1 = \mu\mathbf{s}_2$ for some $\mu \Rightarrow \mathbf{M}\mathbf{s}_1 = \mu\mathbf{M}\mathbf{s}_2 \Rightarrow \ldots \Rightarrow \lambda_1 = \lambda_2.$]

The definitions of eigenvalue and eigenvector apply also to larger square matrices. The characteristic equation $\det(\mathbf{M} - \lambda \mathbf{I}) = 0$ of a real 3×3 matrix $\mathbf{M}$ is a cubic equation, and must therefore have at least one real root. Therefore every real 3×3 matrix has at least one real eigenvector, and every linear transformation of $\mathbb{R}^3$ has at least one invariant line. The procedure for finding the eigenvalues and eigenvectors of a 3×3 matrix is just the same as before, though except in the simplest cases the working is lengthy. There are also various approximate methods which can be used.

Exercise 9G

1. Find the eigenvalues of $\begin{pmatrix} 6 & -2 \\ 1 & 3 \end{pmatrix}$ and the corresponding unit eigenvectors. Which lines are invariant under the transformation with this matrix?

2. Find the equations of the lines which are invariant under the transformation with matrix $\begin{pmatrix} 2 & 1 \\ 3 & 4 \end{pmatrix}$. Describe how the other lines through the origin move in relation to these invariant lines.

3. Find the equation of the line onto which $x + y = 0$ is mapped by the transformation with matrix $\begin{pmatrix} 5 & 3 \\ 12 & 5 \end{pmatrix}$. Find also the values of k for which the line $y = kx$ is mapped onto itself.

4. Reduce $\begin{pmatrix} 3 & 4 \\ 5 & 2 \end{pmatrix}$ to diagonal form.

5. Find the eigenvalues and eigenvectors of $\mathbf{M} = \begin{pmatrix} 0.2 & 0.6 \\ 0.8 & 0.4 \end{pmatrix}$. Hence
   (i) reduce $\mathbf{M}$ to diagonal form; (ii) calculate $\mathbf{M}^n$.
   What is the limit of $\mathbf{M}^n$ as $n \to \infty$?

6. Explain geometrically why $\begin{pmatrix} \cos\theta & -\sin\theta \\ \sin\theta & \cos\theta \end{pmatrix}$ has no real eigenvalues unless $\theta = n\pi$, where n is an integer. If $\theta \neq n\pi$, find the complex eigenvalues, and show that the eigenvectors are independent of θ.

7. Show that the eigenvalues of $\begin{pmatrix} \cos 2\theta & -k\sin 2\theta \\ -\dfrac{1}{k}\sin 2\theta & -\cos 2\theta \end{pmatrix}$ are ± 1, and find the corresponding eigenvectors. For what values of k are these eigenvectors orthogonal? Describe the transformation geometrically when k has these values.

8. Prove that the eigenvalues of the real symmetric matrix $\begin{pmatrix} a & b \\ b & c \end{pmatrix}$ are real, and that the eigenvectors are orthogonal. Are there any exceptional cases?

9. Find the eigenvalues and eigenvectors of the matrix $\mathbf{A} = \begin{pmatrix} 25 & 40 \\ -12 & -19 \end{pmatrix}$.
Express the vector $\begin{pmatrix} 3 \\ -2 \end{pmatrix}$ as a linear combination of the eigenvectors and
deduce that $\mathbf{A}^n \begin{pmatrix} 3 \\ -2 \end{pmatrix} = 5^n \begin{pmatrix} -2 \\ 1 \end{pmatrix} + \begin{pmatrix} 5 \\ -3 \end{pmatrix}$.
[L]

10. A linear transformation of the plane has eigenvectors $\begin{pmatrix} -1 \\ 2 \end{pmatrix}$ and $\begin{pmatrix} 3 \\ 1 \end{pmatrix}$,
with corresponding eigenvalues -1 and 2. Write down the images under
the transformation of $\begin{pmatrix} -1 \\ 2 \end{pmatrix}$ and $\begin{pmatrix} 3 \\ 1 \end{pmatrix}$, and deduce from these the images
of $\begin{pmatrix} 1 \\ 0 \end{pmatrix}$ and $\begin{pmatrix} 0 \\ 1 \end{pmatrix}$. Hence write down the matrix of the transformation.
[SMP]

11. A sequence of rational numbers $\dfrac{p_0}{q_0}, \dfrac{p_1}{q_1}, \ldots, \dfrac{p_n}{q_n}, \ldots$ is such that
$\begin{pmatrix} p_{n+1} \\ q_{n+1} \end{pmatrix} = \begin{pmatrix} 1 & k \\ 1 & 1 \end{pmatrix}\begin{pmatrix} p_n \\ q_n \end{pmatrix}$, where $k > 0$.

   (i) Taking $k = 3$ obtain the first six terms of the sequence
       (a) when $p_0 = q_0 = 1$ (b) when $p_0 = 10, q_0 = 1$.
       Evaluate these terms to 3 D.P.
   (ii) Find the eigenvalues λ_1, λ_2 (with $|\lambda_1| > |\lambda_2|$) and eigenvectors
        s_1, s_2 of $\begin{pmatrix} 1 & k \\ 1 & 1 \end{pmatrix}$ in terms of k.
   (iii) If $\begin{pmatrix} p_0 \\ q_0 \end{pmatrix} = \alpha s_1 + \beta s_2$ show that $\dfrac{p_n}{q_n} = \sqrt{k}\left(\dfrac{\lambda_1^n \alpha - \lambda_2^n \beta}{\lambda_1^n \alpha + \lambda_2^n \beta} \right)$.

        Deduce that, whatever the values of p_0 and q_0, $\dfrac{p_n}{q_n} \to \sqrt{k}$ as $n \to \infty$.

        [A method like this for finding square roots was used by Theon of
        Smyrna, $c.125$ A.D.]

12. A transformation matrix $\mathbf{M}$ is given by $\mathbf{M} = \begin{pmatrix} 1 & 0 & 4 \\ 0 & 5 & 4 \\ 4 & 4 & 3 \end{pmatrix}$. If O is the
origin and A the point $(1, 2, 2)$, what are the coordinates of the image
of A under the transformation $\mathbf{M}$? If A' is the image of A, what is the
ratio of the lengths of OA and OA'?
A second point B has its image B' under $\mathbf{M}$ such that $\mathbf{OB'} = 3\mathbf{OB}$. Find a
set of possible coordinates of B.
A third point C, distinct from A and B, is related to its image C' under
$\mathbf{M}$ by $\mathbf{OC'} = k\,\mathbf{OC}$, where k is a scalar. Find the value of k and a set of
possible coordinates of C.
[MEI]

13. Show that $\begin{pmatrix} 1 \\ 1 \\ 0 \end{pmatrix}$ is an eigenvector of $\begin{pmatrix} 1 & 4 & -1 \\ -1 & 6 & -1 \\ 2 & -2 & 4 \end{pmatrix}$ and find the

corresponding eigenvalue. Hence solve the characteristic equation and
find the other eigenvectors.

14. Find the eigenvalues of the matrix $\mathbf{A} = \begin{pmatrix} 2 & -2 & 3 \\ 1 & 1 & 1 \\ 1 & 3 & -1 \end{pmatrix}$.

    Describe the effect on the line $x = y = z$ of the linear transformation

    defined by $\begin{pmatrix} x_2 \\ y_2 \\ z_2 \end{pmatrix} = \mathbf{A} \begin{pmatrix} x_1 \\ y_1 \\ z_1 \end{pmatrix}$, and find the matrix of the inverse transform-

    mation. Explain the nature of the transformation with matrix $(\mathbf{A} - 3\mathbf{I})$,
    where $\mathbf{I}$ is the unit matrix of order 3.   [L]

15. (i) A 3×3 magic square consists of the numbers 1 to 9 arranged so
        that the digits in each row, column and diagonal have the same
        sum. Give an example of such a square.

    (ii) If a 3×3 magic square is regarded as a matrix, prove that $\begin{pmatrix} 1 \\ 1 \\ 1 \end{pmatrix}$ is an

        eigenvector. Find the corresponding eigenvalue.
    (iii) Generalize (ii) for an $n \times n$ magic square containing the numbers
        1 to n^2.

16. (i) Show that if λ is an eigenvalue of $\mathbf{M}$ then λ^n is an eigenvalue of $\mathbf{M}^n$.
    (ii) A matrix $\mathbf{M}$ is *nilpotent* if $\mathbf{M}^n = \mathbf{0}$ for some n. What can be said
        about the eigenvalues of a nilpotent matrix?

17. Matrices $\mathbf{A}, \mathbf{D}$ are defined by $\mathbf{A} = \begin{pmatrix} 1 & -3 & -3 \\ -8 & 6 & -3 \\ 8 & -2 & 7 \end{pmatrix}$, $\mathbf{D} = \begin{pmatrix} 1 & 0 & 0 \\ 0 & 4 & 0 \\ 0 & 0 & 9 \end{pmatrix}$.

    For each of the eigenvalues 1, 4, 9 of $\mathbf{A}$ find a corresponding eigenvector.
    Write down a matrix $\mathbf{P}$ such that $\mathbf{P}^{-1}\mathbf{A}\mathbf{P} = \mathbf{D}$, and calculate $\mathbf{P}^{-1}$. Write
    down a matrix $\mathbf{C}$ such that $\mathbf{C}^2 = \mathbf{D}$. Hence, or otherwise, find a matrix
    $\mathbf{B}$ such that $\mathbf{B}^2 = \mathbf{A}$.   [SMP]

18. The 3×3 matrix $\mathbf{A}$ is said to be *similar* to the 3×3 matrix $\mathbf{B}$ if there
    exists a matrix $\mathbf{P}$ such that $\mathbf{A} = \mathbf{P}^{-1}\mathbf{B}\mathbf{P}$. Prove that
    (i) every 3×3 matrix is similar to itself
    (ii) if $\mathbf{A}$ is similar to $\mathbf{B}$ then $\mathbf{B}$ is similar to $\mathbf{A}$
    (iii) if $\mathbf{A}$ is similar to $\mathbf{B}$ and $\mathbf{B}$ is similar to $\mathbf{C}$ then $\mathbf{A}$ is similar to $\mathbf{C}$
    (iv) similar matrices have the same characteristic equation.

19. The *trace* $\mathrm{tr}(\mathbf{X})$ of a 3×3 matrix $\mathbf{X}$ with real elements is the sum of the
    elements in the main diagonal of $\mathbf{X}$. If $\mathbf{X}, \mathbf{Y}$ are such matrices and α, β
    are scalars, show that (i) $\mathrm{tr}(\alpha\mathbf{X} + \beta\mathbf{Y}) = \alpha\,\mathrm{tr}(\mathbf{X}) + \beta\,\mathrm{tr}(\mathbf{Y})$;
    (ii) $\mathrm{tr}(\mathbf{X}\mathbf{Y}) = \mathrm{tr}(\mathbf{Y}\mathbf{X})$; (iii) $\mathbf{X}\mathbf{Y} - \mathbf{Y}\mathbf{X}$ cannot equal the identity matrix.
    Suppose there exists a non-singular matrix $\mathbf{P}$ such that

    $\mathbf{P}^{-1}\mathbf{X}\mathbf{P} = \mathbf{D} = \begin{pmatrix} d_1 & 0 & 0 \\ 0 & d_2 & 0 \\ 0 & 0 & d_3 \end{pmatrix}$. Show that $\mathrm{tr}(\mathbf{X}) = d_1 + d_2 + d_3$ and

    deduce that $\mathrm{tr}(\mathbf{X})$ is the sum of the eigenvalues of $\mathbf{X}$. Prove that if
    $\mathrm{tr}(\mathbf{X}^k) = 0$ for every positive integer k, then $\mathbf{D}$ is the zero matrix and
    all the eigenvalues of $\mathbf{X}$ are 0.   [MEI]

20. Calculate the eigenvalues of the matrix $\mathbf{A} = \begin{pmatrix} 1 & 0 & 1 \\ 0 & 2 & 0 \\ 1 & 0 & 3 \end{pmatrix}$, and find three

eigenvectors, $\mathbf{k}_1, \mathbf{k}_2, \mathbf{k}_3$ of unit length, one for each eigenvalue. If $\mathbf{S}$ denotes the 3×3 matrix whose columns are the vectors $\mathbf{k}_1, \mathbf{k}_2, \mathbf{k}_3$, verify that (i) $\mathbf{S}^{\mathrm{T}}\mathbf{S} = \mathbf{S}\mathbf{S}^{\mathrm{T}} = \mathbf{I}$; (ii) $\mathbf{S}^{\mathrm{T}}\mathbf{A}\mathbf{S}$ is a diagonal matrix $\mathbf{D}$. Prove that for any integer $n \geqslant 1$, $\mathbf{A}^n = \mathbf{S}\mathbf{D}^n\mathbf{S}^{\mathrm{T}}$. [MEI]

21. Let $\mathbf{M} = \begin{pmatrix} 0 & 1 & 0 \\ 1 & 0 & 1 \\ 0 & 1 & 0 \end{pmatrix}$. Find a matrix $\mathbf{P}$ such that $\mathbf{P}^{\mathrm{T}}\mathbf{M}\mathbf{P}$ is diagonal and

$\mathbf{P}^{\mathrm{T}}\mathbf{P} = \mathbf{I}$.

22. (M) A unit mass P is suspended from a fixed point by a light spring. A second unit mass Q is suspended from P by a second light spring. For the upper spring, tension = $5 \times$ extension; for the lower spring, tension = $6 \times$ extension. The system is set in motion with P and Q moving vertically. At a later time t the displacements of P and Q from their equilibrium positions are x and y, as shown in the diagram.

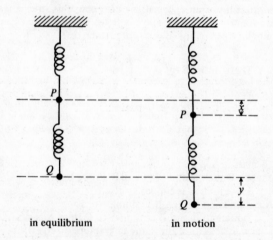

in equilibrium   in motion

(i)   Write down the equations of motion for P and Q.

(ii) Show that these can be written in the form $\ddot{\mathbf{r}} = \mathbf{A}\mathbf{r}$, where

$\mathbf{r} = \begin{pmatrix} x \\ y \end{pmatrix}$ and $\mathbf{A} = \begin{pmatrix} -11 & 6 \\ 6 & -6 \end{pmatrix}$.

(iii) We now seek solutions of $\ddot{\mathbf{r}} = \mathbf{A}\mathbf{r}$ of the form $\mathbf{r} = \mathbf{c}\cos(\omega t + \epsilon)$,

where $\mathbf{c} = \begin{pmatrix} a \\ b \end{pmatrix}$ with a, b, ω, ϵ constant. Show that for solutions of

this form these constants must satisfy $\mathbf{A}\mathbf{c} = -\omega^2\mathbf{c}$. Hence find the possible positive values of ω (it can be shown as in §7.4 that no generality is lost by taking $\omega > 0$), and the corresponding vectors $\mathbf{c}$.

(iv) Deduce that the only solutions of this form are

$$\begin{pmatrix} x \\ y \end{pmatrix} = \begin{pmatrix} 2k_1 \\ 3k_1 \end{pmatrix} \cos(\sqrt{2} . t + \epsilon_1) + \begin{pmatrix} 3k_2 \\ -2k_2 \end{pmatrix} \cos(\sqrt{15} . t + \epsilon_2).$$

23. A certain species of wheat occurs in three genetic strains A, B and C.
When these strains are grown under conditions which ensure self-
pollination, the A and C strains always breed true-to-type (that is, an A
plant will produce seed only of type A and a C plant will produce seed
only of type C), whereas the B strain produces seed of types A, B, C in
the proportions $1 : 2 : 1$. At any given time the proportions of the three
strains of wheat in the population are written as the three components
of the vector $\mathbf{x}$. Write down the matrix $\mathbf{P}$ which, when applied to $\mathbf{x}$,
gives the proportions of each strain in the next generation of wheat
plants (assume that every plant, whatever its strain, produces the same
number of seeds).

Show that $\begin{pmatrix} 1 \\ 0 \\ 0 \end{pmatrix}$ and $\begin{pmatrix} 0 \\ 0 \\ 1 \end{pmatrix}$ are both eigenvectors of $\mathbf{P}$ which correspond to

the eigenvalue 1, and give an interpretation of this result in the context
of the population of wheat plants. By finding another eigenvalue and a
corresponding eigenvector, or otherwise, prove that

$$\mathbf{P}^n = \begin{pmatrix} 1 & \frac{1}{2} - (\frac{1}{2})^{n+1} & 0 \\ 0 & (\frac{1}{2})^n & 0 \\ 0 & \frac{1}{2} - (\frac{1}{2})^{n+1} & 1 \end{pmatrix}.$$

What is the least number of generations that would be required before a
population in which the strains A, B, C occurred in equal proportions
became one with less than 1% of strain B?   [SMP]

24. (P) A system can be in one of three states 1, 2. 3. It changes state at
regular intervals in a random manner so that
Pr (change to state i given that it is in state j) $= p_{ij}$.
Show that $p_{1j} + p_{2j} + p_{3j} = 1$ for all j.
Hence show that one of the eigenvalues of the matrix p_{ij} is unity.
A maintenance team divides its time between 3 places, its base A, a
nearby plant at B and a more distant plant at C, spending a whole number
of days in each. After a day at A they go to B with probability $\frac{3}{4}$. After a
day at B they go to C with probability $\frac{2}{3}$. After a day at C they go back to
B with probability $\frac{1}{2}$. They never go directly from A to C or *vice versa*
and they never spend two consecutive days at B.
(a) Write down the transition matrix for this Markov chain (i.e. the
matrix p_{ij}).
(b) If the team spends Monday at A, find the probability of their being
at C on Thursday.
(c) By finding one of the eigenvectors, or otherwise, find the proportion
of its time that the team spends in each place, over a long period.
                                                                    [O & C]

10 Coordinate geometry

10.1 Polar coordinates

Polar coordinates are an alternative way of describing the position of a point in a plane, first used by Newton in 1671. Let O be a fixed point in the plane, called the *pole*, from which is drawn a half-line called the *initial line*.

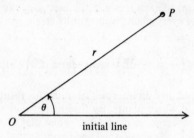

The initial line is usually drawn to the right across the page, like the positive x-axis. The position of any other point P in the plane is determined if we know the length OP, called r, and the angle θ which OP makes with the initial line. The numbers (r, θ) are called the *polar coordinates* of P. As usual the angle θ is positive if the sense of rotation is anti-clockwise from the initial line. For the pole itself $r = 0$ and θ may take any value.

Any pair of numbers (r, θ) will define a unique point, but the converse is not true. For example, the polar coordinates $\left(3, \frac{\pi}{4}\right)$, $\left(3, \frac{9\pi}{4}\right)$, $\left(3, -\frac{7\pi}{4}\right)$ all describe the same point.

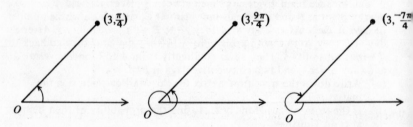

It is sometimes convenient to let r take negative values, by agreeing that the point $(-r, \theta)$ is the same as $(r, \theta + \pi)$. Then the point $\left(3, \frac{\pi}{4}\right)$ can also be described as $\left(-3, -\frac{3\pi}{4}\right)$ or $\left(-3, \frac{5\pi}{4}\right)$.

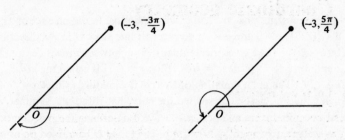

When it is necessary to specify unique polar coordinates of a point, we choose those for which $r > 0$ and $-\pi < \theta \leqslant \pi$; these are called the *principal* polar coordinates.

Qu. 1. Plot the points

(i) $\left(2, \dfrac{5\pi}{4}\right)$ (ii) $\left(-3, \dfrac{5\pi}{6}\right)$ (iii) $\left(1, -\dfrac{4\pi}{3}\right)$ (iv) $\left(-4, \dfrac{3\pi}{2}\right)$.

Qu. 2. Give the principal polar coordinates of the points in Qu. 1.

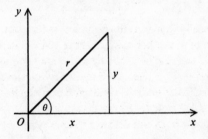

The relation between the polar and Cartesian coordinates of the same point is easily found from the diagram. Taking Ox as the initial line we have

$$
\begin{aligned}
x &= r\cos\theta, &\qquad y &= r\sin\theta \\
r &= \sqrt{(x^2 + y^2)}, &\qquad \theta &= \tan^{-1}\frac{y}{x} \\
& & &\text{or } \tan^{-1}\frac{y}{x} \pm \pi.
\end{aligned}
$$

When finding θ from x and y, care is needed to choose the correct quadrant.

Qu. 3. Find the Cartesian coordinates of the points of Qu. 1.

Qu. 4. Find the principal polar coordinates of the points whose Cartesian coordinates are

(i) $(-\sqrt{8}, \sqrt{8})$ (ii) $(1, -\sqrt{3})$ (iii) $(-5, 0)$ (iv) $(12, 5)$.

By placing restrictions on r and θ, we obtain loci in the plane. For example $\{(r, \theta) : r > 0 \text{ and } \theta = \alpha\}$ gives a half-line from O in the direction α, and $\{(r, \theta) : 2 < r < 3\}$ gives the annular region between circles centre O, radii 2 and 3. If $C = \{(r, \theta) : r = f(\theta)\}$ then $r = f(\theta)$ is called the *polar equation* of C. The polar equation of a curve is sometimes simpler than the Cartesian equation, particularly when the curve has rotational symmetry about O.

Example 1: Find the Cartesian equation of $r = 6 \sin \theta$, and sketch the curve.

Solution: If $r \neq 0$, $\quad r = 6 \sin \theta \Leftrightarrow r^2 = 6r \sin \theta$
$$\Leftrightarrow x^2 + y^2 = 6y.$$

If $r = 0$ then $x = y = 0$, which also satisfies $x^2 + y^2 = 6y$. Therefore the Cartesian equation is $x^2 + y^2 = 6y$, or $x^2 + (y - 3)^2 = 9$, which shows that the curve is a circle of radius 3 with centre at the Cartesian point $(0, 3)$. This can also be seen from the diagram where, from right-angled triangle OPQ, we have $r = 6 \sin \theta$.

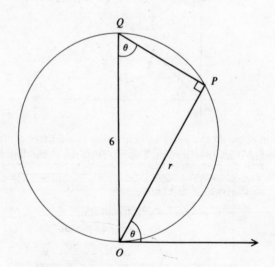

Qu. 5. Describe carefully how the point (r, θ) describes the circle $r = 6 \sin \theta$ as θ increase from 0 to 2π.

Example 2: Draw the curve whose polar equation is $r = 10 \cos 2\theta$. Find the Cartesian equation.

Solution:

θ	0	$\dfrac{\pi}{12}$	$\dfrac{\pi}{6}$	$\dfrac{\pi}{4}$	$\dfrac{\pi}{3}$	$\dfrac{5\pi}{12}$	$\dfrac{\pi}{2}$	$\dfrac{7\pi}{12}$	$\dfrac{2\pi}{3}$	$\dfrac{3\pi}{4}$	$\dfrac{5\pi}{6}$	$\dfrac{11\pi}{12}$	π
r	10	8.7	5	0	−5	−8.7	−10	−8.7	−5	0	5	8.7	10

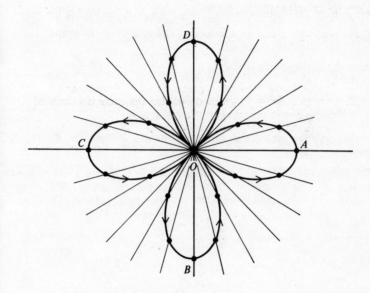

Plotting these points gives the curve $AOBOC$. As θ increases from π to 2π these values of r are repeated, giving the curve $CODOA$. The complete curve is a four-petal rhodonea (rose-curve). To find its Cartesian equation we have

$$r = 10 \cos 2\theta$$
$$= 10 \left(\cos^2 \theta - \sin^2 \theta \right)$$
$$= 10 \left(\frac{x^2}{r^2} - \frac{y^2}{r^2} \right)$$

so
$$r^3 = 10 (x^2 - y^2),$$

i.e.
$$(x^2 + y^2)^{\frac{3}{2}} = 10 (x^2 - y^2),$$

which is not nearly so simple to work with.   □

Exercise 10A

[Polar graph paper should be used in 1−4, with $\frac{1}{2}$ cm as unit.]

1. Tabulate values of $10\cos\theta$ for θ from 0 to π at intervals of $\frac{\pi}{12}$ (i.e. 15°).

   Hence draw the curve $r = 10\cos\theta$. Use the polar equation to prove that this curve is a circle, and check this by finding the Cartesian equation.

2. Draw on a single graph the curves
   (i) $r = 5(\frac{3}{2} + \cos\theta)$ (ii) $r = 5(1 + \cos\theta)$ (iii) $r = 5(\frac{1}{2} + \cos\theta)$.
   These curves are called *limaçons* (snail-shaped); the heart-shaped curve (ii) is also called the *cardioid*.

3. Draw the curve $r = 10\cos 3\theta$.

4. Draw on a single graph a *spiral of Archimedes* $r = \dfrac{9\theta}{\pi}$

   (i) for $0 \leqslant \theta \leqslant 2\pi$ (ii) for $-2\pi \leqslant \theta \leqslant 0$.

5. Prove that $r = a\sec\theta$ is the polar equation of a straight line. Show this line in a sketch.

6. The perpendicular from O to a straight line l is of length p and makes an angle α with the initial line. Prove that the polar equation of l is $r\cos(\theta - \alpha) = p$. Deduce from this the Cartesian equation of l.

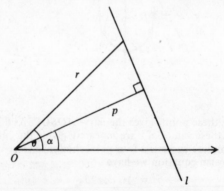

7. Prove that the polar equation of the straight line through the points A, B with polar coordinates (a, α), (b, β) can be written in the form

$$\frac{\sin(\beta - \alpha)}{r} = \frac{\sin(\beta - \theta)}{a} + \frac{\sin(\theta - \alpha)}{b}.$$

[Hint: let $P \equiv (r, \theta)$ be a point on AB, and consider areas AOB, AOP, BOP using '$\frac{1}{2}ab\sin C$'.]

8. Find the polar equation of the circle with radius b whose centre has polar coordinates (a, α). Check that your equation gives the correct result
   (i) when $a = 0$ (ii) when $a = b$.

9. Show that the polar equation $r \cos \left(\theta + \dfrac{\pi}{3} \right) + 1 = 0$ represents a straight line. Calling this line l, find

   (i) the polar coordinates of the point P in which l meets the line $r \cos \theta - 1 = 0$

   (ii) the polar equation of the line through the point $\left(\sqrt{3}, \dfrac{\pi}{2} \right)$ parallel to to l.

   Show that the polar equation of the circle with centre at the point P and radius 2 units is $r = 4 \cos \left(\theta - \dfrac{\pi}{3} \right)$.   [MEI, adapted]

10. Find the polar equations corresponding to these Cartesian equations.

    (i) $3x + 4y = 7$ (ii) $x^2 + y^2 = 8y$ (iii) $x^2 = 6y$

    (iv) $(x^2 + y^2)^2 = x^2 - y^2$.

11. Find the Cartesian equations corresponding to these polar equations.

    (i) $\theta = \dfrac{\pi}{3}$ (ii) $r^2 = 10 + 4r \cos \theta$ (iii) $r^2 = \sec \theta \ \mathrm{cosec} \ \theta$

    (iv) $r = \sqrt{\sin 2\theta}$.

12. Find the equation, in rectangular Cartesian coordinates, of the curve whose equation in polar coordinates is $r = a/(1 + \cos \theta)$, where $a > 0$. Sketch the curve.   [MEI]

13. Sketch in the same diagram the curves with polar equations $r = 2a \cos \theta$, $2r(1 + \cos \theta) = 3a$ and find the polar coordinates of their points of intersection. What is the polar equation of the common chord of the two curves?   [O & C]

14. If C is the curve $r = f(\theta)$ then the *inverse* of C is the curve $r = 1/f(\theta)$. Prove that

    (i) the inverse of a straight line through O is the same straight line

    (ii) the inverse of a straight line not through O is a circle through O

    (iii) the inverse of a circle through O is a straight line not through O

    (iv) the inverse of a circle not through O is a circle not through O.

15. (i) Describe the geometrical transformations which map the curve $r = f(\theta)$ onto (a) the curve $r = kf(\theta)$; (b) the curve $r = f(\theta - \alpha)$, where k and α are constants.

    (ii) Curves of the form $r = ab^\theta$ (a, b positive constants) are called *equiangular spirals*. Sketch an equiangular spiral.

    (iii) Show that the curve obtained by rotating an equiangular spiral S through angle α about O is precisely the same as the curve obtained by enlarging S from O with scale factor $b^{-\alpha}$.

    (iv) Deduce that the angle which the tangent at a point P of S makes with OP produced is constant for all positions of P. This accounts for the name equiangular. [These spirals occur frequently in nature; see D'Arcy Thompson: *On Growth and Form*, Chapter XI.]

16.

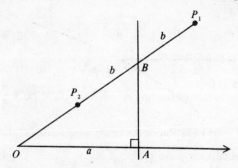

In about 200 B.C. the Greek Nicomedes invented a family of curves called *conchoids*, which are defined as follows.

Let A be the foot of the perpendicular from O to a fixed straight line, with $OA = a$. If B is any point on this line, then points P_1, P_2 are marked on OB at a fixed distance b from B. The locus of P_1 and P_2 is a conchoid, which has two branches. Taking $a = 3$ cm, draw conchoids with
(i) $b = 2$ cm   (ii) $b = 3$ cm   (iii) $b = 4$ cm.
Show that the polar equation of the conchoid is $(r - a \sec \theta)^2 = b^2$, and use it to find the Cartesian equation. [Nicomedes used conchoids to trisect angles: see D. E. Smith: *History of Mathematics, II*, p.298 for details.]

10.2 The parabola

In Vol. 1, §3.7 we showed that all curves with quadratic equations $y = ax^2 + bx + c$ $(a \neq 0)$ have the same shape; these curves are called parabolas. Like the circle, the parabola has a wealth of geometrical properties, some of which we now consider. It is more convenient to start from a different definition of a parabola, and then show later that this gives a curve with a quadratic Cartesian equation.

A *parabola* is defined as the locus of a point P whose distance from a fixed point S equals its distance from a fixed straight line l not through S,

i.e. $\{P : PS = PM\}$,

where M is the foot of the perpendicular from P to l. The fixed point S is called the *focus* of the parabola, and the fixed line l is called the *directrix*.

Qu. 1.   Mark on squared paper a straight line l and a point S 2 cm from l. Draw a circle of radius r cm and a line parallel to l on the same side as S and r cm from l, and mark their points of intersection for $r = 1, 2, 3, \ldots$. Hence draw the parabola with focus S and directrix l.

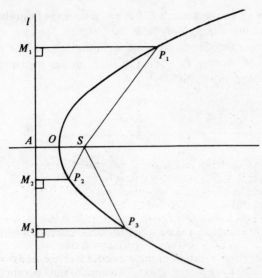

If A is the foot of the perpendicular from S to l, it is clear from the definition that the parabola is symmetrical about SA, which is called the *axis* of the parabola. The midpoint O of SA is equidistant from S and l, and so lies on the parabola; O is called the *vertex* of the parabola. The graph of $y = ax^2 + bx + c$ has its axis parallel to the y-axis, with vertex up if $a < 0$ and vertex down if $a > 0$. In this section however we shall usually draw the axis across the page, with the vertex to the left.

The defining property $PS = PM$ leads directly to an important fact about the tangent to the curve at the point P. Consider the point P to be moving along the curve. Then the direction of the tangent at P is the direction of the

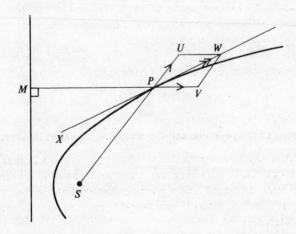

velocity of P. Since PS always equals PM, P is moving away from the focus at the same rate as it is moving away from the directrix. The velocity of P is therefore the vector sum of equal components in the directions SP and MP, the parallelogram of velocities $PUWV$ is a rhombus, and so the tangent XPW bisects angle UPV and angle SPM.

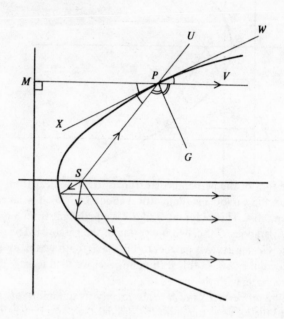

A useful property follows from this. With the notation of the figure,

$$\hat{SPX} = \tfrac{1}{2}\hat{SPM}$$
$$= \tfrac{1}{2}\hat{UPV}$$
$$= \hat{WPV}.$$

Therefore SP and PV are equally inclined to the tangent XPW, and also to the normal PG. A ray of light starting from S and reflected at P on the curve will emerge along PV, which is parallel to the axis. This is called the *parabolic reflector property*. It explains why the parabolic reflectors of car headlights, searchlights, electric fires, etc., product parallel beams. Conversely, incoming rays parallel to the axis are concentrated at the focus by a parabolic reflector; such reflectors are essential parts of reflecting and radio telescopes.

The following questions use the notation of the figure. You should draw your own diagrams, copying only those parts which are needed as you go along.

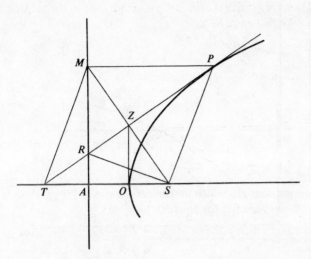

Qu. 2.   The tangent at P meets SM at Z. Prove that PZ is the perpendicular bisector of SM.

Qu. 3.   Mark a point S about 4 cm from a straight edge l of a piece of paper. Fold the paper so that l passes through S, and make a sharp crease. Do this repeatedly, changing the position of the crease slightly each time. Prove that all the creases touch a parabola.

Qu. 4.   Prove that Z lies on the tangent at the vertex O.

Qu. 5.   Draw a straight line m and a point S not on it. Place a set square with the right angle on m and one arm of the right angle through S. Draw the other arm of the right angle. Do this repeatedly, moving the set square slightly each time. Use Qu. 2 and Qu. 4 to prove that all the lines you draw touch a parabola.

Qu. 6.   The tangent meets the directrix at R. Prove that
(i) $R\hat{S}P$ is a right angle   (ii) RP bisects $M\hat{R}S$.

Qu. 7.   The tangent meets the axis at T. Prove that $STMP$ is a rhombus.

In order to apply the powerful machinery of coordinate geometry to the parabola we introduce coordinate axes, taking the vertex O as origin and the axis of the parabola as the x-axis. Then the focus S has coordinates

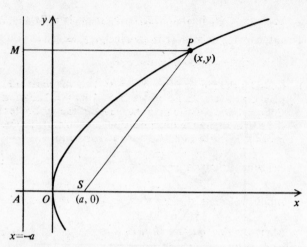

$(a, 0)$, where a is a constant which determines the size of the parabola, and since $AO = OS$ the directrix has equation $x = -a$.

The point $P(x, y)$ lies on the parabola $\Leftrightarrow PS = PM$

$$\Leftrightarrow PS^2 = PM^2$$

$$\Leftrightarrow (x - a)^2 + y^2 = (x + a)^2$$

$$\Leftrightarrow y^2 = 4ax.$$

Therefore the equation of the parabola referred to these axes is $y^2 = 4ax$. The fact that x is a quadratic function of y shows that the focus-directrix definition does give the same curve as the original definition.

There is a simple parametric representation for points on the parabola $y^2 = 4ax$. If (x, y) is any such point, let $t = \dfrac{y}{2a}$; then $x = \dfrac{y^2}{4a} = \dfrac{4a^2 t^2}{4a} = at^2$.

Therefore any point on the parabola can be given in the form $(at^2, 2at)$, and it is easy to check that every point of this form lies on the parabola.

Qu. 8.   Mark on a diagram the position of $(at^2, 2at)$ for $t = -3, -2, -1, 0, 1, 2, 3$.

Example 1: Find the equation of the chord joining the points $P_1(at_1^2, 2at_1)$ and $P_2(at_2^2, 2at_2)$ of the parabola $y^2 = 4ax$.

Solution (i):   The gradient of $P_1 P_2$ is $\dfrac{2at_1 - 2at_2}{at_1^2 - at_2^2} = \dfrac{2(t_1 - t_2)}{t_1^2 - t_2^2} = \dfrac{2}{t_1 + t_2}$.

The line through P_1 with this gradient has equation

$$y - 2at_1 = \frac{2}{t_1 + t_2}(x - at_1^2)$$

or $\qquad (t_1 + t_2)y - 2at_1(t_1 + t_2) = 2x - 2at_1^2$

or $\qquad 2x - (t_1 + t_2)y + 2at_1t_2 = 0.$

This is the equation of chord P_1P_2.   □

Solution (ii): Consider the line $lx + my + n = 0$. This will meet the parabola at $(at^2, 2at)$ if $lat^2 + 2mat + n = 0$. The roots of this quadratic equation are the values of the parameter t at the points where the line meets the parabola. If the line is the chord P_1P_2 these roots are t_1, t_2. In this case,

$$\text{sum of roots} = -\frac{2ma}{la} = t_1 + t_2 \Rightarrow m = -\frac{t_1 + t_2}{2}l$$

$$\text{product of roots} = \frac{n}{la} = t_1t_2 \Rightarrow n = at_1t_2l.$$

Therefore the equation of P_1P_2 is $lx - \dfrac{t_1 + t_2}{2}ly + at_1t_2l = 0$

or $2x - (t_1 + t_2)y + 2at_1t_2 = 0$, as before.   □

Qu. 9. By letting $t_2 \to t_1$ show that
(i) the gradient of the tangent at P_1 is $1/t_1$
(ii) the equation of this tangent is $x - t_1y + at_1^2 = 0.$
Prove these results by an alternative method, using calculus.

Qu. 10. P_1P_2 is called a *focal chord* if it passes through S. Show that in this case $t_1t_2 = -1$.

Example 2: Prove that the tangents at the ends of a focal chord meet at right angles on the directrix.

Solution (i): Let the tangents at P_1, P_2 meet the directrix at R_1, R_2 respectively. By Qu. 6 (i) $R_1S \perp SP_1$ and $R_2S \perp SP_2$. But P_1SP_2 is a straight line. Therefore R_1S and R_2S coincide, and both tangents meet the directrix at the same point R. By Qu. 6(ii)

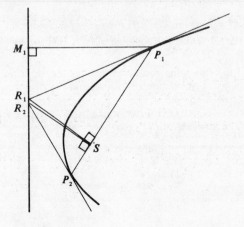

RP_1 is the internal bisector of angle M_1RS and RP_2 is the external bisector. Therefore $RP_1 \perp RP_2$.   □

Solution (ii):   By Qu. 9 (i) the gradients of the tangents at P_1, P_2 are $\dfrac{1}{t_1}, \dfrac{1}{t_2}$, and by Qu. 10, $t_1 t_2 = -1$. The product of the gradients is -1, and so the tangents are perpendicular. The equations of the tangents are

$$x - t_1 y + at_1^2 = 0 \qquad \text{(A)}$$
$$x - t_2 y + at_2^2 = 0 \qquad \text{(B)}$$

$t_2 \times \text{(A)} - t_1 \times \text{(B)}: \quad (t_2 - t_1)x + at_1^2 t_2 - at_2^2 t_1 = 0$

$\Leftrightarrow \quad (t_2 - t_1)x = at_1 t_2 (t_2 - t_1)$

$\Leftrightarrow \quad x = at_1 t_2 = -a$ since $t_1 t_2 = -1$.

Therefore the tangents meet on the directrix.   □

As these examples show, there is often more than one way to solve a problem, so flexibility is important. In the exercise that follows you may use any suitable method. Always draw a diagram, and try not to let the algebra swamp the geometry.

A parabola is the locus $\{P : PS = PM\}$.

With vertex and axis as origin and x-axis its equation is $y^2 = 4ax$ with focus $(a, 0)$ and directrix $x = -a$.

The parametric equations are $x = at^2, y = 2at$.

The equation of the chord $P_1 P_2$ is $2x - (t_1 + t_2)y + 2at_1 t_2 = 0$.

The equation of the tangent at P is $x - ty + at^2 = 0$. The tangent bisects angle SPM.

Exercise 10B

1. The chord of a parabola through the focus perpendicular to the axis is called the *latus rectum* (upright side). Show that the length of the latus rectum of $y^2 = 4ax$ is $4a$.

2. For each of the following parabolas find the length of the latus rectum, the coordinates of the vertex and focus, and the equation of the directrix.
   (i) $y^2 = 16x$ (ii) $y^2 = -x$ (iii) $(y - 3)^2 = 4(x + 2)$
   (iv) $x^2 = 6y$ (v) $(y + 1)^2 = -8(x - 5)$ (vi) $(x + 4)^2 = 14 - 2y$.

3. Find the equation of the parabola
   (i)   with focus $(3, 4)$ and directrix $x = -1$
   (ii) with focus $(-5, 6)$ and vertex $(-5, 2)$
   (iii) with vertex $(2, -1)$ and directrix $x = 5$.

4. The reflector of a radio telescope is a parabolic dish of diameter 85 metres which is 10 metres deep at the centre. Find the distance of the aerial at the focus from the deepest point of the dish.

5. Write down the quadratic equation giving the x-coordinates of the points where the line $y = mx + c$ meets the parabola $y^2 = 4ax$. Find the condition for this to have coincident roots, and hence show that

$$y = mx + \frac{a}{m}$$ touches the parabola for all $m \, (\neq 0)$. Find the coordinates

of the point of contact in terms of a and m.

6. Find the coordinates of the point where the tangents to $y^2 = 4ax$ at $(ap^2, 2ap)$ and $(aq^2, 2aq)$ intersect.

7. Prove that the locus of the midpoints of parallel chords of a parabola is a line parallel to the axis (such a line is called a *diameter* of the parabola). Prove that the tangent at the point where the diameter meets the parabola is parallel to the chords, and that the tangents at the ends of any one of the chords meet on the diameter produced.

8. P, Q, R, P', Q', R' are points on a parabola such that PQ' is parallel to $P'Q$ and QR' is parallel to $Q'R$. Prove that RP' is parallel to $R'P$.

9. The tangents from the point (X, Y) touch the parabola $y^2 = 4ax$ at $P_1(at_1^2, 2at_1)$ and $P_2(at_2^2, 2at_2)$. Explain why t_1, t_2 are the roots of the equation $at^2 - Yt + X = 0$, and hence find $t_1 + t_2$ and $t_1 t_2$ in terms of X, Y, a. Deduce that the equation of the chord of contact $P_1 P_2$ is $yY = 2a(x + X)$. What is this line when (X, Y) is on the parabola?

10. Prove that a circle which has a focal chord as diameter touches the directrix.

11. P is the point $(at^2, 2at)$ on the parabola $y^2 = 4ax$ and OQ is the chord passing through the origin O and parallel to the tangent at P. Find the coordinates of the point of intersection, R, of the tangents at P and Q. Give a reason to show that the locus of R is another parabola. If S is the midpoint of OQ, prove that PSR is a right-angled triangle, and that the area of triangle PQR is $\frac{1}{2} a^2 t^3$.   [O & C]

12. Points $P(ap^2, 2ap)$ and $Q(aq^2, 2aq)$ lie on the parabola $y^2 = 4ax$ and O is the origin. Show that if the angle POQ is a right angle, then $pq = -4$. If P and Q vary so that this condition is satisfied, find the equation of the locus of the midpoint of PQ. State the nature of the locus.   [L]

13. Let P_0 be the parabola $y^2 = 4ax$ and let S_0 be its focus. Show that the midpoints of the chords of P_0 which pass through S_0 lie on a parabola P_1 and that its focus S_1 is the point $(3a/2, 0)$.
The points $S_2, S_3, \ldots, S_n, \ldots$ are derived similarly from $P_1, P_2, \ldots, P_{n-1}, \ldots$ respectively. Find x_n, the x-coordinate of S_n, and determine $\lim_{n \to \infty} x_n$.   [MEI]

14. The tangent and normal at a point P of a parabola meet the axis at T and G respectively, and N is the foot of the perpendicular from P to the axis. Prove that TN is bisected by the vertex, and that NG is constant.

15. Show that the equation of the normal to the parabola $y^2 = 4ax$ at the point $(at^2, 2at)$ is $y + tx = 2at + at^3$.

If this normal meets the parabola again at the point $(aT^2, 2aT)$, show that $t^2 + tT + 2 = 0$, and deduce that T^2 cannot be less than 8.

The line $3y = 2x + 4a$ meets the parabola at the points P and Q. Show that the normals at P and Q meet on the parabola.   [L]

16. Circles C_1, C_2 of radii r_1, r_2 respectively $(r_1 < r_2)$ have their centres on the axis of the parabola $y^2 = 4ax$ $(a > 0)$ and touch each other externally and the parabola internally. Prove that $r_2 - r_1 = 4a$.   [MEI]

17.

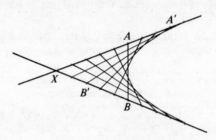

A common exercise in 'curve stitching' is shown in the diagram. The points which are joined are at equal steps along the lines XA and XB, and $XA = XB$.

  (i) Taking AB as the y-axis and the perpendicular from X to AB as the x-axis, show that the equations of XA and XB can be written as $y = hx + k$ and $y = -hx - k$, where h, k are constants.

 (ii) Show that the coordinates of corresponding points which are joined, such as A', B', can be taken as $(t, ht + k)$, $(-t, ht - k)$, where t is a parameter. Find the equation of the line joining these points.

(iii) Use the result of Qu. 5 to show that the curve obtained is a parabola, and find the coordinates of its focus.

 (iv) What happens if the steps on each line continue on AX produced and BX produced?

18. (M) A boy sets out to swim a river of width w. He always aims at the point on the opposite bank directly opposite his starting point. If his speed in still water is the same as the speed of the current, show that his path is a parabola, and that he will approach a point $\frac{1}{2}w$ downstream from his target.

19. The sides VW, WU, UV of a triangle touch $y^2 = 4ax$ at the points $P(ap^2, 2ap)$, $Q(aq^2, 2aq)$, $R(ar^2, 2ar)$ respectively.

  (i) Find the equation of the line through W perpendicular to UV in terms of p, q, r.

 (ii) Find the coordinates of the point where this line meets the directrix, and deduce that the orthocentre of triangle UVW lies on the directrix.

20. (M) Referred to horizontal and vertical axes through the point of projection, the initial velocity of a projectile is $\begin{pmatrix} u \\ v \end{pmatrix}$. Show that its position after time t is $\begin{pmatrix} ut \\ vt - \frac{1}{2}gt^2 \end{pmatrix}$ and hence that the equation of its path is $y = \dfrac{vx}{u} - \dfrac{gx^2}{2u^2}$. Show that this can be written as

$\left(x - \dfrac{uv}{g} \right)^2 = -\dfrac{2u^2}{g}\left(y - \dfrac{v^2}{2g} \right)$, and deduce the position of the vertex and the length of the latus rectum of this parabola.

Show that the directrix is at a height $\dfrac{V^2}{2g}$ above the point of projection, where $V = \sqrt{(u^2 + v^2)}$ is the initial speed. Notice that the height of the directrix does not depend on the direction of projection.

21.

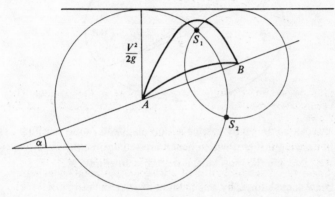

(M) A particle is projected with speed V from a point A on a sloping plane so as to hit the plane at a point B further up the slope. The following geometrical construction can be used to determine the possible paths. Draw the directrix at height $\dfrac{V^2}{2g}$ above A, and construct circles to touch this with centres A and B. If these circles intersect at S_1 and S_2, then S_1 and S_2 are the foci of the two possible parabolic trajectories.

(i)   Explain why this construction works.

(ii) By considering what happens as B moves up the slope, prove that
   (a) the greatest range up the slope occurs when the focus of the trajectory lies on the slope
   (b) the direction of projection then bisects the angle between AB and the vertical
   (c) the direction of motion at B is then perpendicular to the direction of motion at A
   (d) the greatest range up the slope is $\dfrac{V^2}{g(1 + \sin \alpha)}$, where α is the angle of slope.

10.3 The ellipse

There are several ways of defining an ellipse. We start with a definition which immediately gives a simple practical method for drawing ellipses: an *ellipse* is the locus of a point P the sum of whose distances from two fixed points S, S' is constant,

i.e. $\{P : PS + PS' = \text{constant}\}.$

Each fixed point, S or S', is called a *focus* of the ellipse.

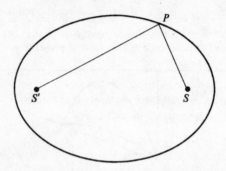

An ellipse can be drawn by passing a loop of thread round pins fixed at S and S', and drawing this taut with a pencil pressed against the paper. As the pencil moves, keeping the loop taut, it describes an ellipse.

Qu. 1.   Draw some ellipses by this method. Notice the effect of
      (i) changing the separation SS' without changing the length of
          the loop
      (ii) changing the length of the loop without changing SS'.

It is immediately clear from the definition that the ellipse is a closed curve symmetrical about both SS' and the perpendicular bisector of SS'. With the notation of the figure, AA' is called the *major axis*, and BB' is the

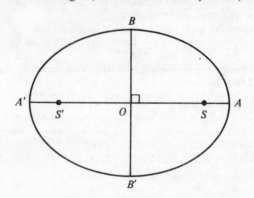

minor axis. These meet at the *centre O.* Any chord through O is called a *diameter*, and is bisected at O. If $OA = a$, then for any point P of the ellipse

$$SP + S'P = SA + S'A \qquad \text{using the definition}$$
$$= S'A' + S'A \qquad \text{since } SA = S'A' \text{ by symmetry}$$
$$= AA'$$
$$= 2a,$$

so the constant length in the definition equals the length of the major axis.

Suppose that $\dfrac{OS}{OA} = e$, so that $OS = ae$. The number e, which is less than 1, is called the *eccentricity*, since it gives a measure of how far each focus is from the centre.

Qu. 2. How does the shape of an ellipse change as e increases from 0 to 1? What 'ellipses' are obtained when $e = 0$ and when $e = 1$? Why is it impossible to have an ellipse with $e > 1$?

Qu. 3. Sketch the families of ellipses obtained by
  (i) keeping e fixed and changing a
 (ii) keeping a fixed and changing e.
In each case assume that the centre and direction of the major axis remain fixed.

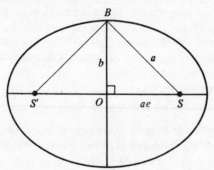

Since B is on the ellipse and is equidistant from S and S', $BS = BS' = a$. Therefore if $OB = b$ then from right-angled triangle OSB $a^2 e^2 + b^2 = a^2$, so that $b^2 = a^2(1 - e^2)$.

Qu. 4. What is the eccentricity of an ellipse with major and minor axes of lengths (i) 10 cm, 6 cm (ii) 25 cm, 24 cm (iii) 2 m, 1 m?

Qu. 5. The chord through a focus perpendicular to the major axis is called a *latus rectum*. Prove that the length of each latus rectum is $2b^2/a$.

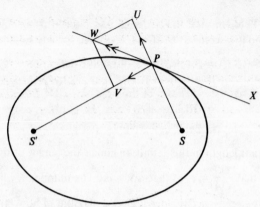

Like the parabola, the ellipse has a reflector property which follows directly from the definition. As the point P moves round the ellipse, $SP + S'P$ remains constant, so the rate at which SP increases equals the rate at which $S'P$ decreases. The velocity of P is therefore the vector sum of equal components in the directions SP and PS', the parallelogram of velocities $PUWV$ is a rhombus, and so the tangent XPW bisects angle UPV. Since angle UPW = angle SPX, the tangent XPW is equally inclined to SP and $S'P$. The normal at P is also equally inclined to SP and $S'P$, so that a ray of light starting from a focus S and reflected at P will pass through the other focus S'.

This is the reason for the term 'focus', which was introduced by Kepler in 1604. In 1609 Kepler published his discovery that the planets move in elliptical orbits, with the Sun at one focus (hence the traditional use of S for a focus). Over seventy years later, Newton proved that the elliptical orbit follows theoretically from his inverse square law of gravitation; this proof was eventually published in his *Principia* of 1687.

Qu. 6.   The perpendicular from S to the tangent at P meets that tangent at Q and meets $S'P$ produced at H. Prove that $HP = SP$, and deduce that the locus of H is a circle centre S'. What is the radius of this circle?

Qu. 7.   Mark a point S on a circular piece of paper (e.g. a filter paper). Fold the paper so that the circumference passes through S, and make a sharp crease. Do this repeatedly, changing the position of the crease slightly each time. Prove that all the creases touch an ellipse with one focus at S. Where is the other focus?

Qu. 8.   Use Qu. 6 to show that an ellipse can be defined as the locus of a point equidistant from a fixed circle and a fixed point inside the circle. Compare this with the definition of a parabola.

Qu. 9. With Q as in Qu. 6, prove that $OQ = a$ and deduce that Q lies on the circle (called the *auxiliary circle*) which has diameter AA'.

Qu. 10. Draw a circle and mark a point S inside it. Place a set square with the right angle on the circle and one arm of the right angle through S. Draw the other arm of the right angle. Do this repeatedly, moving the set square slightly each time. Use Qu. 9 to prove that all the lines you draw touch an ellipse.

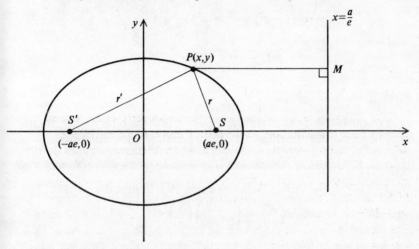

We now introduce coordinates, taking the major and minor axes as x- and y-axes respectively, so that the foci are $(\pm ae, 0)$. Let $P(x, y)$ be a point of the ellipse with $PS = r$, $PS' = r'$.

Applying the distance formula for coordinates,

$$r^2 = (x - ae)^2 + y^2, \ r'^2 = (x + ae)^2 + y^2.$$

Therefore $r^2 - r'^2 = (x - ae)^2 - (x + ae)^2 = -4aex.$
But $\quad r^2 - r'^2 = (r + r')(r - r') = 2a(2r - 2a)$ since $r + r' = 2a$

so that $\quad r - a = -ex$ and $r = e\left(\dfrac{a}{e} - x\right).$

Now $\dfrac{a}{e} - x$ is the distance PM of P from the line $x = \dfrac{a}{e}$. This line is

called the *directrix* corresponding to the focus S, and we have shown that the distance of P from S is e times the distance of P from this directrix. Thus the ellipse is the locus $\{P : PS = e\,PM\}$, where $e < 1$; compare this with the definition of a parabola given on page 282.

By symmetry the line $x = -\dfrac{a}{e}$ is also a directrix, corresponding to the

focus S', with $PS' = e\,PM'$.

The focus-directrix property leads easily to the equation of the ellipse. For $P \equiv (x,y)$ is on the ellipse $\Leftrightarrow PS = e\,PM$

$$\Leftrightarrow PS^2 = e^2\,PM^2$$

$$\Leftrightarrow (x - ae)^2 + y^2 = e^2\left(\frac{a}{e} - x\right)^2$$

$$\Leftrightarrow x^2 - 2aex + a^2e^2 + y^2 = a^2 - 2aex + e^2x^2$$

$$\Leftrightarrow x^2(1 - e^2) + y^2 = a^2(1 - e^2)$$

$$\Leftrightarrow \frac{x^2}{a^2} + \frac{y^2}{a^2(1 - e^2)} = 1$$

$$\Leftrightarrow \frac{x^2}{a^2} + \frac{y^2}{b^2} = 1.$$

The resemblance of this to the equation of a circle suggests another fruitful way of considering the ellipse. The linear transformation with matrix $\begin{pmatrix} 1 & 0 \\ 0 & b/a \end{pmatrix}$ is a one-way stretch from O in the y-direction with scale factor b/a (since $b/a < 1$, the 'stretch' is in fact a 'squash'). It maps the point (X, Y) to the point (x,y), where $\begin{pmatrix} x \\ y \end{pmatrix} = \begin{pmatrix} 1 & 0 \\ 0 & b/a \end{pmatrix}\begin{pmatrix} X \\ Y \end{pmatrix} = \begin{pmatrix} X \\ bY/a \end{pmatrix}$,

i.e. $x = X, \; y = \dfrac{bY}{a}$ or $Y = \dfrac{ay}{b}.$

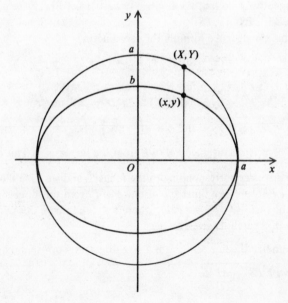

f (X, Y) lies on the auxiliary circle

hen
$$X^2 + Y^2 = a^2$$

o that
$$x^2 + \frac{a^2 y^2}{b^2} = a^2$$

or
$$\frac{x^2}{a^2} + \frac{y^2}{b^2} = 1,$$

and therefore (x, y) lies on the ellipse. The ellipse is thus a squashed circle, formed by shortening in a given ratio all the chords of a circle perpendicular to one diameter. When we look at circular objects, we almost always see ellipses because of our oblique viewpoint.

Qu. 11. Explain how the ellipse can be obtained by stretching the circle $X^2 + Y^2 = b^2$.

Qu. 12. What is the determinant of $\begin{pmatrix} 1 & 0 \\ 0 & b/a \end{pmatrix}$? Prove that the area of the ellipse is πab.

We have seen in Exercise 9A that under every non-singular linear transformation straight lines map to straight lines, parallel lines map to parallel lines, and the ratio of parallel lengths is preserved (in particular, midpoints map to midpoints). These invariants enable us to deduce many facts about the ellipse by transforming the corresponding properties of the circle. These facts will not involve focus or directrix, since these have no counterpart in the circle. It is convenient to draw the circle and the ellipse in separate diagrams, and to let the points $P_1, Q_1, \ldots$ in the circle diagram correspond to the points $P, Q, \ldots$ in the ellipse diagram.

Consider for example the set of chords of the circle parallel to a diameter $P_1 O_1 Q_1$. The midpoints of these chords lie on the diameter $U_1 V_1$ perpendicular to $P_1 Q_1$. The tangents to the circle at U_1 and V_1 are perpendicular to $U_1 V_1$ and therefore parallel to $P_1 Q_1$.

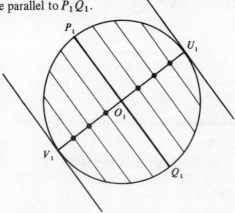

Transforming this figure, we see that the midpoints of the chords of an ellipse parallel to a diameter *POQ* lie on another diameter *UV* of the ellipse, and that the tangents at *U* and *V* are parallel to *PQ*.   *PQ* and *UV* are called *conjugate diameters* of the ellipse; notice that they are in general *not* perpendicular.

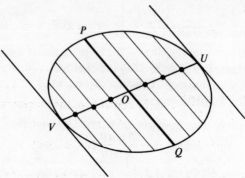

Qu. 13. Prove that *PQ* bisects all the chords parallel to *UV*.

Qu. 14. Prove that the area of the parallelogram formed by the tangents at *P, U, Q, V* is 4*ab*.

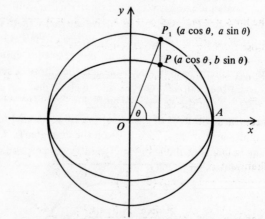

This transformation also gives a very useful parametric representation of points on the ellipse. If P_1 is the point on the auxiliary circle with angle $AOP_1 = \theta$ then P_1 has coordinates $(a \cos \theta, a \sin \theta)$ and the corresponding point P on the ellipse has coordinates $\left(a \cos \theta, \dfrac{b}{a}. a \sin \theta\right)$,

i.e. $(a \cos \theta, b \sin \theta)$.

As θ varies from 0 to 2π the point $(a \cos \theta, b \sin \theta)$ moves once round the ellipse. The angle θ is called the *eccentric angle* of the point P; notice that this is *not* angle *AOP*.

Qu. 15. One end of a diameter of an ellipse has eccentric angle θ. Prove that the ends of the conjugate diameter have eccentric angles $\theta \pm \dfrac{\pi}{2}$.

The equation of the tangent to the ellipse at $(a\cos\theta, b\sin\theta)$ can be found as follows

$$x = a\cos\theta \Rightarrow \frac{dx}{d\theta} = -a\sin\theta$$

$$y = b\sin\theta \Rightarrow \frac{dy}{d\theta} = b\cos\theta.$$

Therefore $\dfrac{dy}{dx} = -\dfrac{b\cos\theta}{a\sin\theta}$ and the equation of the tangent is

$$y - b\sin\theta = -\frac{b\cos\theta}{a\sin\theta}(x - a\cos\theta)$$

$$\Leftrightarrow \quad ay\sin\theta - ab\sin^2\theta = -bx\cos\theta + ab\cos^2\theta$$

$$\Leftrightarrow \quad bx\cos\theta + ay\sin\theta = ab(\cos^2\theta + \sin^2\theta)$$

$$\Leftrightarrow \quad \frac{x\cos\theta}{a} + \frac{y\sin\theta}{b} = 1.$$

Example 1: The lines $y = mx$ and $y = m'x$ are conjugate diameters of the ellipse $\dfrac{x^2}{a^2} + \dfrac{y^2}{b^2} = 1$. Prove that $mm' = -\dfrac{b^2}{a^2}$.

Solution: Let one end of the diameter $y = mx$ be $(a\cos\theta, b\sin\theta)$, so that $m = \dfrac{b\sin\theta}{a\cos\theta}$. By Qu. 15 one end of the conjugate diameter has eccentric angle $\theta + \dfrac{\pi}{2}$.

Therefore $m' = \dfrac{b\sin\left(\theta + \dfrac{\pi}{2}\right)}{a\cos\left(\theta + \dfrac{\pi}{2}\right)} = -\dfrac{b\cos\theta}{a\sin\theta}.$

Therefore $mm' = \dfrac{b\sin\theta}{a\cos\theta}\left(-\dfrac{b\cos\theta}{a\sin\theta}\right) = -\dfrac{b^2}{a^2}.$ $\square$

Summary

An ellipse is the locus $\{P : PS + PS' = 2a\}$.

The tangent and normal at P are equally inclined to PS and PS'.

With major and minor axes as coordinate axes, its equation is

$\dfrac{x^2}{a^2} + \dfrac{y^2}{b^2} = 1$, with foci $(\pm ae, 0)$ and directrices $x = \pm \dfrac{a}{e}$,

where $0 < e < 1$ and $b^2 = a^2(1 - e^2)$.

The focus-directrix property is $PS = e\,PM$, $PS' = e\,PM'$.

The parametric equations are $x = a\cos\theta$, $y = b\sin\theta$.

The tangent at $(a\cos\theta, b\sin\theta)$ is $\dfrac{x\cos\theta}{a} + \dfrac{y\sin\theta}{b} = 1$.

Exercise 10C

In this exercise, E is the ellipse $\dfrac{x^2}{a^2} + \dfrac{y^2}{b^2} = 1$.

1. By plotting points, draw the ellipse $\dfrac{x^2}{25} + \dfrac{y^2}{16} = 1$ accurately on graph paper, taking 1 cm as 1 unit. Calculate the eccentricity, and draw the foci and directrices. Verify the properties (i) $SP + S'P = 2a$ (ii) $SP = e\,PM$ (iii) $S'P = e\,PM'$ by measurement from your diagram for several points P.

2. The orbit of the planet Pluto is an ellipse with major axis of length 1.18×10^{10} km and eccentricity $\frac{1}{4}$. Calculate the length of the minor axis.

3. A point moves round an ellipse. Prove that the ratio of its least and greatest distances from a focus is $1 - e : 1 + e$.

4. A satellite's orbit is an ellipse with the centre of the Earth at one focus. The Earth may be treated as a sphere of radius 6400 km, and the least and greatest heights of the satellite above the Earth's surface are 400 km and 800 km. Show that the eccentricity of the orbit is $1/35$.

5. The Earth moves in an elliptic orbit with the Sun at one focus. The distance of the Earth from the Sun is inversely proportional to the angle which the Sun's diameter (assumed constant) subtends at a point on the Earth. During one year this angle varies from $31'32''$ to $32'36''$. Calculate the eccentricity of the Earth's orbit.

6. The circle C_1 lies entirely inside the circle C_2. A variable circle C touches C_1 externally and touches C_2 internally. Prove that the locus of the centre of C is an ellipse.

7. Find the coordinates of the centre and foci and the equations of the directrices of the following ellipses:
   (i) $2(x-3)^2 + 11(y-1)^2 = 22$ (ii) $x^2 + 2y^2 + 4x - 8y + 4 = 0$.

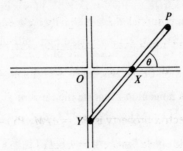

8. The *elliptic trammel* is a mechanical device for drawing ellipses. It consists of a straight rod with a pencil at P and pegs at X, Y which run in perpendicular grooves OX, OY. Prove that if OX, OY are taken as x- and y-axes, with $PX = b$ and $PY = a$, then the locus of P is the ellipse $\frac{x^2}{a^2} + \frac{y^2}{b^2} = 1$. [Hint: use the angle θ shown in the diagram.]

9. A ladder of fixed length rests against a vertical wall. A pot of paint is hooked onto the ladder, not at either end. Show that the locus of this pot as the foot of the ladder slides horizontally away from the wall is part of an ellipse.

10. Two circles have a common centre O. A variable radius cuts them in L and M. Through L and M are drawn lines parallel and perpendicular respectively to a fixed radius. Prove that the locus of the point of intersection of these lines is an ellipse.

11. The feet of the perpendiculars from the foci S, S' to a tangent of E are Q, Q' respectively. Prove that $SQ \cdot S'Q' = b^2$.

12. PQ and UV are conjugate diameters of E. Prove that
    $PQ^2 + UV^2 = 4(a^2 + b^2)$.

13. Prove that just one square can be inscribed in E, and that its area is
    $\frac{4a^2 b^2}{a^2 + b^2}$.

14. Prove that the tangents to an ellipse at the end of a focal chord meet on the corresponding directrix.

15. Prove that the two tangents of E with gradient m have equations
    $y = mx \pm \sqrt{(m^2 a^2 + b^2)}$.

16. Use Qu. 15 to find the gradients and hence the equations of the tangents to the ellipse $\frac{x^2}{6} + \frac{y^2}{3} = 1$ from the point $(-2, 5)$. Find the coordinates of the point of contact of each tangent.

17. Prove that the gradients of the tangents from the point (X, Y) to E are the roots m_1, m_2 of the equation $m^2(a^2 - X^2) + 2mXY + b^2 - Y^2 = 0$. Deduce that the locus of the point of intersection of perpendicular tangents of E is the circle $x^2 + y^2 = a^2 + b^2$. This is called the *director circle* of the ellipse.

18. An elliptical disc slides between two fixed perpendicular straight lines. Prove that the locus of its centre is an arc of a circle.

19. Prove that the equation of the tangent to E at the point $P_1(x_1, y_1)$ is $\frac{xx_1}{a^2} + \frac{yy_1}{b^2} = 1$. The tangent at P_1 meets the tangent at $P_2(x_2, y_2)$ at T. Show that the line $\frac{xx_1}{a^2} + \frac{yy_1}{b^2} = \frac{xx_2}{a^2} + \frac{yy_2}{b^2}$ passes through T and through the midpoint of $P_1 P_2$. Prove that if $P_1 T P_2$ is a right angle, then $\frac{x_1 x_2}{a^4} + \frac{y_1 y_2}{b^4} = 0$.   [L]

20. With the notation of Qu. 19, show that if T has coordinates (x_3, y_3) then $P_1 P_2$ has equation $\frac{xx_3}{a^2} + \frac{yy_3}{b^2} = 1$.

21. Show that the equation of the tangent and normal to the ellipse $x = a \cos\theta, y = b \sin\theta$ at the point $\theta = t$ are $xb \cos t + ya \sin t = ab$ and $xa \sin t - yb \cos t = (a^2 - b^2) \sin t \cos t$ respectively.
The normal at P intersects the x-axis at N, and the perpendicular drawn from the origin O to the tangent at P intersects the tangent at M. Prove that $OM \cdot PN = b^2$.   [L]

Questions 22–26 are intended to be tackled by the 'squashed circle' method of page 297.

22. PQ is a diameter and R is any other point of an ellipse. Prove that the diameters parallel to PR and QR are conjugate.

23. The tangents at points P and Q of an ellipse with centre O meet at T. Prove that OT bisects PQ.

24. A parallelogram is incribed in an ellipse. Show that its sides are parallel to a pair of conjugate diameters.

25. The fixed point L lies inside an ellipse. Find the locus of the midpoint of a variable chord through L.

26. (i) Prove that the greatest possible area of a triangle inscribed in E is $\frac{3\sqrt{3}\,ab}{4}$.

    (ii) Prove that there are infinitely many inscribed triangles with this area, and that they all touch a similar concentric ellipse half as long as E.

    (iii) Prove that the tangent at any vertex of one of these triangles is parallel to the opposite side.

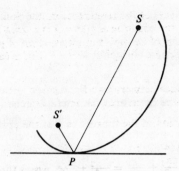

27. (M) A piece of light inextensible flexible string is attached to fixed points
    S, S'. The length of the string exceeds SS'. A small smooth ring slides on
    the string.
    (i)   What is the locus of the ring?
    (ii) The ring comes to rest at P. Use the potential energy of the system
          to explain why the tangent to the locus at P is horizontal.
    (iii) By resolving the forces on the ring show that $SP, S'P$ are equally
          inclined to the horizontal.
    (iv) Deduce the reflector property of the ellipse.

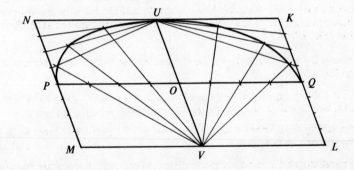

28. The points A, B, C, D have coordinates $(0, a), (0, -a), (\lambda, 0), (a, a - \lambda)$
    respectively, where a is a positive constant and $0 \leqslant \lambda \leqslant a$. The lines AD
    and BC produced meet at P. Prove that, as λ increases from 0 to a, P
    describes a quarter circle of radius a. Use this to explain the following
    draughtsman's construction for an ellipse given a pair of conjugate
    diameters PQ, UV meeting at O.
    Draw the parallelogram $KLMN$ with sides parallel to the diameters, as in
    the figure. Divide each of PQ, KL, MN into the same number of equal
    parts. The intersections of lines from U to points of KQ with lines from
    V to points of OQ give points on the top right part of the ellipse. The
    other parts are obtained by lines (a) from U to NP and from V to OP,
    (b) from V to PM and from U to OP, (c) from V to LQ and from U to
    OQ. Draw an ellipse by this method.

10.4 The hyperbola

A *hyperbola* is the locus of a point P the difference of whose distances from two fixed points S, S' is constant,

i.e. $\{P : |PS - PS'| = \text{constant}\}.$

S and S' are the *foci* of the hyperbola. This definition is closely related to the definition of an ellipse, so it is not surprising that there are many similarities between the properties of hyperbolas and ellipses. There are also some striking differences.

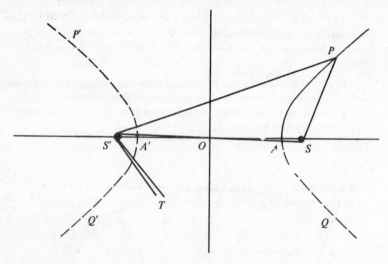

Part of a hyperbola may be drawn by knotting a piece of thread to a pencil P and passing the thread round pins S and S', as shown in the diagram, holding the two ends together at T. As $S'T$ changes, the lengths $S'P$ and SP change by equal amounts, and so their difference remains constant as required.

Qu. 1.   Draw parts of some hyperbolas by this method.

This method will produce only the section PA shown solid in the diagram, but the reflection AQ of this in SS' also satisfies the definition, and PAQ, extending infinitely far, is called one *branch* of the hyperbola. By symmetry there is another branch $P'A'Q'$, the reflection of PAQ in the perpendicular bisector of SS'. The line SS' is called the *transverse axis*, and its perpendicular bisector is the *conjugate axis* (which does not meet the curve). The axes cross at the *centre* O. Any chord through O is a *diameter*, and is bisected at O. Each of the points A, A' is a *vertex*. The eccentricity e of the hyperbola is the ratio OS/OA, so that if $OA = OA' = a$ then $OS = OS' = ea$.

Qu. 2.   Prove that the constant length in the definition equals $2a$.

Qu. 3.   Describe (a) $\{P : PS - PS' = 2a\}$ (b) $\{P : PS' - PS = 2a\}$.

Qu. 4.   Use the triangle inequality $|PS - PS'| \leqslant SS'$ to show that $e \geqslant 1$.
Describe the hyperbola when $e = 1$, i.e. $\{P : |PS - PS'| = SS'\}$.

Qu. 5.   How does the shape of a hyperbola change as e increases?

Hyperbolas are used in a modern method of navigating. Two radio beacons S, S' transmit simultaneous radio pulses, and the time lag between their arrivals at a ship P is measured. From this the difference $PS - PS'$ can be calculated, and the ship located on one branch of a hyperbola with foci S, S'. The same procedure is applied to pulses from S and a third beacon S'', giving one branch of a second hyperbola. The intersection of these two branches on a specially drawn chart fixes the position of the ship.

As the point P moves along the hyperbola, SP and $S'P$ increase at the same rate since their difference is constant. It follows, by the same argument that has been used for the parabola and the ellipse, that the tangent to the hyperbola at P is equally inclined to SP and $S'P$. This is the reflector property of the hyperbola; the details of the proof are left for the reader to supply.

Qu. 6. P is a point on a hyperbola with foci S, S'. The perpendicular from S to the tangent at P meets that tangent at Q and, when produced, meets $S'P$ at H. Prove that $HP = SP$, and deduce that H lies on the circle centre S', radius $2a$. [Draw separate diagrams showing P on each of the two branches.]

Qu. 7.   Use Qu. 6 to show that the locus of a point equidistant from a fixed circle and a fixed point outside the circle is one branch of a hyperbola. Compare this with Qu. 8 on page 294.

Qu. 8.   With Q as in Qu. 6 prove that $OQ = a$ and deduce that Q lies on the circle (called the *auxiliary circle*) which has diameter AA'.

Qu. 9.   Draw a circle and mark a point S outside it. Place a set square with the right angle on the circle and one arm of the right angle through S. Draw the other arm of the right angle. Do this repeatedly, moving the set square slightly each time. Use Qu. 8 to show that all the lines you draw touch a hyperbola.

The set square construction of Qu. 9 is worth closer investigation. As the right angle Q moves round the circle anticlockwise from A, the tangent PQ turns clockwise at first, and the point of contact P moves away from A along the curve. But when Q has passed T (the point of contact of the tangent from S to the auxiliary circle), the tangent PQ turns anticlockwise, and the point of contact moves along the other branch. This continues until Q reaches U, the point of contact of the other tangent from S to the auxiliary circle. After this, PQ is again turning clockwise and P completes its journey along the first branch, moving towards A.

The two lines OT and OU are the limiting positions of the tangent as the point P moves to infinity along the curve. These lines are called the *asymptotes* of the hyperbola; they do not meet the curve, but they approach arbitrarily close to it.

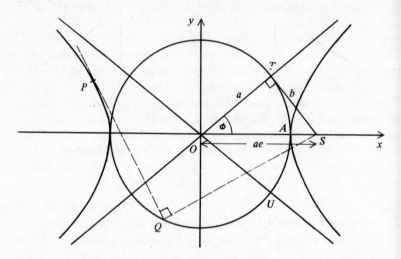

In triangle OTS let $TS = b$. Then since $OT = OA = a$ and $OS = ae$,

$$b^2 = a^2 e^2 - a^2 = a^2(e^2 - 1).$$

If we take the transverse and conjugate axes of the hyperbola as x- and y-axes respectively, then with the angle ϕ as in the diagram the gradients of the asymptotes are $\pm \tan \phi = \pm b/a$, and their equations are $y = \pm \dfrac{b}{a} x$.

Qu. 10. Sketch hyperbolas (i) with $b < a$ (ii) with $b > a$.

As with the ellipse, the lines $x = \pm \dfrac{a}{e}$ are called directrices. Notice that each directrix of a hyperbola lies between the corresponding focus and the centre, since $e > 1$.

With the notation of the figure, the proof that $PS = e\,PM$ is very similar to the proof on page 295; the details are left to the reader.

If P' is the point where PM produced meets the other branch, then

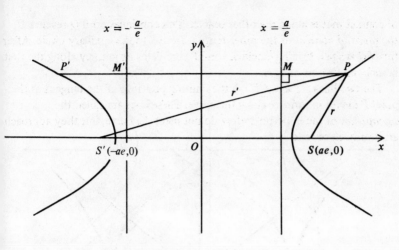

$$P'M = P'M' + M'M$$

$$= PM + \frac{2a}{e} \quad \text{since } PM = P'M' \text{ by symmetry}$$

$$= \frac{PS}{e} + \frac{2a}{e}.$$

Therefore $eP'M = PS + 2a$

$$= PS' \qquad \text{using the hyperbola definition}$$

$$= P'S \qquad \text{by symmetry.}$$

Therefore the focus-directrix property $PS = ePM$ holds for points on both branches of the hyperbola. By symmetry we also have $PS' = ePM'$. The equation of the hyperbola follows easily from this. For $P \equiv (x, y)$ is on the hyperbola $\Leftrightarrow$

$$PS = ePM$$

$$\Leftrightarrow \qquad PS^2 = e^2 PM^2$$

$$\Leftrightarrow (x - ae)^2 + y^2 = e^2 \left(x - \frac{a}{e}\right)^2$$

$$\Leftrightarrow \frac{x^2}{a^2} - \frac{y^2}{a^2(e^2 - 1)} = 1$$

$$\Leftrightarrow \quad \frac{x^2}{a^2} - \frac{y^2}{b^2} = 1.$$

Example 1: Through any point P of the hyperbola $\dfrac{x^2}{a^2} - \dfrac{y^2}{b^2} = 1$ lines are

drawn parallel to the asymptotes, meeting them at K, L. Prove that $4PK \cdot PL = a^2 + b^2$.

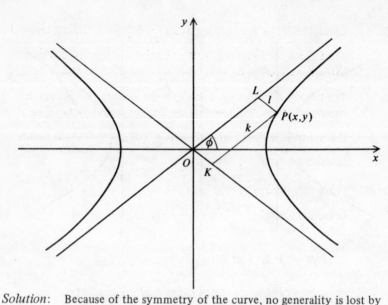

Solution: Because of the symmetry of the curve, no generality is lost by
taking P in the first quadrant as in the diagram. Let $PK = k$,
$PL = l$, and let the angle between the x-axis and the asymptote
$y = \dfrac{b}{a}x$ be ϕ. Then $x = k\cos\phi + l\cos\phi$ so that $k + l = x\sec\phi$,
and $y = k\sin\phi - l\sin\phi$ so that $k - l = y\operatorname{cosec}\phi$.
Therefore $(k + l)^2 - (k - l)^2 = x^2\sec^2\phi - y^2\operatorname{cosec}^2\phi$.
But $(k + l)^2 - (k - l)^2 = 4kl$ and

$$\sec\phi = \frac{\sqrt{(a^2 + b^2)}}{a}, \quad \operatorname{cosec}\phi = \frac{\sqrt{(a^2 + b^2)}}{b}$$

Therefore $4kl = \dfrac{a^2 + b^2}{a^2}x^2 - \dfrac{a^2 + b^2}{b^2}y^2$

$$= (a^2 + b^2)\left(\frac{x^2}{a^2} - \frac{y^2}{b^2}\right)$$

$$= a^2 + b^2, \text{ since } \frac{x^2}{a^2} - \frac{y^2}{b^2} = 1. \qquad \square$$

There are various ways of expressing the coordinates of a point on
$\dfrac{x^2}{a^2} - \dfrac{y^2}{b^2} = 1$ in terms of a parameter. For many purposes the most convenient
is $x = a\sec\theta, y = b\tan\theta$.

Qu. 11. Check that $(a \sec\theta, b \tan\theta)$ satisfies $\dfrac{x^2}{a^2} - \dfrac{y^2}{b^2} = 1$. Describe how this point traces the hyperbola as θ increases from 0 to 2π, paying particular attention to what happens when θ is near $\pi/2$ or $3\pi/2$.

Qu. 12. The tangent at $V(a\cos\theta, a\sin\theta)$ on the auxiliary circle cuts the transverse axis at N. Show that the line through N perpendicular to the transverse axis cuts the hyperbola $\dfrac{x^2}{a^2} - \dfrac{y^2}{b^2} = 1$ at the points $(a\sec\theta, \pm b\tan\theta)$.

Qu. 13. Prove that the equation of the tangent to $\dfrac{x^2}{a^2} - \dfrac{y^2}{b^2} = 1$ at $(a\sec\theta, b\tan\theta)$ is $\dfrac{x\sec\theta}{a} - \dfrac{y\tan\theta}{b} = 1$.

The equation of the tangent at $P(a\sec\theta, b\tan\theta)$ given in Qu. 13 can be rearranged by multiplying through by $\cos\theta$: $\dfrac{x}{a} - \dfrac{y\sin\theta}{b} = \cos\theta$. As $\theta \to \dfrac{\pi}{2}$, P moves to infinity and this tangent tends to the asymptote $\dfrac{x}{a} - \dfrac{y}{b} = 0$. Similarly the other asymptote $\dfrac{x}{a} + \dfrac{y}{b} = 0$ is obtained by letting $\theta \to \dfrac{3\pi}{2}$.

The point (x, y) lies on one or other asymptote if and only if $\left(\dfrac{x}{a} - \dfrac{y}{b}\right)\left(\dfrac{x}{a} + \dfrac{y}{b}\right) = 0 \Leftrightarrow \dfrac{x^2}{a^2} - \dfrac{y^2}{b^2} = 0$. This is the equation of the pair of asymptotes. Notice its close relation to the equation of the hyperbola, $\dfrac{x^2}{a^2} - \dfrac{y^2}{b^2} = 1$.

Qu. 14. Show that the hyperbolas $\dfrac{x^2}{a^2} - \dfrac{y^2}{b^2} = \lambda$, where a, b are fixed and λ is a variable parameter, all have the same asymptotes. What happens (i) when $\lambda = 0$ (ii) when λ is negative?

If $a = b$, the asymptotes are the lines $y = \pm x$, which are perpendicular, and the curve is called a *rectangular hyperbola*.

Qu. 15. Show that all rectangular hyperbolas have eccentricity $\sqrt 2$, and hence that they are all the same shape.

The rectangular hyperbola $x^2 - y^2 = a^2$ has many properties in addition to those of all hyperbolas (just as the circle $x^2 + y^2 = a^2$ has properties in addition to those of all ellipses).

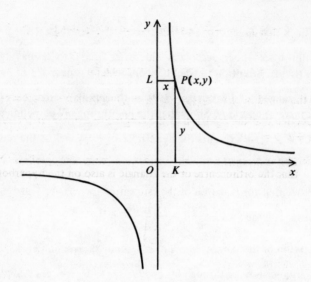

Since the asymptotes of a rectangular hyperbola are perpendicular we can use them as coordinate axes. Let $P(x, y)$ be a point on a rectangular hyperbola referred to the axes, with one branch in the first quadrant. The lines through P parallel to the asymptotes are in this case the perpendiculars PK, PL from P to the coordinate axes. As shown in Example 1,

$4PK \cdot PL = a^2 + b^2 = 2a^2$ (since $b = a$). Therefore $4yx = 2a^2$ or $xy = \dfrac{a^2}{2}$.

Writing c^2 for $\dfrac{a^2}{2}$, we have the equation $xy = c^2$ for a rectangular hyperbola referred to its asymptotes as coordinate axes. In particular, the familiar curve of reciprocals $y = 1/x$ is now seen to be a rectangular hyperbola.

Points on $xy = c^2$ have the very useful parametric equations $x = ct, y = c/t \, (t \neq 0)$.

Qu. 16. Check that $(ct, c/t)$ lies on $xy = c^2$ and that every point of the curve corresponds to one and only one non-zero value of t. Describe how $(ct, c/t)$ traces the curve as t increases from $-\infty$ to ∞.

To find the equation of the chord joining $(ct_1, c/t_1)$, $(ct_2, c/t_2)$, consider the line $lx + my + n = 0$.

The point $\left(ct, \dfrac{c}{t}\right)$ lies on this $\Leftrightarrow lct + mc/t + n = 0$

$$\Leftrightarrow lct^2 + nt + mc = 0.$$

If $lx + my + n = 0$ is the required chord, then the roots of this quadratic are t_1, t_2.

Therefore $t_1 + t_2 = -\dfrac{n}{lc}$, so that $n = -lc(t_1 + t_2)$,

and $t_1 t_2 = \dfrac{mc}{lc}$, so that $m = l\,t_1 t_2$.

Therefore the chord is $lx + l\,t_1 t_2 y - lc\,(t_1 + t_2) = 0$ or $x + t_1 t_2 y = c(t_1 + t_2)$.

Qu. 17. Deduce that the equation of the tangent to $xy = c^2$ at $(ct, c/t)$ is
$x + t^2 y = 2ct$.

Example 2: The vertices of a triangle are on a rectangular hyperbola. Prove
that the orthocentre of the triangle is also on the hyperbola.

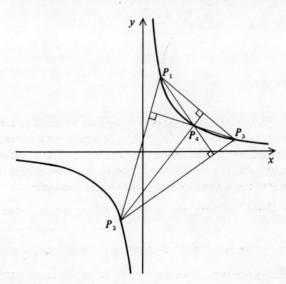

Solution: Let the triangle be $P_1 P_2 P_3$, and let the perpendicular from P_3 to
$P_1 P_2$ meet the hyperbola again at P_4. Let the parameters of P_1, P_2,
P_3, P_4 be t_1, t_2, t_3, t_4. The equation of $P_1 P_2$ is $x + t_1 t_2 y = c(t_1 + t_2)$,
so the gradient of $P_1 P_2$ is $-\dfrac{1}{t_1 t_2}$.
Similarly the gradient of $P_3 P_4$ is $-\dfrac{1}{t_3 t_4}$.

Since $P_1 P_2 \perp P_3 P_4$, $\left(-\dfrac{1}{t_1 t_2}\right)\left(-\dfrac{1}{t_3 t_4}\right) = -1$, and so $t_1 t_2 t_3 t_4 = -1$.

Since this condition is symmetrical in t_1, t_2, t_3, t_4, $P_1 P_4 \perp P_2 P_3$.
Therefore P_4 lies on the perpendicular from P_1 to $P_2 P_3$, and so P_4
is the orthocentre of triangle $P_1 P_2 P_3$.
Similarly each of P_1, P_2, P_3 is the orthocentre of the triangle
formed by the other three points. $\square$

Summary.

A hyperbola is the locus $\{P : |PS - PS'| = 2a\}$. Referred to its

transverse and conjugate axes its equation is $\dfrac{x^2}{a^2} - \dfrac{y^2}{b^2} = 1$,

with foci $(\pm ae, 0)$ and directrices $x = \pm \dfrac{a}{e}$,

where $e > 1$ and $b^2 = a^2(e^2 - 1)$.

The tangent at P is equally inclined to PS and PS'.

The focus-directrix property is $PS = e\,PM$.

The asymptotes are $y = \pm\dfrac{b}{a}x$ or $\dfrac{x^2}{a^2} - \dfrac{y^2}{b^2} = 0$.

One parametric form for the hyperbola is $x = a \sec\theta, y = b \tan\theta$.
The tangent at this point is $\dfrac{x \sec\theta}{a} - \dfrac{y \tan\theta}{b} = 1$.

A rectangular hyperbola has perpendicular asymptotes.
Referred to these its equation is $xy = c^2$,

with parametric equations $x = ct, \ y = \dfrac{c}{t}$.

The chord joining points with parameters t_1, t_2 is
$x + t_1 t_2 y = c(t_1 + t_2)$.
The tangent at the point with parameter t is $x + t^2 y = 2ct$.

Exercise 10D

In this exercise H is the hyperbola $\dfrac{x^2}{a^2} - \dfrac{y^2}{b^2} = 1$.

1. Draw the hyperbola $\dfrac{x^2}{16} - \dfrac{y^2}{9} = 1$ carefully by plotting points, taking

   1 cm as unit. Draw the asymptotes. Find the eccentricity, and put the
   foci and directrices on your diagram. Check each of the properties
   (i) $|PS - PS'| = 2a$ (ii) $PS = e\,PM$ (iii) $PS' = e\,PM'$
   by measurement for several positions of P on the hyperbola.

2. The rectangle $TUVW$ has vertices $(a, b), (-a, b), (-a, -b), (a, -b)$
   respectively. Show that   (i) H touches TW and UV at their midpoints
                              (ii) TV and UW are the asymptotes of H
                              (iii) the circle with TV as diameter meets the
                                    x-axis at the foci of H
                              (iv) the directrices pass through the points
                                   where the auxiliary circle meets the
                                   asymptotes.

3. Starting with the rectangle $TUVW$ of Qu. 2 and using its properties, draw quick sketches of the following hyperbolas with their asymptotes, foci and directrices. (i) $\dfrac{x^2}{9} - \dfrac{y^2}{16} = 1$ (ii) $\dfrac{x^2}{4} - y^2 = 1$ (iii) $x^2 - y^2 = 50$.

4. Find the equation of the hyperbola with foci $(\pm 6, 0)$ and directrices $x = \pm 1\frac{1}{2}$.

5. Find the equation of the hyperbola with asymptotes $3x + y = 0$ and $3x - y = 0$ which passes through $(2, 4)$.

6. Find the equation of the tangent and normal to $x^2 - 5y^2 = 1$ at $(9, 4)$.

7. The line $y = mx + c$ meets H at P_1, P_2 and meets the asymptotes at Q_1, Q_2.
    (i) Write down the quadratic equation whose roots are the x-coordinates of P_1, P_2.
    (ii) Write down the quadratic equation whose roots are the x-coordinates of Q_1, Q_2.
    (iii) Hence show that P_1P_2 and Q_1Q_2 have the same midpoint.
    (iv) Deduce that $P_1Q_1 = P_2Q_2$.

8. The tangent at a point P on H meets the asymptotes at Q_1, Q_2. Prove that P is the midpoint of Q_1Q_2.

9. If a given line l meets the hyperbola $\dfrac{x^2}{a^2} - \dfrac{y^2}{b^2} = \lambda$ at P_1 and P_2, show that that the midpoint M of P_1P_2 is independent of λ. If the coordinates of M are (x_0, y_0), find the equation of l.   [O]

10. Prove that a variable tangent to a hyperbola forms with the asymptotes a triangle of constant area. Prove also that the locus of the centroid of this triangle is a similar hyperbola.

11. A set of parallel chords of H have gradient m. Prove that their midpoints lie on the diameter of gradient m', where $mm' = b^2/a^2$.

12. The hyperbolas H and $\dfrac{x^2}{a^2} - \dfrac{y^2}{b^2} = -1$ are said to be *conjugate*. Sketch both hyperbolas in the same diagram. If the eccentricities of conjugate hyperbolas are e, e', show that $e^2 e'^2 = e^2 + e'^2$.

13. Prove that if the line $y = mx + c$ touches H then $a^2 m^2 = b^2 + c^2$. Investigate whether the converse is true.

14. Adapt the method of Ex. 10C, Qu. 17 to show that the locus of the point of intersection of perpendicular tangents of H is the circle $x^2 + y^2 = a^2 - b^2$ if $a > b$, and that there are no perpendicular tangents if $a < b$. What happens when $a = b$?

15. Prove that $x = a \cosh\phi, y = b \sinh\phi$ are parametric equations for one branch of H, and suggest similar equations for the other branch.

16. The points $P_1 (a \cos\theta, a \sin\theta)$ and $P_2 (a \cosh\phi, a \sinh\phi)$ lie on the circle $x^2 + y^2 = a^2$ and the rectangular hyperbola $x^2 - y^2 = a^2$ respectively.

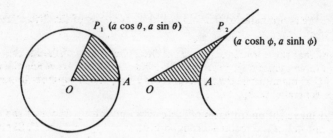

Show that area OAP_1 is proportional to θ and that area OAP_2 is proportional to ϕ, where O is the origin and A is $(a, 0)$. This accounts for the descriptions *circular* function for cosine and sine, and *hyperbolic* function for cosh and sinh.

17. Prove that the lines $\dfrac{x}{a} + \dfrac{y}{b} = t$, $\dfrac{x}{a} - \dfrac{y}{b} = \dfrac{1}{t}$ $(t \neq 0)$, which are parallel to the asymptotes, meet on H. Deduce the parametric equations $x = \dfrac{a}{2}\left(t + \dfrac{1}{t}\right)$, $y = \dfrac{b}{2}\left(t - \dfrac{1}{t}\right)$ for H.

18. Show that the equation of the tangent to H at $\left(\dfrac{a}{2}\left(t + \dfrac{1}{t}\right), \dfrac{b}{2}\left(t - \dfrac{1}{t}\right)\right)$ is $\dfrac{x}{a}(t^2 + 1) - \dfrac{y}{b}(t^2 - 1) = 2t$. Find the coordinates of the points where this tangent meets the asymptotes, and deduce the result of Qu. 8 again.

19. The same point of H has coordinates $(a \sec \theta, b \tan \theta)$, $(a \cosh \phi, b \sinh \phi)$, $\left(\dfrac{a}{2}\left(t + \dfrac{1}{t}\right), \dfrac{b}{2}\left(t - \dfrac{1}{t}\right)\right)$. Show that $t = \tan\left(\dfrac{\theta}{2} + \dfrac{\pi}{4}\right)$, and find t in terms of ϕ.

20. $P_1(x_1, y_1)$ and $P_2(x_2, y_2)$ are points on the hyperbola $\dfrac{x^2}{a^2} - \dfrac{y^2}{b^2} = 1$. Prove that $\dfrac{y_2 - y_1}{x_2 - x_1} = \dfrac{b^2}{a^2} \dfrac{(x_1 + x_2)}{(y_1 + y_2)}$ and deduce that the equation of $P_1 P_2$ may be written $y - \eta = \dfrac{b^2}{a^2} \cdot \dfrac{\xi}{\eta}(x - \xi)$ where (ξ, η) is the midpoint of $P_1 P_2$.

Hence, or otherwise, show that the locus of midpoints (ξ, η) of the chords passing through a fixed point (u, v) on the hyperbola is

$$\dfrac{(x - \tfrac{1}{2}u)^2}{a^2} - \dfrac{(y - \tfrac{1}{2}v)^2}{b^2} = \tfrac{1}{4}.$$ [MEI]

21. Show that the equation of the tangent at $(ct, c/t)$ to the rectangular hyperbola $xy = c^2$ is $t^2 y + x = 2ct$.
The point P is on the curve $xy = 4$ and the tangent at P to this curve is also a tangent to the hyperbola $4x^2 - 9y^2 = 36$. Find the coordinates of the two possible positions of P. Find also the equations of the tangents to the curve $xy = 4$ at these two points and the coordinates of the points in which these two tangents touch the other hyperbola.   [L]

22. Find the coordinates of the foci and the equations of the directrices of $xy = c^2$.

23. Show that the normal to $xy = c^2$ at the point $P(ct, c/t)$ meets the hyperbola again at $Q(-c/t^3, -ct^3)$. Find the point R at which the circle on PQ as diameter meets the hyperbola again, and show that the normal at R is parallel to PQ.

24. Write down the coordinates of the centre and the equations of the asymptotes of the rectangular hyperbola $(x - h)(y - k) = c^2$. Sketch the hyperbolas $2x(y - 2) = 3$ and $2y(x - 1) = 3$, and find the coordinates of the points P and Q in which they intersect. Show that the tangents to the hyperbolas at P and Q form a parallelogram.

10.5 The family of conics

Many links between the parabola, ellipse and hyperbola have now appeared, for example the focus-directrix and reflector properties. These curves were originally studied by the Greeks, particularly Apollonius of Perga (247–205 B.C.), as sections of a right circular cone. For this reason they are known as *conic sections* or just *conics*.

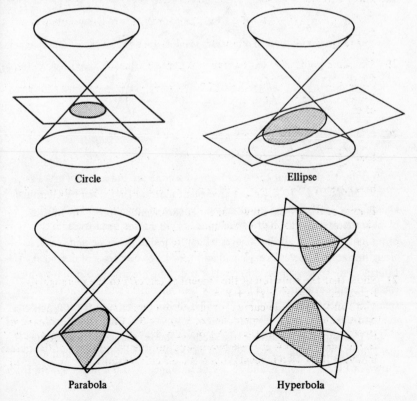

Circle

Ellipse

Parabola

Hyperbola

Consider a double cone standing with its axis vertical. A horizontal plane not through the vertex cuts the cone in a circle. When the plane is tilted slightly the section is an ellipse. As the angle of tilt increases, the section becomes more elongated until, when the plane is parallel to a generator of the cone, the section is a parabola. With further tilting, the plane cuts the other half of the cone too, and the section is a hyperbola. The parabola is thus the borderline case, separating ellipses from hyperbolas.

Qu. 1. Explain how a pair of straight lines can be obtained as a conic section.

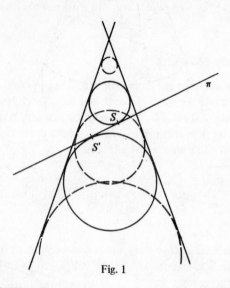

Fig. 1

In order to show that these sections of a cone are indeed the curves as defined in §§10.2–10.4 we use an elegant method first given by Germinal Dandelin in 1822, taking for example the case of the ellipse. The cutting plane π meets only one half of the cone, so we do not need to draw the other half. Think of all the spheres which lie inside the cone and touch it along horizontal circles. Very small and very large spheres will not meet π. Intermediate spheres meet π in circles, and there are just two spheres which touch π, one between π and the vertex touching at S, and one on the other side of π touching at S' (see Fig. 1 which shows the vertical section perpendicular to π).

Let P be any point on the conic section, and let the generator through P meet the horizontal circles of contact of these two spheres in K and K', as shown in Fig. 2. Then PK and PS are both tangents to the upper sphere from

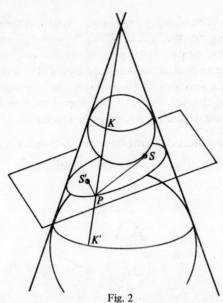

Fig. 2

P, and so $PS = PK$. Similarly $PS' = PK'$. Therefore
$PS + PS' = PK + PK' = KK'$. But KK' is the distance along a generator
between the two circles of contact, and this is constant for all positions of P.
Therefore the locus of P in the plane π is $\{P : PS + PS' = \text{constant}\}$, which
is an ellipse with foci S and S'.

Qu. 2. Adapt this method to show that a hyperbola is a conic section.
[The parabola is dealt with in Ex. 10E, Qu. 12.]

Qu. 3. By considering how Fig. 1 changes as the tilt and distance from the
vertex of π vary, show that ellipses of all shapes and sizes can be
obtained as sections of a single cone. Does the same apply to
hyperbolas?

Ellipses and hyperbolas are called *central* conics. Their standard Cartesian
equations $\dfrac{x^2}{a^2} \pm \dfrac{y^2}{b^2} = 1$ are very similar, but the standard equation of the
parabola, $y^2 = 4ax$, is different since the parabola has no centre which can
be taken as origin. The unity of the conic sections is more apparent in their
standard polar equations. To find these, we take one focus as pole and use
the focus-directrix property $PS = e\,PM$, where $e < 1$ for an ellipse, $e = 1$ for
a parabola, $e > 1$ for a hyperbola. In Fig. 3, S is a focus of the conic, SK is
the initial line which is perpendicular to the directrix MK corresponding to
S, and $SK = k$.

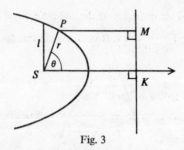

Fig. 3

The point P with polar coordinates (r, θ) lies on the conic

$$\Leftrightarrow \qquad PS = ePM$$

$$\Leftrightarrow \qquad r = e(k - r\cos\theta)$$

$$\Leftrightarrow \quad r(1 + e\cos\theta) = ek.$$

If $r = l$ when $\theta = \dfrac{\pi}{2}$ then $l(1 + e.0) = ek$, so that $ek = l$. The length l is called the *semi-latus rectum*. Thus

> the polar equation of the conic with eccentricity e and semi-latus rectum l is $\dfrac{l}{r} = 1 + e\cos\theta$,
>
> where the pole is at a focus and the initial line is perpendicular to the directrix.

Qu. 4. What curve does this polar equation give when $e = 0$?

Exercise 10E

1. Draw the hyperbola $\dfrac{3}{r} = 1 + 2\cos\theta$ accurately by plotting points. What is the angle between the asymptotes?

2. Sketch on the same diagram the conics $\dfrac{2}{r} = 1 + e\cos\theta$ for $e = 0, \frac{1}{2}, 1, \frac{3}{2}$.

3. The chord PQ passes through the focus S of a conic of semi-latus rectum l. Prove that $\dfrac{1}{PS} + \dfrac{1}{QS} = \dfrac{2}{l}$.

4. A square is drawn with SP as diagonal, where P is any point on a conic with focus S. Prove that the loci of the other vertices of the square are similar conics.

5. Show that the polar coordinates of the other focus of the conic
$\dfrac{l}{r} = 1 + e \cos\theta$ are $\left(\dfrac{2el}{1-e^2}, \pi\right)$. Describe how this focus moves as e
varies from 0 to ∞.

6. Find the polar equation of the directrix of the conic $\dfrac{l}{r} = 1 + e \cos\theta$
which corresponds to the focus at the pole.

7. PP', QQ' are perpendicular focal chords of a conic. Prove that
$\dfrac{1}{PP'} + \dfrac{1}{QQ'}$ is constant.

8. PP', QQ' are perpendicular focal chords of a conic. Show that, if
$PP' = QQ'$ for all positions of P, then the conic is either a circle or a
rectangular hyperbola.

9. Using the result of Ex. 10A, Qu. 7, show that the polar equation of the
chord joining the points of the conic $\dfrac{l}{r} = 1 + e \cos\theta$ where $\theta = \alpha$ and
$\theta = \beta$ is $\dfrac{l \sin(\beta - \alpha)}{r} = (1 + e \cos\alpha)\sin(\beta - \theta) + (1 + e\cos\beta)\sin(\theta - \alpha)$,
and that this reduces to $\dfrac{l}{r} = \cos\left(\theta - \dfrac{\alpha + \beta}{2}\right)\sec\left(\dfrac{\beta - \alpha}{2}\right) + e\cos\theta$.
Deduce that the tangent at the point where $\theta = \alpha$ has polar equation
$\dfrac{l}{r} = \cos(\theta - \alpha) + e\cos\theta$.

10. Prove that the tangents at the end of a focal chord of a conic meet on
the corresponding directrix.

11. The tangents from a point T to a conic touch the conic at P and Q.
Prove that TP and TQ subtend equal angles at a focus.

12. *Proof that a parabola is a conic section.*
   (i) Show that if the plane π is parallel to a generator then there is just
       one Dandelin sphere touching the cone in a circle and also touching
       π. Let this sphere touch π at S, and let the plane through the circle
       of contact meet π in the line l.
   (ii) P is any point on the conic section. The generator through P meets
        the circle of contact at K. Explain why $PS = PK$.
   (iii) The foot of the perpendicular from P to l is M. Explain why
         $PM = PK$.
   (iv) Deduce that $PS = PM$, and hence that the conic section is the
        parabola in π with focus S and directrix l.

13. The fixed point C and a variable point P have position vectors $\mathbf{c}, \mathbf{r}$
respectively in three-dimensional Euclidean space. Describe in geometrical
terms the locus Γ defined by $\mathbf{r} \cdot \mathbf{c} = |\mathbf{r}||\mathbf{c}| \cos\alpha$ where α is a given acute
angle. If $\mathbf{c} = (a, b, c)$ and $\mathbf{r} = (x, y, z)$, express the equation of Γ in
Cartesian form and show that, for all real numbers λ,
$$(x, y, z) \in \Gamma \implies (\lambda x, \lambda y, \lambda z) \in \Gamma.$$

Taking the special case $c = (4, 0, 3)$, $\cos \alpha = \frac{4}{5}$, show that Γ meets the plane $z = 6$ in a parabola and that the sphere $|r - c| = 3$ touches the plane at the focus of this parabola.   [MEI]

14. Show that the coordinates (x, y, z) of a point on the line joining the points (x_0, y_0, z_0), (f, g, h) may be expressed as

$$(x, y, z) = (f, g, h) + \lambda(x_0 - f, y_0 - g, z_0 - h).$$

Hence or otherwise show that the cone which passes through the ellipse $\dfrac{x^2}{a^2} + \dfrac{y^2}{b^2} = 1$, $z = 0$ and has its vertex at (f, g, h), $h \neq 0$, has equation

$$\frac{(fz - hx)^2}{a^2} + \frac{(gz - hy)^2}{b^2} = (z - h)^2.$$

What is the significance of the condition $h \neq 0$?
Show that this cone meets the plane $x = 0$ in a circle if the vertex lies on a certain hyperbola in the plane $y = 0$.

15. The conic C_λ has equation $\dfrac{x^2}{a^2 + \lambda} + \dfrac{y^2}{b^2 + \lambda} = 1$, where $a > b > 0$.

   (i)   Show that if $\lambda > -b^2$ then C_λ is an ellipse. Can C_λ be a circle?
   (ii) Show that if $-a^2 < \lambda < -b^2$ then C_λ is a hyperbola. Can C_λ be a rectangular hyperbola?
   (iii) Describe C_λ if $\lambda < -a^2$.
   (iv) Show that all the ellipses and hyperbolas C_λ have the same foci. The set of curves C_λ is said to form a *confocal system* of ellipses and hyperbolas.
   (v)   Use focal distance properties to show that there is a unique ellipse and a unique hyperbola of the system through each point of the plane except for points on the axes.
   (vi) Use reflector properties to show that each ellipse of the system meets each hyperbola of the system orthogonally, i.e. that the tangents at the intersection are perpendicular. Sketch this confocal system of orthogonal curves.
   (vii) Describe the system of orthogonal curves which is obtained in the limit if the distance between the foci tends to zero.

16. Prove that the locus of points from which an elliptical disc appears to be circular is a hyperbola.

11 Approximate methods II

11.1 Errors

Most practical applications of mathematics lead to numerical solutions. Only in the simplest cases are these solutions likely to be exact, since (i) usually the information on which the calculations are based is itself not exact, and (ii) it may not be possible or worthwhile to solve the resulting equations etc. by exact methods. An approximate solution is often acceptable, but it is important to give some indication of the degree to which it is accurate.

We first distinguish between *mistakes* and *errors*. Mistakes are made by the person doing the calculation, and can be put right if they are noticed. Of course all possible steps should be taken to avoid mistakes; orderly setting out of calculations, use of checks, and care in reading tables and copying numbers will help. Errors are inaccuracies which cannot be avoided. They may be caused by inexact information (data error, as for example when experimental data are used), by the mathematical process (method error, as for example when taking $\cos \theta$ to be $1 - \frac{1}{2}\theta^2$ for small θ), or by truncating or rounding the numbers being used. The value of π is $3.141\,592\,65 \ldots$; *truncated* to 4 D.P. this is 3.1415 with truncation error $0.000\,092\,65 \ldots$. The value of π *rounded* to 4 D.P. is 3.1416 with rounding error $-0.000\,007\,34 \ldots$.

Qu. 1. Solve $5x + 3 = 2x + 9$. Find also the greatest and least possible solutions if all the coefficients have been rounded to the nearest integer.

Qu. 2. Find the limits of the error when a positive number is
(i) truncated to n D.P. (ii) rounded to n D.P.

If y is an approximation to the number x, then the error ϵ involved in using y instead of x is $x - y$, so that $x = y + \epsilon$. The *absolute error* is $|\epsilon|$, and the *relative error* is $\left|\dfrac{\epsilon}{x}\right|$.

Since $\dfrac{\epsilon}{x} = \dfrac{\epsilon}{y + \epsilon} = \dfrac{\epsilon}{y}\left(1 + \dfrac{\epsilon}{y}\right)^{-1} = \dfrac{\epsilon}{y} - \dfrac{\epsilon^2}{y^2} + \ldots \approx \dfrac{\epsilon}{y}$ if $\dfrac{\epsilon}{y}$ is small, we often use $\left|\dfrac{\epsilon}{y}\right|$ instead of $\left|\dfrac{\epsilon}{x}\right|$ for the relative error. This is more convenient since y is known whereas in general x is not.

If $x_1 = y_1 + \epsilon_1$ and $x_2 = y_2 + \epsilon_2$ then $x_1 + x_2 = y_1 + y_2 + \epsilon_1 + \epsilon_2$. The absolute error in using $y_1 + y_2$ instead of $x_1 + x_2$ is $|\epsilon_1 + \epsilon_2| \leqslant |\epsilon_1| + |\epsilon_2|$. Similarly the absolute error in using $y_1 - y_2$ instead of $x_1 - x_2$ is $|\epsilon_1 - \epsilon_2| \leqslant |\epsilon_1| + |\epsilon_2|$. Notice that both cases use the *sum* of the absolute errors.

Qu. 3.   Evaluate $3.52 + 4.713 - 0.2147 - 1.86$, where each number has been rounded, giving your answer to the greatest accuracy that can be guaranteed.

With the same notation, $\begin{aligned}[t] x_1 x_2 &= (y_1 + \epsilon_1)(y_2 + \epsilon_2) \\ &= y_1 y_2 + y_1 \epsilon_2 + y_2 \epsilon_1 + \epsilon_1 \epsilon_2. \end{aligned}$

Neglecting the product $\epsilon_1 \epsilon_2$, we see that the error in using $y_1 y_2$ instead of $x_1 x_2$ is $y_1 \epsilon_2 + y_2 \epsilon_1$. The relative error is approximately

$$\left| \frac{y_1 \epsilon_2 + y_2 \epsilon_1}{y_1 y_2} \right| = \left| \frac{\epsilon_1}{y_1} + \frac{\epsilon_2}{y_2} \right| \leqslant \left| \frac{\epsilon_1}{y_1} \right| + \left| \frac{\epsilon_2}{y_2} \right|.$$

To deal with quotients, we use a binomial expansion:

$$\frac{x_1}{x_2} = \frac{y_1 + \epsilon_1}{y_2 + \epsilon_2} = \frac{y_1 + \epsilon_1}{y_2 \left(1 + \dfrac{\epsilon_2}{y_2}\right)} = \frac{y_1 + \epsilon_1}{y_2} \left(1 - \frac{\epsilon_2}{y_2} + \ldots\right)$$

$$\approx \frac{y_1}{y_2} + \frac{\epsilon_1}{y_2} - \frac{\epsilon_2 y_1}{y_2^2} \quad \text{neglecting products of } \epsilon\text{'s}$$

$$= \frac{y_1}{y_2} + \frac{y_2 \epsilon_1 - y_1 \epsilon_2}{y_2^2}.$$

The error in using $\dfrac{y_1}{y_2}$ for $\dfrac{x_1}{x_2}$ is approximately $\dfrac{y_2 \epsilon_1 - y_1 \epsilon_2}{y_2^2}$ (notice the similarity with the derivative of a quotient). The relative error is

approximately $\left| \dfrac{y_2 \epsilon_1 - y_1 \epsilon_2}{y_2^2} \div \dfrac{y_1}{y_2} \right| = \left| \dfrac{\epsilon_1}{y_1} - \dfrac{\epsilon_2}{y_2} \right| \leqslant \left| \dfrac{\epsilon_1}{y_1} \right| + \left| \dfrac{\epsilon_2}{y_2} \right|.$

The maximum possible absolute error in the sum or difference of two quantities is the sum of the absolute errors in the quantities.

The maximum possible relative error in the product or quotient of two quantities is taken to be the sum of the relative errors in the quantities.

Qu. 4.   The breaking load L of a rectangular beam is proportion to bd^2/s, where b, d, s are the breadth, depth and span of the beam. If b, d, s are subject to relative errors of $3, 2, 1$ per cent respectively, what is the maximum possible percentage error in L?

An error in x will usually produce an error in $f(x)$, where f is any function. If $x = y + \epsilon$ then $f(x) = f(y + \epsilon) \approx f(y) + \epsilon f'(y)$, using the first Taylor approximation centred at y. Therefore the error in using $f(y)$ instead of $f(x)$ is approximately $\epsilon f'(y)$.

Qu. 5. Illustrate this, using a graph of f.

Example 1: A student measures the side a and the angles A, B of a triangular field ABC, and uses these values to calculate b by the Sine Formula. He finds that $a = 243$ m, $A = 74°$, $B = 29°$. The length may be in error by up to 2%, and the angles by up to $\frac{1}{2}°$. Estimate the maximum possible percentage error in his answer.

Solution: Taking $f(x) \equiv \sin x$, the *relative* error in using $\sin y$ instead of $\sin x$ is $\dfrac{\epsilon \cos y}{\sin y} = \epsilon \cot y$.

In this case $\epsilon = \frac{1}{2}° = \dfrac{\pi}{360}$ rad.

Since $b = \dfrac{a \sin B}{\sin A}$, the greatest possible percentage error in b is

$$2 + \frac{\pi}{360} \times \cot 29° \times 100 + \frac{\pi}{360} \times \cot 74° \times 100$$

$$= 2 + \frac{\pi}{3.6} (1.804 + 0.287) \approx 3.8. \qquad \square$$

Qu. 6. If the student in Example 1 had time to measure just one of the angles more accurately, which one should he choose?

Finally we introduce the idea of *differencing*, which has widespread applications in numerical work, though there is not space to pursue it far here. We are interested in the values of $f(x)$ for the equally spaced values $x_0, x_0 + h, x_0 + 2h, \ldots, x_0 + rh, \ldots$, where x_0 and h are constants. We denote the corresponding values of $f(x)$ by $f_0, f_1, f_2, \ldots, f_r, \ldots$, so that $f_r = f(x_0 + rh)$. Having tabulated these values we also tabulate the differences Δf between successive entries, and then the differences $\Delta^2 f$ between these differences, and so on, as in the following example where $x_0 = 0$ and $h = 1$.

x	f	Δf	$\Delta^2 f$	$\Delta^3 f$	$\Delta^4 f$
0	−18				
		7			
1	−11		−6		
		1		6	
2	−10		0		0
		1		6	
3	−9		6		0
		7		6	
4	−2		12		0
		19		6	
5	17		18		
		37			
6	54				

Qu. 7. Prove that the sum of the entries in any difference column equals the difference between the last and first entries of the preceding column. This is a useful check when compiling a table of differences.

We notice that in this case the third differences $\Delta^3 f$ are constant, and the fourth differences $\Delta^4 f$ are zero. This is because this $f(x)$ is a cubic polynomial (in fact $(x-2)^3 - 10$). For if $f(x)$ is a polynomial of degree n, $f(x) \equiv c_0 x^n + c_1 x^{n-1} + \ldots + c_n$ say, then

$$\Delta f(x) \equiv f(x+h) - f(x)$$

$$\equiv c_0(x+h)^n + c_1(x+h)^{n-1} + \ldots + c_n - (c_0 x^n + c_1 x^{n-1} + \ldots + c_n)$$

$$\equiv c_0 x^n + c_0 nh x^{n-1} + \ldots + c_1 x^{n-1} + \ldots - c_0 x^n - c_1 x^{n-1} - \ldots$$

$$\equiv c_0 nh \, x^{n-1} + \ldots,$$

a polynomial of degree $n-1$. Each differencing reduces the degree of the polynomial by 1, so that in the nth difference column the values are those of a polynomial of degree zero, i.e. a non-zero constant. We can use this to extend the table of values of the polynomial, using the entries along the lowest diagonal line of the table to produce new values, shown in heavy type.

x	f	Δf	$\Delta^2 f$	$\Delta^3 f$
				6
			18	
		37		6
6	54		24	
		61		6
7	115		30	
		91		
8	206			

Qu. 8. The fourth-degree polynomial $f(x)$ has values $f(0) = 7, f(1) = 4$,
$f(2) = -2, f(3) = 10, f(4) = 5$. Form a difference table, and deduce
the values of $f(5), f(6)$ and $f(-1)$.

The following difference table is for $f(x) \equiv e^x$ for $x = 1.0\,(0.1)\,1.9$.
All the entries are to 4 D.P., but it is not necessary to write the decimal point;
for example the 33 written in the $\Delta^3 f$ column is really 0.0033.

x	f	Δf	$\Delta^2 f$	$\Delta^3 f$	$\Delta^4 f$	$\Delta^5 f$	$\Delta^6 f$	$\Delta^7 f$
1.0	2.7183							
		2859						
1.1	3.0042		300					
		3159		33				
1.2	3.3201		333		1			
		3492		34		4		
1.3	3.6693		367		5		-6	
		3859		39		-2		11
1.4	4.0552		406		3		5	
		4265		42		3		-10
1.5	4.4817		448		6		-5	
		4713		48		-2		9
1.6	4.9530		496		4		4	
		5209		52		2		
1.7	5.4739		548		6			
		5757		58				
1.8	6.0496		606					
		6363						
1.9	6.6859							

The differences do not become constant in this case, since e^x is not a polynomial function. After reaching their lowest values in the $\Delta^5 f$ column, they
start to grow again; this is because of the accumulation of rounding errors.
The effect of a single error ϵ on successive differences is shown by this table.

f	Δf	$\Delta^2 f$	$\Delta^3 f$	$\Delta^4 f$
0				
	0			
0		0		
	0		0	
0		0		ϵ
	0		ϵ	
0		ϵ		-4ϵ
	ϵ		-3ϵ	
ϵ		-2ϵ		6ϵ
	$-\epsilon$		3ϵ	
0		ϵ		-4ϵ
	0		$-\epsilon$	
0		0		ϵ
	0		0	
0		0		
	0			
0				

This pattern, with its alternating binomial coefficients, can be used to detect a mistake in a table, as in the following example.

Example 2: For equally spaced values of x, a function f is said to take the values 1.669, 1.811, 1.971, 2.151, 2.352, 2.557, 2.828, 3.108, 3.418. Draw up a table of differences, and comment on what it shows.

Solution:	f	Δf	$\Delta^2 f$	$\Delta^3 f$	$\Delta^4 f$
	1.669				
		142			
	1.811		18		
		160		2	
	1.971		20		−1
		180		1	
	2.151		21		−18
		201		−17	
	2.352		4		79
		205		62	
	2.557		66		−119
		271		−57	
	2.828		9		78
		280		21	
	3.108		30		•
		310			
	3.418				

$$\begin{array}{cc} \epsilon & -18 \\ -4\epsilon & 79 \\ 6\epsilon & -119 \\ -4\epsilon & 78 \\ \epsilon & \bullet \end{array}$$

Comparison of the $\Delta^4 f$ patterns   6ϵ and   −119   shows that

there is probably an error −20 (i.e. −0.020) in the entry 2.557, which should therefore be 2.577. Repetition of the wrong figure in a number is a common mistake. Another slip to guard against is the transposition of two figures, such as 3.148 instead of 3.418. □

Exercise 11A

1. The sides of a rectangle are found to be 51.7 cm and 82.2 cm to 3 S.F. Show that the area of the rectangle can be determined from this information to only 1 S.F.

2. In the equation $x^2 + 4.2x - 3.9 = 0$ the coefficients 4.2 and −3.9 are correct to 1 D.P. Solve the equation, giving your solutions to the greatest accuracy that can be guaranteed.

3. A physics book states that 'to obtain the viscosity of water, η, from
   Poiseuille's formula $\eta = \dfrac{\pi p a^4 t}{8 l Q}$ to an accuracy of 1% the radius a must
   be measured with an accuracy of at least 1 in 400'. Explain why.

4. Three quantities, x, y and z, are measured and found to be 1.61, 2.23
   and 4.66 respectively, correct to 2 D.P. Find x^2, $(y - z)^2$ and
   $x^2 - (y - z)^2$ to 2 D.P. and in each case calculate the maximum possible
   error involved.   [MEI]

5. Find the values of x, between $-90°$ and $90°$, which satisfy the equation
   $2 \cos^2 x = b \sin x + c$, where $b = -1, c = 2$. The values of b and c are
   only known within an error of ± 0.01, so that $b = -1 \pm 0.01, c = 2 \pm 0.01$.
   Find the bounds within which the larger value of x must lie.   [MEI]

6. The mass m of an electron moving with velocity v is given by
   $$m = m_0 \left(1 - \frac{v^2}{c^2}\right)^{-\frac{1}{2}},$$ where the rest-mass m_0 and the speed of light c are
   constants. Show that if v/c is small then the relative error in m is approxi-
   mately v^2/c^2 times the relative error in v.

7. The magnetic field H at a point on the axis of a coil of radius a carrying
   a constant current is $k(a^2 + x^2)^{-3/2}$, where x is the distance of the point
   from the centre of the coil and k is a constant. If the error in measuring
   x is ϵ, find an expression for the corresponding error in H. If ϵ is
   constant, find the values of x for which the absolute error in H is
   (i) least   (ii) greatest.

8. If the logarithm of x to base e is known to lie between $l + \epsilon$ and $l - \epsilon$,
   derive a value for the maximum relative error in taking x to be e^l. Show
   that if $\log_{10} x$ is known to within 5×10^{-6}, the maximum relative error
   in x is about 1.2×10^{-5}.   [MEI]

9. The table gives some values of the cubic polynomial $p(x)$

x	0.2	0.4	0.6	0.8
$p(x)$	0.625	0.704	1.597	1.234

   Use a difference table to find $p(0)$ and $p(1)$.

10. Copy and complete this difference table.

x	f	Δf	$\Delta^2 f$	$\Delta^3 f$	$\Delta^4 f$
0.1	?				
		?			
0.2	3.0544		?		
		?		-240	
0.3	3.0909		-250		-24
		?		?	
0.4	?		?		
		?			
0.5	?				

11. The following table contains an error. By constructing a difference table, locate the error and correct the original table.

x	1	2	3	4	5	6	7	8	9	10
y	7	12	21	34	51	70	97	126	159	196

[MEI]

12. Estimate the maximum possible error in the fifth differences due to a rounding off error in the least significant digit of one value of an arbitrary function.

13. (i) Write down the first three orders of differences for the following table of values. Locate and correct a probable error in the table and suggest how this type of error may occur.

x	7	8	9	10	11	12	13	14
y	103.32	110.62	118.04	125.62	133.04	141.42	149.72	158.34

(ii) If the variable y in (i) is a cubic polynomial in x, use the table of differences to express y in the form
$a + b(x - 7) + c(x - 7)(x - 8) + d(x - 7)(x - 8)(x - 9)$.
Hence evaluate y to 2 D.P. when $x = 9.4$, and compare the value obtained from the cubic with that obtained by linear interpolation.

[MEI]

14. The values of a polynomial u at $x = a, a + h, \ldots, a + nh$ are denoted by $u_a, u_{a+h}, \ldots, u_{a+nh}$ respectively. Prove that
$$u_{a+nh} = u_n + \binom{n}{1} \Delta u_a + \binom{n}{2} \Delta^2 u_a + \ldots + \binom{n}{r} \Delta^r u_a + \ldots + \Delta^n u_a,$$
where $\Delta u_a \equiv u_{a+h} - u_a$, $\Delta^r u_a \equiv \Delta^{r-1} u_{a+h} - \Delta^{r-1} u_a \, (r = 2, 3, \ldots, n)$.

The following table gives the values u_x of a polynomial of the fourth degree for certain integral values of x. Calculate the value of the polynomial when $x = 7.5$.

x	5	6	7	8	9
u_x	6.195	5.919	5.630	5.326	5.006

[MEI]

11.2 Iteration

Here is a well-known and efficient method of calculating square roots.

Let x_1 be a positive approximation to $\sqrt{X}$. If $x_1 > \sqrt{X}$ then $\dfrac{X}{x_1} < \dfrac{X}{\sqrt{X}} = \sqrt{X}$, while if $x_1 < \sqrt{X}$ then $\dfrac{X}{x_1} > \sqrt{X}$; in both cases $\sqrt{X}$ lies between x_1 and $\dfrac{X}{x_1}$, so we expect to get closer to $\sqrt{x}$ by taking $x_2 = \frac{1}{2}\left(x_1 + \dfrac{X}{x_1}\right)$. Repeating this

process produces a sequence of approximations $x_1, x_2, x_3, \ldots$, where
$x_{n+1} = \frac{1}{2}\left(x_n + \dfrac{X}{x_n}\right)$. For example, to find $\sqrt{20}$ we take $X = 20$, $x_1 = 4$.

Then $x_2 = \frac{1}{2}\left(4 + \dfrac{20}{4}\right) = 4.5$

$x_3 = \frac{1}{2}\left(4.5 + \dfrac{20}{4.5}\right) = \frac{1}{2}(4.5 + 4.\dot{4}) = 4.47\dot{2}$

$x_4 = \frac{1}{2}\left(4.47\dot{2} + \dfrac{20}{4.47\dot{2}}\right) = \frac{1}{2}(4.47\dot{2} + 4.472\,049\,7)$

$= 4.472\,136\,0$ by calculator, working to 7 D.P.

and this is in fact $\sqrt{20}$ correct to 7 D.P., found with remarkable ease.

Qu. 1.   Find $\sqrt{20}$ by this method, starting with (i) $x_1 = 5$ (ii) $x_1 = 100$.

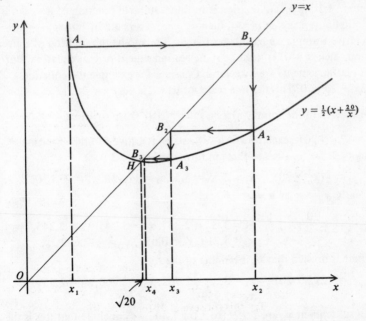

This process can be illustrated graphically. The diagram shows the line
$y = x$ and the curve $y = \frac{1}{2}\left(x + \dfrac{20}{x}\right)$ for $x > 0$; the curve is in fact one

branch of a hyperbola with asymptotes $x = 0$, $y = \frac{1}{2}x$, but for clearness the

diagram is not to scale. The line and curve meet at H, where $x = \frac{1}{2}\left(x + \dfrac{20}{x}\right)$

$\Rightarrow x = \dfrac{20}{x} \Rightarrow x^2 = 20 \Rightarrow x = \sqrt{20}.$

Let the line through $(x_1, 0)$ parallel to the y-axis meet the curve at A_1. The coordinates of A_1 are $\left(x_1, \frac{1}{2}\left(x_1 + \frac{20}{x_1}\right)\right)$, i.e. (x_1, x_2). The line $y = x_2$ through A_1 parallel to the x-axis meets the line $y = x$ at $B_1(x_2, x_2)$, and the line through B_1 parallel to the y-axis meets the x-axis at $(x_2, 0)$. Now the whole process can be repeated, drawing lines up to the curve (A_2), across to $y = x$ (B_2) and down to $(x_3, 0)$, and so on. The zig-zag line $A_1 B_1 A_2 B_2 \ldots$ approaching H indicates clearly how the numbers $x_1, x_2, x_3, \ldots$ approach $\sqrt{20}$.

This method is an example of *iteration*, in which we approach a solution by repeated calculation, using each approximation to produce the next. The formula $x_{n+1} = \frac{1}{2}\left(x_n + \frac{X}{x_n}\right)$ is called an *iterative* formula. Iterative methods are often used in computing, where it is simple to program a loop to perform the repeated calculation.

The general iterative formula is of the form $x_{n+1} = \phi(x_n)$. The basic problem in setting up an iteration is to find a suitable function ϕ. If x_n tends to a finite limit, α say, as $n \to \infty$, then $x_{n+1} \to \alpha$ also, and by letting $n \to \infty$ in the iterative formula we have $\alpha = \phi(\alpha)$. So if α is to be the solution of a given equation, then $x = \phi(x)$ must be some rearrangement of that equation. But not all rearrangements are successful. Consider for example the equation $x^3 - 2x - 20 = 0$. This has one real root which is near $x = 3$. The equation can be written in the forms (a) $x = \frac{1}{2}(x^3 - 20)$; (b) $x = \frac{20}{x^2 - 2}$; (c) $x = (2x + 20)^{\frac{1}{3}}$, each of which suggests an iterative formula. Starting with $x_1 = 3$, we try each of these in turn, working to 3 D.P.

(a) $x_{n+1} = \frac{1}{2}(x_n^3 - 20)$. $x_1 = 3, x_2 = 3.5, x_3 = 11.438, x_4 = 738.106, \ldots$
    Clearly $x_n \to \infty$ as $n \to \infty$.

(b) $x_{n+1} = \dfrac{20}{x_n^2 - 2}$. $x_1 = 3, x_2 = 2.857, x_3 = 3.245, x_4 = 2.345,$
                             $x_5 = 5.719, x_6 = 0.651, x_7 = -12.692, \ldots$

There is no sign that x_n is tending to the root.

(c) $x_{n+1} = (2x_n + 20)^{\frac{1}{3}}$. $x_1 = 3, x_2 = 2.962, x_3 = 2.960, x_4 = 2.959,$
                             $x_5 = 2.959.$

All subsequent x_n are 2.959 (to 3 D.P.), and we conclude that this is the root to 3 D.P.

It is instructive to see what happens graphically in each case. As before we draw $y = x$ and $y = \phi(x)$, and then follow the zig-zag path from $(3, 0)$, alternately moving up to the curve and across to $y = x$. For clearness the diagrams are not to scale. We see how in (a) and (b) the zig-zag leads away from the root, while in (c) it leads towards it.

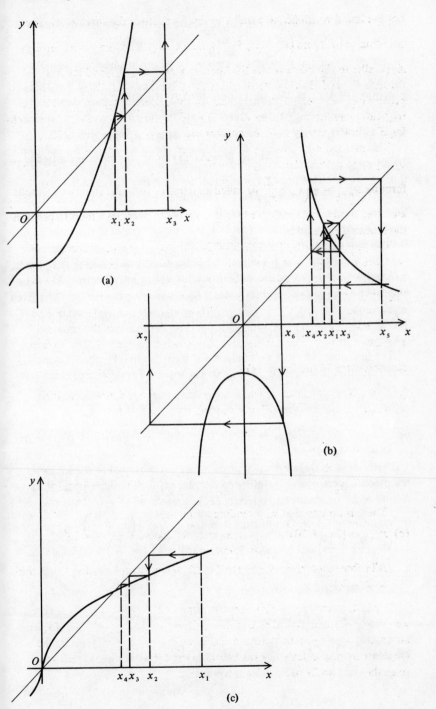

(a)

(b)

(c)

Qu. 2.   Draw diagrams for (a), (b), (c) starting with x_1 just less than the root. Notice that (c) is again the only successful iteration.

It is clearly desirable to have some way of telling in advance whether an iteration is going to converge to a solution. Suppose that α is a root of the equation $x = \phi(x)$, and let the error in the approximation x_n be ϵ_n, so that

$$x_n + \epsilon_n = x_{n+1} + \epsilon_{n+1} = \alpha.$$

Since $x_{n+1} = \phi(x_n)$, $\alpha - \epsilon_{n+1} = \phi(\alpha - \epsilon_n) \approx \phi(\alpha) - \epsilon_n \phi'(\alpha)$,

using the first Taylor approximation centred at α.

But $\alpha = \phi(\alpha)$. Therefore $\epsilon_{n+1} \approx \epsilon_n \phi'(\alpha)$. So if $|\phi'(\alpha)| < 1$, and if x_n is sufficiently close to α for the Taylor expansion to be a good approximation, then $|\epsilon_{n+1}| < |\epsilon_n|$, and the approximations move closer to the root at each stage. On the other hand if $|\phi'(\alpha)| > 1$, the error increases in magnitude. If $\phi'(\alpha)$ is positive, the errors are of constant sign; if $\phi'(\alpha) < 0$, the sign of the error alternates. This is summarized in the four diagrams opposite.

Since we do not know the numerical value of α, it may be necessary to estimate $\phi'(\alpha)$, for example by finding $\phi'(x_1)$.

Qu. 3.   Find $\phi'(3)$ for each of the iterations (a), (b), (c) above, and verify that the sequences behave as predicted.

The best possible value we can have for $\phi'(\alpha)$ is zero. The first Taylor approximation then gives $\epsilon_{n+1} \approx \epsilon_n . 0$. The second Taylor approximation is

$$\alpha - \epsilon_{n+1} \approx \phi(\alpha) - \epsilon_n \phi'(\alpha) + \frac{\epsilon_n^2}{2} \phi''(\alpha)$$

or
$$\epsilon_{n+1} \approx -\frac{\epsilon_n^2}{2} \phi''(\alpha) \quad \text{since } \alpha = \phi(\alpha) \text{ and } \phi'(\alpha) = 0.$$

The error at each stage is then approximately proportional to the square of the preceding error, and therefore convergence will be rapid. This is called *second-order* or *quadratic* convergence, whereas if $\phi'(\alpha) \neq 0$ the convergence is *first-order* or *linear*. Our original iteration $x_{n+1} = \frac{1}{2}\left(x_n + \frac{X}{x_n}\right)$ has $\phi(x) \equiv \frac{1}{2}\left(x + \frac{X}{x}\right)$ and $\alpha = \sqrt{X}$, so that $\phi'(\alpha) = \frac{1}{2}\left(1 - \frac{X}{\alpha^2}\right) = 0$. Thus this method has second order convergence, which is why it is so efficient.

The Newton-Raphson method of §2.4 is a general iterative method for solving $f(x) = 0$ in which $x_{n+1} = x_n - \frac{f(x_n)}{f'(x_n)}$, so that $\phi(x) \equiv x - \frac{f(x)}{f'(x)}$.

Therefore $\phi'(x) \equiv 1 - \frac{f'(x) . f'(x) - f(x) . f''(x)}{[f'(x)]^2}$

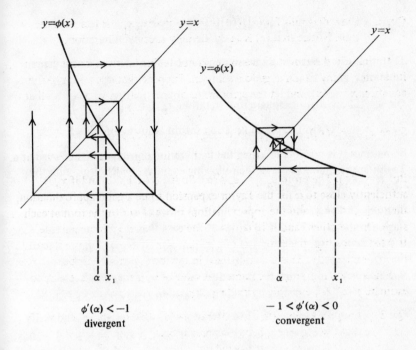

$\phi'(\alpha) < -1$
divergent

$-1 < \phi'(\alpha) < 0$
convergent

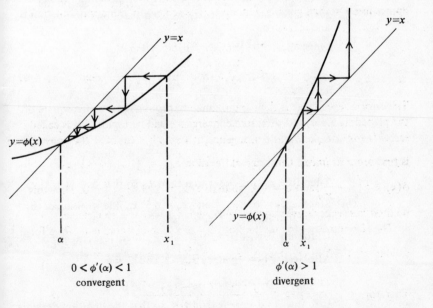

$0 < \phi'(\alpha) < 1$
convergent

$\phi'(\alpha) > 1$
divergent

and
$$\phi'(\alpha) = 1 - \frac{[f'(\alpha)]^2 - 0 \cdot f''(\alpha)}{[f'(\alpha)]^2} = 0.$$

Therefore every Newton-Raphson iteration has second-order convergence (at least).

Qu. 4. Show that the square root iteration $x_{n+1} = \frac{1}{2}\left(x_n + \frac{X}{x_n}\right)$ is in fact the Newton-Raphson iteration for the solution of $x^2 - X = 0$.

Sometimes it is possible to show that an iteration converges without using Taylor approximations, as in the following example, which deals with the iteration (c) used above.

Example 1: The root of $x^3 - 2x - 20 = 0$ is α, and $\alpha = x_n + \epsilon_n$ where $x_{n+1} = (2x_n + 20)^{1/3}$, $x_1 = 3$. Prove that

(i) $x_{n+1}^3 - \alpha^3 = 2(x_n - \alpha)$ (ii) $\alpha < x_{n+1} < x_n$ (iii) $|\epsilon_n| < \frac{1}{6^{n-1}}$.

Solution: (i) $x_{n+1}^3 - \alpha^3 = 2x_n + 20 - \alpha^3$

$\qquad\qquad\qquad = 2x_n + 20 - 2\alpha - 20$ since $\alpha^3 - 2\alpha - 20 = 0$

$\qquad\qquad\qquad = 2(x_n - \alpha)$.

(ii) $2 < \alpha < 3$ since $2^3 - 2 \times 2 - 20 < 0$ and $3^3 - 2 \times 3 - 20 > 0$.
Now $x_{n+1} - \alpha$ has the same sign as $x_{n+1}^3 - \alpha^3$, which has the same sign as $x_n - \alpha$ by (i). Therefore all the x_r are on the same side of α, and since $x_1 = 3 > \alpha$ it follows that $x_n > \alpha$ and $x_{n+1} > \alpha$.

Since $x_{n+1}^3 - \alpha^3 \equiv (x_{n+1} - \alpha)(x_{n+1}^2 + \alpha x_{n+1} + \alpha^2)$ we have

$$0 < x_{n+1} - \alpha = \frac{2(x_n - \alpha)}{x_{n+1}^2 + \alpha x_{n+1} + \alpha^2}$$

$$< \frac{2(x_n - \alpha)}{3\alpha^2} \qquad \text{since } x_{n+1} > \alpha$$

$$< \frac{x_n - \alpha}{6} \qquad \text{since } \alpha > 2.$$

Therefore x_{n+1} is closer to α than x_n, so that $\alpha < x_{n+1} < x_n$.

(iii) Since $\epsilon_{n+1} = \alpha - x_{n+1}$ and $\epsilon_n = \alpha - x_n$, the working of (ii) shows that $|\epsilon_{n+1}| < \frac{1}{6}|\epsilon_n|$.
If, for any particular k, $|\epsilon_k| < \frac{1}{6^{k-1}}$ then $|\epsilon_{k+1}| < \frac{1}{6} \cdot \frac{1}{6^{k-1}} = \frac{1}{6^k}$.
Also $|\epsilon_1| < 1 = \frac{1}{6^0}$ since $x_1 = 3$ and $2 < \alpha < 3$.
Therefore by induction $|\epsilon_n| \leqslant \frac{1}{6^{n-1}}$ for all $n \geqslant 1$.

It follows that $\epsilon_n \to 0$ as $n \to \infty$, so that the iteration does converge to α.   □

Exercise 11B

1. Use the iteration $x_{n+1} = (6x_n + 3)^{\frac{1}{4}}$ to find to 3 S.F. the root of $x^4 - 6x - 3 = 0$ near $x = 2$.

2. Find the root of the equation $x^3 - 10x + 8 = 0$ near $x = 1$ to 3 S.F.

3. If the iteration $x_{n+1} = \dfrac{20}{x_n}$, $x_1 = 4$ converges what will be its limit?

   Does the iteration converge? Illustrate what is happening by means of a diagram. [This shows one possibility when $\phi'(\alpha) = -1$.]

4. Devise simple examples of iterations $x_{n+1} = \phi(x_n)$ with $\phi'(\alpha) = 1$, where $\alpha = \phi(\alpha)$, such that
   (i) the iteration converges to α (ii) the iteration diverges.

5. Show that the only possible limits of the iteration $x_{n+1} = x_n^2$ are 0 and 1. Show graphically that the iteration diverges if $x_1 > 1$, and find what happens if $0 < x_1 < 1$.

6. By finding the points of intersection of the graph of the function $f(x) = 5 - 5/x^2$ with a suitable straight line, find to 2 S.F. the positive roots of the equation $x = 5 - 5/x^2$.
   The iterative method based on the relation $x_{n+1} = f(x_n)$, using an initial approximation $x_0 > 2$, is to be employed to evaluate one of these roots. By inspecting your graph, or otherwise, find to which root the iteration will converge. Perform one iteration to improve the approximation you have found for this root.   [L]

7. Show that the equation $x^{10} = 10x - 7$ has one root between 0 and 1 and one root between 1 and 2. Decide to which of these roots the iteration $x_{n+1} = (10x_n - 7)^{1/10}$ will converge, and find this root to 2 D.P. Devise a suitable iteration for the other root, and use it to find that root to 2 D.P.

8. By using a suitable iteration find the root of $x^2 = \sin x$ near $\dfrac{\pi}{4}$ to 2 D.P. (or to 5 D.P. if you have a scientific calculator).

9. The flow diagram is intended to represent an iterative process for finding

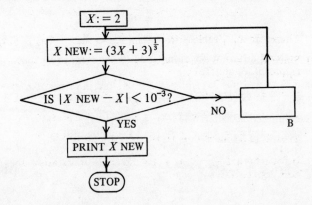

an approximation to the root of
$$x^3 - 3x - 3 = 0$$
that lies near $x = 2$.
   (i)   What should be inserted in the blank box B?
   (ii) Use the iteration to find the root of $x^3 - 3x - 3 = 0$ correct to
        3 S.F.
   (iii) Sketch the graphs $y = 3x + 3$ and $y = x^3$ and show geometrically
        how the iteration converges.   [SMP]

10. The iterative formula $x_{n+1} = a + \dfrac{b}{x_n}$ $(a, b > 0), (n = 1, 2, 3, \ldots)$, can
    be used to obtain successive approximations to a root of the quadratic
    equation $x^2 = ax + b$. For the equation $2x^2 - x - 20 = 0$, write down
    suitable values of a and b and, starting at $x_1 = 3$, find the values of x_2,
    x_3, x_4. Write a flow chart, using the above iteration, which, starting with
    $x_1 = 3$, will generate successive approximations to a root of
    $2x^2 - x - 20 = 0$, and will terminate, either when two successive
    approximations first agree correct to two places of decimals, or when
    x_{20} has been found, whichever first occurs.   [C]

11. Show that both the iterations (a) $x_{n+1} = \frac{1}{3}(2x_n + 12/x_n^2)$, $x_1 = 2$
    (b) $x_{n+1} = \frac{1}{2}(x_n + 12/x_n^2)$, $x_1 = 2$ give better approximations to $\sqrt[3]{12}$ as
    n increases. Which set of approximations is likely to approach $\sqrt[3]{12}$ more
    quickly? Evaluate x_2, x_3 in each case, and compare their values with
    $\sqrt[3]{12} = 2.289 \ldots$.

12. In the early days of computing, division had to be done iteratively; the
    iteration $x_{n+1} = x_n(2 - ax_n)$ was used to find $1/a$.
   (i)   Show that this is the Newton-Raphson iteration for the solution of
        $1/x - a = 0$.
   (ii) Show that $1 - ax_{n+1} = (1 - ax_n)^2$, and deduce that $\epsilon_{n+1} = a\,\epsilon_n^2$,
        where $x_n + \epsilon_n = \dfrac{1}{a}$.
   (iii) State the range of values of x_1 for which the iteration converges.

13. Take a long rectangular strip of paper, with ends A and B. Make a crease
    X_1 straight across the strip somewhere. Then make the following sequence
    of creases:
    $$Y_n = \text{the crease when } B \text{ is folded to } X_n,$$
    $$X_{n+1} = \text{the crease when } A \text{ is folded to } Y_n.$$
    Try this for several different positions of X_1. What seem to be the
    limiting positions of X_n and Y_n?
    If the length of the strip is $3a$, and $x_1 = a + \epsilon$, find in terms of a, ϵ, n
    the distances x_n and y_n of X_n and Y_n from A.

14. The positive numbers a_n satisfy the relation $a_{n+1} = (a_n + 3)^{\frac{1}{2}}$
    $(n = 1, 2, 3, \ldots)$, and $a_n \to l$ as $n \to \infty$. Find the value of l.
    Prove that $a_{n+1}^2 - l^2 = a_n - l$, and that $a_n > l$ implies $a_n > a_{n+1} > l$.
    Show that, when $a_1 = \frac{5}{2}$, $a_n - l < \dfrac{1}{2^{2n-1}}$.
                                                                         [MEI]

15. The positive numbers $x_1, x_2, x_3, \ldots$ satisfy the relation
    $x_{r+1} = \dfrac{2(6x_r + 1)}{3(3x_r + 5)}$. Prove that, if x_r tends to a limit as r tends to infinity,
    that limit must be $\frac{1}{3}$. Prove also that, if u_r is written for $x_r - \frac{1}{3}$, then
    $u_{r+1} = \dfrac{u_r}{2 + u_r}$, and deduce, by induction or otherwise, that, if $u_1 = 1$,
    $u_r = \dfrac{1}{2^r - 1}$. Is it established that, in this case, $x_r \to \frac{1}{3}$?   [MEI]

16. A sequence of real numbers a_n ($n = 1, 2, 3, \ldots$) satisfies the relation
    $a_{n+1} = \dfrac{3a_n + 5}{a_n + 3}$. Show that, if $a_1 = 2$, then for all $n \geqslant 1$
    (i) $a_n \geqslant 2$ (ii) $a_n^2 < 5$ (iii) $5 - a_{n+1}^2 < \frac{1}{5}(5 - a_n^2)$
    (iv) $a_{n+1} > a_n$ (v) $\sqrt{5} - a_{n+1} < \frac{1}{4} \times 5^{-n}$.
    Hence find a rational number which differs from $\sqrt{5}$ by less than
    5×10^{-4}.   [MEI]

17. The two sequences $\alpha_1, \alpha_2, \alpha_3, \ldots$ and $\beta_1, \beta_2, \beta_3, \ldots$ are obtained from
    the relations $\alpha_n + \beta_n = p$, $\alpha_{n+1}\beta_n = q$, with $\alpha_1 = 1$. Show that if
    $\alpha_n \to \alpha$ and $\beta_n \to \beta$ as $n \to \infty$, then α, β are the solutions of
    $x^2 - px + q = 0$. See what happens when this method is used with
    (i) $x^2 - 17x + 5 = 0$ (ii) $x^2 - 5x + 17 = 0$ (iii) $x^2 - 17 = 0$.

11.3 Convergence of series

From any infinite sequence $u_1, u_2, u_3, \ldots, u_n, \ldots$ we can form a new
sequence, called the sequence of *partial sums*, by taking $S_1 = u_1$, $S_2 = u_1 + u_2$,
$S_3 = u_1 + u_2 + u_3, \ldots, S_n = \sum\limits_{r=1}^{n} u_r, \ldots$. These partial sums can also be
defined inductively by the relations $S_{n+1} = S_n + u_{n+1}$, $S_1 = u_1$. If S_n tends to
a fixed finite limit, S say, as $n \to \infty$ then we say that the series
$u_1 + u_2 + u_3 + \ldots$ *converges* to S, and write $\sum\limits_{r=1}^{\infty} u_r = S$. In all other cases the
series *diverges*.

Sometimes it is possible to find S_n explicitly, and then tell whether or not
the series converges, as in the following examples.

(a) If $u_n = \dfrac{1}{n(n+1)}$ then $S_n = \sum\limits_{r=1}^{n} \dfrac{1}{r(r+1)} = \sum\limits_{r=1}^{n} \left(\dfrac{1}{r} - \dfrac{1}{r+1}\right)$

$$= (1 - \tfrac{1}{2}) + (\tfrac{1}{2} - \tfrac{1}{3}) + \ldots + \left(\dfrac{1}{n} - \dfrac{1}{n+1}\right)$$

$$= 1 - \frac{1}{n+1}.$$

As $n \to \infty$, $S_n \to 1$, so $\sum_{r=1}^{\infty} \frac{1}{r(r+1)} = 1$.

(b) If $u_n = n(n+1)$ then $S_n = \sum_{r=1}^{n} r(r+1) = \frac{1}{3}n(n+1)(n+2)$

(Ex. 6A, Qu. 11). S_n can be made as large as we please by making n sufficiently large, so the series diverges to infinity.

(c) If $u_n = (-1)^n$ then $S_n = \begin{cases} -1 & \text{if } n \text{ is odd} \\ 0 & \text{if } n \text{ is even} \end{cases}$, which can be written more compactly as $S_n = \frac{1}{2}((-1)^n - 1)$. S_n does not tend to a limit, so the series diverges; it is said to oscillate finitely.

(d) If $u_n = x^{n-1}$ then $S_n = \sum_{r=1}^{n} x^{r-1} = \frac{1-x^n}{1-x}$, the sum of a geometric progression. If $|x| < 1$ the series converges to $\frac{1}{1-x}$; if $|x| \geqslant 1$ the series diverges.

Qu. 1. Describe how series (d) diverges when
   (i) $x > 1$ (ii) $x = 1$ (iii) $x = -1$ (iv) $x < -1$.

If $S_n \to S$ as $n \to \infty$ then $S_{n-1} \to S$ also, and $u_n = S_n - S_{n-1} \to S - S = 0$. Therefore if the series converges, then $u_n \to 0$. The converse is *not* true, as can be seen by taking $u_n = \frac{1}{\sqrt{n}}$. For then $u_n \to 0$ but

$$S_n = 1 + \frac{1}{\sqrt{2}} + \frac{1}{\sqrt{3}} + \ldots + \frac{1}{\sqrt{n}}$$

$$> n \cdot \frac{1}{\sqrt{n}} \qquad \text{replacing each term by the smallest}$$

$$= \sqrt{n} \to \infty,$$

so the series diverges.

> The condition $u_n \to 0$ is necessary but not sufficient for the convergence of $\sum_{r=1}^{\infty} u_r$.

This supplies us with a test for divergence: if $u_n \not\to 0$ as $n \to \infty$ then $\sum_{r=1}^{\infty} u_r$ is divergent. Thus although we cannot easily evaluate $\sum_{r=1}^{n} \sin r$ we can be sure that the series diverges, since $\sin n \not\to 0$ as $n \to \infty$.

If all the terms of a sequence are positive, $u_n > 0$ for all n, then $S_{n+1} = S_n + u_{n+1} > S_n$, so the partial sums form a sequence of positive increasing terms. If there exists a number, K say, such that $S_n < K$ for all n, we say that $\{S_n\}$, the sequence of partial sums, is *bounded above*. If all the u_n are positive there are just two possibilities.

(1) If $\{S_n\}$ is not bounded above, then for each K however large there is a particular n, N say, such that $S_N \geqslant K$, and then, since $\{S_n\}$ is increasing, $S_n > K$ for all $n > N$. This is just what we mean by saying that $S_n \to \infty$ as $n \to \infty$.

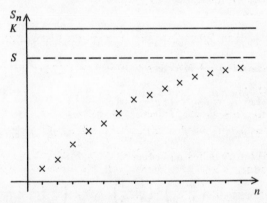

(2) If $\{S_n\}$ is bounded above by K then geometrical intuition suggests that S_n tends to a limit, S say, as $n \to \infty$, with $S \leqslant K$. This is in fact true, and will be assumed from now on. To prove it would require a thorough investigation of the nature of the real number system, which we cannot undertake here. This is indicated by the fact that the result is not true within the more restricted rational number system. For example, the positive increasing sequence of rational numbers $3, 3.1, 3.14, 3.141, 3.1415, 3.14159, \ldots$ is bounded above (by 4) but does not have a rational limit, since the limit is π which is irrational. We can summarize (1) and (2) by saying that

> a series of *positive* terms converges if its partial sums are bounded above, and diverges to infinity if they are not.

Q. 2.   State the corresponding results for a series of negative terms.

Example 1: Show that (i) $1 + \frac{1}{2} + \frac{1}{3} + \frac{1}{4} + \frac{1}{5} + \frac{1}{6} + \ldots$ diverges

(ii) $1 - \frac{1}{2} + \frac{1}{3} - \frac{1}{4} + \frac{1}{5} - \frac{1}{6} + \ldots$ converges.

Solution: (i) Consider the partial sums S_n for which n is a power of 2. For example

$$S_{2^5} = S_{32} = 1 + \frac{1}{2} + (\frac{1}{3} + \frac{1}{4}) + (\frac{1}{5} + \frac{1}{6} + \frac{1}{7} + \frac{1}{8})$$
$$+ (\frac{1}{9} + \ldots + \frac{1}{16}) + (\frac{1}{17} + \ldots + \frac{1}{32})$$
$$\text{[8 terms]} \qquad\qquad \text{[16 terms]}$$

$$> 1 + \frac{1}{2} + (\frac{1}{4} + \frac{1}{4}) + (\frac{1}{8} + \frac{1}{8} + \frac{1}{8} + \frac{1}{8})$$
$$+ (\frac{1}{16} + \ldots + \frac{1}{16}) + (\frac{1}{32} + \ldots + \frac{1}{32})$$
$$\text{[8 terms]} \qquad\qquad \text{[16 terms]}$$

$$= 1 + \frac{1}{2} + \frac{1}{2} + \frac{1}{2} + \frac{1}{2} + \frac{1}{2}$$

$$= 3\frac{1}{2}.$$

Similarly $S_{2^m} > 1 + \frac{1}{2}m$, which can be made arbitrarily large. Therefore the partial sums are unbounded and the *harmonic* series $\sum\limits_{r=1}^{\infty} \frac{1}{r}$ diverges.

(ii) Let $T_n = S_{2n} = \left(1 - \frac{1}{2}\right) + \left(\frac{1}{3} - \frac{1}{4}\right) + \ldots + \left(\frac{1}{2n-1} - \frac{1}{2n}\right) > 0$

Then $T_{n+1} = T_n + \frac{1}{2n+1} - \frac{1}{2n+2} > T_n$, so that $\{T_n\}$ is a positive increasing sequence.
But

$$T_n = 1 - \left(\frac{1}{2} - \frac{1}{3}\right) - \left(\frac{1}{4} - \frac{1}{5}\right) - \ldots - \left(\frac{1}{2n-2} - \frac{1}{2n-1}\right) - \frac{1}{2n}$$

$$= 1 - \frac{1}{6} - \frac{1}{20} - \ldots - \frac{1}{(2n-2)(2n-1)} - \frac{1}{2n}$$

$$< 1.$$

So $\{T_n\}$ is bounded above by 1. Therefore T_n tends to a limit, T say, as $n \to \infty$. Thus $S_{2n} \to T$ as $n \to \infty$.

Also $S_{2n+1} = S_{2n} + \frac{1}{2n+1} \to T + 0 = T$ as $n \to \infty$. Therefore the alternating harmonic series $\sum\limits_{r=1}^{\infty} \frac{(-1)^{r-1}}{r}$ converges to T.

In Ex. 11C, Qu. 15 we shall show that $T = \ln 2$.   □

One common way to show that the partial sums are bounded is to compare them with the partial sums of a more convenient series. For example, since $\frac{1}{n^2} \leqslant \frac{2}{n(n+1)}$ for all $n \geqslant 1$,

$$\sum_{r=1}^{n} \frac{1}{r^2} \leqslant \sum_{r=1}^{n} \frac{2}{r(r+1)} < 2, \text{ since } \sum_{r=1}^{\infty} \frac{2}{r(r+1)} = 2.$$

Therefore $\sum_{r=1}^{\infty} \frac{1}{r^2}$ converges to a sum between 0 and 2. Notice that this

method does not tell us the value of $\sum_{r=1}^{\infty} \frac{1}{r^2}$; in fact it is $\frac{\pi^2}{6}$, a surprising

result due to Euler.

Qu. 3.   Prove that $\sum_{r=1}^{\infty} \frac{1}{r^\alpha}$ converges if $\alpha > 2$ and diverges if $\alpha < 1$.

Another useful test for the convergence of a series of positive terms uses comparison with a geometric series. This is known as *D'Alembert's ratio test*, though it was first used by the Cambridge mathematician Edward Waring in 1776.

> If for a series of positive terms u_n there is a fixed number k
>
> and an integer N such that $\frac{u_{n+1}}{u_n} < k < 1$ for all $n \geqslant N$
>
> then $\sum_{r=1}^{\infty} u_r$ converges.

For $u_{N+m} < k u_{N+m-1} < k^2 u_{N+m-2} < \ldots < k^m u_N$.

Therefore $\sum_{r=N}^{N+m} u_r = \sum_{r=0}^{m} k^m u_N = u_N \frac{1 - k^{m+1}}{1-k} < \frac{u_N}{1-k}$.

So if $n > N$, $\sum_{r=1}^{n} u_r < \sum_{r=1}^{N-1} u_r + \frac{u_N}{1-k}$, a fixed number. Therefore the

partial sums are bounded, and $\sum_{r=1}^{\infty} u_r$ converges. The statement of the test

should be noted carefully: it is *not* sufficient merely to have $\frac{u_{n+1}}{u_n} < 1$, as

the example $u_n = \frac{1}{\sqrt{n}}$ has already shown.

Example 2: Show that the series $1 + x + \frac{x^2}{2!} + \frac{x^3}{3!} + \ldots + \frac{x^n}{n!} + \ldots$ is

convergent for all positive x.

Solution: Let $u_n = \frac{x^n}{n!}$. Then $\frac{u_{n+1}}{u_n} = \frac{x^{n+1}}{(n+1)!} \cdot \frac{n!}{x^n} = \frac{x}{n+1}$.

If $n > 2x - 1$ then $\frac{u_{n+1}}{u_n} < \frac{1}{2}$, which is a fixed number less than 1.

Therefore by D'Alembert's ratio test the series converges.
(It also converges for all negative x, see Ex. 11C, Qu. 12).   □

Qu. 4. Deduce from Example 2 that, for all x, $\dfrac{x^n}{n!} \to 0$ as $n \to \infty$.

Example 3: Estimate the truncation error involved in taking

$$\sum_{r=1}^{10} \frac{2^r}{r!} \text{ as the value of } \sum_{r=1}^{\infty} \frac{2^r}{r!}.$$

Solution: The truncation error is $\displaystyle\sum_{r=11}^{\infty} \frac{2^r}{r!} = \frac{2^{11}}{11!} + \frac{2^{12}}{12!} + \frac{2^{13}}{13!} + \frac{2^{14}}{14!} + \cdots$

$$= \frac{2^{11}}{11!}\left(1 + \frac{2}{12} + \frac{2^2}{12.13} + \frac{2^3}{12.13.14} + \cdots\right)$$

$$< \frac{2^{11}}{11!}\left(1 + \frac{2}{12} + \frac{2^2}{12^2} + \frac{2^3}{12^3} + \cdots\right)$$

$$= \frac{2^{11}}{11!} \times \frac{1}{1 - \frac{1}{6}} = \frac{6 \times 2^{11}}{5 \times 11!} \approx 6.2 \times 10^{-5}.$$

Therefore the truncation error is less than 6.2×10^{-5}.   □

Exercise 11C

1. Show that $\dfrac{1}{2n-1} - \dfrac{1}{2n+1} \equiv \dfrac{2}{(2n-1)(2n+1)}$. Hence find

   $\displaystyle\sum_{r=1}^{\infty} \frac{1}{(2r-1)(2r+1)}$. How many terms of the series must be taken to

   give this sum with a truncation error less than 10^{-3}?

2. Prove that if u_n is positive and $\dfrac{u_{n+1}}{u_n} < k < 1$ for all n, then the error in

   taking $\displaystyle\sum_{r=1}^{n} u_r$ as the value of $\displaystyle\sum_{r=1}^{\infty} u_r$ is less than $\dfrac{k u_n}{1 - k}$.

3. Prove that $\displaystyle\sum_{r=1}^{\infty} \frac{1}{r \cdot 2^r}$ converges, and that the error in taking the sum of

   the first five terms as the sum of the infinite series is less than 0.01.

4. Let $u_n = nx^{n-1}$, where x is a fixed number between 0 and 1. Prove that

   $\dfrac{u_{n+1}}{u_n} < \frac{1}{2}(1 + x)$ provided that $n > \dfrac{2x}{1-x}$. Deduce that, for

   $0 < x < 1$, $\displaystyle\sum_{r=1}^{\infty} rx^{r-1}$ converges, and that, for $|x| < 1$, $nx^n \to 0$ as $n \to \infty$.

   By differentiating $\displaystyle\sum_{r=1}^{n} x^r$, or otherwise, find the value of $\displaystyle\sum_{r=1}^{\infty} rx^{r-1}$.

5. A sequence of positive real numbers $\{a_r : r = 1, 2, 3, \ldots\}$ is defined by
$$a_r = \frac{3.5.7. \ldots .(2r+1)}{5.8.11. \ldots .(3r+2)}.$$ Prove that $a_r < \left(\frac{2}{3}\right)^r$. Deduce that the sum $\sum_{r=1}^{n} a_r$ does not exceed 2, and that, if $n > 15$, it is within 0.005 of the estimate obtained by summing the first 15 terms of the series.   [MEI]

6. Prove that $\displaystyle\sum_{r=1}^{n} \frac{1}{r(r+1)(r+2)} = \frac{1}{4} - \frac{1}{2(n+1)(n+2)}.$

It is desired to estimate $\displaystyle\sum_{r=1}^{1000} \frac{1}{r(r+1.01)(r+2.001)}$ to within 0.001 by summing only the first N terms. Find a suitable value for N.   [MEI]

7. In the binomial expression of $(1-x)^{\frac{1}{3}}$, where x is positive and less than 1, show that the ratio of the term in x^{n+1} to that in x^n is $\frac{3n-1}{3n+3} x.$

By comparing the sum of all the terms after the first three with a suitable infinite geometric series, prove that the magnitude of the error involved in neglecting these terms is less than $(5x^3/81)(1-x)^{-1}$. Show that, if $x \leqslant 0.1$, this error is, in magnitude, less than 0.000 07 and hence, using the binomial expansion, evaluate $(0.9)^{\frac{1}{3}}$ correct to 3 D.P.   [MEI]

8. If $u_r = r \ln\left(\dfrac{r+2}{r+3}\right)$ find $\displaystyle\sum_{r=1}^{n} u_r$, and decide whether $\displaystyle\sum_{r=1}^{\infty} u_r$ converges.

9. Given that $|x| > 1$, find in its simplest form the sum to infinity of the series $\dfrac{4}{x} + \dfrac{5}{x^2} + \dfrac{6}{x^3} + \dfrac{4}{x^4} + \dfrac{5}{x^5} + \dfrac{6}{x^6} + \ldots + \dfrac{4}{x^{3n-2}} + \dfrac{5}{x^{3n-1}} + \dfrac{6}{x^{3n}} + \ldots .$

10. Find the set of real values of x for which the series $\displaystyle\sum_{r=1}^{\infty} (-1)^r \left(\frac{1+x}{1+x^2}\right)^r$ is convergent. Find the value of x for which the sum to infinity of this series is a maximum.   [L]

11. Use the method of Example 1 (ii), §11.3 to show that if $0 < u_{n+1} < u_n$ and $u_n \to 0$ as $n \to \infty$ then $u_1 - u_2 + u_3 - u_4 + u_5 - u_6 + \ldots + (-1)^{r-1} u_r + \ldots$ converges to a limit between 0 and u_1.

12. Use Qu. 11 to show that
  (i) $1 + x + \dfrac{x^2}{2!} + \dfrac{x^3}{3!} + \ldots$ converges for all negative x
  (ii) $1 + 2x + 3x^2 + 4x^3 + \ldots$ converges for $-1 < x < 0$.

13. (i) The diagrams (overleaf) show the curve $y = \dfrac{1}{x}$ with approximations to the area under the curve from $x = 1$ to $x = n$ by means of upper rectangles (Fig. 1) and lower rectangles (Fig. 2) of unit width. Use

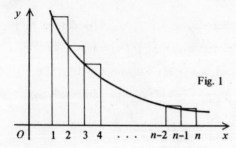

Fig. 1

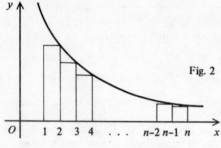

Fig. 2

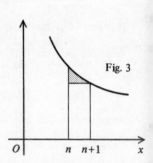

Fig. 3

these to show that $\displaystyle\sum_{r=2}^{n} \frac{1}{r} < \ln n < \sum_{r=1}^{n-1} \frac{1}{r}$. Hence show that if

$d_n = \displaystyle\sum_{r=1}^{n} \frac{1}{r} - \ln n$ then $0 < d_n < 1$.

(ii) Show that $d_n - d_{n+1}$ equals the shaded area in Fig. 3, and hence that $\{d_n\}$ is a decreasing sequence.

(iii) Deduce that d_n tends to a limit, γ say, between 0 and 1 as $n \to \infty$. [The value of γ is approximately $0.557\,256\,6$; γ is called *Euler's constant*.]

14. Use Qu. 13 to estimate $\displaystyle\sum_{r=1}^{10^6} \frac{1}{r}$.

15. The result of Qu. 13 can be written as $\displaystyle\sum_{r=1}^{n} \frac{1}{r} = \ln n + \gamma + \epsilon_n$, where

$\epsilon_n \to 0$ as $n \to \infty$. Let $S_n = 1 - \frac{1}{2} + \frac{1}{3} - \frac{1}{4} + \ldots + (-1)^{n-1} \frac{1}{n}$. Show that

$S_{2n} = 1 + \frac{1}{2} + \frac{1}{3} + \ldots + \frac{1}{2n} - \left(1 + \frac{1}{2} + \frac{1}{3} + \ldots + \frac{1}{n}\right)$ and hence that

$S_{2n} = \ln 2 + \epsilon_{2n} - \epsilon_n$. Deduce that $1 - \frac{1}{2} + \frac{1}{3} - \frac{1}{4} + \frac{1}{5} - \frac{1}{6} + \ldots$ converges to $\ln 2$.

16. Use the method of Qu. 13 with the curve $y = \dfrac{1}{x^\alpha}$, $\alpha > 0$, to show that

$\displaystyle\sum_{r=1}^{n} \frac{1}{r^\alpha} - \int_1^n \frac{1}{x^\alpha}\, dx$ tends to a limit as $n \to \infty$. Deduce that $\displaystyle\sum_{r=1}^{\infty} \frac{1}{r^\alpha}$

converges if $\alpha > 1$ and diverges if $\alpha \leqslant 1$.

17. Find whether $\displaystyle\sum_{r=3}^{\infty} \frac{1}{r \ln r}$ converges.

11.4 Power series

In §2.5 we obtained the Maclaurin or Taylor polynomial approximations to various functions; now we consider in more detail the accuracy of these approximations.

If $f(x) = a_0 + a_1 x + a_2 x^2 + \ldots + a_n x^n + E_n(x)$, then $E_n(x)$ is the truncation error introduced by using the polynomial $a_0 + a_1 x + a_2 x^2 + \ldots + a_n x^n$ in place of $f(x)$. If $E_n(x) \to 0$ as $n \to \infty$, then the infinite series $\sum_{r=0}^{\infty} a_r x^r$ converges to $f(x)$; this series is called the *power series expansion* of $f(x)$.

We start with two special cases obtained from the important identity

$$\frac{1}{1+x} \equiv 1 - x + x^2 - x^3 + \ldots + (-x)^{n-1} + \frac{(-x)^n}{1+x}.$$

Qu. 1. Prove this identity by summing the geometric progression
$$1 - x + x^2 - \ldots + (-x)^{n-1}.$$

Replacing x by t^2 and then integrating from 0 to x gives

$$\int_0^x \frac{1}{1+t^2}\,dt = x - \frac{x^3}{3} + \frac{x^5}{5} + \ldots + (-1)^{n-1}\frac{x^{2n-1}}{2n-1} + \int_0^x \frac{(-1)^n\, t^{2n}}{1+t^2}\,dt.$$

But $\int_0^x \frac{1}{1+t^2}\,dt = \left[\tan^{-1} t\right]_0^x = \tan^{-1} x$. Therefore

$$\tan^{-1} x = x - \frac{x^3}{3} + \frac{x^5}{5} - \frac{x^7}{7} + \ldots + (-1)^n \frac{x^{2n-1}}{2n-1} + R_n(x),$$

where $R_n(x) = \int_0^x \frac{(-1)^n\, t^{2n}}{1+t^2}\,dt = E_{2n-1}(x) = E_{2n}(x)$.

Since $1 + t^2 \geqslant 1$, it follows that $0 \leqslant \dfrac{t^{2n}}{1+t^2} \leqslant t^{2n}$ with equality only

when $t = 0$, and so $|R_n(x)| = \left| \int_0^x \frac{t^{2n}}{1+t^2}\,dt \right| < \left| \int_0^x t^{2n}\,dt \right| = \frac{|x^{2n+1}|}{2n+1}$.

Here we have used the property

$$f(t) < g(t) \text{ for } 0 < t < x \Rightarrow \int_0^x f(t)\,dt < \int_0^x g(t)\,dt; \text{ see } §12.1.$$

If $|x| \leqslant 1$ then $\dfrac{|x^{2n+1}|}{2n+1} \leqslant \dfrac{1}{2n+1} \to 0$ as $n \to \infty$.

If $|x| > 1$, let $|x| = 1 + h$. Then

$\dfrac{|x|^n}{n} = \dfrac{1 + nh + \ldots + h^n}{n} > \dfrac{nh}{n} = h$; therefore the terms do not tend to

zero and so the series does not converge. Therefore

$$\tan^{-1}x \;=\; x - \frac{x^3}{3} + \frac{x^5}{5} - \frac{x^7}{7} + \ldots + (-1)^{n-1}\frac{x^{2n-1}}{2n-1} + \ldots \;\text{ for } |x| \leqslant 1.$$

This is called Gregory's series, after the Scottish mathematician James Gregory, who published it in 1668, well before Newton's and Leibniz's accounts of calculus appeared. The particular case $x = 1$ is of special interest because $\tan^{-1} 1 = \dfrac{\pi}{4}$,

$$\frac{\pi}{4} \;=\; 1 - \tfrac{1}{3} + \tfrac{1}{5} - \tfrac{1}{7} + \tfrac{1}{9} - \ldots,$$

a result first given by Leibniz in 1673. This simple series is not much use for calculating the value of π, since it converges very slowly; about 8000 terms have to be taken to guarantee four-figure accuracy. But Gregory's series can be used to calculate π quite efficiently (see Ex. 11D, Qu. 2).

Gregory obtained his series by integrating the series expansion of $\dfrac{1}{1 + t^2}$. At about the same time Nicolaus Mercator (1620–87), who was born in Denmark but lived for a long time in London, used the same idea to find the power series expansion of $\ln(1 + x)$ by integrating $\dfrac{1}{1 + t}$:

$$\ln(1 + x) \;=\; \int_0^x \frac{1}{1 + t}\, dt$$

$$= \int_0^x \left(1 - t + t^2 - t^3 + \ldots + (-t)^{n-1} + \frac{(-t)^n}{1 + t}\right) dt$$

$$= x - \frac{x^2}{2} + \frac{x^3}{3} - \frac{x^4}{4} + \ldots + (-1)^{n-1}\frac{x^n}{n} + E_n(x),$$

where $\qquad E_n(x) \;=\; \displaystyle\int_0^x \frac{(-t)^n}{1 + t}\, dt.$

If $t > 0$ then $1 + t > 1$ and $\dfrac{t^n}{1 + t} < t^n$. Therefore if $0 \leqslant x \leqslant 1$,

$$|E_n(x)| = \int_0^x \frac{t^n}{1 + t}\, dt < \int_0^x t^n\, dt = \frac{x^{n+1}}{n+1} \leqslant \frac{1}{n+1} \to 0 \text{ as } n \to \infty.$$

If $-1 < x < 0$, let $s = -t$ so that $ds = -dt$ and $s = -x$ when $t = x$.

Then $E_n(x) = (-1)^n \displaystyle\int_0^{-x} \frac{(-s)^n}{1 - s}\,(-ds) = -\int_0^{-x} \frac{s^n}{1 - s}\, ds$. Since x is

negative, $1 - s > 1 + x$ and $\dfrac{s^n}{1 - s} < \dfrac{s^n}{1 + x}$ for $0 < s < -x$, so that

$$|E_n(x)| = \int_0^{-x} \frac{s^n}{1-s}\, ds < \int_0^{-x} \frac{s^n}{1+x}\, ds = \frac{(-x)^{n+1}}{(1+x)(n+1)} < \frac{1}{(1+x)(n+1)}$$

$\rightarrow 0$ as $n \rightarrow \infty$. Therefore $E_n(x) \rightarrow 0$ as $n \rightarrow \infty$ if $-1 < x \leqslant 1$.

If $x > 1$ the series does not converge since x^n/n does not tend to zero, while if $x \leqslant -1$, $\ln(1+x)$ does not exist; so the question of convergence to $\ln(1+x)$ does not arise. This establishes Mercator's series

---

$$\ln(1+x) = x - \frac{x^2}{2} + \frac{x^3}{3} - \frac{x^4}{4} + \ldots + (-1)^{n-1}\frac{x^n}{n} + \ldots \text{ for } -1 < x \leqslant 1.$$

Qu. 2. What familiar result is obtained by putting $x = 1$ in this?

To establish the convergence of the power series for e^x, we use Ex. 4A, Qu. 25. This shows that $e^x = 1 + x + \frac{x^2}{2!} + \ldots + \frac{x^n}{n!} + E_n(x)$

where $\qquad E_n(x) = e^x \int_0^x \frac{t^n}{n!} e^{-t}\, dt.$

If $x > 0$, $0 < e^{-t} < 1$ for $0 < t < x$ and so

$$0 < E_n(x) < e^x \int_0^x \frac{t^n}{n!}\, dt = e^x \frac{x^{n+1}}{(n+1)!}.$$

If $x < 0$, let $s = -t$. Then $dt = -ds$, and $s = -x = |x|$ when $t = x$.

Therefore $|E_n(x)| = e^x \int_0^{|x|} \frac{s^n}{n!} e^s\, ds < e^x \int_0^{|x|} \frac{s^n}{n!} e^{|x|}\, ds = \frac{|x|^{n+1}}{(n+1)!}.$

In both cases $E_n(x) \rightarrow 0$ as $n \rightarrow \infty$, so that

---

$$e^x = 1 + x + \frac{x^2}{2!} + \frac{x^3}{3!} + \ldots + \frac{x^n}{n!} + \ldots \text{ for all } x.$$

We have already seen (§11.3 Example 2; Ex. 11C, Qu. 12) that the exponential series converges for all x. We have now shown that the series converges to the 'right' sum, e^x. There are cases (admittedly rare) in which the Maclaurin series for $f(x)$ converges, but not to $f(x)$; see Ex. 12D, Qu. 11.

Finally we state without proof a general theorem about truncation errors of Maclaurin approximations.

If $f^{(n)}(t)$ exists for all t between 0 and x (inclusive), then
$$f(x) = f(0) + xf'(0) + \frac{x^2}{2!}f''(0) + \ldots + \frac{x^{n-1}}{(n-1)!}f^{(n-1)}(0) + \frac{x^n}{n!}f^{(n)}(\theta x),$$
where $0 < \theta < 1$.

Qu. 3. When $n = 1$, the theorem can be stated in the form
$[f(x) - f(0)]/x = f'(\theta x)$, where $0 < \theta < 1$. Give a geometrical interpretation of this.

Qu. 4. Adapt the method of Ex. 2E, Qu. 25 to give a geometrical interpretation of the case $n = 2$.

Qu. 5. Use the theorem to prove again the convergence of the exponential series.

The range of accuracy of Taylor and Maclaurin expansions is often very small. For computing purposes it is usually much more efficient to use an expansion of the form $f(x) = \sum\limits_{r=0}^{\infty} a_r T_r(x)$, where $T_r(x)$ is the *Chebyshev polynomial* of degree r defined by $T_r(x) = \cos(r \cos^{-1} x)$, though further details are beyond the scope of this book.

Exercise 11D

1. Use the first three terms of Gregory's series to find approximations to
   (i) $\tan^{-1} 0.2$ (ii) $\tan^{-1}(-0.5)$ (iii) $\tan^{-1} 0.9$. Work to 3 D.P. and use tables to find to how many decimal places each answer is correct.

2. Let $\alpha = \tan^{-1}\frac{1}{5}$. Find exact values of (i) $\tan 4\alpha$ (ii) $\tan\left(4\alpha - \dfrac{\pi}{4}\right)$.
   Deduce that $\dfrac{\pi}{4} = 4 \tan^{-1}\frac{1}{5} - \tan^{-1}\frac{1}{239}$.

   Use this and Gregory's series to calculate π correct to 4 D.P., or more if you have a calculator. [This method was first used by John Machin in 1706.]

3. By giving x a suitable value in Gregory's series, prove that

$$\pi = 2\sqrt{3} \sum_{r=0}^{\infty} \frac{(-3)^{-r}}{2r + 1}.$$

4. Prove that if θ is small then $\dfrac{\tan \theta - \theta}{\tan^3 \theta} \approx \frac{1}{3}$.

5. Prove that if $x \geqslant 1$ then $\tan^{-1} x = \dfrac{\pi}{2} - \dfrac{1}{x} + \dfrac{1}{3x^3} - \dfrac{1}{5x^5} + \dfrac{1}{7x^7} - \cdots$.

6. Use the following table of values to plot on a single large graph the curve $y = \ln(1 + x)$ and the successive approximations (i) $y = x$;
   (ii) $y = x - \dfrac{x^2}{2}$; (iii) $y = x - \dfrac{x^2}{2} + \dfrac{x^3}{3}$; (iv) $y = x - \dfrac{x^2}{2} + \dfrac{x^3}{3} - \dfrac{x^4}{4}$;
   for $-0.8 \leqslant x \leqslant 1.2$

x	0.2	0.4	0.6	0.8	1	1.2
$\dfrac{x^2}{2}$	0.020	0.080	0.180	0.320	0.500	0.720
$\dfrac{x^3}{3}$	0.003	0.021	0.072	0.171	0.333	0.576
$\dfrac{x^4}{4}$	0.000	0.006	0.032	0.102	0.250	0.518
$\ln(1+x)$	0.182	0.336	0.470	0.588	0.693	0.788
$\ln(1-x)$	-0.223	-0.511	-0.916	-1.609		

7. Prove that $\ln\left(\dfrac{1}{1-x}\right) = x + \dfrac{x^2}{2} + \dfrac{x^3}{3} + \dfrac{x^4}{4} + \ldots + \dfrac{x^n}{n} + \ldots\,$. For what values of x is this expansion valid?

8. Prove that $\ln\left(\dfrac{1+x}{1-x}\right) = 2\left(x + \dfrac{x^3}{3} + \dfrac{x^5}{5} + \ldots + \dfrac{x^{2n+1}}{2n+1} + \ldots\right)$ if $|x| < 1$.

9. The series expansion of $\frac{1}{2}\ln\left(\dfrac{1+x}{1-x}\right)$ obtained from Qu. 8 is very similar to the series for $\tan^{-1}x$. Prove that if $x = \tanh y$ then $e^{2y} = \dfrac{1+x}{1-x}$, and deduce that $\frac{1}{2}\ln\left(\dfrac{1+x}{1-x}\right) = \tanh^{-1}x$.

10. Find the first four terms in the power series expansions of the following, stating the values of x for which each is valid.

   (i) $\ln(1+3x)$ (ii) $\ln(1-2x)$ (iii) $\ln\left(\dfrac{1+3x}{1-2x}\right)$

   (iv) $\ln(1+x-6x^2)$ (v) $\ln\sqrt{(1+3x)}$ (vi) $\ln\left(\dfrac{1}{\sqrt[3]{(1-2x)}}\right)$

11. Find the expansion of $\ln(1+y+y^2)$ as far as the term in y^4

   (i) by putting $x = y + y^2$ in $\ln(1+x)$ (ii) using $1 + y + y^2 = \dfrac{1-y^3}{1-y}$.

   Use (ii) to give also the coefficient of y^n in the expansion.

12. Find approximations to $\ln 2$
   (i) by putting $x = 1$ in Mercator's series
   (ii) by putting $x = \frac{1}{2}$ in Qu. 7.
   (iii) by putting $x = \frac{1}{3}$ in Qu. 8.
   In each case use four terms and compare your answer with
   $\ln 2 = 0.6931$ (to 4 S.F.)

13 Use Qu. 8 to show that if $y > 0$ then
$$\ln y = 2\left(\frac{y-1}{y+1} + \frac{1}{3}\left(\frac{y-1}{y+1}\right)^3 + \frac{1}{5}\left(\frac{y-1}{y+1}\right)^5 + \ldots\right).$$
Hence calculate $\ln 10 \; (= \ln \frac{5}{4} + 3\ln 2)$ correct to 4 D.P.

14. Prove that $\log_2 e - \log_4 e + \log_8 e - \log_{16} e + \ldots = 1$.

15. Let $0 \leqslant x \leqslant 1$ and define $T_0 = 1$, $T_n = \frac{x}{n} T_{n-1}$, $(n \geqslant 1)$
$$S_0 = 1, \quad S_n = S_{n-1} + T_n, \;(n \geqslant 1).$$
Prove that, for every $n \geqslant 1$, $e^x = S_{n-1} + t_n$, where $0 \leqslant t_n \leqslant T_n e$.
Deduce that $T_n < 10^{-6} e^{-1}$ implies $0 \leqslant e^x - S_{n-1} < 10^{-6}$.
(The expansion of e^x in ascending powers of x may be assumed.) Explain
how you would use the foregoing as a program to compute e^x with error
not exceeding 10^{-6} and illustrate your answer by means of a flow
diagram.   [MEI]

16. *Proof that e is irrational*
   (i) Suppose that e is rational, and let $e = a/b$, where a and b are positive
       integers. Show that $b!e$ is a positive integer.
   (ii) Show that $b!e = b!\left(1 + 1 + \frac{1}{2!} + \frac{1}{3!} + \ldots + \frac{1}{b!}\right) + E_b$, where
$$E_b = \frac{1}{b+1} + \frac{1}{(b+1)(b+2)} + \frac{1}{(b+1)(b+2)(b+3)} + \ldots$$
   (iii) By comparing E_b with a suitable geometric series, show that
         $0 < E_b < 1/b$, and hence that $b!e$ is not an integer.
   (iv) Deduce from the contradictory results (i) and (iii) that e is
        irrational.

17 By using integration by parts, show that
$$F(x) = \int_x^\infty \frac{e^{x-t}}{t}\, dt = \frac{1}{x} - \frac{1}{x^2} + \frac{2!}{x^3} - \ldots + (-1)^{n-1}\frac{(n-1)!}{x^n} + E(x,n),$$
where the truncation error $E(x,n)$ satisfies $|E(x,n)| \leqslant (n-1)!\, x^{-n}$.
Show that the terms of the series for $F(x)$ decrease in magnitude when
$n < x$, but increase thereafter. Hence show that, for $x = 10$, n may be
chosen so that the error is approximately -3.6×10^{-5} and evaluate
$F(10)$ to 3 decimal places.   [MEI]

18. Use the theorem of §11.4 to prove that, for all x,
$$\sin x = x - \frac{x^3}{3!} + \frac{x^5}{5!} - \ldots + (-1)^n \frac{x^{2n+1}}{(2n+1)!} + \ldots$$
and $\cos x = 1 - \frac{x^2}{2!} + \frac{x^4}{4!} - \ldots + (-1)^n \frac{x^{2n}}{(2n)!} + \ldots$.

19. Prove that $\displaystyle\int_0^1 \sin(x^2)\, dx = \frac{1}{3} - \frac{1}{7.3!} + \frac{1}{11.5!} - \frac{1}{15.7!} + \ldots$,
and evaluate this to 4 D.P.

12 Calculus II

12.1 Properties of definite integrals

The definite integral $\int_a^b f(x)\,dx$, where $a < b$, is the limit of the sum $\sum_{x=a}^{b} f(x)\,\delta x$ as the number of terms increases and $\delta x \to 0$ (see Vol. 1, §6.8).

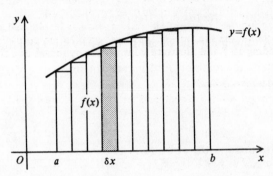

If $f(x) > 0$ for $a \leqslant x \leqslant b$ then $\int_a^b f(x)\,dx$ is positive and gives the area under the curve $y = f(x)$ from $x = a$ to $x = b$.

Though it is possible to evaluate some definite integrals directly as the limits of sums, it is usually much easier to make use of the Fundamental Theorem of Calculus, which states that if $f(x)$ is continuous then

$\dfrac{dA}{dx} = f(x)$, where A is the area $KLPQ$ of the figure (see Vol. 1, §6.6).

From this it follows that $A = F(x) + c$, where F is a function such that $F'(x) = f(x)$; $F(x)$ is called a *primitive* of $f(x)$. When Q is at K, $x = a$ and $A = 0$, so that $0 = F(a) + c$ and therefore $c = -F(a)$.

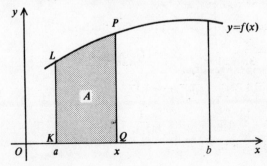

When $x = b, A = F(b) + c = F(b) - F(a)$.

Therefore the area under $y = f(x)$ from $x = a$ to $x = b$ is

$$\int_a^b f(x)\,dx = F(b) - F(a), \text{ where } F'(x) = f(x).$$

The problem of evaluating a definite integral is therefore solved if we can find and evaluate an appropriate primitive.

Interchanging a and b changes the sign of $F(b) - F(a)$, and therefore $\int_b^a f(x)\,dx$ is defined to be $-\int_a^b f(x)\,dx$.

Qu. 1.   Prove that $\int_a^b f(x)\,dx + \int_b^c f(x)\,dx = \int_a^c f(x)\,dx$

(i) by using properties of sums;   (ii) by using primitives.

Qu. 2.   Illustrate the result of Qu. 1 in terms of areas under $y = f(x)$ for the case $a < b < c$.

Example 1: Find $\int_0^2 |x^2 - 2|\,dx$.

Solution: For $0 \leqslant x \leqslant \sqrt{2}, |x^2 - 2| = 2 - x^2$;
for $\sqrt{2} \leqslant x \leqslant 2, |x^2 - 2| = x^2 - 2$.

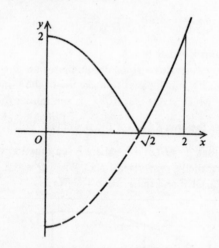

Using the result of Qu. 1,

$$\int_0^2 |x^2 - 2|\,dx = \int_0^{\sqrt{2}} |x^2 - 2|\,dx + \int_{\sqrt{2}}^2 |x^2 - 2|\,dx$$

$$= \int_0^{\sqrt{2}} (2 - x^2)\,dx + \int_{\sqrt{2}}^2 (x^2 - 2)\,dx$$

$$= \left[2x - \frac{x^3}{3}\right]_0^{\sqrt2} + \left[\frac{x^3}{3} - 2x\right]_{\sqrt2}^2$$

$$= 2\sqrt2 - \frac{2\sqrt2}{3} + \frac{8}{3} - 4 - \frac{2\sqrt2}{3} + 2\sqrt2$$

$$= \frac{4}{3}(2\sqrt2 - 1). \qquad\qquad\qquad \square$$

If f, g are functions and λ, μ are constants then

$$\sum_{x=a}^{b} [\lambda f(x) + \mu g(x)]\, \delta x \;=\; \lambda \sum_{x=a}^{b} f(x)\, \delta x + \mu \sum_{x=a}^{b} g(x)\, \delta x.$$

It follows by taking the limit as $\delta x \to 0$ that

$$\int_a^b [\lambda f(x) + \mu g(x)]\, dx \;=\; \lambda \int_a^b f(x)\, dx + \mu \int_a^b g(x)\, dx. \qquad (A)$$

Qu. 3.   Prove equation (A) by using primitives.

Taking $\lambda = -1, \mu = 0$ in equation (A) gives $\displaystyle \int_a^b [-f(x)]\, dx = -\int_a^b f(x)\, dx$.

If $f(x) < 0$ for $a < x < b$ then

$$\int_a^b f(x)\, dx = -\int_a^b [-f(x)]\, dx$$

$$= -\int_a^b |f(x)|\, dx$$

$$< 0 \quad \text{since} \quad |f(x)| > 0 \quad \text{for} \quad a < x < b.$$

In order to preserve the connection between definite integrals and areas we must therefore say that if $f(x)$ is negative for $a < x < b$, then the area between $y = f(x)$ and the x-axis from $x = a$ to $x = b$ is also negative.

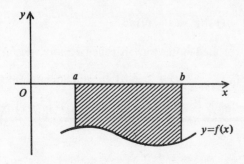

Taking $\lambda = 1, \mu = -1$ in equation (A) gives

$$\int_a^b [f(x) - g(x)] dx = \int_a^b f(x) dx - \int_a^b g(x).$$

If $f(x) > g(x)$ for $a < x < b$, then $f(x) - g(x) > 0$ for $a < x < b$ and therefore $\int_a^b [f(x) - g(x)] dx > 0$. Thus

$$f(x) > g(x) \text{ for } a < x < b \Rightarrow \int_a^b f(x) dx > \int_a^b g(x) dx.$$

Qu. 4.   Interpret this in terms of areas. Is the converse true?

This result can be used to estimate the value of a definite integral which we cannot find exactly.

Example 2: Prove that $0.5 < \int_0^{\frac{1}{2}} \dfrac{1}{\sqrt{(1 - x^3)}} dx < 0.53$.

Solution: $0 < x < \frac{1}{2} \Rightarrow 0 < x^3 < x^2$

$$\Rightarrow 1 > 1 - x^3 > 1 - x^2$$

$$\Rightarrow 1 < \frac{1}{\sqrt{(1 - x^3)}} < \frac{1}{\sqrt{(1 - x^2)}}.$$

Therefore $\int_0^{\frac{1}{2}} \dfrac{1}{\sqrt{(1 - x^3)}} dx$ lies between $\int_0^{\frac{1}{2}} 1 . dx = 0.5$ and

$$\int_0^{\frac{1}{2}} \frac{1}{\sqrt{(1 - x^2)}} dx = [\sin^{-1} x]_0^{\frac{1}{2}} = \frac{\pi}{6} \approx 0.524 < 0.53.$$

Therefore $0.5 < \int_0^{\frac{1}{2}} \dfrac{1}{\sqrt{(1 - x^3)}} dx < 0.53$. $\square$

Since the variable x does not appear in the final expression $F(b) - F(a)$ for $\int_a^b f(x) dx$, we can replace it by any other variable:

$$\int_a^b f(x) dx = \int_a^b f(y) dy = \int_a^b f(\theta) d\theta = \dots .$$

Letters $x, y, \theta, \dots$ used in this way are called *dummy variables*; the variables r in $\sum_{r=1}^{4} r^3$ and z in $\{z : z^2 + 4z + 3 = 0\}$ are also dummy variables.

For any function f, $\int_{-a}^{a} f(x) dx = \int_{-a}^{0} f(x) dx + \int_{0}^{a} f(x) dx$.

Substituting $u = -x$,

$$\int_{-a}^{0} f(x)\, dx = \int_{a}^{0} f(-u) \cdot (-du)$$

$$= -\int_{a}^{0} f(-u)\, du$$

$$= \int_{0}^{a} f(-u)\, du, \text{ interchanging the limits,}$$

$$= \int_{0}^{a} f(-x)\, dx, \text{ since } u \text{ is a dummy variable.}$$

Therefore $\int_{-a}^{a} f(x)\, dx = \int_{0}^{a} f(-x)\, dx + \int_{0}^{a} f(x)\, dx = \int_{0}^{a} [f(-x) + f(x)]\, dx.$

In particular, if f is an odd function, $f(-x) = -f(x)$ and so $\int_{-a}^{a} f(x)\, dx = 0$;

if f is an even function $f(-x) = f(x)$ and so $\int_{-a}^{a} f(x)\, dx = 2 \int_{0}^{a} f(x)\, dx.$

Qu. 5.   Illustrate these results about odd and even functions in terms of areas.

Example 3: Find $\int_{-3}^{3} \dfrac{(x + 3)^3}{x^2 + 3}\, dx.$

Solution: $\int_{-3}^{3} \dfrac{(x + 3)^3}{x^2 + 3}\, dx = \int_{-3}^{3} \dfrac{x^3 + 9x^2 + 27x + 27}{x^2 + 3}\, dx$

$$= \int_{-3}^{3} \frac{x^3 + 27x}{x^2 + 3}\, dx + \int_{-3}^{3} \frac{9x^2 + 27}{x^2 + 3}\, dx.$$

The first of these integrals is zero since $x \mapsto \dfrac{x^3 + 27x}{x^2 + 3}$ is an odd function.

Therefore $\int_{-3}^{3} \dfrac{(x + 3)^3}{x^2 + 3}\, dx = \int_{-3}^{3} \dfrac{9x^2 + 27}{x^2 + 3}\, dx = \int_{-3}^{3} 9\, dx = 54.$

$\square$

One further useful result is obtained by substituting $u = a - x$ in
$\int_{0}^{a} f(x)\, dx.$ When $x = 0, u = a$ and when $x = a, u = 0$; also $x = a - u$ and
$dx = -du.$ Therefore $\int_{0}^{a} f(x)\, dx = \int_{a}^{0} f(a - u)(-du)$

$$= \int_{0}^{a} f(a - u)\, du.$$

Since u is a dummy variable, we may replace it by x. Therefore

$$\int_{0}^{a} f(x)\, dx = \int_{0}^{a} f(a - x)\, dx.$$

Qu. 6.   Sketch $y = f(x)$ and $y = f(a - x)$ on the same axes, and hence give
a geometric interpretation of $\displaystyle\int_0^a f(x)\,dx = \int_0^a f(a - x)\,dx$.

Example 4: Find (i) $\displaystyle\int_0^{\pi/2} \sin^2 \theta\, d\theta$ (ii) $\displaystyle\int_0^{\pi} \sin^2 \theta\, d\theta$ (iii) $\displaystyle\int_0^{\pi} \theta \sin^2 \theta\, d\theta$.

Solution: In the following solutions notice at (A) the use of the fact that
$I = J \Rightarrow I = \frac{1}{2}(I + J)$.

(i) $\displaystyle\int_0^{\pi/2} \sin^2 \theta\, d\theta = \int_0^{\pi/2} \sin^2 \left(\frac{\pi}{2} - \theta\right) d\theta$

$\displaystyle\qquad\qquad\quad = \int_0^{\pi/2} \cos^2 \theta\, d\theta$

$\displaystyle\qquad\qquad\quad = \tfrac{1}{2}\int_0^{\pi/2} (\sin^2 \theta + \cos^2 \theta)\, d\theta \qquad\qquad \text{(A)}$

$\displaystyle\qquad\qquad\quad = \tfrac{1}{2}\int_0^{\pi/2} 1 \cdot d\theta = \frac{\pi}{4}.$

(ii) $\displaystyle\int_0^{\pi} \sin^2 \theta\, d\theta = 2 \int_0^{\pi/2} \sin^2 \theta\, d\theta \quad$ since the graph $y = \sin^2 \theta$ is

$\displaystyle\qquad\qquad\quad = 2 \times \frac{\pi}{4} = \frac{\pi}{2}. \qquad\qquad$ symmetrical about $\theta = \dfrac{\pi}{2}$

(iii) $\displaystyle\int_0^{\pi} \theta \sin^2 \theta\, d\theta = \int_0^{\pi} (\pi - \theta) \sin^2 (\pi - \theta)\, d\theta$

$\displaystyle\qquad\qquad\qquad = \int_0^{\pi} (\pi - \theta) \sin^2 \theta\, d\theta$

$\displaystyle\qquad\qquad\qquad = \tfrac{1}{2}\int_0^{\pi} (\theta + \pi - \theta) \sin^2 \theta\, d\theta \qquad\qquad \text{(A)}$

$\displaystyle\qquad\qquad\qquad = \frac{\pi}{2}\int_0^{\pi} \sin^2 \theta\, d\theta = \frac{\pi^2}{4}. \qquad\qquad \square$

Summary

$$\int_a^b f(x)\,dx = \lim_{\delta x \to 0} \sum_{x=a}^{b} f(x)\,\delta x = F(b) - F(a), \text{ where } F'(x) = f(x).$$

$$\int_b^a f(x)\,dx = -\int_a^b f(x)\,dx; \int_a^b f(x)\,dx + \int_b^c f(x)\,dx = \int_a^c f(x)\,dx.$$

$$f(x) > g(x) \text{ for } a < x < b \Rightarrow \int_a^b f(x)\,dx > \int_a^b g(x)\,dx.$$

$$\int_{-a}^a f(x)\,dx = \begin{cases} 0 \text{ if } f \text{ is odd} \\ 2\displaystyle\int_0^a f(x)\,dx \text{ if } f \text{ is even;} \end{cases} \int_0^a f(x)\,dx = \int_0^a f(a-x)\,dx.$$

Exercise 12A

1–8. Evaluate these integrals.

1. $\displaystyle\int_{-1}^{2} (3x + |x|)\, dx$ 2. $\displaystyle\int_{3}^{7} |6 + 5x - x^2|\, dx$ 3. $\displaystyle\int_{0}^{\pi} |1 - 2 \sin x|\, dx$

4. $\displaystyle\int_{1/e}^{e} e^{|\ln x|}\, dx$ 5. $\displaystyle\int_{-\pi/2}^{\pi/2} \sin^9 \theta\, d\theta$ 6. $\displaystyle\int_{-a}^{a} \frac{x^7}{a^4 + x^4}\, dx$

7. $\displaystyle\int_{-1/2}^{1/2} \frac{5x^3 - 12x + 4}{\sqrt{(1 - x^2)}}\, dx$ 8. $\displaystyle\int_{-\pi/2}^{\pi/2} (\cos \theta + 2 \sin \theta)^3\, d\theta$

9. Find all the values of c for which $\displaystyle\int_{0}^{c} (x - 1)(x - 4)\, dx = 0$. Explain the geometrical significance of these values with the help of a sketch graph of $y = (x - 1)(x - 4)$.

10. Prove that $\displaystyle\int_{0}^{\pi} x \sin x\, dx = \int_{0}^{\pi} (\pi - x) \sin x\, dx$, and hence evaluate the first integral.

11. Prove that $\displaystyle\int_{0}^{\pi} \frac{x \sin x}{1 + \cos^2 x}\, dx = \frac{\pi^2}{4}$.

12. Find $\displaystyle\int_{0}^{\pi/2} x \left(\frac{\pi}{2} - x\right) \cos^2 x\, dx$.

13. Prove that $\displaystyle\int_{0}^{\pi/4} \ln(1 + \tan \theta)\, d\theta = \int_{0}^{\pi/4} \ln \left(\frac{2}{1 + \tan \theta}\right) d\theta$.

    Deduce that $\displaystyle\int_{0}^{\pi/4} \ln(1 + \tan \theta)\, d\theta = \frac{\pi}{8} \ln 2$.

14. By means of a suitable substitution, show that

    $\displaystyle\int_{a}^{2a} f(x)\, dx = \int_{0}^{a} f(2a - x)\, dx$.

    Deduce that $\displaystyle\int_{0}^{2a} f(x)\, dx = \int_{0}^{a} f(x)\, dx + \int_{0}^{a} f(2a - x)\, dx$.

    Hence evaluate the integrals $\displaystyle\int_{0}^{2\pi} \sin^5 \theta\, d\theta$ and $\displaystyle\int_{0}^{\pi} \cos^3 \theta\, d\theta$.   [SMP]

15. Prove that (i) $\displaystyle\int_{a}^{b} f(x)\, dx = \int_{a}^{b} f(a + b - x)\, dx$

    (ii) $\displaystyle\int_{0}^{na} f(x)\, dx = n \int_{0}^{a} f(nx)\, dx$. Give geometric interpretations.

16. Show that, when $0 < x < 1$, $x^{\frac{1}{2}}$ is greater than $x^{\frac{2}{3}}$. If $I = \displaystyle\int_{0}^{1} \frac{2\, dx}{x^{\frac{1}{2}} + x^{\frac{2}{3}}}$, show that $\displaystyle\int_{0}^{1} x^{-\frac{1}{2}}\, dx < I < \int_{0}^{1} x^{-\frac{2}{3}}\, dx$, and deduce that the value of I lies between 2 and 3.   [L]

17. Establish the inequalities, when $0 \leqslant x \leqslant 1$,
$$2 \geqslant \sqrt{(4 - x^2 + x^4)} \geqslant \sqrt{(4 - x^2)} > 0.$$
    Deduce that $0.5 < \int_0^1 \dfrac{dx}{\sqrt{(4 - x^2 + x^4)}} < 0.524.$

    Use Simpson's rule with four strips to evaluate the integral and verify that your answer satisfies the above inequalities.   [MEI]

18. Use the fact that $x - \dfrac{x^3}{6} < \sin x < x$ for $x > 0$ to show that
$$x^{\frac{1}{2}}\left(1 - \frac{x^2}{12}\right) < \sqrt{\sin x} < x^{\frac{1}{2}} \text{ for } 0 < x < 1.$$
    Deduce that $\frac{9}{14} < \int_0^1 \sqrt{\sin x} \, dx < \frac{2}{3}.$

19. Find the sixth-degree polynomial $P(x)$ and the constant A such that
    $x^4(1 - x)^4 \equiv (1 + x^2)P(x) + A.$

    Hence show that $\int_0^1 \dfrac{x^4(1 - x)^4}{1 + x^2} dx = \frac{22}{7} - \pi.$

    Use the fact that $\frac{1}{2} < \dfrac{1}{1 + x^2} < 1$ for $0 < x < 1$ to deduce that
    $$\frac{22}{7} - \frac{1}{630} < \pi < \frac{22}{7} - \frac{1}{1260}.$$

20. By considering the derivative of $\sin\theta/\theta$, or otherwise, prove that
    $\dfrac{2}{\pi} < \dfrac{\sin\theta}{\theta} < 1$ if $0 < \theta < \frac{1}{2}\pi$, and deduce that, if $\lambda > 0$, then
    $$\frac{1}{\lambda}(1 - e^{-\frac{1}{2}\pi\lambda}) \leqslant \int_0^{\frac{1}{2}\pi} e^{-\lambda\sin\theta} \, d\theta \leqslant \frac{\pi}{2\lambda}(1 - e^{-\lambda}).$$

    Obtain corresponding bounds for $\int_0^\pi e^{-\lambda\sin\theta} \, d\theta.$ [MEI]

12.2 Mean values

We know that the (arithmetic) mean of the n numbers $y_1, y_2, \ldots, y_n$ is

$\bar{y} = \dfrac{y_1 + y_2 + \ldots + y_n}{n} = \dfrac{1}{n} \sum\limits_{i=1}^{n} y_i$. We now consider how to find the mean

value of a variable y which depends on x and varies continuously as x varies
from a to b.

Suppose that $y = f(x)$. Divide the region under the curve $y = f(x)$ into n

strips of width $\delta x = \dfrac{b - a}{n}$ by means of the lines $x = a, x = x_1, x = x_2, \ldots,$

$x = x_n = b$, where $x_i = a + i \, \delta x \qquad (i = 1, 2, \ldots, n).$
Let $y_i = f(x_i)$. Then the mean of the n values $y_1, y_2, \ldots, y_n$ is

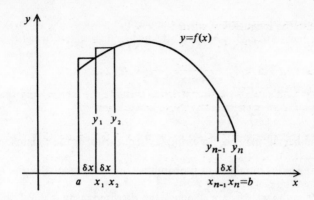

$$\bar{y} = \frac{1}{n} \sum_{i=1}^{n} y_i = \frac{1}{n} \sum_{i=1}^{n} f(x_i) = \frac{1}{n\,\delta x} \sum_{i=1}^{n} f(x_i)\,\delta x = \frac{1}{b-a} \sum_{i=1}^{n} f(x_i)\,\delta x.$$

As n increases, the calculation of $\bar{y}$ takes into account more and more values of y, and so becomes more and more representative of the overall behaviour of the function for $a \leqslant x \leqslant b$. Also, as $n \to \infty$, $\sum_{i=1}^{n} f(x_i)\,\delta x \to \int_a^b f(x)\,dx$.

It is therefore natural to make the following definition.

> The *mean value* of $f(x)$ for $a \leqslant x \leqslant b$ is $\dfrac{1}{b-a} \displaystyle\int_a^b f(x)\,dx$.

The mean value is the height of the rectangle bounded by $x = a, x = b$, $y = 0$ whose area equals the area under $y = f(x)$ from $x = a$ to $x = b$.

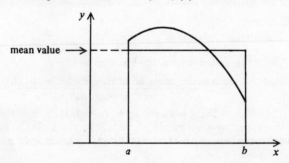

In particular, if a particle moving in a straight line has displacement s metres and velocity $v\,\mathrm{m\,s^{-1}}$ after t seconds then by this definition the mean value of v for $t_1 \leqslant t \leqslant t_2$ is $\dfrac{1}{t_2-t_1} \displaystyle\int_{t_1}^{t_2} v\,dt = \dfrac{s_2-s_1}{t_2-t_1}$, which is simply the average velocity as defined in Vol. 1, §5.1.

We often use some form of weighted mean in place of the ordinary arithmetic mean: to each value y_i is attached a non-negative 'weight' w_i, and

the mean is then $\dfrac{\sum\limits_{i=1}^{n} w_i y_i}{\sum\limits_{i=1}^{n} w_i}$.

The numbers w_i may be, for example, frequencies, or the weights used in constructing index numbers, or probabilities $\left(\text{in which case } \sum\limits_{i=1}^{n} w_i = 1\right)$ or masses. When the weights are all equal, the weighted mean is simply the arithmetic mean.

In the continuous case the weights w_i are replaced by a *weighting function* w such that $w(x) \geqslant 0$ for $a \leqslant x \leqslant b$.

> The weighted mean of $f(x)$ for $a \leqslant x \leqslant b$ with weighting
>
> function w is $\dfrac{\displaystyle\int_a^b w(x) f(x)\, dx}{\displaystyle\int_a^b w(x)\, dx}$.

Example 1: The village of Long Standing consists of one straight street 2 km long. The bus stop is at one end of this street. The population is concentrated towards the church and pub, $\frac{4}{3}$ km from the bus stop, in such a way that the number of people living between x and $(x + \delta x)$ km from the bus stop $\approx kx^2(2-x)\,\delta x$, $(0 \leqslant x \leqslant 2)$. Find the mean distance of the villagers from the bus stop.

Solution: The total population is approximately $\sum\limits_{x=0}^{2} kx^2(2-x)\,\delta x$.

The mean distance from the bus stop is

$$\frac{\text{sum of distances of all villagers from bus stop}}{\text{total population}}.$$

The villagers living between x and $(x + \delta x)$ km from the bus stop contribute approximately $x \cdot kx^2(2-x)\,\delta x$ to the sum of distances. Therefore the mean distance $\bar{x}$ km is approximately

$$\sum_{x=0}^{2} x \cdot kx^2(2-x)\,\delta x \left/ \sum_{x=0}^{2} kx^2(2-x)\,\delta x \right.$$

Taking the limit as $\delta x \to 0$ we have

$$\bar{x} = \int_0^2 x \cdot kx^2(2-x)\, dx \left/ \int_0^2 kx^2(2-x)\, dx \right.;$$

this is the weighted mean of the distance, with the population density function as weighting function.

$$\text{Thus } \bar{x} = \frac{\displaystyle\int_0^2 (2x^3 - x^4)\,dx}{\displaystyle\int_0^2 (2x^2 - x^3)\,dx} = \frac{\left[\frac{1}{2}x^4 - \frac{1}{5}x^5\right]_0^2}{\left[\frac{2}{3}x^3 - \frac{1}{4}x^4\right]_0^2} = \frac{\frac{8}{5}}{\frac{4}{3}} = 1\frac{1}{5}.$$

The mean distance from the bus stop is $1\frac{1}{5}$ km.   □

Weighted means of continuous functions are used in probability theory to find expectations and variances, and in mechanics to find centroids and moments of inertia.

Exercise 12B

1. Find the mean value of x^3 for $0 \leqslant x \leqslant 6$.

2. Find the mean value of $\sin^2 x$ for $0 \leqslant x \leqslant 2\pi$.

3. Prove that

   f is a linear function $\Rightarrow$ the mean value of $f(x)$ for $a \leqslant x \leqslant b$

   $$\text{is } f\left(\frac{a+b}{2}\right).$$

   Give a counterexample to show that the converse is false.

4. (i) Find to 3 S.F. the value of a such that

   $$e^a = \text{ mean value of } e^x \text{ for } 0 \leqslant x \leqslant 4.$$

   (ii) Find to 3 S.F. the value of b such that

   $$e^4 = \text{ mean value of } e^x \text{ for } 0 \leqslant x \leqslant b.$$

5. The mean value of $f(x)$ for $a \leqslant x \leqslant b$ is k. Prove that the mean value of $f(x) - k$ for $a \leqslant x \leqslant b$ is zero.

6. The voltage in an electrical circuit is $E = 4 \sin 3t$ at time t.
   (i) Sketch the graph of E against t. What is the period of this graph?
   (ii) Show that the mean value of E for $0 \leqslant t \leqslant 2\pi/3$ is zero.
   (iii) In order to avoid the cancelling which produces the zero mean value we may find either
       (a) the *mean absolute value*, i.e. the mean value of $|E|$, for $0 \leqslant t \leqslant 2\pi/3$
       or (b) the *root mean square value*, i.e. the square root of the mean value of E^2, for $0 \leqslant t \leqslant 2\pi/3$.
       Find both these values.

7. The alternating current I from the mains is given at time t by
   $I = I_0 \sin \omega t$, where I_0, ω are constants. Show that the root mean square
   current over one period is $\dfrac{1}{\sqrt{2}} I_0$.

8. The voltage E and current I in an electrical circuit at time t are given by
   $E = 120 \cos t$, $I = 3 \cos \left(t - \dfrac{\pi}{3}\right)$. The power is the product of the
   voltage and the current. Find the mean value of the power over one
   period.

9. A crowd, stretching for 300 m along a straight road from the entrance to
   a football ground, is most dense near the gates, so that at distance
   $(300 - x)$ m from the entrance there are kx^2 people per metre of road.
   Find the mean distance of the crowd from the entrance to the ground.

10. If $y = \sin x + 5c \sin 5x$, where c is a constant, and if M and N denote the
    mean values of y and y^2, respectively, over the interval $0 \leqslant x \leqslant \pi$, prove
    that $M = 2(1 + c)/\pi$, $N = \frac{1}{2}(1 + 25c^2)$. Verify that $N > M^2$ for all
    values of c.   [MEI]

11.

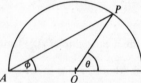

Points are uniformly distributed round a semicircle of radius R, and it is
required to find the mean distance of these points from the end A of the
diameter.
   (i) Show that the distance AP is $2R \sin \phi = 2R \sin \dfrac{\theta}{2}$.

   (ii) Which of the expressions

       (a) $\dfrac{2}{\pi} \displaystyle\int_0^{\pi/2} 2R \sin \phi \, d\phi$;   (b) $\dfrac{1}{\pi} \displaystyle\int_0^{\pi} 2R \sin \dfrac{\theta}{2} \, d\theta$

       gives the mean distance correctly? Explain why.
   (iii) Find the mean distance.

12. Prove that if $f(x)$ lies between h and k for $a \leqslant x \leqslant b$, then so does every
    weighted mean of $f(x)$ for $a \leqslant x \leqslant b$.

13. Find the mean distance of points on a circular disc of radius 8 cm
    (i) from the centre of the disc
    (ii) from a point on the axis of the disc 6 cm from the plane of the disc.

14. Find the mean distance from the centre of points inside a sphere of
    radius R.

15. Show that a cone of semi-vertical angle α with its vertex at O cuts out on
    a sphere of radius a with its centre at O an area of $2\pi a^2 (1 - \cos \alpha)$.
    Show that the mean distance of points inside a unit sphere from a fixed
    point on its surface is 1.2.   [MEI]

16. (P) A random variable x has the probability density function p where
$$p(x) = \begin{cases} c\,x^2(2-x) & \text{for } 0 \leqslant x \leqslant 2 \\ 0 & \text{otherwise.} \end{cases}$$
Determine the value of c and find the mean and standard deviation of this distribution.

17. (P) A probability density function of a random variable x is defined as follows:
$$f(x) = \begin{cases} x(x-1)(x-2) & \text{for } 0 \leqslant x < 1 \\ \lambda & \text{for } 1 \leqslant x \leqslant 3 \\ 0 & \text{otherwise} \end{cases}$$
where λ is a suitable constant. Calculate the expectation μ of x. What is the probability that x is less than or equal to μ?   [MEI]

18. (M) The density of a thin straight rod AB of length 2 units is $\dfrac{\rho}{1+x^2}$ at a distance x from A $(0 \leqslant x \leqslant 2)$. Find the position of the centroid of the rod.

19. (M) Find a mass density distribution so that the centroid of a straight rod of length L will be at a distance $L/5$ from one end of the rod.

20. (M) A uniform solid occupies the volume of revolution obtained by rotating $y = x^{-3/2}$, $1 \leqslant x \leqslant k$, about the x-axis. Find the coordinates of the centroid, and describe what happens to the centroid as $k \to \infty$.

12.3 Finding the area of a sector

In order to find the area of the section OAB bounded by the curve $r = f(\theta)$ and the lines $\theta = \alpha$, $\theta = \beta$ $(\alpha < \beta)$ we first divide it into elementary sectors OPQ, where P and Q have coordinates (r, θ) and $(r + \delta r, \theta + \delta\theta)$ as in Fig. 1.

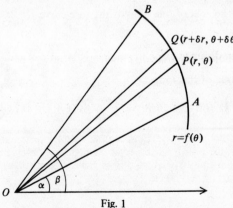

Fig. 1

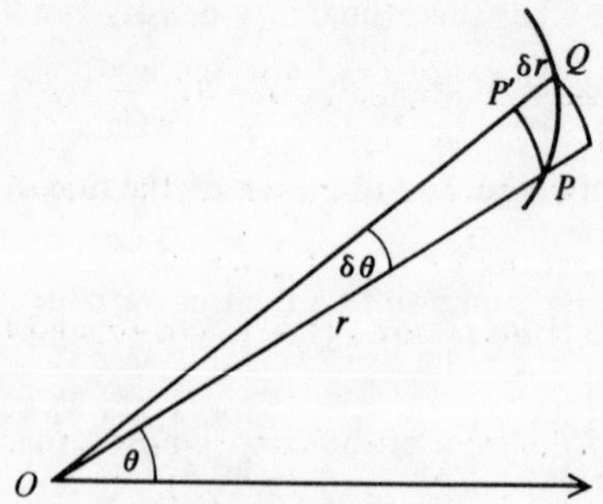

Fig. 2

Suppose first for simplicity that r increases with θ for $\alpha < \theta < \beta$. Then the area of sector OPQ lies between the areas of the circular sectors OPP', OQQ' (Fig. 2), i.e.

$$\tfrac{1}{2}r^2\,\delta\theta < \text{area } OPQ < \tfrac{1}{2}(r + \delta r)^2\,\delta\theta.$$

Therefore

$$\sum_{\theta=\alpha}^{\beta} \tfrac{1}{2}r^2\,\delta\theta < \text{area } OAB < \sum_{\theta=\alpha}^{\beta} \tfrac{1}{2}(r + \delta r)^2\,\delta\theta.$$

As $\delta\theta \to 0$ these two approximating sums have the common limit

$\int_{\alpha}^{\beta} \tfrac{1}{2}r^2\,d\theta$. A similar argument applies if r decreases as θ increases. Most

curves can be split into parts where r is increasing and parts where r is decreasing; all the curves considered in this book consist of a finite number of such parts. Adding the sector areas of all these parts gives the final result:

the area of the sector bounded by $r = f(\theta)$, $\theta = \alpha$, $\theta = \beta$ is

$$\int_{\alpha}^{\beta} \tfrac{1}{2}r^2\,d\theta.$$

Example: Find the area of one loop of the rose curve $r = 10\cos 2\theta$ drawn on p.279.

Solution: One loop is described as θ increases from $-\dfrac{\pi}{4}$ to $\dfrac{\pi}{4}$.

The area of this loop is

$$\int_{-\pi/4}^{\pi/4} \tfrac{1}{2} \cdot 100 \cos^2 2\theta\, d\theta = \int_{-\pi/4}^{\pi/4} 25(1 + \cos 4\theta)\, d\theta$$

$$= \left[25\theta + \tfrac{25}{4}\sin 4\theta \right]_{-\pi/4}^{\pi/4}$$

$$= \frac{25\pi}{2}. \qquad \qquad \square$$

Exercise 12C

1. Find the area of the sector bounded by $r = \theta(1 + 2\sqrt{\theta})$, $\theta = 1$, $\theta = 4$.

2. Find the areas of the two portions into which the line $\theta = \dfrac{\pi}{2}$ divides the upper half of the cardioid $r = 5(1 + \cos\theta)$.

3. Sketch the *lemniscate* (ribbon bow) $r^2 = a^2 \cos 2\theta$, and find the area of one of its loops.

4. Show that the area of the loop of the curve whose equation in polar coordinates is $r \cos\theta = a \cos 2\theta$ is $a^2(2 - \tfrac{1}{2}\pi)$.   [O & C]

5. A gardener with mathematical tendencies makes two new flower beds. One is circular with radius 2 metres; the other is bounded by the curve $r = 2 + \tfrac{1}{2}\sin 4\theta$, where r is in metres. Sketch the flower beds and without doing any calculation decide which has the greater area. Calculate the area of the second flower bed as a percentage of the area of the circular one.

6. The interior of the circle $r = 3a \cos\theta$ is divided into two parts by the curve $r = a(1 + \cos\theta)$. Find the area of the part whose boundary passes through the origin.

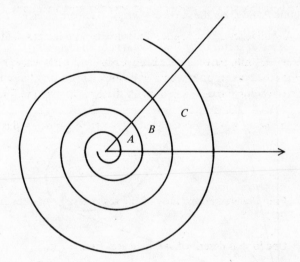

7. The diagram shows the equiangular spiral $r = ae^{k\theta}$ and the lines $\theta = 0$, $\theta = \dfrac{\pi}{4}$. Prove that the areas of the regions $A, B, C, \ldots$ between these lines and successive whorls of the spiral form a geometric sequence.

8. Sketch the two ellipses $\frac{x^2}{a^2} + \frac{y^2}{b^2} = 1$, $\frac{x^2}{b^2} + \frac{y^2}{a^2} = 1$ $(a > b)$ on the same axes. By changing to polar coordinates show that the area of the region common to both curves is $4ab \tan^{-1}\left(\frac{b}{a}\right)$. What happens to this region if b remains fixed and (i) $a \to b$ (ii) $a \to \infty$?

Check that the correct area is obtained from $4ab \tan^{-1}\left(\frac{b}{a}\right)$ in each case.

9. A curve is defined by parametric equations $x = f(t), y = g(t)$. By differentiating the relation $\tan\theta = \frac{y}{x}$ with respect to t show that $r^2 \frac{d\theta}{dt} = x \frac{dy}{dt} - y \frac{dx}{dt}$.

Deduce that if P_1, P_2 are the points for which $t = t_1, t = t_2$ $(t_2 > t_1)$ and θ increases as t increases then the area of the sector bounded by OP_1, OP_2 and the curve $P_1 P_2$ is $\frac{1}{2}\int_{t_1}^{t_2}\left(x \frac{dy}{dt} - y \frac{dx}{dt}\right)dt$.

10. The arc AB is defined by $x = 2 + 3\cos t, y = 4 + \sin t, 0 \leq t \leq \frac{\pi}{2}$.

Find the area of the sector bounded by this arc and the lines OA, OB.

11. The points P and Q on the curve $x = at^2, y = at^3$ are given by $t = T$ and $t = 2T$ respectively. Prove that

$$\text{area of sector } OPQ = 3.1 \times \text{area of triangle } OPQ.$$

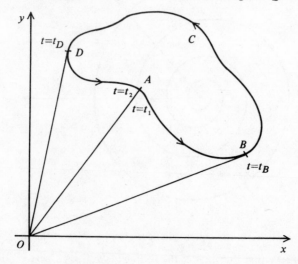

12. The figure shows a loop traced by a point P with coordinates (x, y) which depend on a parameter t. As t increases from t_1 to t_2, P moves once anticlockwise around the loop from A via B (where $t = t_B$), C, D (where $t = t_D$) back to A. Explain why $\frac{1}{2}\int\left(x \frac{dy}{dt} - y \frac{dx}{dt}\right)dt$ is negative

from t_1 to t_B, positive from t_B to t_D, and negative from t_D to t_2. Deduce that the area of the loop is $\frac{1}{2}\int_{t_1}^{t_2}\left(x\frac{dy}{dt}-y\frac{dx}{dt}\right)dt$.

13. Sketch the *deltoid* $x = 2a\cos t + a\cos 2t$, $y = 2a\sin t - a\sin 2t$ and find the area it encloses.

14. Sketch the curve $x = t^4$, $y = t^3 - t + 1$, and find the area of the loop.

12.4 Some important limits

The expressions x^3, $\sqrt{x}$ and $\sqrt{\dfrac{3x^2+1}{x+2}}$ all tend to infinity as $x\to\infty$. Since $\dfrac{\sqrt{x}}{x^3}\to 0$ as $x\to\infty$ we say that $\sqrt{x}\to\infty$ more slowly than x^3. On the other hand, since $\sqrt{\dfrac{3x^2+1}{x+2}}\Big/\sqrt{x} = \sqrt{\dfrac{3x^2+1}{x^2+2x}} = \sqrt{\dfrac{3+1/x^2}{1+2/x}}\to\sqrt{3}$,

a non-zero limit, as $x\to\infty$, we say that $\sqrt{\dfrac{3x^2+1}{x+2}}$ and $\sqrt{x}$ are of the same order of greatness for large x.

It is easy to define a sequence of expressions of decreasing order of greatness, for example $\sqrt{x}, \sqrt[3]{x}, \sqrt[4]{x}, \sqrt[5]{x}, \ldots$, and it might be expected that if $f(x)\to\infty$ as $x\to\infty$, however slowly, then we could find a sufficiently small positive number α such that $x^\alpha\to\infty$ even more slowly.

We can show that this is not so by taking $f(x)\equiv\ln x$.

Qu. 1. Use a calculator to find to 2 S.F. the values of

(i) $\dfrac{\ln x}{x}$ (ii) $\dfrac{\ln x}{x^{0.1}}$ (iii) $\dfrac{\ln x}{x^{0.001}}$ for $x = 10, 10^3, 10^6, 10^{10}, 10^{50}$.

To put this on a sound footing, we first note that since $\dfrac{1}{t} < 1$ for $t > 1$ it follows that for $y > 1$

$$\ln y = \int_1^y \frac{1}{t}\,dt < \int_1^y 1\cdot dt = y - 1.$$

Therefore $$0 < \frac{\ln y}{y^2} < \frac{y-1}{y^2}\to 0 \text{ as } y\to\infty.$$

Thus $\ln y\to\infty$ more slowly than y^2.

Now put $y^2 = x^\alpha$, where α is any fixed positive number.

Then $$\frac{\ln y}{y^2} = \frac{\ln(x^{\alpha/2})}{x^\alpha} = \frac{\alpha}{2}\cdot\frac{\ln x}{x^\alpha}, \text{ so that } \frac{\ln x}{x^\alpha} = \frac{2}{\alpha}\frac{\ln y}{y^2}.$$

As $x \to \infty$, $y \to \infty$ and $\dfrac{\ln y}{y^2} \to 0$. Therefore

> if α is any positive number (however small), $\dfrac{\ln x}{x^\alpha} \to 0$ as $x \to \infty$;
>
> $\ln x \to \infty$ more slowly than any positive power of x.

Qu. 2. Estimate the value of $\ln x / x^{0.001}$
(i) when $x = 10^{1000}$ (ii) when $x = 10^{10\,000}$.

Qu. 3. By writing $\dfrac{(\ln x)^A}{x^\alpha} = \left(\dfrac{\ln x}{x^{\alpha/A}}\right)^A$, prove that if A and α are positive
then, however large A and however small α, $\dfrac{(\ln x)^A}{x^\alpha} \to 0$ as $x \to \infty$.

It is easy to deduce from this the corresponding results about the
behaviour of $\ln x$ as $x \to 0$ and of e^x as $x \to \infty$.

$$\text{If } x = \frac{1}{y} \text{ then } x^\alpha \ln x = \frac{\ln\left(\dfrac{1}{y}\right)}{y^\alpha} = -\frac{\ln y}{y^\alpha}$$

As $x \to 0$ through positive values, $y \to \infty$ and $\dfrac{\ln y}{y^\alpha} \to 0$. Therefore

> $x^\alpha \ln x \to 0$ as $x \to 0$ through positive values;
>
> the tendency of $|\ln x|$ to become large as $x \to 0$ is overcome
> by any positive power of x, however small.

If $x = \ln y$ then $\dfrac{x^A}{e^x} = \dfrac{(\ln y)^A}{y}$. As $x \to \infty$, $y = e^x \to \infty$ and by Qu. 3

(with $\alpha = 1$), $\dfrac{(\ln y)^A}{y} \to 0$. Therefore

> $\dfrac{x^A}{e^x} \to 0$ as $x \to \infty$;
>
> $e^x \to \infty$ faster than any positive power of x.

Qu. 4. Use a calculator to estimate how large x must be if $\dfrac{x^{20}}{e^x} < 0.01$.

Exercise 12D

1. Arrange the following in increasing order of greatness for large x:
$x \ln x$, $\ln \ln x$, e^{x^3}, $\sinh x$, e^{e^x}, $(\ln x)^{20}$, $(\ln x)^x$.

2. Sketch the graphs (i) $y = x^{\frac{1}{2}} \ln x$ (ii) $y = x^2 \ln x$ for $0 < x < 2$, paying particular attention to the shape near the origin.

3. Prove that $P(x) e^{-x} \to 0$ as $x \to \infty$, where $P(x)$ is any polynomial.

4. Find $\dfrac{d}{dx} (x^2 e^{-x^2})$ and determine the intervals in which the function $x^2 e^{-x^2}$ is increasing and those in which it is decreasing. Sketch the curve $y = x^2 e^{-x^2}$.   [MEI]

5. Sketch the curve $y = x^3 e^{-x}$, indicating the points of inflexion. Find the equations of the tangents to the curve that pass through the origin.

   Show that, for $a > 6$, $5 < \displaystyle\int_0^a x^3 e^{-x} \, dx < 6$.   [L]

6. Find the maxima and minima of the function $f(x) = e^{-x^2}(c + cx^2 + x^4)$ where c is a constant, distinguishing the different cases according to the value of c. Sketch the graph of the function in the cases $c = 1$ and $c = 3$.
                                                                          [MEI]

7. When $x > 0$, prove that the function $\dfrac{\ln x}{x}$ has a maximum at $x = e$ and find its value at this point.
   Find any points of inflexion, and sketch the graph of the function.
   How many real roots are there to the equation $\ln x = kx$ where
   (a) $k = \frac{1}{4}$, (b) $k = 4$?
   Use your graph to find how many real positive roots there are to the equation $x^4 = 4 \{x^2 + (\ln x)^2\}$.   [O & C]

8. Sketch the graph of $y = x^{1/x}$ for $x > 0$.

9. Sketch the curve $y = x^n e^{-x}$, (n a positive integer, $n > 1$)
   (i) when n is even;   (ii) when n is odd.
   Prove that the maximum of this curve lies, for all values of n, on the curve $y = x^x e^{-x}$. Prove also that, for positive x, the gradient of the curve $y = x^n e^{-x}$ decreases for values of x lying in an interval of length $2\sqrt{n}$.
                                                                          [C]

10. The function ψ is defined by $\psi(x) \equiv e^{-1/x^2} (x \neq 0)$, $\psi(0) = 0$.
    (i) Prove that $\psi(x) \to 0$ as $x \to 0$ and that $\psi(x) \to 1$ as $x \to \pm \infty$. Sketch the graph of $y = \psi(x)$.
    (ii) Find $\psi'(x)$ and $\psi''(x)$ for $x \neq 0$.
    (iii) Prove by induction that $\psi^{(n)}(x) = P_n\left(\dfrac{1}{x}\right) e^{-1/x^2}$ for $x \neq 0$, where P_n is a polynomial. Deduce that $\psi^{(n)}(x) \to 0$ as $x \to 0$.
    (iv) Use the definition $\psi^{(k+1)}(0) = \lim\limits_{h \to 0} \dfrac{\psi^{(k)}(h) - \psi^{(k)}(0)}{h}$ to prove that $\psi^{(n)}(0) = 0$ for all n.
    [Thus if a particle moves on a straight line with displacement $\psi(t)$ at time t, then the particle is initially at rest at the origin with zero acceleration, zero rate of change of acceleration, and so on—and yet it moves immediately!]

11. Show that the Maclaurin expansion of $\psi(x)$, where ψ is as in Qu. 10, converges for all x, but that the sum of the series equals $\psi(x)$ only when $x = 0$.

12.5 Improper integrals

There are two extensions of the idea of a definite integral which involve infinity and give rise to what are called improper integrals. First consider the area under the curve $y = \dfrac{1}{x^2}$ from $x = 1$ to $x = X$. This area is

$$\int_1^X \frac{1}{x^2}\,dx = \left[-\frac{1}{x}\right]_1^X = 1 - \frac{1}{X}.$$

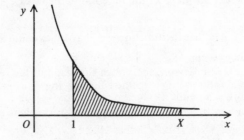

As $X \to \infty$, $1 - \dfrac{1}{X} \to 1$, so although the region bounded by the curve, the line $x = 1$, and the positive x-axis extends infinitely far, it has only a finite area. We say that $\displaystyle\int_1^\infty \frac{1}{x^2}\,dx$ *converges* to the value 1.

The curve $y = \dfrac{1}{x}$ also has the x-axis as an asymptote, but since

$$\int_1^X \frac{1}{x}\,dx = [\ln x]_1^X = \ln X,$$ which does not tend to a finite limit as $X \to \infty$, we say that $\displaystyle\int_1^\infty \frac{1}{x}\,dx$ *diverges*.

The general definitions are similar.

If $\displaystyle\int_a^X f(x)\,dx \to L$ as $X \to \infty$, where L is a fixed finite number, we say that $\displaystyle\int_a^\infty f(x)\,dx$ converges and that $\displaystyle\int_a^\infty f(x)\,dx = L$.

In all other cases we say that $\displaystyle\int_a^\infty f(x)\,dx$ diverges.

Example 1: Investigate the convergence of (i) $\int_0^\infty \sin x \, dx$ (ii) $\int_0^\infty e^{-x} \sin x \, dx$.

Solution: (i) $\int_0^X \sin x \, dx = [-\cos x]_1^X = 1 - \cos X.$

As $X \to \infty$, $\cos X$ oscillates between ± 1, but does not tend to a limit. Therefore $\int_0^\infty \sin x \, dx$ diverges.

(ii) Integrating by parts twice,

$$\int_0^X e^{-x} \sin x \, dx = [-e^{-x} \cos x]_0^X - \int_0^X e^{-x} \cos x \, dx$$

$$= -e^{-X} \cos X + 1 - [e^{-x} \sin x]_0^X - \int_0^X e^{-x} \sin x \, dx,$$

so that $\int_0^X e^{-x} \sin x \, dx = \frac{1}{2}(1 - e^{-X} \cos X - e^{-X} \sin X).$

As $X \to \infty$, $e^{-X} \to 0$ while $|\cos X|$ and $|\sin X|$ never exceed 1.

Therefore $e^{-X} \cos X \to 0$ and $e^{-X} \sin X \to 0$.

Therefore $$\int_0^\infty e^{-x} \sin x \, dx = \frac{1}{2}.$$ □

Similarly $\int_{-\infty}^b f(x) \, dx$ is defined to be the limit as $Y \to \infty$ of $\int_{-Y}^b f(x) \, dx$, if this limit exists, and $\int_{-\infty}^\infty f(x) \, dx$ is defined to be $\int_{-\infty}^a f(x) \, dx + \int_a^\infty f(x) \, dx$, where a is any number, provided that both these integrals converge separately. Such integrals occur frequently in probability theory.

The other type of improper integral arises when we consider the area between a curve and a vertical asymptote. For example the area under the curve $y = x^{-1/3}$ from $x = h$ to $x = 1$, where $0 < h < 1$, is

$\int_h^1 x^{-1/3} \, dx = [\frac{3}{2} x^{2/3}]_h^1 = \frac{3}{2} - \frac{3}{2} h^{2/3}$. As $h \to 0$, $\frac{3}{2} - \frac{3}{2} h^{2/3} \to \frac{3}{2}$, so the area

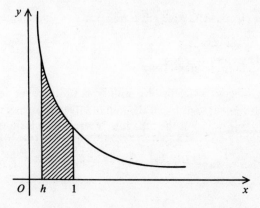

of the region bounded by the curve, the axes and the line $x = 1$ is $\frac{3}{2}$. We say that $\int_0^1 x^{-1/3}\,dx$ converges to $\frac{3}{2}$; it is necessary to use a limiting process since the integrand $x^{-1/3}$ is undefined when $x = 0$. Similarly

---

if $f(x)$ is defined for $a \leqslant x \leqslant h < b$, but undefined for $x = b$, and if $\int_a^h f(x)\,dx \to L$ as $h \to b$, we say that $\int_a^b f(x)\,dx$ converges and that $\int_a^b f(x)\,dx = L$. Otherwise we say that $\int_a^b f(x)\,dx$ diverges.

Example 2: Find (i) $\displaystyle\int_0^1 \frac{x}{\sqrt{(1 - x^2)}}\,dx$ (ii) $\displaystyle\int_0^1 \frac{x^2}{\sqrt{(1 - x^2)}}\,dx$.

Solution: In each case the integrand is undefined when $x = 1$.

(i) $\displaystyle\int_0^h \frac{x}{\sqrt{(1 - x^2)}}\,dx = \left[-\sqrt{(1 - x^2)}\right]_0^h$

$$= 1 - \sqrt{(1 - h^2)}$$

$$\to 1 \text{ as } h \to 1.$$

Therefore $\displaystyle\int_0^1 \frac{x}{\sqrt{(1 - x^2)}}\,dx = 1$.

(ii) Let $x = \sin u$ and $h = \sin \alpha$.

Then $\displaystyle\int_0^h \frac{x^2}{\sqrt{(1 - x^2)}}\,dx = \int_0^\alpha \frac{\sin^2 u}{\cos u} \cdot \cos u\,du$

$$= \int_0^\alpha \tfrac{1}{2}(1 - \cos 2u)\,du$$

$$= \left[\frac{u}{2} - \frac{\sin 2u}{4}\right]_0^\alpha = \frac{\alpha}{2} - \frac{\sin 2\alpha}{4}.$$

As $h \to 1$, $\alpha \to \dfrac{\pi}{2}$ and $\dfrac{\alpha}{2} - \dfrac{\sin 2\alpha}{4} \to \dfrac{\pi}{4}$.

Therefore $\displaystyle\int_0^1 \frac{x^2}{\sqrt{(1 - x^2)}}\,dx = \frac{\pi}{4}$. □

To deal with integrals which involve both types of improper behaviour, we subdivide the domain so as to get the sum of integrals each of which is improper in only one way. For the function f whose graph is shown,

$$\int_{-\infty}^\infty f(x)\,dx = \int_{-\infty}^a f(x)\,dx + \int_a^b f(x)\,dx + \int_b^c f(x)\,dx + \int_c^\infty f(x)\,dx,$$

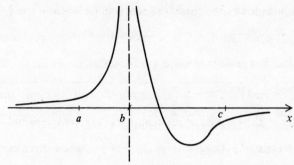

provided that these four integrals all converge separately. The particular
values of a and c do not matter, provided that $a < b < c$.

Exercise 12E

1. Find the values of m for which   (i) $\displaystyle\int_1^\infty x^m \, dx$ converges

   (ii) $\displaystyle\int_0^1 x^m \, dx$ converges   (iii) $\displaystyle\int_0^\infty x^m \, dx$ converges.

2–10. Investigate these integrals, evaluating those that converge.

2. $\displaystyle\int_0^\infty \frac{1}{(x+2)^3} \, dx$ 3. $\displaystyle\int_1^\infty \frac{1}{3+x^2} \, dx$ 4. $\displaystyle\int_{-\infty}^0 \frac{1}{\sqrt{(4-x)}} \, dx$

5. $\displaystyle\int_0^\infty x \cdot e^{-x^2} \, dx$ 6. $\displaystyle\int_3^\infty \frac{1}{x(x+3)} \, dx$ 7. $\displaystyle\int_0^1 (1-x)^{-1/5} \, dx$

8. $\displaystyle\int_0^3 \frac{1}{x(x+3)} \, dx$ 9. $\displaystyle\int_0^{\pi/2} \frac{\cos x}{\sqrt{\sin x}} \, dx$ 10. $\displaystyle\int_{-\infty}^\infty \frac{1}{x^2+4x+8} \, dx$

11. Explain the fallacy in this argument.

$\displaystyle\int_{-1}^1 \frac{1}{x^2} \, dx = \left[-\frac{1}{x}\right]_{-1}^1 = -2.$ But $\dfrac{1}{x^2} > 0$ for all x. Therefore $-2 > 0$.

12. Show that, if $a > 0$, $\displaystyle\int_0^\infty e^{-ax} \cos bx \, dx = \frac{a}{a^2+b^2}$.

13. (i) Use the substitution $u = \dfrac{1}{x}$ to show that $\displaystyle\int_{1/X}^1 \frac{\ln x}{x} \, dx = -\int_1^X \frac{\ln x}{x} \, dx$,
and hence that $\displaystyle\int_{1/X}^X \frac{\ln x}{x} \, dx = 0$.

(ii) Deduce that if $\displaystyle\int_0^\infty \frac{\ln x}{x} \, dx$ converges, its value is zero.

(iii) Use the substitution $v = \ln x$ to find whether $\displaystyle\int_0^\infty \frac{\ln x}{x} \, dx$ does
converge.

14. Use the substitution $x = a \cos^2\theta + b \sin^2\theta$ to show that
$$\int_a^b \frac{1}{\sqrt{\{(x-a)(b-x)\}}}\, dx = \pi \qquad (0 < a < b).$$

15. (P) A probability density function f is a function for which $f(x) \geqslant 0$ for all x and $\int_{-\infty}^{\infty} f(x)\, dx = 1$.

   (i) For the Laplace distribution $f(x) \equiv A e^{-\lambda|x|}$, where A and λ are positive constants. Find A in terms of λ.

   (ii) For the Cauchy distribution $f(x) \equiv \dfrac{B}{1+x^2}$, where B is a constant. Find B.

16. (P) The variance of a symmetrical probability distribution is $\int_{-\infty}^{\infty} x^2 f(x)\, dx$, where f is the probability density function. Find the variances of the Laplace and Cauchy distributions given in Qu. 15.

17. Sketch the graph of the function $\ln \sin y$, $0 \leqslant y \leqslant \pi$.
   If $A = \int_0^{\pi/2} \ln(2\cos x)\, dx$ and $B = \int_0^{\pi/2} \ln(2\sin x)\, dx$, prove that $A + B = B$. Hence evaluate $\int_0^{\pi/2} \ln \cos x\, dx$. [Assume that all these integrals converge.] [MEI, adapted]

18. If $\sec\theta = \cosh u$, with $u > 0$ and $0 < \theta < \dfrac{\pi}{2}$, express (i) $\tan\theta$; (ii) $\dfrac{d\theta}{du}$ in terms of u. Hence, or otherwise, evaluate $\int_0^{\infty} \dfrac{1}{\cosh u}\, du$. [SMP]

19. A function f is periodic, with period a, that is to say $f(x + a) = f(x)$ for all x. Show that, if k is any integer,
$$\int_{ka}^{(k+1)a} e^{-pt} f(t)\, dt = e^{-kpa} \int_0^a e^{-pt} f(t)\, dt,$$ assuming that these integrals exist.
   By subdividing the interval $0 \leqslant x \leqslant (n+1)a$ suitably, and using the result above, show that
$$\int_0^{(n+1)a} e^{-pt} f(t)\, dt = \frac{1 - e^{-(n+1)pa}}{1 - e^{-pa}} \int_0^a e^{-pt} f(t)\, dt, \text{ where } p > 0.$$
   If $f(t) = \begin{cases} t, & 0 \leqslant t \leqslant 1 \\ 2 - t, & 1 < t < 2 \end{cases}$ with $a = 2$, show that
$$\int_0^{\infty} e^{-pt} f(t)\, dt = \frac{e^p - 1}{p^2(e^p + 1)}. \qquad \text{[MEI]}$$

12.6 Reduction formulae

It may be possible to relate an integral whose integrand involves a positive integer n to a similar integral with $n - 1$ (or $n - 2, n - 3, \ldots$) in place of n. For example

$$\int x^n e^{-x} dx = x^n(-e^{-x}) - \int nx^{n-1}(-e^{-x}) dx$$

$$= n \int x^{n-1} e^{-x} dx - x^n e^{-x},$$

or $\qquad\qquad I_n = n I_{n-1} - x^n e^{-x}$, where $I_n = \int x^n e^{-x} dx$.

This relation is called a *reduction formula*. By using it repeatedly we can find $\int x^n e^{-x} dx$ for any particular n. Thus

$$\int x^3 e^{-x} dx = I_3 = 3I_2 - x^3 e^{-x}$$

$$= 3(2I_1 - x^2 e^{-x}) - x^3 e^{-x}$$

$$= 3(2(I_0 - xe^{-x}) - x^2 e^{-x}) - x^3 e^{-x}.$$

But $\qquad\qquad I_0 = \int e^{-x} dx = -e^{-x} + c.$

Therefore $\quad \int x^3 e^{-x} dx = -e^{-x}[3(2(1 + x) + x^2) + x^3] + c'$

$$= -e^{-x}[6 + 6x + 3x^2 + x^3] + c'.$$

The search for a reduction formula usually starts with integration by parts. Definite integrals can lead to simpler reduction formulae, since some terms may become zero when evaluated between the limits of integration, as in the next example.

Example 1: Find a reduction formula for $\int_0^{\pi/2} \sin^n \theta \, d\theta$, and evaluate

(i) $\int_0^{\pi/2} \sin^6 \theta \, d\theta$ (ii) $\int_0^{\pi/2} \sin^7 \theta \, d\theta.$

Solution: Let $I_n = \int_0^{\pi/2} \sin^n \theta \, d\theta.$

Then $I_n = \int_0^{\pi/2} \sin^{n-1} \theta \, . \, d(-\cos \theta)$

$$= [-\sin^{n-1} \theta \, \cos \theta]_0^{\pi/2} + \int_0^{\pi/2} (n-1)\sin^{n-2} \theta \, \cos^2 \theta \, d\theta$$

$$= 0 + (n-1) \int_0^{\pi/2} \sin^{n-2} \theta \, (1 - \sin^2 \theta) \, d\theta$$

$$= (n-1)(I_{n-2} - I_n).$$

Therefore $n I_n = (n-1)I_{n-2}$, or $I_n = \dfrac{n-1}{n} I_{n-2}$ $(n > 1).$

In particular (i) $I_6 = \dfrac{5}{6} I_4 = \dfrac{5}{6} \cdot \dfrac{3}{4} I_2 = \dfrac{5}{6} \cdot \dfrac{3}{4} \cdot \dfrac{1}{2} I_0$

$$= \dfrac{5}{6} \cdot \dfrac{3}{4} \cdot \dfrac{1}{2} \cdot \dfrac{\pi}{2} \text{ since } I_0 = \int_0^{\pi/2} 1 \, . \, d\theta = \dfrac{\pi}{2}$$

$$= \dfrac{5\pi}{32}.$$

(ii) $I_7 = \dfrac{6}{7} I_5 = \dfrac{6}{7} \cdot \dfrac{4}{5} \cdot I_3 = \dfrac{6}{7} \cdot \dfrac{4}{5} \cdot \dfrac{2}{3} I_1$

$$= \dfrac{6}{7} \cdot \dfrac{4}{5} \cdot \dfrac{2}{3} \cdot 1 \text{ since } I_1 = \int_0^{\pi/2} \sin \theta \, d\theta = [-\cos \theta]_0^{\pi/2} = 1$$

$$= \dfrac{16}{35}. \qquad\qquad\qquad\qquad\qquad\qquad\qquad\qquad\qquad\qquad \square$$

Qu. 1. Prove that $\displaystyle\int_0^{\pi/2} \cos^n \theta \, d\theta = \dfrac{n-1}{n} \int_0^{\pi/2} \cos^{n-2} \theta \, d\theta \quad (n > 1)$.

Example 2: Find a reduction formula for $\displaystyle\int_0^\infty \dfrac{1}{(1 + x^2)^n} \, dx$, and hence evaluate this integral.

Solution: Let $I_n = \displaystyle\int_0^X \dfrac{1}{(1 + x^2)^n} \, dx$.

Then $I_n = \left[\dfrac{1}{(1 + x^2)^n} \cdot x \right]_0^X - \displaystyle\int_0^X (-n) \dfrac{2x}{(1 + x^2)^{n+1}} \cdot x \, dx$

$$= \dfrac{X}{(1 + X^2)^n} + 2n \int_0^X \dfrac{x^2 + 1 - 1}{(1 + x^2)^{n+1}} \, dx$$

$$= \dfrac{X}{(1 + X^2)^n} + 2n \, (I_n - I_{n+1}).$$

In this case, integration by parts has increased n instead of reducing it, so we replace n by $n - 1$:

$$I_{n-1} = \dfrac{X}{(1 + X^2)^{n-1}} + 2(n - 1) \, (I_{n-1} - I_n)$$

so that $I_n = \dfrac{1}{2n - 2} \cdot \dfrac{X}{(1 + X^2)^{n-1}} + \dfrac{2n - 3}{2n - 2} \, I_{n-1}, \quad n > 1. \quad$ (A)

Now let $X \to \infty$, assuming for the moment that

$J_n = \displaystyle\int_0^\infty \dfrac{1}{(1 + x^2)^n} \, dx$ converges for $n > 1$. Then since

$\dfrac{X}{(1 + X^2)^{n-1}} \to 0$, $J_n = \dfrac{2n - 3}{2n - 2} \, J_{n-1}$. Applying this repeatedly,

$$J_n = \frac{2n-3}{2n-2} \cdot \frac{2n-5}{2n-4} \cdot \frac{2n-7}{2n-6} \cdot \ldots \cdot \frac{1}{2} \cdot \frac{\pi}{2}, \text{ since}$$

$$J_1 = \lim_{X \to \infty} I_1 = \lim_{X \to \infty} \tan^{-1} X = \frac{\pi}{2}.$$

Note that the convergence of J_n for $n > 1$ follows from the convergence of J_1 by induction, using equation (A). The expression for J_n can be written more compactly by introducing extra factors ($\uparrow$) in the numerator and denominator:

$$J_n = \frac{2n-2}{2n-2} \cdot \frac{2n-3}{2n-2} \cdot \frac{2n-4}{2n-4} \cdot \frac{2n-5}{2n-4} \cdot \frac{2n-6}{2n-6} \cdot \frac{2n-7}{2n-6} \cdot \ldots \cdot \frac{2}{2} \cdot \frac{1}{2} \cdot \frac{\pi}{2}$$

$$\qquad\;\; \uparrow \qquad\qquad\quad \uparrow \qquad\qquad\quad \uparrow \qquad\qquad\qquad \uparrow$$

$$= \frac{(2n-2)!}{[(2n-2)(2n-4)(2n-6) \ldots 2]^2} \cdot \frac{\pi}{2}$$

$$= \frac{(2n-2)!}{[2^{n-1}(n-1)!]^2} \cdot \frac{\pi}{2} = \frac{(2n-2)!\,\pi}{2^{2n-1}[(n-1)!]^2}. \qquad\qquad \square$$

Exercise 12F

1. Use the results of Example 1 and Qu. 1 to evaluate

(i) $\displaystyle\int_0^{\pi/2} \sin^4 \theta \, d\theta$ (ii) $\displaystyle\int_0^{\pi/2} \sin^5 \theta \, d\theta$ (iii) $\displaystyle\int_0^{\pi/2} \cos^9 \theta \, d\theta$

(iv) $\displaystyle\int_0^{\pi} \cos^6 \theta \, d\theta$ (v) $\displaystyle\int_0^{\pi} \sin^3 \theta \, d\theta$ (vi) $\displaystyle\int_0^{\pi} \cos^{11} \theta \, d\theta$

(vii) $\displaystyle\int_{\pi}^{3\pi/2} \sin^7 \theta \, d\theta$

2. If $I_n = \int (x^2 + 3)^n \, dx$, prove that $(2n+1)I_n = x(x^2+3)^n + 6n\,I_{n-1}$.

3. Find a reduction formula for $\int (\ln x)^n \, dx$. Evaluate $\displaystyle\int_1^e (\ln x)^4 \, dx$.

4. Find a reduction formula for $\displaystyle\int_0^{\pi} x^n \cos x \, dx$.

5. Let $I_n = \displaystyle\int_0^{\pi/2} \cos^n x \cos nx \, dx$.

   Use $\cos(n-1)x = \cos nx \cos x + \sin nx \sin x$ to prove that $I_n = \frac{1}{2} I_{n-1}$.
   Evaluate I_5.

6. Let $I_n = \displaystyle\int \frac{\sin nx}{\sin x} \, dx$. Prove that $I_n - I_{n-2} = \frac{2\sin(n-1)x}{n-1}$.

   Deduce that $\displaystyle\int_0^{\pi} \frac{\sin nx}{\sin x} \, dx$ is 0 if n is even and π if n is odd.

7. If $I_n = \int_0^{\pi/4} \tan^n x\, dx$ and $n \geqslant 2$, prove that $I_n + I_{n-2} = \dfrac{1}{n-1}$

and hence prove by induction that

$$1 - \frac{1}{3} + \frac{1}{5} - \frac{1}{7} + \ldots + \frac{(-1)^{n-1}}{2n-1} = \frac{\pi}{4} + (-1)^{n-1} I_{2n}.$$

Deduce that $1 - \frac{1}{3} + \frac{1}{5} - \frac{1}{7} + \ldots + \frac{1}{501}$ differs from $\pi/4$ by less than $\frac{1}{1000}$.
                                                                                    [MEI]

8. Let $I_{m,n} = \int_0^{\pi/2} \sin^m \theta \cos^n \theta\, d\theta$.   (i) Prove that $I_{m,n} = I_{n,m}$.

   (ii) By differentiating $\sin^{m-1} \theta \cos^{n+1} \theta$ prove that $I_{m,n} = \dfrac{m-1}{m+n} I_{m-2,n}$.

9. Use Qu. 8 to evaluate

   (i) $\displaystyle\int_0^{\pi/2} \sin^3 \theta \cos^4 \theta\, d\theta$ (ii) $\displaystyle\int_0^{\pi/2} \sin^5 \theta \cos^7 \theta\, d\theta$

   (iii) $\displaystyle\int_0^{\pi/2} \sin^6 \theta \cos^4 \theta\, d\theta$ (iv) $\displaystyle\int_0^1 x^6 (1-x^2)^{3/2}\, dx$.

10. If $B(p,q) = \displaystyle\int_0^1 x^{p-1} (1-x)^{q-1}\, dx$, where $p > 0$, $q > 0$, show that:

   (i) $B(p,q) = B(q,p)$
   (ii) $p B(p,q) = (q-1) B(p+1, q-1)$, for $q > 1$
   (iii) $B(p,q) = \dfrac{q-1}{p+q-1} B(p, q-1)$, for $q > 1$
   (iv) $B(p,q) = 2 \displaystyle\int_0^{\pi/2} \sin^{2p-1} \theta \cos^{2q-1} \theta\, d\theta$
   (v) If p, q are integers, $B(p,q) = \dfrac{(p-1)!\,(q-1)!}{(p+q-1)!}$.

$B(p,q)$ is called Euler's beta function; it has important applications in probability theory. You may assume that $B(p,q)$ converges for $0 < p < 1$, $0 < q < 1$.

11. (i) Use Example 1 to show that if $I_n = \displaystyle\int_0^{\pi/2} \sin^n \theta\, d\theta$ then

   $$I_{2m} = \frac{2m-1}{2m} \cdot \frac{2m-3}{2m-2} \cdot \ldots \cdot \frac{1}{2} \cdot \frac{\pi}{2} \text{ and}$$

   $$I_{2m+1} = \frac{2m}{2m+1} \cdot \frac{2m-2}{2m-1} \cdot \ldots \cdot \frac{2}{3}.$$

   (ii) By considering the graphs $y = \sin^n \theta$ for $n = 2m-1, 2m, 2m+1$,

   $0 \leqslant \theta \leqslant \dfrac{\pi}{2}$, show that $I_{2m-1} > I_{2m} > I_{2m+1}$, and hence that

   $$\frac{I_{2m-1}}{I_{2m+1}} > \frac{I_{2m}}{I_{2m+1}} > 1.$$

(iii) Show that $\dfrac{I_{2m-1}}{I_{2m+1}} \to 1$ as $m \to \infty$.

(iv) Deduce that $\dfrac{\pi}{2} = \lim\limits_{m\to\infty} \left(\dfrac{2}{1} \cdot \dfrac{2}{3} \cdot \dfrac{4}{3} \cdot \dfrac{4}{5} \cdot \ldots \cdot \dfrac{2m}{2m-1} \cdot \dfrac{2m}{2m+1} \right)$.

This is called Wallis's product, after John Wallis who gave it in 1655 in the form $\dfrac{4}{\pi} = \dfrac{3}{2} \cdot \dfrac{3}{4} \cdot \dfrac{5}{4} \cdot \dfrac{5}{6} \cdot \dfrac{7}{6} \cdot \dfrac{7}{8} \cdot \ldots$; it converges very slowly.

12. By considering the derivative of $x^p e^{-x^2}$ or otherwise, prove that, if for $n \geqslant 0$ $I_n = \displaystyle\int_0^\infty x^n e^{-x^2}\, dx$, then $I_{n+2} = \tfrac{1}{2}(n+1) I_n$. Deduce that, if $\displaystyle\int_0^\infty e^{-x^2}\, dx = \tfrac{1}{2}\sqrt{\pi}$, then $\displaystyle\int_0^\infty x^6 e^{-x^2}\, dx = \tfrac{15}{16}\sqrt{\pi}$. [MEI]

13. If $I_n = \displaystyle\int_0^\phi \dfrac{\sin^n\theta}{\cos\theta}\, d\theta$, prove that $I_n = I_{n-2} - \dfrac{\sin^{n-1}\phi}{n-1}$, for $n = 2, 3, 4, \ldots$

Hence obtain expressions for I_{2n} and I_{2n-1} for $n \geqslant 1$.

By considering the limit as $n \to \infty$ for an appropriate value of ϕ, show that

$$\ln(1+\sqrt{2}) = \sum_{r=1}^{\infty} \dfrac{1}{(2r-1)} \cdot \dfrac{1}{2^{r-\frac{1}{2}}} \quad \text{and} \quad \ln 2 = \sum_{r=1}^{\infty} \dfrac{1}{r} \cdot \dfrac{1}{2^r}. \qquad \text{[O \& C]}$$

14. *Proof that π is irrational.*

(i) Let $I_n = \displaystyle\int_{-1}^{1} (1-x^2)^n \cos\dfrac{\pi x}{2}\, dx$. By integrating by parts twice, prove that $\left(\dfrac{\pi}{2}\right)^2 I_n = 2n(2n-1) I_{n-1} - 4n(n-1) I_{n-2}$, $n \geqslant 2$.

(ii) Show that $\dfrac{\pi}{2} \cdot I_0 = 2$, $\left(\dfrac{\pi}{2}\right)^3 I_1 = 4$, and prove by induction that $\left(\dfrac{\pi}{2}\right)^{2n+1} I_n = n!\, P_n\left(\dfrac{\pi}{2}\right)$, where P_n is a polynomial of degree $\leqslant n$ with integer coefficients.

(iii) Suppose that $\dfrac{\pi}{2}$ is rational, and let $\dfrac{\pi}{2} = \dfrac{a}{b}$ where a and b are integers. Prove that $\dfrac{a^{2n+1}}{n!} I_n = b^{2n+1} P_n\left(\dfrac{a}{b}\right)$ and deduce that $\dfrac{a^{2n+1}}{n!} I_n$ is an integer for all n.

(iv) Show that $0 < (1-x^2)^n \cos\dfrac{\pi x}{2} < 1$ for $-1 < x < 1$, and deduce that $0 < I_n < 2$. Use this and the fact that $\dfrac{a^{2n+1}}{n!} \to 0$ as $n \to \infty$ (see §11.3, Qu. 4) to show that $0 < \dfrac{a^{2n+1}}{n!} I_n < 1$ if n is sufficiently large.

(v) Deduce from the contradictory results of (iii) and (iv) that π is irrational.

13 Complex numbers II

13.1 Roots of complex numbers

If r is a positive real number and q is a positive integer, then there is just one positive real number s such that $s^q = r$. We call s the positive qth root of r, and we write $s = r^{1/q}$.

Suppose now that z is a given complex number, and that we wish to find w such that $w^q = z$. It is best to work in the polar form, so let $z = r(\cos\theta + j\sin\theta)$ and $w = s(\cos\phi + j\sin\phi)$. Then

$$w^q = z \Leftrightarrow s^q(\cos\phi + j\sin\phi)^q = r(\cos\theta + j\sin\theta)$$

$$\Leftrightarrow s^q(\cos q\phi + j\sin q\phi) = r(\cos\theta + j\sin\theta),$$

using De Moivre's theorem with the integer q. Now two complex numbers in polar form can be equal only if they have the same moduli and if their arguments are equal or differ by an integer multiple of 2π. Thus

$$s^q = r \quad \text{and} \quad q\phi = \theta + 2k\pi, \quad \text{where } k \text{ is an integer.}$$

Therefore $s = r^{\frac{1}{q}}$ (since r and s are positive) and $\phi = \dfrac{\theta + 2k\pi}{q}$, so that

$$w = r^{\frac{1}{q}}\left(\cos\left(\frac{\theta + 2k\pi}{q}\right) + j\sin\left(\frac{\theta + 2k\pi}{q}\right)\right).$$

If k_1 and k_2 are two values of k differing by a multiple of q, then $\dfrac{\theta + 2k_1\pi}{q}$ and $\dfrac{\theta + 2k_2\pi}{q}$ differ by a multiple of 2π, and so produce the same w. By letting k take the values $0, 1, 2, \ldots, q-1$, we obtain q different complex numbers w. Since any other integer differs from one of $0, 1, 2, \ldots, q-1$ by a multiple of q, there can be no further distinct values of w.

Therefore the non-zero complex number $r(\cos\theta + j\sin\theta)$ has precisely q different qth roots. These are

$$r^{\frac{1}{q}}\left(\cos\left(\frac{\theta + 2k\pi}{q}\right) + j\sin\left(\frac{\theta + 2k\pi}{q}\right)\right)$$

where $k = 0, 1, 2, \ldots, q-1$.

Example 1: Find the four fourth roots of -4.

Solution: Since $-4 = 4 (\cos \pi + j \sin \pi)$, the four fourth roots all have modulus $4^{1/4} = \sqrt{2}$. The fourth roots are

$$\sqrt{2} \left(\cos \frac{\pi}{4} + j \sin \frac{\pi}{4} \right) = \sqrt{2} \left(\frac{1}{\sqrt{2}} + j \cdot \frac{1}{\sqrt{2}} \right) = 1 + j$$

$$\sqrt{2} \left(\cos \left(\frac{\pi + 2\pi}{4} \right) + j \sin \left(\frac{\pi + 2\pi}{4} \right) \right) = -1 + j$$

$$\sqrt{2} \left(\cos \left(\frac{\pi + 4\pi}{4} \right) + j \sin \left(\frac{\pi + 4\pi}{4} \right) \right) = -1 - j$$

$$\sqrt{2} \left(\cos \left(\frac{\pi + 6\pi}{4} \right) + j \sin \left(\frac{\pi + 6\pi}{4} \right) \right) = 1 - j. \qquad \square$$

Qu. 1. Check that $(\pm 1 \pm j)^4 = -4$.

Example 2: Represent $-3 + 3j$ and its five fifth roots on an Argand diagram.

Solution: Since $-3 + 3j = 18^{\frac{1}{2}} \left(\cos \frac{3\pi}{4} + j \sin \frac{3\pi}{4} \right)$, the fifth roots all have modulus $18^{\frac{1}{10}} \approx 1.335$.

Their arguments are $\dfrac{3\pi}{20}, \dfrac{3\pi}{20} + \dfrac{2\pi}{5}, \dfrac{3\pi}{20} + \dfrac{4\pi}{5}, \dfrac{3\pi}{20} + \dfrac{6\pi}{5}, \dfrac{3\pi}{20} + \dfrac{8\pi}{5}$

or (taking principal arguments) $\dfrac{3\pi}{20}, \dfrac{11\pi}{20}, \dfrac{19\pi}{20}, -\dfrac{13\pi}{20}, -\dfrac{5\pi}{20}$.

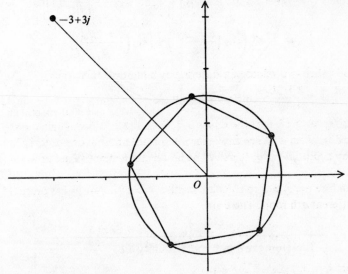

The fifth roots are represented by the vertices of a regular pentagon inscribed in the circle $|z| = 18^{\frac{1}{10}}$. $\qquad \square$

Qu. 2. Express $18^{\frac{1}{10}}\left(\cos\dfrac{3\pi}{20} + j\sin\dfrac{3\pi}{20}\right)$ in the form $x + jy$, giving x and y to 2 D.P.

The Argand diagram in Example 2 is typical of the general case: the q qth roots of z are represented by the vertices of a regular q-gon inscribed in the the circle with centre O and radius $|z|^{1/q}$.

One special case is worth considering: when $z = 1$ we see that the equation $w^q = 1$ has the q distinct solutions

$$\cos\frac{2k\pi}{q} + j\sin\frac{2k\pi}{q},\ k = 0, 1, 2, \ldots, q-1.$$

These are called the qth *roots of unity*. They are represented in the Argand diagram by the vertices of the regular q-gon inscribed in the unit circle and having one vertex at the point 1.

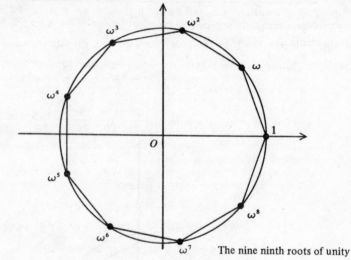

The nine ninth roots of unity

If $\omega = \cos\dfrac{2\pi}{q} + j\sin\dfrac{2\pi}{q}$ then (by De Moivre's theorem)

$$\omega^k = \cos\frac{2k\pi}{q} + j\sin\frac{2k\pi}{q}.$$

> Therefore the q q-th roots of unity are
>
> $1, \omega, \omega^2, \ldots, \omega^{q-1}$, where $\omega = \cos\dfrac{2\pi}{q} + j\sin\dfrac{2\pi}{q}$.

Qu. 3. Write down (i) the two square roots of unity (ii) the three cube roots of unity (iii) the four fourth roots of unity.

Qu. 4. Prove that if α is one qth root of z, then the others are

$$\alpha\omega, \alpha\omega^2, \ldots, \alpha\omega^{q-1}, \text{ where } \omega = \cos\frac{2\pi}{q} + j\sin\frac{2\pi}{q}.$$

Qu. 5. Prove that $(\omega^r)^* = \omega^{q-r}$.

The sum of all the qth roots of unity is

$$1 + \omega + \omega^2 + \ldots + \omega^{q-1} = \frac{1-\omega^q}{1-\omega} \quad \text{since } \omega \neq 1$$

$$= 0 \quad \text{since } \omega^q = 1.$$

> Therefore the sum of the q qth roots of unity is zero.

Many problems involving regular polygons can be solved by using the Argand diagram and the roots of unity.

Example 3: Three points are equally spaced around a circle of radius a. Prove that the sum of the squares of their distances from any diameter is $\frac{3a^2}{2}$.

Solution: Take the diameter as the imaginary axis and the circle as $|z| = a$. Let one of the points be α, so that the other two are $\alpha\omega$ and $\alpha\omega^2$, where ω is a complex cube root of 1.
The square of the distance of the point z from the imaginary axis is $(\text{Re}(z))^2 = \frac{1}{4}(z + z^*)^2$, so the required sum of squares is

$$\tfrac{1}{4}(\alpha + \alpha^*)^2 + \tfrac{1}{4}(\alpha\omega + \alpha^*\omega^*)^2 + \tfrac{1}{4}(\alpha\omega^2 + \alpha^*\omega^{*2})^2$$

$$= \tfrac{1}{4}\{(\alpha + \alpha^*)^2 + (\alpha\omega + \alpha^*\omega^2)^2 + (\alpha\omega^2 + \alpha^*\omega)^2\} \quad \text{using Qu. 5}$$

$$= \tfrac{1}{4}\{\alpha^2 + 2\alpha\alpha^* + \alpha^{*2} + \alpha^2\omega^2 + 2\alpha\alpha^*\omega^3 + \alpha^{*2}\omega^4$$
$$+ \alpha^2\omega^4 + 2\alpha\alpha^*\omega^3 + \alpha^{*2}\omega^2\}$$

$$= \tfrac{1}{4}\{\alpha^2(1 + \omega + \omega^2) + 6\alpha\alpha^* + \alpha^{*2}(1 + \omega + \omega^2)\}$$
$$\text{since } \omega^3 = 1 \text{ and } \omega^4 = \omega$$

$$= \tfrac{3}{2}\alpha\alpha^* \quad \text{since } 1 + \omega + \omega^2 = 0$$

$$= \tfrac{3}{2}a^2 \quad \text{since } \alpha\alpha^* = |\alpha|^2 = a^2. \qquad \square$$

If $z = r(\cos\theta + j\sin\theta)$ then $z^p = r^p(\cos p\theta + j\sin p\theta)$ for all integers p by De Moivre's theorem. The complex number z^p has q qth roots

$$r^{p/q}\left(\cos\left(\frac{p\theta + 2k\pi}{q}\right) + j\sin\left(\frac{p\theta + 2k\pi}{q}\right)\right), \quad k = 0, 1, 2, \ldots, q-1.$$

384   Roots of complex numbers   Ex. 13A

One of these is $r^{p/q}\left(\cos\dfrac{p\theta}{q}+j\sin\dfrac{p\theta}{q}\right)$, and if we now define $z^{p/q}$ to mean

this qth root of z^p, we can extend De Moivre's theorem to rational powers also, since (taking $r=1$) we have

$$(\cos\theta+j\sin\theta)^{p/q} = \left(\cos\frac{p\theta}{q}+j\sin\frac{p\theta}{q}\right).$$

The index notation $z^{p/q}$ has to be used with care. For example,

$$j = \cos\frac{\pi}{2}+j\sin\frac{\pi}{2} \quad\text{so}\quad j^{\frac{1}{2}} = \cos\frac{\pi}{4}+j\sin\frac{\pi}{4}$$

and $(j^{\frac{1}{2}})^3 = \cos\dfrac{3\pi}{4}+j\sin\dfrac{3\pi}{4} = \dfrac{1}{\sqrt2}(-1+j).$

But $(j^3)^{\frac{1}{2}} = \left(\cos\left(-\dfrac{\pi}{2}\right)+j\sin\left(-\dfrac{\pi}{2}\right)\right)^{\frac{1}{2}} = \cos\left(-\dfrac{\pi}{4}\right)+j\sin\left(-\dfrac{\pi}{4}\right) = \dfrac{1}{\sqrt2}(1$

Therefore $(j^{\frac{1}{2}})^3 \neq (j^3)^{\frac{1}{2}}$. We always have to bear in mind the many-valued nature of complex roots: the two cubes of the square roots of j are the same as the two square roots of j^3, but the notation $(j^{\frac{1}{2}})^3$ refers to one and $(j^3)^{\frac{1}{2}}$ to the other.

Qu. 6.   Which of these is $j^{\frac{3}{2}}$?

Qu. 7.   Show that $(j^{\frac{1}{3}})^2 = (j^2)^{\frac{1}{3}}$.

Qu. 8.   Find a condition involving p and $\arg z$ which ensures that $(z^{1/q})^p = (z^p)^{1/q}$.

Exercise 13A

1. Find both square roots of $7-4j$, giving your answer in the form $x+yj$ with x and y to 2 D.P.

2. Find all three cube roots of -8, giving your answers in the form $x+yj$. Hence write z^3+8 as the product of three linear factors.

3. Find the four fourth roots of (i) 9; (ii) -16; (iii) $-\dfrac{1}{2}+\dfrac{\sqrt3}{2}j$.

4. Express the following in polar form.

   (i) $\left(\cos\dfrac{2\pi}{5}+j\sin\dfrac{2\pi}{5}\right)^{1/6}$ 　　　 (ii) $\left(32\left(\cos\dfrac{\pi}{3}+j\sin\dfrac{\pi}{3}\right)\right)^{3/5}$

   (iii) $\left(8\left(\cos\dfrac{3\pi}{8}+j\sin\dfrac{3\pi}{8}\right)\right)^{-2/3}$ 　　 (iv) $\left(81\left(\cos\left(-\dfrac{\pi}{3}\right)+j\sin\left(-\dfrac{\pi}{3}\right)\right)\right)^{-7}$

5. One fourth root of z is $3-2j$. What is z, and what are its other fourth roots?

6. Represent the six solutions of the equation $(z - j)^6 = 64$ in the Argand diagram.

7. A regular pentagon in the Argand diagram has centre $1 + 4j$ and one vertex at $5 + 4j$. Write down the equation whose solutions are represented by the vertices of this pentagon.

8. Explain the fallacy in the following argument:

$$ j = \sqrt{(-1)} = \sqrt{\left(\frac{1}{-1}\right)} = \frac{\sqrt{1}}{\sqrt{(-1)}} = \frac{1}{j}, \text{ so } j^2 = 1. \text{ But } j^2 = -1. $$

Therefore $1 = -1$.'

9. Prove that if $|z| = |w|$ then $z + w = \lambda\sqrt{(zw)}$, where λ is real.

10. A, B, C, D are four points on the circle $|z| = 1$, representing the complex numbers a, b, c, d. Use Qu. 9 to show that

$$ AB \text{ is perpendicular to } CD \iff ab = -cd. $$

Find the point of intersection of AB and CD in terms of a, b, c, d when AB is perpendicular to CD.

11. If ω is a complex cube root of unity show that

(i) $1 + \omega = \dfrac{1}{1 + \omega^2}$

(ii) $1 + \omega$ and $1 + \omega^2$ are complex cube roots of -1

(iii) $a^3 + b^3 = (a + b)(a + \omega b)(a + \omega^2 b)$

(iv) $a^3 + b^3 + c^3 - 3abc = (a + b + c)(a + b\omega + c\omega^2)(a + b\omega^2 + c\omega).$

12. A regular hexagon is inscribed in the unit circle. One vertex is α. Give the other vertices in terms of α and ω, where ω is a complex cube root of unity.

13. Three points are equally spaced round a circle of radius a. Prove that the sum of the squares of their distances from any tangent is $\dfrac{9a^2}{2}$.

[Use the method of Example 3, taking the line $\text{Re}(z) = a$ as the tangent. You may use the result of Example 3 to simplify the working.]

14. (i) (a) Draw an Argand diagram showing the points $1, \omega, \omega^2, \omega^3, \omega^4$, where $\omega = \cos\dfrac{2\pi}{5} + j\sin\dfrac{2\pi}{5}$.

(b) If $\alpha = \omega^2$ show that the points $1, \alpha, \alpha^2, \alpha^3, \alpha^4$ are the same as the points in (a), but in a different order. Indicate this order by joining the successive points on your diagram.

(c) Repeat (b) with α replaced by β, where $\beta = \omega^3$.

(ii) Repeat the procedure of (i) taking $\omega = \cos\dfrac{2\pi}{6} + j\sin\dfrac{2\pi}{6}$ and considering the points $1, \omega, \omega^2, \omega^3, \omega^4, \omega^5$ and $1, \alpha, \alpha^2, \alpha^3, \alpha^4, \alpha^5$.

(iii) Do likewise for the seventh and eighth roots of unity.

(iv) If $\omega = \cos\dfrac{2\pi}{n} + j\sin\dfrac{2\pi}{n}$ and $\alpha = \omega^m$, in what circumstances does $\{1, \omega, \omega^2, \ldots, \omega^{n-1}\} = \{1, \alpha, \alpha^2, \ldots, \alpha^{n-1}\}$?

15. The vertices A_1, A_2, A_3, A_4, A_5 of a regular pentagon lie on a circle of unit radius with centre at the point O. A_1 is the midpoint of OP. Prove that

(i) $PA_1 . PA_2 . PA_3 . PA_4 . PA_5 = 31$ (ii) $\sum_{n=1}^{5} (PA_n)^2 = 25$

(iii) $A_1 A_2 . A_1 A_3 . A_1 A_4 . A_1 A_5 = 5$.   [CS]

16. Show that Example 3 and Qu. 13 can be generalized by taking $n \ (\geqslant 3)$ points equally spaced round a circle, the sums of squares in this case being $\dfrac{na^2}{2}$ and $\dfrac{3na^2}{2}$. (M) Deduce the moments of inertia of a uniform hoop about a diameter and about a tangent.

17. The Polish mathematician Wronski (1778–1853) once wrote that

$$\pi = \frac{2\infty}{\sqrt{-1}} \left\{ (1 + \sqrt{-1})^{\frac{1}{\infty}} - (1 - \sqrt{-1})^{\frac{1}{\infty}} \right\}.$$

Show that this is not as mad as it looks. (Start by replacing ∞ by n.)

13.2 Solving polynomial equations

Complex numbers were originally invented in order to provide solutions for quadratic equations which have real coefficients but no real roots. But, as the previous sections have shown, their use goes far beyond this limited objective. If we concentrate on solving polynomial equations, we can first remove the restriction that the coefficients should be real. For if $\alpha z^2 + \beta z + \gamma = 0$, where α, β, γ are complex and $\alpha \neq 0$, then completing the square in the usual way gives $\left(z + \dfrac{\beta}{2\alpha}\right)^2 = \dfrac{\beta^2 - 4\alpha\gamma}{4\alpha^2}$. By §13.1 the complex number $\dfrac{\beta^2 - 4\alpha\gamma}{4\alpha^2}$ has two complex square roots, so that the equation has the two complex roots $\dfrac{-\beta \pm (\beta^2 - 4\alpha\gamma)^{\frac{1}{2}}}{2\alpha}$. Secondly we can find complex roots of some equations of higher degree. For example the equation $(3 + 2j)z^7 + 4 - 5j = 0$ has seven complex roots, the seven seventh roots of $\dfrac{-4 + 5j}{3 + 2j}$.

It is surprising that in both cases we can solve more complicated polynomial equations without having to invent a more complicated number system of 'super-complex' numbers. As early as 1629 the Flemish mathematician Girard conjectured that within the set of complex numbers every polynomial equation of degree n has precisely n roots, counting repetitions if necessary. (For example $(z - 2)^3 (z + j)^2 = 0$ is said to have three roots equal to 2 and two roots equal to $-j$.) Strenuous efforts were made to prove this, but as so often happened it was Gauss who produced the first satisfactory proof, in 1799. The chief difficulty is to establish the existence of *one* root, and the fact that every polynomial equation with complex coefficients has a complex root is called the *Fundamental Theorem of Algebra*. This is too difficult to prove here, but see Ex. 13B, Qu. 17.

The Fundamental Theorem shows that the development of the number system to solve various types of polynomial equations, as outlined in §5.1, is completed by the complex numbers. But we are still faced with the practical problem of finding the roots. For all equations of degree four or less and for some equations of higher degree there are exact formulae for the roots. In other cases approximate methods have to be used. Even when the equation can be solved exactly it may be quicker to find the roots to the required accuracy by approximation.

If the coefficients of the equation are *real*, there is a useful general result about the roots.

> If α is a complex root of a polynomial equation with real coefficients, then so is α^*.

The proof of this depends on the properties of conjugates. Let the equation be be

$$p(z) \equiv c_0 z^n + c_1 z^{n-1} + c_2 z^{n-2} + \ldots + c_n = 0,$$

where the coefficients c_r are real. Taking conjugates,

$$
\begin{aligned}
(p(z))^* &= (c_0 z^n + c_1 z^{n-1} + \ldots + c_n)^* \\
&= (c_0 z^n)^* + (c_1 z^{n-1})^* + \ldots + c_n{}^* \quad \text{since } (z_1 + z_2)^* = z_1{}^* + z_2{}^* \\
&= c_0{}^*(z^n)^* + c_1{}^*(z^{n-1})^* + \ldots + c_n{}^* \quad \text{since } (z_1 z_2)^* = z_1{}^* z_2{}^* \\
&= c_0(z^n)^* + c_1(z^{n-1})^* + \ldots + c_n \quad \text{since the } c_r \text{ are real} \\
&= c_0(z^*)^n + c_1(z^*)^{n-1} + \ldots + c_n \quad \text{since } (z^r)^* = (z^*)^r \\
&= p(z^*)
\end{aligned}
$$

Therefore α is a root of $p(z) = 0 \iff p(\alpha) = 0$

$$\iff p(\alpha^*) = (p(\alpha))^* = 0^* = 0$$

$$\iff \alpha^* \text{ is a root of } p(z) = 0.$$

Qu. 1.   Show that the equation $z^2 - 3jz - 2 = 0$ has roots j and $2j$, which are *not* conjugate.

Since the non-real roots of a real polynomial equation occur in conjugate pairs, the equation must have an even number of non-real roots. Therefore if the equation is of odd degree, there must be an odd number of real roots, and so at least one real root.

Qu. 2.   Give a graphical argument to show that a real polynomial equation
         of odd degree must have at least one real root.

If $z = \alpha$ is a non-real root of $p(z) = 0$ then $z - \alpha$ is a factor of $p(z)$.
If $p(z)$ has real coefficients then $z - \alpha^*$ is also a factor, and is distinct from
$z - \alpha$. Therefore $p(z)$ is divisible by

$$(z - \alpha)(z - \alpha^*) = z^2 - (\alpha + \alpha^*)z + \alpha\alpha^*$$
$$= z^2 - 2\,\mathrm{Re}(\alpha)z + |\alpha|^2, \text{ a real quadratic polynomial.}$$

So by pairing together the complex factors $(z - \alpha)$ and $(z - \alpha^*)$, we can
express any real polynomial as the product of real linear or quadratic factors,
where the quadratic factors are *irreducible*, i.e. they cannot themselves be
split into real linear factors.

Qu. 3.   If the real polynomial equation $p(z) = 0$ of degree n has r real roots,
         how many irreducible real quadratic factors does $p(z)$ have?

Example 1: Show that $2 - 4j$ is a root of $z^4 - 10z^3 + 41z^2 - 108z - 60 = 0$
             and find the other roots.

Solution: The coefficients are real, so if $2 - 4j$ is a root, then so is $2 + 4j$,
             and $(z - 2 + 4j)(z - 2 - 4j) = z^2 - 4z + 20$ is a factor. To test
             whether this is so, we divide by $z^2 - 4z + 20$.

$$
\begin{array}{r}
z^2 - 6z - 3 \\
\hline
z^2 - 4z + 20 \overline{)\, z^4 - 10z^3 + 41z^2 - 108z - 60} \\
z^4 - 4z^3 + 20z^2 \\
\hline
-6z^3 + 21z^2 - 108z \\
-6z^3 + 24z^2 - 120z \\
\hline
-3z^2 + 12z - 60 \\
-3z^2 + 12z - 60
\end{array}
$$

$z^4 - 10z^3 + 41z^2 - 108z - 60 \equiv (z^2 - 4z + 20)(z^2 - 6z - 3)$.
Therefore $2 \pm 4j$ are roots, and the other roots are found by
solving $z^2 - 6z - 3 = 0$.

By the formula $z^2 - 6z - 3 = 0 \Leftrightarrow z = \dfrac{6 \pm \sqrt{48}}{2} = 3 \pm \sqrt{12}$.

So the other three roots are $2 + 4j, 3 + \sqrt{12}, 3 - \sqrt{12}$.   □

Example 2: Prove that all the roots of $(1 + jz)^n = (1 - jz)^n$ are real.
             Find them when n is odd.

Solution: $(1 + jz)^n = (1 - jz)^n$

       $\Leftrightarrow$ $1 + jz = (1 - jz)\alpha$, where α is an nth root of unity

       $\Leftrightarrow$ $1 - jz^* = (1 + jz^*)\alpha^*$ (taking conjugates)

$$\Rightarrow (1 + jz)(1 - jz^*) = (1 - jz)(1 + jz^*)\, \alpha\alpha^*$$

$$\Leftrightarrow 1 + j(z - z^*) + zz^* = 1 - j(z - z^*) + zz^* \quad \text{(since } \alpha\alpha^* = 1)$$

$$\Leftrightarrow z - z^* = 0 \Leftrightarrow z = z^* \Leftrightarrow z \text{ is real.}$$

Now $1 + jz = (1 - jz)\alpha$

$$\Leftrightarrow jz(\alpha + 1) = \alpha - 1$$

$$\Leftrightarrow z = \frac{1}{j}\frac{\alpha - 1}{\alpha + 1} \quad \text{(since } n \text{ is odd, } \alpha \neq -1, \text{ so } \alpha + 1 \neq 0).$$

Let $\alpha = \cos\theta + j\sin\theta$, where $\theta = \dfrac{2k\pi}{n}$, $k = 0, 1, 2, \ldots, n-1$.

Then $\alpha - 1 = \cos\theta - 1 + j\sin\theta = -2\sin^2\dfrac{\theta}{2} + 2j\sin\dfrac{\theta}{2}\cos\dfrac{\theta}{2}$

$$= 2j\sin\frac{\theta}{2}\left(\cos\frac{\theta}{2} + j\sin\frac{\theta}{2}\right)$$

and $\alpha + 1 = \cos\theta + 1 + j\sin\theta = 2\cos^2\dfrac{\theta}{2} + 2j\sin\dfrac{\theta}{2}\cos\dfrac{\theta}{2}$

$$= 2\cos\frac{\theta}{2}\left(\cos\frac{\theta}{2} + j\sin\frac{\theta}{2}\right).$$

Therefore $z = \dfrac{1}{j}\dfrac{j\sin\dfrac{\theta}{2}}{\cos\dfrac{\theta}{2}} = \tan\dfrac{\theta}{2} = \tan\dfrac{k\pi}{n}$, $k = 0, 1, 2, \ldots, n-1$. $\quad\square$

Qu. 4. What is the degree of the equation when n is even? What are the roots in this case?

Exercise 13B

1. Verify that $1 + j$ is a root of $z^3 + 4z^2 - 10z + 12 = 0$, and find the other roots.

2. Verify that $2 + \sqrt{3}j$ is a root of $z^4 - z^3 + z + 35 = 0$, and find the other roots.

3. Find an equation of the smallest possible degree with real coefficients which has roots 4, $2 + j$, $5 - 3j$.

4. Prove that if p, q, r, s are real and non-zero, then the equation $pz^3 + qz^2 + rz + s = 0$ has a pure imaginary root if and only if $ps = qr$ and $pr > 0$.
   Solve the equation $3z^3 - 2z^2 + 6z - 4 = 0$.

5. Prove that $z^q - 1 = (z - 1)(z - \omega)(z - \omega^2) \ldots (z - \omega^{q-1})$, where $\omega = \cos\dfrac{2\pi}{q} + j\sin\dfrac{2\pi}{q}$.

6. Solve the equation $z^4 + z^3 + z^2 + z + 1 = 0$. Hence write $z^4 + z^3 + z^2 + z + 1$ as the product of two real quadratic factors.

7. Solve the equation $243z^5 + 81z^4 + 27z^3 + 9z^2 + 3z + 1 = 0$.

8. Express $z^6 + 1$ as the product of three real quadratic factors.

9. Solve the equation $z^7 + z^4 - z^3 = 1$.

10. Show that all the roots of $(z - 1)^n = z^n$ have real part $\frac{1}{2}$.

11. Solve $(z + j)^n + (z - j)^n = 0$.

12. Show that $z^{2n} - 2z^n \cos n\alpha + 1 = 0 \Leftrightarrow z^n = \cos n\alpha \pm j \sin n\alpha$ and hence solve the first equation completely.
    Deduce that $z^{2n} - 2z^n \cos n\alpha + 1$ is the product of the n quadratic

    factors $z^2 - 2z \cos \left(\alpha + \dfrac{2k\pi}{n} \right) + 1$, $k = 0, 1, 2, \ldots, n - 1$.

    [This was first proved by De Moivre in 1730.]

13. The real polynomial $p(x)$ is positive for all real values of x. Prove that $p(x)$ can be expressed as the sum of the squares of two real polynomials. Find two such polynomials when

    $$p(x) \equiv (x^2 + 4)(x^2 - 4x + 5)(x^2 + 6x + 13).$$

    [Hint: let the roots of $p(x) = 0$ be $\alpha, \alpha^*, \beta, \beta^*, \ldots, \lambda, \lambda^*$, and let

    $$(x - \alpha)(x - \beta) \ldots (x - \lambda) \equiv u(x) + jv(x).]$$

14. Verify that $(5 + j)^2 = 24 + 10j$, and use this to solve the equation $z^2 + (1 + 3j)z - 8 - j = 0$.
    By putting $z = x + jy$ in this equation, find the real points of intersection of the curves $x^2 - y^2 + x - 3y - 8 = 0$
    and $2xy + 3x + y - 1 = 0$.

15. Use the equation $z^3 = -8$ to find all the real points of intersection of the curves

    $$x^3 - 3xy^2 + 8 = 0$$
    $$3x^2y - y^3 = 0.$$

16. Deduce from the Fundamental Theorem of Algebra that the two curves

    $$x^4 - 6x^2y^2 + y^4 + 3x^2 - 3y^2 + x - 4 = 0$$
    $$4x^3y - 4xy^3 + 6xy + y = 0$$

    have at least one real point of intersection.

17. Assuming the Fundamental Theorem of Algebra, prove by induction that in the field of complex numbers every polynomial equation of degree n has precisely n roots.

18. (i) Let $f(z) \equiv z^n + a_1 z^{n-1} + a_2 z^{n-2} + \ldots + a_n$. Show that
       $$|f(z)| \geqslant r^n - |a_1 z^{n-1} + a_2 z^{n-2} + \ldots + a_n|, \text{ where } r = |z|.$$
    (ii) Let M be the greatest value $|a_i|$ for $i = 1, 2, \ldots, n$. Show that
       $$|a_1 z^{n-1} + a_2 z^{n-2} + \ldots + a_n| \leqslant M(r^{n-1} + r^{n-2} + \ldots + 1).$$

(iii) Deduce that $|f(z)| \geqslant r^n - \dfrac{M(r^n - 1)}{r - 1}$ and that if $r - 1 > M$ then $|f(z)| > 1$.

(iv) Hence show that all the roots of the equation $f(z) = 0$ lie in the circular region $|z| \leqslant M + 1$.

13.3 Functions of a complex variable

The crucial role of expressions of the form $\cos \theta + j \sin \theta$ in the theory of complex numbers can be seen in a surprisingly different way by using the power series for cosine and sine. If y is any real number then, as in Ex. 11D, Qu. 18,

$$\cos y = 1 - \frac{y^2}{2!} + \frac{y^4}{4!} - \frac{y^6}{6!} + \ldots$$

and

$$\sin y = y - \frac{y^3}{3!} + \frac{y^5}{5!} - \frac{y^7}{7!} + \ldots.$$

Therefore

$$\cos y + j \sin y = 1 + jy - \frac{y^2}{2!} - \frac{jy^3}{3!} + \frac{y^4}{4!} + \frac{jy^5}{5!} - \frac{y^6}{6!} - \frac{jy^7}{7!} + \ldots$$

$$= 1 + jy + \frac{(jy)^2}{2!} + \frac{(jy)^3}{3!} + \frac{(jy)^4}{4!} + \frac{(jy)^5}{5!} + \frac{(jy)^6}{6!} + \frac{(jy)^7}{7!} + \ldots$$

(assuming that two infinite series can be added in this way).
Now this final series is what we get by writing jy in place of x in the exponential series $1 + x + \dfrac{x^2}{2!} + \dfrac{x^3}{3!} + \ldots$ for e^x.

So far we have not attached any meaning to an expression such as e^z with a complex exponent, but this series, together with the basic exponential property $e^{a+b} = e^a \cdot e^b$ which we naturally wish to preserve, suggest that it is sensible to make the following *definition* of e^z.

> If $z = x + jy$ then $e^z = e^x(\cos y + j \sin y)$.

Notice that when $y = 0$, $e^z = e^x$, so that when z is real this definition of e^z agrees with the exponential function we have used until now. Also, taking $x = 0$, $e^{jy} = \cos y + j \sin y$, agreeing with the result indicated by the power series.

The law of exponents still holds for complex exponentials, since

$$e^{z_1} \cdot e^{z_2} = e^{x_1}(\cos y_1 + j \sin y_1) e^{x_2}(\cos y_2 + j \sin y_2)$$

$$= e^{x_1} e^{x_2}(\cos y_1 + j \sin y_1)(\cos y_2 + j \sin y_2)$$

$$= e^{x_1+x_2}(\cos(y_1 + y_2) + j \sin(y_1 + y_2))$$

$$= e^{x_1+x_2+j(y_1+y_2)}$$

$$= e^{z_1+z_2}.$$

In particular $e^z \cdot e^{-z} = e^0 = 1$, so that $e^{-z} = \dfrac{1}{e^z}$. The polar form of a complex number can now be written compactly as $re^{j\theta}$, and De Moivre's theorem becomes the simple statement

$$(e^{j\theta})^n = e^{jn\theta} \text{ for all rational } n.$$

An equivalent of this definition of e^z was given in 1714 by Roger Cotes (1682–1716), after whose early death Newton remarked 'If Cotes had lived we might have known something'. The exponential form was made widely known by Euler in an influential book published in 1748. The particular case $z = j\pi$ is specially interesting, since $e^{j\pi} = \cos \pi + j \sin \pi = -1$,

i.e. $$e^{j\pi} + 1 = 0.$$

This remarkable statement links the five fundamental numbers $0, 1, e, \pi, j$, the three fundamental operations of addition, multiplication, exponentiation, and the fundamental relation of equality.

If θ is real then, by definition, $e^{j\theta} = \cos \theta + j \sin \theta$

and $e^{-j\theta} = \cos(-\theta) + j \sin(-\theta) = \cos\theta - j \sin\theta$

Therefore $$\cos \theta = \frac{e^{j\theta} + e^{-j\theta}}{2} \text{ and } \sin \theta = \frac{e^{j\theta} - e^{-j\theta}}{2j}.$$

By using complex numbers in this way, we are able to find a simple link between the hitherto very different worlds of the trigonometric functions and the exponential function.

Qu. 1.   Prove that $e^{z+2n\pi j} = e^z$. This means that e^z has the imaginary period $2\pi j$.

All the standard trigonometrical results can be obtained directly from these formulae. For example

$$\sin \theta \cos \phi + \cos \theta \sin \phi = \frac{e^{j\theta} - e^{-j\theta}}{2j} \cdot \frac{e^{j\phi} + e^{-j\phi}}{2} + \frac{e^{j\theta} + e^{-j\theta}}{2} \cdot \frac{e^{j\phi} - e^{-j\phi}}{2j}$$

$$= \frac{1}{4j} (e^{j(\theta+\phi)} + e^{j(\theta-\phi)} - e^{j(-\theta+\phi)} - e^{-j(\theta+\phi)}$$
$$+ e^{j(\theta+\phi)} - e^{j(\theta-\phi)} + e^{j(-\theta+\phi)} - e^{-j(\theta+\phi)})$$

$$= \frac{e^{j(\theta+\phi)} - e^{-j(\theta+\phi)}}{2j}$$

$$= \sin(\theta + \phi).$$

Qu. 2.   Prove by similar methods
(i) $\cos^2\theta + \sin^2\theta = 1$ (ii) $\cos 2\theta = 1 - 2\sin^2\theta$.

Qu. 3.   Prove that $\tan\theta = \dfrac{e^{2j\theta} - 1}{j(e^{2j\theta} + 1)}$.

Similarity of definition makes it clear now why the corresponding results hold for the hyperbolic functions $\cosh\theta = \dfrac{e^\theta + e^{-\theta}}{2}$ and $\sinh\theta = \dfrac{e^\theta - e^{-\theta}}{2}$.

The presence of the j in the denominator of $\sin\theta$ explains why, when we change a general relation between trigonometric functions into the corresponding relation between hyperbolic functions, each product (or implied product) of two sines gives a reversal of sign; this is Osborn's rule, which was noted in §4.7.

If t is real then the function $t \mapsto e^{jt}$ is an example of a complex valued function of a real variable since it has domain $\mathbb{R}$ (the real numbers) and codomain $\mathbb{C}$ (the complex numbers). If f is any such function then $f(t)$ is a complex number, $f(t) = u(t) + jv(t)$ say. If t changes by δt then $f(t + \delta t) = u(t + \delta t) + jv(t + \delta t)$. Therefore

$$\frac{f(t+\delta t) - f(t)}{\delta t} = \frac{u(t+\delta t) - u(t)}{\delta t} + j\frac{v(t+\delta t) - v(t)}{\delta t}.$$

Taking limits as $\delta t \to 0$, $f'(t) = u'(t) + jv'(t)$, provided that $u'(t)$ and $v'(t)$ exist for the value of t in question.

Qu. 4.   Discuss how to represent a complex valued function of a real variable graphically.

Example 1: If α is a fixed complex number and $f(t) = e^{\alpha t}$ prove that $f'(t) = \alpha e^{\alpha t}$.

Solution: Let $\alpha = a + bj$. Then

$$f(t) = e^{at} \cdot e^{jbt} = e^{at}\cos bt + je^{at}\sin bt$$
$$\text{and } f'(t) = ae^{at}\cos bt - be^{at}\sin bt + aje^{at}\sin bt + bje^{at}\cos bt$$
$$= ae^{at}(\cos bt + j\sin bt) + be^{at}(j\cos bt - \sin bt)$$
$$= (a + bj)e^{at}(\cos bt + j\sin bt)$$
$$= \alpha e^{\alpha t}. \qquad \square$$

There are also functions of a complex variable. The functions $z \mapsto |z|$ and $z \mapsto \arg z$ are familiar examples of real-valued function of a complex variable, mapping from $\mathbb{C}$ to $\mathbb{R}$. The functions $z \mapsto z^2$ and $z \mapsto e^z$ are complex valued functions of a complex variable, mapping from $\mathbb{C}$ to $\mathbb{C}$.

Qu. 5.   Discuss how to represent real-valued and complex-valued functions of a complex variable graphically.

It is natural to extend the domains of the trigonometric and hyperbolic functions by making the following definitions

$$\cos z = \frac{e^{jz} + e^{-jz}}{2}, \quad \sin z = \frac{e^{jz} - e^{-jz}}{2j}$$

$$\cosh z = \frac{e^z + e^{-z}}{2}, \quad \sinh z = \frac{e^z - e^{-z}}{2}.$$

If z is real, these agree with the corresponding functions of a real variable.

Qu. 6.   Prove the following results from these definitions.
    (i) $\cos jz = \cosh z$ (ii) $\sin jz = j \sinh z$
    (iii) $\cosh jz = \cos z$ (iv) $\sinh jz = j \sin z$.

Qu. 7.   Prove from these definitions that $\cos z$ and $\sin z$ have period 2π and that $\cosh z$ and $\sinh z$ have period $2\pi j$.

The results of Qu. 6 make it possible to find the real and imaginary parts of these trigonometric or hyperbolic functions. For example

$$\begin{aligned}
\cos z &= \cos(x + jy) \\
&= \cos x \cos jy - \sin x \sin jy \\
&= \cos x \cosh y - j \sin x \sinh y.
\end{aligned}$$

Qu. 8.   Find the real and imaginary parts of $\cosh(x + jy)$.

Example 2: Find the values of z for which $\cos z = 5$.

Solution: $\cos(x + jy) = 5 \Leftrightarrow \cos x \cosh y - j \sin x \sinh y = 5$
                          $\Leftrightarrow \cos x \cosh y = 5$ and $\sin x \sinh y = 0$.

Now $\sin x \sinh y = 0 \Leftrightarrow \sin x = 0$ or $\sinh y = 0$. If $\sinh y = 0$ then $y = 0$, so $\cosh y = 1$ and $\cos x = 5$, which is impossible since x is real.
If $\sin x = 0$ then $x = n\pi$, $\cos x = (-1)^n$ and $(-1)^n \cosh y = 5$. This is impossible if n is odd, but if n is even we have $\cosh y = 5$, so $y = \pm \cosh^{-1} 5 \approx \pm 2.292$.
Thus $\cos z = 5 \Leftrightarrow z = 2k\pi \pm j \cosh^{-1} 5$. $\square$

Since they were first studied systematically by Cauchy in the early nineteenth century, functions of a complex variable have turned out to have connections with many branches of pure mathematics, and to be of enormous practical value for physicists and engineers. The ideas involved in differentiating and integrating such functions need careful consideration, which cannot be given here.

Exercise 13C

1. Express in the form $x + jy$ (i) $e^{-j\pi}$ (ii) $e^{j\pi/4}$ (iii) $e^{(5+j\pi)/6}$

2. Find the real and imaginary parts of e^{2+3j} to 3 S.F.

3. Represent on Argand diagrams the sets of points for which
   (i) $e^z = 1$ (ii) $e^z = je$.

4. Find all the complex numbers z for which $e^z = \dfrac{1 - \sqrt{3}j}{2e^3}$.

5. If $f(t) = Ae^{jmt}$, where m is real, show that $f''(t) + m^2 f(t) = 0$.

6. The position z of a point moving on a circle of radius a is given at time t by $z = ae^{j\theta}$.
   Prove that $\dot{z} = a\dot{\theta}je^{j\theta}$ (where the dot denotes the derivative with respect to t) and state what this shows about the magnitude and direction of the velocity of the point.
   Prove also that $\ddot{z} = -a\dot{\theta}^2 e^{j\theta} + a\ddot{\theta} je^{j\theta}$, and interpret this in terms of the components of the acceleration of the point.

7. The position z of a point P at time t is given by $z = re^{j\theta}$. Find $\dot{z}$ and $\ddot{z}$. Deduce the radial and transverse components of the velocity and acceleration of P. (The radial direction is the direction of $\mathbf{OP}$, the transverse direction is perpendicular to $\mathbf{OP}$.)

8. Let $C = \int e^{3x} \cos 2x \, dx$ and $S = \int e^{3x} \sin 2x \, dx$. Show that
   $$C + jS = \frac{e^{(3+2j)x}}{3 + 2j} + A \quad \text{(where } A \text{ is a constant). Hence find } C \text{ and } S.$$

9. Use the method of Qu. 8 to find $\int e^{ax} \cos bx \, dx$ and $\int e^{ax} \sin bx \, dx$.

10. If m and n are integers prove that
    $$\int_0^{2\pi} e^{jnx} e^{-jmx} \, dx = \begin{cases} 0 & \text{if } m \neq n \\ 2\pi & \text{if } m = n. \end{cases}$$
    Deduce that if $m^2 \neq n^2$,
    $$\int_0^{2\pi} \cos nx \cos mx \, dx = \int_0^{2\pi} \cos nx \sin mx \, dx = \int_0^{2\pi} \sin nx \sin mx \, dx = 0$$
    and that $\displaystyle\int_0^{2\pi} \cos^2 nx \, dx = \int_0^{2\pi} \sin^2 nx = \pi, \quad (n \neq 0)$.

11. Find the real and imaginary parts of $\sinh(x + jy)$.

12. Prove that $\cosh(\pi j - \theta) = -\cosh\theta$ and $\sinh\left(\dfrac{j\pi}{2} - \theta\right) = j\cosh\theta$.

13. If $z = \sin(x + jy)$
    (i) find $\mathrm{Re}(z)$ and $\mathrm{Im}(z)$ (ii) prove that $|z|^2 = \frac{1}{2}(\cosh 2y - \cos 2x)$
    (iii) prove that $\tan(\arg z) = \cot x \tanh y$.

14. Find z if $\cosh z = -1$.

15. Prove that if $k > 1$, $\sin z = k \Leftrightarrow z = \dfrac{4n+1}{2}\pi \pm j\cosh^{-1}k$.

    State the corresponding result if $k < -1$.

16. Prove that $w = \ln|z| + j\arg z$ is one solution of the equation $e^w = z$,
    and find all the other solutions in terms of z. (The given solution is
    called the *principal logarithm* of z, and we write
    $\ln z = \ln|z| + j\arg z$.)

17. Find the principal logarithm of
    (i) -1 (ii) j (iii) 4 (iv) -4 (v) $-3j$ (vi) $4 + 3j$.

18. By writing $\tan z$ in terms of e^{jz}, prove that
    $\tan^{-1}x = \dfrac{1}{2j}\ln\dfrac{1+jx}{1-jx}$. Prove the same result by using partial fractions
    to find $\displaystyle\int\frac{1}{1+x^2}\,dx$.

19. A complex power of a complex number is defined as follows:
    $$z^w = e^{w\ln z}\ (z \neq 0).$$
    Prove that j^j is real, and find its value to 3 S.F.

20. Find to 3 S.F.   (i) 1^j (ii) $(-1)^j$ (iii) $(1-j)^{-1/j}$ (iv) $(1+2j)^{3+4j}$.

21. (i) Prove that if $z \neq 0$ then $z^{w_1}z^{w_2} = z^{w_1+w_2}$.
    (ii) By considering $z_1 = z_2 = -1$, $w = j$ show that $(z_1 z_2)^w$ need not
    equal $z_1{}^w z_2{}^w$. Find conditions on z_1 and z_2 which ensure that
    $(z_1 z_2)^w = z_1{}^w z_2{}^w$.

13.4 The complex number field

When we started work on complex numbers in §5.2 we noted the
inconvenient fact that $\sqrt{-1}$ is not a real number, then introduced a symbol
j for $\sqrt{-1}$, and proceeded on the assumption that j can be treated as a real
number.

To illustrate that this 'bold assumption' approach can run into difficulty,
consider what happens if we try to remove another mathematical incon-
venience, the ban on dividing by zero. We know that $1/0$ is not a real number,

but let us introduce the symbol ∞ for $1/0$, and proceed on the assumption that ∞ can be treated as a real number. Then since $\infty = 1/0$ it follows that $\infty \times 0 = 1$. We can immediately deduce a contradiction from the fact that $0 \times 2 = 0 \times 3$. For

$$0 \times 2 = 0 \times 3 \Rightarrow \infty \times (0 \times 2) = \infty \times (0 \times 3)$$
$$\Rightarrow (\infty \times 0) \times 2 = (\infty \times 0) \times 3, \text{ by associativity}$$
$$\Rightarrow \qquad 1 \times 2 = 1 \times 3$$
$$\Rightarrow \qquad\qquad 2 = 3, \text{ which is absurd.}$$

So the assumption that $1/0$ can be treated as a real number has to be abandoned.

To make certain that the complex numbers will never lead to a similar disaster, we have to find a definition which ensures that they obey the 'usual laws of arithmetic'. It is notoriously difficult to say what numbers are, but rather simpler to say what they can do. In the real number system we have a set $\mathbb{R}$ of elements with two closed, associative and commutative binary operations, addition $+$ and multiplication $\times$, each with an identity element (the distinct numbers 0 and 1 respectively). Each number a has an inverse $-a$ for addition, and each number a except 0 has an inverse a^{-1} for multiplication. Multiplication is distributive over addition. Thus if a, b, c are arbitrary real numbers, not necessarily distinct, we have

Closure   (A1) $a + b \in \mathbb{R}$; (M1) $a \times b \in \mathbb{R}$
Associativity   (A2) $(a+b)+c = a+(b+c)$; (M2) $(a \times b) \times c = a \times (b \times c)$
Commutativity (A3) $a + b = b + a$; (M3) $a \times b = b \times a$
Identity   (A4) $a + 0 = a$; (M4) $a \times 1 = a$ $(1 \neq 0)$
Inverse   (A5) $a + (-a) = 0$; (M5) $a \times a^{-1} = 1$ $(a \neq 0)$
Distributivity   (MA) $a \times (b + c) = (a \times b) + (a \times c)$.

Subtraction and division are defined by $a - b = a + (-b)$ and $a \div b = a \times b^{-1}$; from now on we shall often write ab instead of $a \times b$.

> A set with two binary operations which have all the properties corresponding to A1-5, M1-5, MA is said to be a *field*.

Qu. 1. Check that the set of rational numbers with addition and multiplication is a field.

Qu. 2. Explain why the following sets with addition and multiplication fail to be fields: (i) non-negative real numbers   (ii) integers.

Qu. 3. Is the set of real 2×2 matrices with matrix addition and multiplication a field? Give reasons.

The rational numbers and the real numbers both have all the field properties, and yet some statements (for example 'there is a number x such that $x^2 = 2$') are true for real numbers but false for rational numbers. This shows that these properties are not sufficient to determine the behaviour of a set of numbers completely. However, all the standard algebraic processes of simplifying, expanding, factorizing and solving equations (but not extracting roots or using inequalities) can be derived from these field properties.

Example 1: Prove that $a \times 0 = 0$.

Solution:

$$1 + 0 = 1 \qquad \text{by (A4)}$$
$$\Rightarrow \qquad a \times (1 + 0) = a \times 1$$
$$\Rightarrow \qquad a \times 1 + a \times 0 = a \times 1 \qquad \text{by (MA)}$$
$$\Rightarrow \qquad a + a \times 0 = a \qquad \text{by (M4)}$$
$$\Rightarrow \qquad a \times 0 + a = a \qquad \text{by (A3)}$$
$$\Rightarrow (a \times 0 + a) + (-a) = a + (-a)$$
$$\Rightarrow a \times 0 + (a + (-a)) = a + (-a) \qquad \text{by (A2)}$$
$$\Rightarrow \qquad a \times 0 + 0 = 0 \qquad \text{by (A5)}$$
$$\Rightarrow \qquad a \times 0 = 0 \qquad \text{by (A4)} \qquad \square$$

Qu. 4.   Prove that (i) $ab = 0 \Rightarrow a = 0$ or $b = 0$ (ii) $(-a)b = -(ab)$, stating the field properties that you use at each step.

Although the set of all real 2×2 matrices with matrix addition and multiplication is not a field (see Qu. 3), the subset of enlargement matrices $\begin{pmatrix} a & 0 \\ 0 & a \end{pmatrix}$, $a \in \mathbb{R}$, does have all the field properties; we leave it for the reader to check this. There is an obvious one-to-one correspondence between real numbers and enlargement matrices, defined by

$$a \leftrightarrow \begin{pmatrix} a & 0 \\ 0 & a \end{pmatrix}.$$

This correspondence is such that if we add two real numbers and add the corresponding matrices, then the resulting real number corresponds to the resulting matrix. The same applies if we multiply in each set instead of adding. This is summarized in the diagram on the next page.

We say that the correspondence $a \leftrightarrow \begin{pmatrix} a & 0 \\ 0 & a \end{pmatrix}$ *preserves the structure* of addition and multiplication in the two sets. Such a correspondence is called an *isomorphism*, and these two fields are said to be *isomorphic* (meaning 'of the same form').

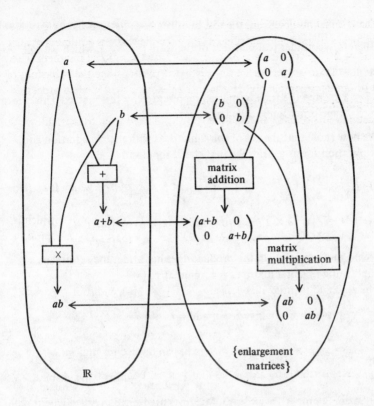

Qu. 5. Show that (i) the correspondence $a \leftrightarrow \begin{pmatrix} -a & 0 \\ 0 & -a \end{pmatrix}$ preserves

addition but not multiplication (ii) the correspondence

$a \leftrightarrow \begin{pmatrix} a+1 & 0 \\ 0 & a+1 \end{pmatrix}$ preserves neither addition nor multiplication.

Qu. 6. What, if anything, is preserved by the correspondence in which

$a \leftrightarrow \begin{pmatrix} 1/a & 0 \\ 0 & 1/a \end{pmatrix}$ for $a \neq 0$ and $0 \leftrightarrow \begin{pmatrix} 0 & 0 \\ 0 & 0 \end{pmatrix}$?

The system of enlargement matrices reproduces all the properties of the

real number system (for example, there is a matrix $\begin{pmatrix} x & 0 \\ 0 & x \end{pmatrix}$ such that

$\begin{pmatrix} x & 0 \\ 0 & x \end{pmatrix}^2 = \begin{pmatrix} 2 & 0 \\ 0 & 2 \end{pmatrix}$), though expressed in a different, less convenient notation.

But we now have the possibility of expanding the system, by making use of

the 'north-east' and 'south-west' elements in the matrices. A glance back at

§5.6 suggests that it may be fruitful to follow our enlargement by a rotation. We therefore consider real matrices of the form $\begin{pmatrix} x & -y \\ y & x \end{pmatrix}$, which we call *spiral* matrices.

Qu. 7. Describe the transformation represented by this matrix. This transformation is called a *spiral similarity*.

We now show that the set of spiral matrices with matrix addition and multiplication forms a field. We first check for closure:

(A1) $\begin{pmatrix} x_1 & -y_1 \\ y_1 & x_1 \end{pmatrix} + \begin{pmatrix} x_2 & -y_2 \\ y_2 & x_2 \end{pmatrix} = \begin{pmatrix} x_1 + x_2 & -(y_1 + y_2) \\ y_1 + y_2 & x_1 + x_2 \end{pmatrix}$, which is spiral;

(M1) $\begin{pmatrix} x_1 & -y_1 \\ y_1 & x_1 \end{pmatrix}\begin{pmatrix} x_2 & -y_2 \\ y_2 & x_2 \end{pmatrix} = \begin{pmatrix} x_1 x_2 - y_1 y_2 & -(x_1 y_2 + y_1 x_2) \\ x_1 y_2 + y_1 x_2 & x_1 x_2 - y_1 y_2 \end{pmatrix}$, which is spiral.

Interchanging subscripts in the product does not affect the result, so multiplication of spiral matrices is commutative (M3).

The zero and identity matrices are spiral (A4, M4), and so is $\begin{pmatrix} -x & y \\ -y & -x \end{pmatrix}$, which is the inverse for addition of $\begin{pmatrix} x & -y \\ y & x \end{pmatrix}$: (A5). Since $\begin{vmatrix} x & -y \\ y & x \end{vmatrix} = x^2 + y^2$, each spiral matrix has an inverse for multiplication unless $x = y = 0$; this inverse is $\dfrac{1}{x^2 + y^2}\begin{pmatrix} x & y \\ -y & x \end{pmatrix}$, which is again spiral: (M5). The remaining properties (A2, A3, M2, MA) hold since they are true for all 2×2 matrices. Therefore the set of spiral matrices forms a field.

This field has the set of enlargement matrices as a *subfield* (i.e. a subset which is itself a field), isomorphic to the real numbers. It also has the crucial property which the real numbers lack. For if $\mathbf{J}$ is the special spiral matrix $\begin{pmatrix} 0 & -1 \\ 1 & 0 \end{pmatrix}$ then $\mathbf{J}^2 = \begin{pmatrix} -1 & 0 \\ 0 & -1 \end{pmatrix} = -\mathbf{I}$.

Qu. 8. What transformation has matrix $\mathbf{J}$?

Every spiral matrix can be written as an expression of the form $x\mathbf{I} + y\mathbf{J}$. These expressions can be manipulated by 'ordinary' algebra, using $\mathbf{I}^2 = -\mathbf{J}^2 = \mathbf{I}$, $\mathbf{IJ} = \mathbf{JI} = \mathbf{J}$. If we abbreviate still further by writing 1 for $\mathbf{I}$ and j for $\mathbf{J}$, with $j^2 = -1$, we obtain the familiar complex numbers $x + yj$, which we now regard as merely shorthand for spiral matrices. Thus the statement

$$t = 2 + 11j \text{ is a solution of } t^2 - 4t + 125 = 0$$

is shorthand for $\mathbf{T} = \begin{pmatrix} 2 & -11 \\ 11 & 2 \end{pmatrix}$ is a solution of the matrix equation

$$\mathbf{T}^2 - \begin{pmatrix} 4 & 0 \\ 0 & 4 \end{pmatrix}\mathbf{T} + \begin{pmatrix} 125 & 0 \\ 0 & 125 \end{pmatrix} = \begin{pmatrix} 0 & 0 \\ 0 & 0 \end{pmatrix}.$$

By defining complex numbers as spiral matrices in this way we have shown that complex numbers form a field, and so we can be confident that our work with them will not produce contradictions.

Exercise 13D

1. Prove that if $a \neq 0$ then
$$ax + b = c \Leftrightarrow x = (c - b) \div a,$$
stating which field property is used at each step.

2. Prove that the numbers $p + q\sqrt{2}$, where p, q are rational, with ordinary addition and multiplication form a field.

3. Prove that every subfield of the real numbers contains all the rational numbers.

4. The set Z consists of all integers (positive, zero and negative). Operations of 'addition', $\oplus$, and 'multiplication', $\otimes$, are defined on Z as follows:
$$m \oplus n = m + n - 1, \quad m \otimes n = m + n - mn.$$
Show that the system $(Z, \oplus, \otimes)$ is closed, commutative, and associative with respect to each operation, and that 'multiplication' is distributive over 'addition'. Show also that each operation has an identity element in Z, and find them. Determine whether each element of Z has an inverse with respect to each operation. [MEI]

5. Write the matrix equation $\begin{pmatrix} 2 & -1 \\ 1 & 2 \end{pmatrix}\mathbf{Z} + \begin{pmatrix} 4 & 5 \\ -5 & 4 \end{pmatrix} = \begin{pmatrix} 1 & -3 \\ 3 & 1 \end{pmatrix}\mathbf{Z}$ in terms of complex numbers, and hence find the matrix $\mathbf{Z}$.

6. By solving the equivalent equation in complex numbers, find the two matrices $\mathbf{Z}$ for which
$$\mathbf{Z}^2 - \begin{pmatrix} 10 & 0 \\ 0 & 10 \end{pmatrix}\mathbf{Z} + \begin{pmatrix} 29 & 0 \\ 0 & 29 \end{pmatrix} = \begin{pmatrix} 0 & 0 \\ 0 & 0 \end{pmatrix}.$$

7. Find two matrices $\mathbf{Z}$ for which
$$\begin{pmatrix} 1 & -1 \\ 1 & 1 \end{pmatrix}\mathbf{Z}^2 + \begin{pmatrix} 4 & 0 \\ 0 & 4 \end{pmatrix}\mathbf{Z} + \begin{pmatrix} 1 & 1 \\ -1 & 1 \end{pmatrix} = \begin{pmatrix} 0 & 0 \\ 0 & 0 \end{pmatrix}.$$

8. Find $\mathbf{P}^6$ if $\mathbf{P} = \begin{pmatrix} \frac{\sqrt{3}}{2} & -\frac{1}{2} \\ \frac{1}{2} & \frac{\sqrt{3}}{2} \end{pmatrix}$.

9. What operation on complex numbers corresponds to transposing spiral matrices?

10. Establish the result $zz^* = |z|^2$ by means of matrices.

11. What property of determinants corresponds to the fact that $|z_1 z_2| = |z_1||z_2|$?

12. If z, w are complex, which of the following sets of matrices (with the usual matrix operations) form a field?

(i) $\begin{pmatrix} z & 0 \\ 0 & z \end{pmatrix}$ (ii) $\begin{pmatrix} z & -w \\ w & z \end{pmatrix}$ (iii) $\begin{pmatrix} z & -w^* \\ w & z^* \end{pmatrix}$.

13. (i) If $z = x + yj$, $w = u + vj$, prove that

$$\det \begin{pmatrix} z & -w^* \\ w & z^* \end{pmatrix} = x^2 + y^2 + u^2 + v^2.$$

(ii) By using the product of two matrices of this type prove that, if each of two integers is the sum of four perfect squares, then so is their product.

(iii) Express $(1^2 + 3^2 + 4^2 + 6^2)(2^2 + 2^2 + 5^2 + 7^2)$ as the sum of four squares in two different ways.

[In fact every integer can be expressed as the sum of at most four squares; the first proof of this was published by J. L. Lagrange in 1770.]

13.5 Groups

In mathematics we are often interested in a set of elements with one or more binary operations (i.e. rules for combining two elements). The study of the properties of such a system is called the algebra of the system. Familiar examples are real-number algebra, matrix algebra, vector algebra, and the algebra of sets. These algebras may differ; for example multiplication of real numbers is closed and commutative, the vector product of two vectors is closed but not commutative, while the scalar product is commutative but not closed. But we cannot get far in any algebra without using (or having to avoid) some basic properties: much of the work in §9.4 depends on the associativity of matrix multiplication, and we cannot progress from the definition $\mathbf{a} \cdot \mathbf{b} = |\mathbf{a}||\mathbf{b}| \cos \theta$ to the very useful component form $\mathbf{a} \cdot \mathbf{b} = a_1 b_1 + a_2 b_2 + a_3 b_3$ until we have shown that the scalar product is distributive over vector addition.

During the nineteenth century the interest of mathematicians gradually switched from particular algebras to *abstract algebra*. In abstract algebra we assume certain statements (called axioms) about unspecified objects and operations, and then see what can be deduced from them. This abstract approach has various advantages. We are free to choose the 'rules of the game' as we please; some sets of axioms turn out to be more fruitful than others. One abstract theorem can be interpreted in many different ways by taking particular objects and operations which satisfy the axioms. Thus abstract methods have great generality and often reveal connections between properties of apparently quite separate systems.

One of the simplest sets of axioms defines a *group*.

> A group $(S, *)$ is a set S with a binary operation $*$ such that
>
> (C) $*$ is *closed* in S, i.e. $a * b \in S$ for all $a, b \in S$
>
> (A) $*$ is *associative*, i.e. $(a * b) * c = a * (b * c)$ for all $a, b, c \in S$
>
> (N) there is an identity or *neutral* element $e \in S$ such that
> $$a * e = e * a = a \text{ for all } a \in S$$
>
> (I) each $a \in S$ has an *inverse* element, a^{-1}, such that
> $$a * a^{-1} = a^{-1} * a = e, \text{ where } a^{-1} \in S.$$

It is amazing that such a simple set of axioms has led over the last 150 years to a vast theory of groups, penetrating into almost every corner of mathematics.

There are many familiar examples of groups.

1. Properties A1, A2, A4, A5 of §13.4 show that the set of real numbers with addition forms a group, and properties M1, M2, M4, M5 show that the set of non-zero real numbers with multiplication also forms a group. Both these groups have infinitely many elements, and in both cases the operation is commutative. A group with a commutative operation is said to be *Abelian*, after the Norwegian Niels Abel (1802–29), one of the pioneers of group theory.

2. The numbers $\{1, -1, j, -j\}$ give this multiplication table

$\times$	1	-1	j	$-j$
1	1	-1	j	$-j$
-1	-1	1	$-j$	j
j	j	$-j$	-1	1
$-j$	$-j$	j	1	-1

We see that the system is closed (whereas with $\{1, -1, j\}$ it would not be), that 1 is the identity element, and that each element has an inverse (since the identity appears once in each row and column). Multiplication of complex numbers is associative. Therefore we have a finite group with four distinct elements; this is called a group of *order* four.

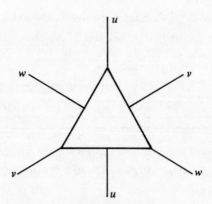

3. A hole in the shape of an equilateral triangle is cut in a piece of card. Axes u, v, w are drawn on the card. An equilateral triangle fits into the hole, and can be moved in the following ways.

P : rotation $120°$ anticlockwise about the axis through the centre perpendicular to the plane of the card

Q : rotation $120°$ clockwise about the same axis

U : half turn about axis u

V : half turn about axis v

W : half turn about axis w.

One side of the triangle is marked ∗ so that we can show when it has been turned over.

Performing P and then U we have

But 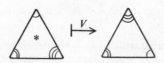 . So P followed by U is equivalent

to V, and we write $UP = V$.

Qu. 1.   Show that $PU \neq UP$. What single transformation is equivalent to PU?

These five transformations, together with the identity transformation I, give the following table. Notice that the entry for UP is written in row U and column P.

	first transformation					
followed by	I	P	Q	U	V	W
I	I	P	Q	U	V	W
P	P	Q	I	W	U	V
Q	Q	I	P	V	W	U
U	U	V	W	I	P	Q
V	V	W	U	Q	I	P
W	W	U	V	P	Q	I

(Rows labelled on the left: *second transformation*)

The identity is I, and each element has an inverse. The operation is a composition of mappings, which is always associative (see §9.1). We therefore have a non-Abelian group of order six. It is called the *symmetry group* of the equilateral triangle.

The following example shows how some simple general properties can be deduced from the group axioms. For brevity we now write ab instead of $a * b$.

Example 1: Prove that in every group
      (i) the identity is unique;   (ii) the inverse of each element is unique;
      (iii) $(a^{-1})^{-1} = a$;   (iv) $ax = ay \Rightarrow x = y$;
      (v) the equation $ax = b$ has the unique solution $x = a^{-1} b$.

Solution: (i) If e and f are both identity elements then

$$e = ef, \text{ since } f \text{ is an identity}$$
$$= f, \text{ since } e \text{ is an identity}.$$

  (ii) $ab = e \Rightarrow a^{-1}(ab) = a^{-1}e = a^{-1}$
$$\Rightarrow (a^{-1}a)b = a^{-1}$$
$$\Rightarrow \quad eb = a^{-1}$$
$$\Rightarrow \quad\quad b = a^{-1}$$

Similarly $ba = e \Rightarrow b = a^{-1}$.

  (iii) Since $a^{-1}a = aa^{-1} = e$, a is an inverse of a^{-1}. By (ii) there can be no other inverse, so that $(a^{-1})^{-1} = a$.

  (iv) $ax = ay \Rightarrow a^{-1}(ax) = a^{-1}(ay)$
$$\Rightarrow (a^{-1}a)x = (a^{-1}a)y$$
$$\Rightarrow \quad ex = ey$$
$$\Rightarrow \quad\quad x = y. \text{ This is called the left cancellation law.}$$

(v) $ax = b \Rightarrow a^{-1}(ax) = a^{-1}b$

$\Rightarrow (a^{-1}a)x = a^{-1}b$

$\Rightarrow ex = a^{-1}b$

$\Rightarrow x = a^{-1}b.$

When $x = a^{-1}b$, $ax = a(a^{-1}b) = (aa^{-1})b = eb = b.$ □

Qu. 2. State and prove the right cancellation law.

Qu. 3. Prove that $(ab)^{-1} = b^{-1}a^{-1}$.

Example 2: If $S = \{2, 4, 6, 8\}$ and
$a * b = $ the remainder when ab is divided by 10,
show that $(S, *)$ is a group isomorphic to the group
$(\{1, -1, j, -j\}, \times)$ given above.

Solution: Since $(ab)c = a(bc)$, the remainders when each side is divided by 10 are equal, so $*$ is associative. The table shows that $(S, *)$ is a group in which, perhaps surprisingly, the identity is 6.

*	2	4	6	8
2	4	8	2	6
4	8	6	4	2
6	2	4	6	8
8	6	2	8	4

To establish that $(S, *)$ is isomorphic to $(\{1, -1, j, -j\}, \times)$ we have to find a one-to-one correspondence between the sets which preserves the group structure. Clearly the identity elements must correspond, so that $1 \leftrightarrow 6$. The elements -1 and 4 are both self-inverse, so we take $-1 \leftrightarrow 4$. There is no obvious reason for pairing the remaining elements in any particular way, so we try $j \leftrightarrow 2$ and $-j \leftrightarrow 8$. The combination tables are then

×	1	−1	j	−j
1	1	−1	j	−j
−1	−1	1	−j	j
j	j	−j	−1	1
−j	−j	j	1	−1

*	6	4	2	8
6	6	4	2	8
4	4	6	8	2
2	2	8	4	6
8	8	2	6	4

Both tables have the same pattern:

	■	●	□	○
■	■	●	□	○
●	●	■	○	□
□	□	○	●	■
○	○	□	■	●

This shows that the correspondence preserves the structure, and so the groups are isomorphic.   □

Qu. 4. Show that $1 \leftrightarrow 6$, $-1 \leftrightarrow 4$, $j \leftrightarrow 8$, $-j \leftrightarrow 2$ is also an isomorphism.

A subset of S which is itself a group with the same operation $*$ is called a *subgroup* of $(S, *)$. For example, {even integers} is a subgroup of $(\mathbb{R}, +)$, and

$\{1, -1\}$ is a subgroup of $(\{1, -1, j, -j\}, \times)$. Every group $(S, *)$ has the trivial subgroups S and $\{e\}$; subgroups other than these are called *proper* subgroups.

The symmetry group of the equilateral triangle given above has four proper subgroups: $\{I, P, Q\}, \{I, U\}, \{I, V\}, \{I, W\}$. Let $H = \{I, P, Q\}$. If we follow each transformation of H by U we obtain

$$UI = U, \quad UP = V, \quad UQ = W.$$

We write this as $UH = U\{I, P, Q\} = \{U, V, W\}$.

Similarly $VH = V\{I, P, Q\} = \{V, W, U\} = \{U, V, W\}$

and $PH = P\{I, P, Q\} = \{P, P^2, PQ\} = \{P, Q, I\} = H.$

A set formed by combining each of the elements of H with one fixed element in this way is called a *left coset.* Similarly
$HV = \{I, P, Q\}V = \{IV, PV, QV\} = \{V, U, W\}$ is a right coset.

Qu. 5.   Write down the cosets (i) QH (ii) HQ (iii) WH (iv) HU.

Qu. 6.   If $K = \{I, U\}$ write down all the left cosets of K.

Qu. 7.   Show that $PK \neq KP$.

With $K = \{I, U\}$ the left cosets are in three equal pairs,

$$IK = UK = \{I, U\}, \quad PK = WK = \{P, W\}, \quad QK = VK = \{Q, V\},$$

which partition the six elements of S into three disjoint subsets, each with two elements.

This behaviour is typical of the general case. Suppose that $(S, *)$ is a finite group of order n, and that H is a subgroup of order m, $H = \{h_1, h_2, \ldots, h_m\}$ say, where the h_i are elements of S (one of them being e).

(i)   In the left coset $aH = \{ah_1, ah_2, \ldots, ah_m\}$, all the elements are distinct, for $ah_i = ah_j \Rightarrow h_i = h_j$ by the left cancellation law. Therefore aH also has m elements.

(ii)   Since $e \in H$, $ae = a \in aH$. Therefore every element of S belongs to at least one left coset.

(iii)   Two left cosets, aH and bH, are either disjoint or equal. Suppose that aH and bH are not disjoint, i.e. that $aH \cap bH \neq \emptyset$. Let $x \in aH \cap bH$. Then $x = ah_i = bh_j$ for some particular i, j, and so $a = bh_j h_i^{-1}$.

     Then $y \in aH \Rightarrow y = ah_k = bh_j h_i^{-1} h_k = bh_l,$
                               since $h_j, h_i, h_k \in H$, which is a subgroup
     $\Rightarrow y \in bH.$

     Therefore every element of aH is an element of bH. Similarly every element of bH is an element of aH. Therefore $aH = bH$.

Results (i), (ii), (iii) show that the left cosets partition the whole set S into disjoint subsets, each containing m elements. If there are p subsets, then the total number of elements is pm, so that $pm = n$, and therefore m is a factor of n. This proves that

> the order of any subgroup of a finite group
> is a factor of the order of the group.

This is known as Lagrange's Theorem, but although Lagrange did publish some related work in 1770 the term 'group' was first used in its modern sense in 1830 by the brilliant Evariste Galois, who was killed in a duel at the age of twenty in 1832.

Exercise 13E

1. A binary operation, $\circ$, is defined on the set $\mathbb{R}$ of real numbers by

$$a \circ b = \begin{cases} \text{maximum of } a \text{ and } b \text{ if } a \neq b \\ a \text{ if } a = b. \end{cases}$$

   State whether $\circ$ is (i) closed, (ii) associative, giving your reasons. Show that $\mathbb{R}$ does not contain an identity element for this operation. Suggest a set S on which this operation can be defined which does contain an identity. Does every element of your set S have an inverse in S?   [MEI]

2. Let $\mathbb{R}$ be the set of real numbers and let S be the subset of $\mathbb{R}$ consisting of all the real numbers except 1. A commutative composition '$*$' is defined by the rule $a * b = a + b - ab$.
   (i)   Is the system $(\mathbb{R}, *)$ closed and associative?
   (ii) Does $(\mathbb{R}, *)$ have an identity element?
   (iii) Does every element of $(\mathbb{R}, *)$ have an inverse?
   Answer the same three questions for the sub-system $(S, *)$.
   Solve the equation $2 * (x * (5 * x)) = 2 * x$ in the system $(S, *)$.   [MEI]

3. The operation $\circ$ is defined on the number pairs $A \equiv (a_1, a_2)$ and $B \equiv (b_1, b_2)$ so that $A \circ B = (a_1 + b_2, a_2 + b_1)$ and $A = B$ if and only if $a_1 = b_1$ and $a_2 = b_2$. Find whether the operation is associative. Find P such that $A \circ P = A$, and Q such that $Q \circ A = A$, and determine whether I, J can be found such that, for all A, $A \circ I = A$ and $J \circ A = A$. Given that $A^* = (-a_2, -a_1)$, discuss the following:

$$B \circ A = C \circ A$$
$$\Rightarrow B \circ A \circ A^* = C \circ A \circ A^*$$
$$\Rightarrow \quad B \circ (0, 0) = C \circ (0, 0)$$
$$\Rightarrow \qquad\qquad B = C.$$

                                                                          [MEI]

4. In the combination table of a group with identity e the following rectangles occur:

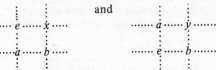

Express x and y in terms of a and b.

5.

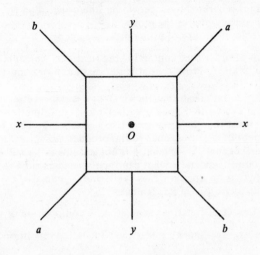

(i) Write out the combination table of the group of symmetries of a square, using the transformations I for identity
R, H, S for rotations 90° anticlockwise, 180°, 90° clockwise about O respectively.
X, Y, A, B for half turns about axes x, y, a, b respectively.

(ii) Find two subgroups of order four which are not isomorphic, giving a simple reason for the lack of isomorphism.

(iii) Let $K = \{I, A\}$. Write down all the left cosets of K, and verify that they partition the set of transformations into four distinct subsets.

6. Let $(S, *)$ be a finite group and let $a \in S$ be an element other than the identity. Let C be the set of all positive integral powers of a,
$C = \{a, a^2, a^3, \ldots\}$, where a^k means $a * a * \ldots * a$ with k factors. Show that C is a subgroup. This subgroup C is said to be *generated* by a.

7. A group which can be generated by a single element is said to be *cyclic*.

(i) Prove that $(\{1, -1, j, -j\}, \times)$ is cyclic, and name two possible generators.

(ii) Describe the cyclic group generated by $\omega = \cos\dfrac{2\pi}{7} + j \sin\dfrac{2\pi}{7}$

under the multiplication of complex numbers.

(iii) The produce mod 7 of a and b is the remainder when ab is divided by 7. Prove that $(\{1, 2, 3, 4, 5, 6\}, \times \text{ mod } 7)$ is cyclic with generator 5. Does any other element generate the whole group?

8. Given that elements e, a, b form a group under a law of combination $*$, complete the table. Hence explain why groups consisting of three elements are (i) all isomorphic to each other; (ii) cyclic.

$*$	e	a	b
e	e	a	b
a	a	.	.
b	b	.	.

[SMP]

9. (i) The *order* (or period) of an element a of a finite group is the smallest integer k such that $a^k = e$, the identity. By considering the subgroup generated by a, prove that the order of an element is a factor of the order of the group.

   (ii) Prove that all groups of prime order are cyclic.

10. G is a group of order four containing the (distinct) elements e (identity element), a, b and c. Prove that at least one of a, b or c must be its own inverse.

    Suppose, for definiteness, that the element a is self-inverse. Prove, by showing that every other possibility leads to a contradiction, that the product ab must then be equal to c. Use a similar argument to find which element of G corresponds to the product ba.

    Suppose further that the element b is not self-inverse (a still being assumed to be self-inverse). What can then be said about b^2?

    Hence show that, in this case, G is cyclic and is generated by b. Is G generated by any other of its elements?   [SMP]

11. Show that the set of functions generated by composition of the functions $a : x \mapsto \dfrac{1}{x}$ and $b : x \mapsto 1 - x$ (including repetitions) forms a group of order six. Is this isomorphic to the symmetry group of the equilateral triangle?

12. If A and B are sets, the symmetric difference $A \, \Delta \, B$ is defined by $A \, \Delta \, B = (A \cap B') \cup (A' \cap B)$, as in the Venn diagram. Prove that the set of subsets of a set $\&$ forms an Abelian group under Δ, with each element except the identity having order 2.

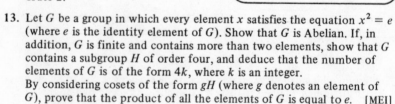

13. Let G be a group in which every element x satisfies the equation $x^2 = e$ (where e is the identity element of G). Show that G is Abelian. If, in addition, G is finite and contains more than two elements, show that G contains a subgroup H of order four, and deduce that the number of elements of G is of the form $4k$, where k is an integer.

    By considering cosets of the form gH (where g denotes an element of G), prove that the product of all the elements of G is equal to e.   [MEI]

14. Let G be a group and let H be any non-empty subset of the elements of G. Prove that H is a subgroup of G if and only if

$$xy^{-1} \in H \quad \text{for all } x, y \in H.$$

The centre C of a group G is defined to be the set of all elements of G which commute with every element of G (that is, $c \in C$ if and only if $gc = cg$ for all $g \in G$). Prove that C is a subgroup of G.

Show that the set of matrices of the form $\begin{pmatrix} 1 & p & q \\ 0 & 1 & r \\ 0 & 0 & 1 \end{pmatrix}$, where p, q, r are

rational numbers, forms a group under matrix multiplication. Find the centre of this group.   [SMP]

5. G is a group and $D(G)$ is the set of ordered pairs of elements of the form (g, ϵ) where g is an element of G and $\epsilon = \pm 1$. Show that $D(G)$ together with the binary operation

$$(g, \epsilon) . (h, \delta) = (gh^\epsilon, \epsilon\delta)$$

is a group if and only if G is Abelian. (Here (h, δ) denotes another element of $D(G)$.)

Find the condition on G which makes $D(G)$ an Abelian group and show that, if this condition is satisfied, then $D(D(G))$ is also an Abelian group.   [MEI]

6. A subgroup H of a finite group G is said to be a *normal subgroup* if, for each element $h \in H$ and each element $g \in G$, the element $g^{-1}hg \in H$. Prove that
   (i) G and $\{e\}$ (where e is the identity element of G) are normal subgroups of G;
   (ii) if G is a commutative group, then every subgroup of G is a normal subgroup;
   (iii) H is a normal subgroup if and only if every left coset of H is a right coset of H (that is, $gH = Hg$, for all $g \in G$);
   (iv) if the order of G is twice the order of H, then H is a normal subgroup of G.

   Give (and justify) an example of a finite group G and a subgroup H which is *not* a normal subgroup.   [MEI]

Miscellaneous exercise

1 to 5 on chapter 1

1. If A and B are non-empty sets of real numbers, then $A \times B$ is defined to be the subset of the real plane consisting of all the pairs (a, b), with $a \in A$, $b \in B$. Decide whether each of the following statements is true or false, and justify your answers with proofs or counter-examples as appropriate:
   (i) $A \times B = B \times A$;
   (ii) $A \times (B \cap C) = (A \times B) \cap (A \times C)$;
   (iii) if $A \times B \subseteq A \times C$, then $B \subseteq C$;
   (iv) $(A \times B) \cup (C \times D) = (A \cup C) \times (B \cup D)$.
   (You may assume throughout that $A, B, C, B \cap C$ and D are all non-empty.)
   [MEI]

2. If $\mathbf{A} = \begin{pmatrix} 1 & 1 \\ 0 & 2 \end{pmatrix}$, find $\mathbf{A}^2$, $\mathbf{A}^3$ and conjecture a form for $\mathbf{A}^n$ where n is any positive integer. Prove the truth of your conjecture by mathematical induction.
   [O & C]

3. By induction, or otherwise, prove that, for all positive integers n,
$$1^2 + 2^2 + 3^2 + \ldots + n^2 = \tfrac{1}{6}n\,(n+1)\,(2n+1).$$

   Hence determine the sum of the squares of the first n even integers i.e.
$$2^2 + 4^2 + \ldots + (2n)^2.$$
   [JMB]

4. Prove by induction that, for all positive integers n,
$$1.2 + 2.3 + 3.4 + \ldots + n(n+1) = \tfrac{1}{3}n\,(n+1)(n+2).$$
   [L]

5. Show that, for every positive integer k,
$$3^{2(k+1)} + 7 = (8 \times 3^{2k}) + (3^{2k} + 7)$$

   and hence, by mathematical induction or otherwise, prove that, for all positive integers n, $3^{2n} + 7$ has the factor 8.
   [MEI]

6 to 10 on chapter 2

6. Show that the curve $y = x\,e^{-x^2/2}$ has two stationary points and investigate their nature. Sketch the curve. (You may assume that $x\,e^{-x^2/2} \to 0$ as $x \to \infty$.)
   [L]

7. Show that the turning points of the function $y = f(x)$ are given by roots of the equation $f'(x) = 0$. Show further that, if x_0 is a root of this equation, and if $f''(x_0) < 0$, then x_0 gives a maximum value of the function. Find the values of x which give stationary values of the function
$$f(x) = \frac{a^2}{1+x} + \frac{b^2}{1-x} \quad (0 < a < b).$$

Determine the maximum and minimum values of the function and give a rough sketch of the graph of $y = f(x)$. [O & C]

8. In the interval $0 \leqslant x \leqslant a$ the gradient of the curve $y = f(x)$ is positive and increasing. The curve crosses the x-axis at a point P between the origin and the point $Q(a, 0)$. Show graphically that the point on the x-axis given by $x = a - f(a)/f'(a)$ lies between P and Q.
By drawing the curve $y = 4e^{-x}$ and the line $y = x$, estimate to one decimal place the root of the equation $x - 4e^{-x} = 0$. Use the Newton-Raphson method once to obtain a second approximation to the root, giving your answer to three decimal places. State with reasons whether this second approximation is greater or less than the root. [L]

9. Establish the formula $\log_b a = \dfrac{\log_c a}{\log_c b}$. Find the first two terms in the expansion of $\log_{10}(10 + x)$ in ascending powers of x, and state the coefficient of x^n. [JMB]

10. Find, using Simpson's rule with five ordinates, an approximate value for the area of the finite region bounded by the curve $y = \sin x$ and the x-axis between the values $x = 0$ and $x = \pi$. Find the percentage error in your result by comparing it with the exact value obtained by integration. [MEI]

11 to 15 on chapter 3

11. The points P and Q have position vectors $\mathbf{p}$ and $\mathbf{q}$ respectively relative to an origin O, which does not lie on PQ. Three points R, S, T have respective position vectors $\mathbf{r} = \frac{1}{4}\mathbf{p} + \frac{3}{4}\mathbf{q}$, $\mathbf{s} = 2\mathbf{p} - \mathbf{q}$, $\mathbf{t} = \mathbf{p} + 3\mathbf{q}$. Show in one diagram the positions of O, P, Q, R, S and T. [JMB]

12. The position vectors, relative to the origin O, of points A and B are respectively $\mathbf{a}$ and $\mathbf{b}$. State, in terms of $\mathbf{a}$ and $\mathbf{b}$, the position vector of the point T which lies on AB and is such that $AT = 2TB$. (Give reasons.) Find the position vector of the point M on OT produced such that BM and OA are parallel. If AM is produced to meet OB produced in K, determine the ratio $OB : BK$. [O & C]

13. The position vectors of the points A and B with respect to the origin O are $\mathbf{a} = 2\mathbf{i} + 6\mathbf{j} + 9\mathbf{k}$ and $\mathbf{b} = 6\mathbf{i} + 6\mathbf{j} + 7\mathbf{k}$ respectively. Find the vector equation of the internal bisector of the angle AOB. Obtain the relation satisfied by the vectors $\mathbf{p}$ and $\mathbf{q}$ if the lines $\mathbf{r} = \mathbf{p} + \lambda\mathbf{a}$, $\mathbf{r} = \mathbf{q} + \mu\mathbf{b}$ intersect. Find the position vector of their common point when $\mathbf{p} = 2\mathbf{i}$, $\mathbf{q} = \mathbf{k}$. [L]

14. A, B, C are the points $(8, 9, -2)$, $(3, -1, -7)$, $(-5, 3, 4)$ respectively. P is the point on AB such that $AP : PB = 2 : 3$. Find the equations of the line through P parallel to BC, giving your answer in the form
$$\frac{x - a}{l} = \frac{y - b}{m} = \frac{z - c}{n}.$$ [NI]

15. Kepler postulated in *Mysterium Cosmographicum* (1596) that the radii r_1, r_2, r_3 of the orbits of the planets Saturn, Jupiter and Mars were radii

of spheres related in the following way. Inside a sphere of radius r_1 is inscribed a cube. Inside this cube is inscribed a sphere of radius r_2. Inside this sphere is inscribed a regular tetrahedron and inside this tetrahedron is inscribed a sphere of radius r_3. Calculate the ratios $r_1 : r_2 : r_3$.   [OS]

16 to 20 on chapter 4

16. Evaluate, correct to two decimal places, $\int_0^1 (1-x) \sin x \, dx$.   [JMB]

17. Functions f_1, f_2, f_3 are defined by

$$f_1(x) = \frac{1}{1+x^2}, \qquad x \in \mathbb{R},$$

$$f_2(x) = \frac{1}{\sqrt{(1-x^2)}}, \quad -1 < x < 1, x \in \mathbb{R},$$

$$f_3(x) = \ln(1+x), \quad x > -1, x \in \mathbb{R}.$$

For each function state the range.
In each case find whether an inverse function exists. If the inverse function f^{-1} does exist, give an expression for $f^{-1}(x)$; if not, suggest one way in which a sub-domain could be chosen so that f, restricted to the sub-domain, would have an inverse.   [L]

18. Prove that $\dfrac{d}{dx} \sin^{-1} x = \dfrac{1}{(1-x^2)^{\frac{1}{2}}}$.

Given that the variables x and y satisfy the equation

$$\sin^{-1} 2x + \sin^{-1} y + \sin^{-1}(xy) = 0,$$

find dy/dx when $x = y = 0$.   [JMB]

19. Give a precise description (including a diagram) of the set of points (x, y) obtained as t takes all real values
   (i) when $x = 1 + 2 \sin t$, $y = 3 + 2 \cos t$;
   (ii) when $x = 1 + 2 \sin^2 t$, $y = 3 + 2 \cos^2 t$.   [NI]

20. Show that, for all real values of x, $1 + \frac{1}{2}x^2 > x$.
By using the series expansion for $\cosh x$, or otherwise, deduce that $\cosh x > x$ for all real values of x.
Sketch, on the same diagram, the line $y = x$ and the curves $y = \cosh x$, $y = \cosh^{-1} x$.
Prove that the point on the curve $y = \cosh x$ which is closest to the line $y = x$ has coordinates $(\ln(1 + \sqrt{2}), \sqrt{2})$.   [L]

21 to 30 on chapters 1, 2, 3, 4

21. Given that $y = e^x \sin x$, show that

$$\frac{dy}{dx} = 2^{\frac{1}{2}} e^x \sin\left(x + \frac{\pi}{4}\right),$$

and prove, by mathematical induction or otherwise, that

$$\frac{d^n y}{dx^n} = 2^{n/2} e^x \sin\left(x + \frac{n\pi}{4}\right),$$

where $\dfrac{d^n y}{dx^n}$ denotes the nth derivative of y with respect to x. Hence

(i) find the values of x in the interval $0 \leqslant x \leqslant 2\pi$ for which y has stationary values, distinguishing between maximum and minimum values;

(ii) find the first three non-zero terms in the Maclaurin expansion of $e^x \sin x$.   [MEI]

22. (i) State sufficient conditions for a function f of a real variable x to have a maximum or a minimum at $x = x_0$. If $f'(x_0) = f''(x_0) = 0$, how would you determine the nature of the stationary value?

(ii) Find values of x for which the function

$$(x^2 + 1)\, e^{ax}$$

has stationary values in the three cases $0 < a < 1, a = 1, a > 1$.
Find the type of the stationary values and draw a *rough* sketch of the graph of the function in each case.   [OS]

23. Sketch the graph of $f(x) = \cos \pi x - 3x$ for x taking values from 0 to $\tfrac{1}{2}$. Explain the effect of the following flow chart.

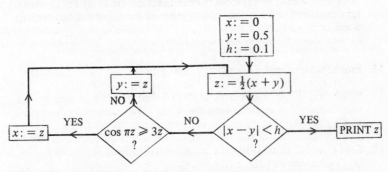

Find the first three values of z and the printout.
Describe ways in which the flow chart could be generalized. What are the advantages and disadvantages of using this method to solve an equation by computer?   [MEI]

24. A curve is defined by the parametric equations

$$x = t^2, y = t^3, t \geqslant 0.$$

Show that the equation of the tangent to the curve at the point $P(p^2, p^3)$ is $2y = 3px - p^3,$

and find the equation of the normal at P.
The tangent at P meets the x-axis at A; the normal at P meets the y-axis at B. Find the coordinates of A and B. The mid-point C of AB has coordinates (X, Y). Show that

$$3Y = p(9X + 1).$$

Hence find the Cartesian equation satisfied by the coordinates of C for all positions of P on the given curve.   [JMB]

25. (i) State the expansion of $\cosh x$ as a series of ascending powers of x and hence sum the series $\dfrac{x}{2!} + \dfrac{x^2}{4!} + \ldots + \dfrac{x^n}{(2n)!} + \ldots$.

    (ii) Evaluate $\displaystyle\int_e^{e^2} \ln \sqrt{x} \, dx$.   [L]

26. Explain, by reference to a general cubic polynomial, the method of 'nested multiplication' for the evaluation of a polynomial. Show that, for the polynomial $A_n x^n + A_{n-1} x^{n-1} + \ldots + A_0$, where n is a positive integer and $A_n, A_{n-1}, \ldots, A_0$ are real numbers, evaluation by the method of nested multiplication requires a smaller number of multiplication operations than evaluation term-by-term.
Draw a flow diagram for the evaluation of this polynomial by the method of nested multiplication.   [MEI]

27. The curve $y = e^{-ax} \cos(x + b)$ is denoted by S and the curves $y = e^{-ax}$ and $y = -e^{-ax}$ by E_1 and E_2 respectively $(x > 0, a > 0)$. Show that E_1 and E_2 each touch S at every point at which they meet S. Show that the y-coordinates of the consecutive turning points of S are in geometric progression and find their common ratio. Show also that, if a is small, the x-coordinate of each point of contact of S with E_1 and E_2 differs by approximately a from the x-coordinate of a corresponding turning point of S.   [JMB]

28. A transformation from x to y is given by the relation $y = \dfrac{ax + b}{cx + d}$, all numbers being real. It is written briefly in the form $y = T(x)$. The 'inverse operation', expressing x in terms of y, is written $x = T^{-1}(y)$. Prove that T^{-1} is of the same form as T and find the corresponding numbers a', b', c' and d'. Prove also that, if T_1, given by a_1, b_1, c_1, d_1, and T_2, given by a_2, b_2, c_2, d_2, are two such transformations, then $T_1\{T_2(x)\}$ is also of the same type. Find the conditions satisfied by a, b, c, d if $T\{T(x)\} \equiv x$, identically for all values of x.   [CS]

29. Define, in terms of e^x, the functions $\sinh x$ and $\tanh x$ and prove that, for $x > 0$, $\sinh x > \tanh x$.

    (a) Sketch, in two separate diagrams, the graphs of
$$y = \sinh x, \quad y = \tanh x.$$

    (b) Prove that, for small x, the function

$$f(x) = \frac{x}{\sinh x} - \frac{\tanh x}{x} \quad \text{is positive.} \qquad \text{[O \& C]}$$

30. Show that, if a, b, c, and d are constants, the equations
$$u_{n+2} - (a + b) u_{n+1} + ab u_n = 0, \quad n \geqslant 0,$$
determine u_n uniquely for all $n \geqslant 2$, when it is given that $u_0 = c$ and $u_1 = d$. By induction, or otherwise, prove that u_n is either of the form

$$\lambda a^n + \mu b^n \quad \text{for all } n \geqslant 0$$
$$\text{or} \quad (\lambda + n\mu) a^n \quad \text{for all } n \geqslant 2.$$

Find the functions $u_n(t)$ for $|t| \leqslant 1$ which satisfy $u_0(t) = 0$ and $u_1(t) = 1$ and

$$u_{n+2}(t) + 2u_{n+1}(t) + t u_n(t) = 0, \quad n \geqslant 0. \qquad \text{[OS]}$$

31 to 35 on chapter 5

31. (a) If $Z_1 = 1 + 3j$ and $Z_2 = 3 - j$, mark on an Argand diagram the points representing $Z_1, Z_2, Z_1 Z_2, Z_1/Z_2$.

   (b) Define the modulus and the argument of a complex number. Prove that if $|z_1 + z_2| = |z_1 - z_2|$, then the arguments of z_1 and z_2 differ by $\frac{1}{2}\pi$ or $\frac{3}{2}\pi$.

   (c) State De Moivre's Theorem for a positive integral exponent. Show that $(1 - j)^{16}$ is a real number. [MEI]

32. (i) If z is the complex number $x + yj$, define the conjugate complex number z^*, and prove that $(z^2)^* = (z^*)^2$.

   (ii) If n is a positive integer and $(3 + j)^n = a_n + b_n j$, where a_n and b_n are real, prove (by using conjugates or otherwise) that $a_n^2 + b_n^2 = 10^n$. [NI]

33. If z_1, z_2 are complex numbers, show by using the Argand diagram (or otherwise) that $|z_1 + z_2| \leqslant |z_1| + |z_2|$. If $z_1 = 6 + 8j$ and $|z_2| = 5$ find
   (i) the particular z_2 for which $|z_1 + z_2| = |z_1| + |z_2|$;
   (ii) the greatest possible value (to the nearest degree) of $\arg(z_1 + z_2)$. [NI]

34. Three points A_1, A_2, A_3 representing the complex numbers z_1, z_2, z_3 in the Argand diagram are equally spaced round the circumference of the unit circle. Show that $z_1 + z_2 + z_3 = 0$. The point P represents the complex number Z and $|Z| = 3$. Show, by using the fact that $zz^* = |z|^2$, or otherwise, that
$$PA_1^2 + PA_2^2 + PA_3^2 = 30.$$ [JMB]

35. (a) Find the values of the complex numbers z_1 and z_2 which satisfy the simultaneous equations
$$z_1 - 2jz_2 = -1, \quad 2z_1 - 5z_2 = 23j,$$
   expressing your results in the form $a + jb$, where a and b are real.

   (b) The points A, B on an Argand diagram represent $z_1 = \cos\theta + j\sin\theta$, $z_2 = 1 + \cos\theta + j\sin\theta$ respectively and O is the origin. Show the triangle OAB in a diagram and prove that it is isosceles. Find the modulus and the argument of z_2. Write down the modulus and the argument of each of the complex numbers
$$\left(1 + \cos\frac{\pi}{3} + j\sin\frac{\pi}{3}\right), \quad \left(1 + \cos\frac{\pi}{3} + j\sin\frac{\pi}{3}\right)^6.$$ [MEI]

36 to 40 on chapter 6

36. (a) Show that, if the equations
$$x^2 + bx + c = 0, \quad x^2 + px + q = 0$$
   have a common root, then
$$(c - q)^2 = (b - p)(cp - bq).$$

   (b) Using the Remainder Theorem, or otherwise, factorize the expression $(x - y)^3 + (y - z)^3 + (z - x)^3$

into linear factors.

(c)   The polynomial equation $f(x) = 0$ has roots a and b. These roots are not repeated and there are no roots between a and b. Show that, if $f(x) = (x - a)(x - b) g(x)$, then $g(a)$ and $g(b)$ have the same sign.
[MEI]

37. If the roots of the equation $x^3 - 9x^2 + 3x - 39 = 0$ are α, β, γ, show that an equation whose roots are $\alpha - 3, \beta - 3$, and $\gamma - 3$ is

$$x^3 - 24x - 84 = 0.$$

Show also that the equation $x^3 - 24x - 84 = 0$ has only one real root, and show that this root lies between 6 and 7. Sketch the two curves $y = x^3 - 9x^2 + 3x - 39$ and $y = x^3 - 24x - 84$ on the same diagram.
[L]

38. Find an equation for each of the asymptotes of the curve

$$y = x + 6 + \frac{4}{x + 1}$$

and sketch this curve, showing clearly how it approaches its asymptotes. Find the set of values of x for which

$$\frac{x^2 + 7x + 10}{x + 1} > 2x + 7.$$
[L]

39. Prove the identity $\dfrac{1}{4 - 5 \sin^2 x} = \dfrac{\sec^2 x}{4 - \tan^2 x}$.

Hence evaluate $\displaystyle\int_0^{\pi/4} \frac{dx}{4 - 5 \sin^2 x}$ to 3 decimal places, by using the substitution $u = \tan x$, or otherwise.
[NI]

40. The domain of the function defined by $f(x) = \dfrac{4(2x - 7)}{(x - 3)(x + 1)}$ is the set of all real values of x other than 3 and -1. Express $f(x)$ in partial fractions. Hence, or otherwise, show that $f'(x)$ is zero for two positive integral values of x. State these values and obtain the corresponding values of $f(x)$. Sketch the graph of $f(x)$. Determine the area of the region bounded by the curve, the x-axis, and the lines $x = 4, x = 6$. (Your result may be given in terms of logarithms.)
[JMB]

41 to 45 on chapter 7

41. Find y in terms of x when $\dfrac{dy}{dx} = \dfrac{(x - 1)\sqrt{(y^2 + 1)}}{xy}$, given that $y = -\sqrt{(e^2 - 2e)}$ when $x = e$.
[L]

42. Solve the differential equation $\dfrac{dy}{dx} - 3y = 2x\, e^{3x}$, given that $y = 1$ when $x = 0$.
[L]

43. (i)   A particle moving in a straight line has its motion described by the equation $\dfrac{d^2x}{dt^2} = -\omega^2 x$, where ω is a constant. Verify that

$x = A \cos(\omega t + \epsilon)$, where A and ϵ are constants, is a solution of this equation.

(ii) At low tide the depth of water over a harbour bar is 2 m and at high tide it is $6\frac{2}{3}$ m. If on a particular day low tide is at 8.00 a.m. and high tide is at 2.15 p.m., during what part of the day can a ship drawing 4 m enter the harbour? (Assume that the motion of the water's surface during its rise and fall is simple harmonic.)   [NI]

44. Find the value of the constant p for which $p\,e^{2x}$ is a solution of the differential equation $\dfrac{d^2y}{dx^2} - 6\dfrac{dy}{dx} + 13y = 10e^{2x}$.

Solve this differential equation completely given that $y = 0$ and $\dfrac{dy}{dx} = 0$ when $x = 0$.   [L]

45. (i) The gradient of a curve which passes through the point $(0, 1)$ is given by
$$\frac{dy}{dx} = \frac{1}{x + y + 4}.$$

Estimate the value of y at $x = 1$ by using iteratively the approximation $y(h) \approx y(0) + \frac{1}{2}h\{y'(0) + y'(h)\}$, taking $h = 1$ and giving your result to three decimal places.

(ii) Given that $\dfrac{d^2y}{dx^2} = \frac{1}{2}x - xy$, and that, when $x = 0$, $y = \frac{1}{2}$ and $\dfrac{dy}{dx} = -2$, show that the first four non-zero terms in the Maclaurin expansion of y are $\frac{1}{2} - 2x + \dfrac{x^4}{6} - \dfrac{x^7}{252}$.   [L]

46 to 55 on chapters 5, 6, 7

46. (i) If $z = x + yj$ prove that (a) $y = \dfrac{1}{2j}(z - z^*)$; (b) $|z|^2 = zz^*$, where z^* is the complex conjugate of z.

(ii) In the Argand diagram the complex number z is represented by the point (x, y). Prove that the area of triangle OAB formed by the points representing $0, 1, z$ is $\pm\dfrac{1}{4j}(z - z^*)$, where the sign is chosen to give a positive result.

(iii) The triangle $OA'B'$ has vertices representing $0, w, wz$, where w is a complex number. Give a full description in terms of w of the geometrical transformation which maps triangle OAB to triangle $OA'B'$. Deduce that the area of triangle $OA'B'$ is $\pm\dfrac{1}{4j}w\,w^*\,(z - z^*)$.

(iv) By putting $v = wz$, or otherwise, prove that the area of the triangle whose vertices represent $0, w, v$ is $\pm\dfrac{1}{4j}\begin{vmatrix} v & w \\ v^* & w^* \end{vmatrix}$.   [NI]

47. (i) By considering $f(r) - f(r + 1)$, where $f(r) = \dfrac{r+2}{r(r+1)}$, or otherwise,

show that $\displaystyle\sum_{r=1}^{n} \frac{r+4}{r(r+1)(r+2)} = \frac{3}{2} - \frac{n+3}{(n+1)(n+2)}$.

(ii) $g(x)$ is the polynomial $x^3 + 7x^2 + 11x + 5$.
$h(x)$ is the polynomial $x^4 - 22x^2 + 24x + k$, $(k > 0)$.
(a) Solve the equation $g(x) = 0$, given that it has integer roots.
(b) The equation $h(x) = 0$ has a repeated positive root and a
negative root. Explain why the fourth root must be negative
and solve the equation given that it has two distinct roots in
common with the equation $g(x) = 0$. State the value of k.   [L]

48. For any complex number $z \neq -1$ define $f(z) = \dfrac{z}{1+z}$. Prove that $f(z)$ is

real if and only if z is real. If $|z| = 1$ show that the real part of $f(z)$ is $\frac{1}{2}$.
Sketch on one diagram the loci $C = \{z : |z| = 1, z \neq -1\}$ and
$L = \{w : w = f(z)$ for some $z \in C\}$ and indicate the way in which f
maps C to L. To what set does f map the set $\{z : |z| > 1\}$? (Give reasons.)
[OS]

49. Sketch the curve given by $y^2 = x^3$. Indicate also the regions for which
the expression $y^2 - x^3$ is (a) positive, (b) negative.

A function y satisfies $\dfrac{dy}{dx} = y^2 - x^3$ such that $y = \alpha$ when $x = 0$.

Without trying to solve this differential equation, show that
(i)   the function cannot have more than one turning point;
(ii) any turning point is a maximum;
(iii) the value of x is positive at such a point;
(iv) if $\alpha = -1$, the function does have a maximum. Sketch the graph of
this function on the same paper as the curve $y^2 = x^3$.   [MEI]

50. Integrate with respect to x.

(i) $\dfrac{x^2 + 1}{x^2 - 1}$;   (ii) $\dfrac{\cos x}{1 + \sin x}$;   (iii) $\cos^{-1} x$;

(iv) $\dfrac{x + 2}{\sqrt{(1 - x^2)}}$;   (v) $e^{-x} \sin x$;   (vi) $\dfrac{1}{x^2 + 4x + 3}$;

(vii) $\dfrac{1}{x^2 + 4x + 5}$;   (viii) $\dfrac{x(x - 2)}{(1 + 2x)(1 + x^2)}$;   (ix) $\dfrac{1 + x^2}{x(1 + 3x^2)}$;

(x) $\operatorname{cosec} x$.

51. The numbers a and b are real and $a \neq b$. Prove that
(i)   the equation $(x - a)^3 + (x - b)^3 = 0$ has precisely one real root;
(ii) the equation $(x - a)^4 + (x - b)^4 = (b - a)^4$
has precisely two real roots;
(iii) the equation $(x - a)^4 + (x - b)^4 = 0$ has no real roots.
Find the values of c for which the equation $(x - a)^4 + (x - b)^4 = c$ has
(i) no real roots, (ii) precisely two real roots.   [OS]

52. Show that the equation $\dfrac{1}{x + 1} = \dfrac{x}{x - 2}$ has no real roots. Using the same

axes, sketch the graphs of (a) $y = \dfrac{1}{x + 1}$ and (b) $y = \dfrac{x}{x - 2}$

distinguishing carefully between (a) and (b). Determine the set of values of x for which $\dfrac{1}{x+1} < \dfrac{x}{x-2}$. Find the number of real roots of the

equation $\left|\dfrac{1}{x+1}\right| = \left|\dfrac{x}{x-2}\right|$.   [L]

53. Prove that any rational root of the equation $x^2 + bx + c = 0$, where b and c are integers, must be an integer. State and prove a similar result for a polynomial equation of degree n. Show that $x^3 - 6x^2 + 6x + 8 = 0$ has an integer root, and hence find all the roots.
[OS]

54. By using an approximate formula for $f'(x)$ in terms of $f(x+h)$, $f(x-h)$ and h, show how one can obtain a similar approximation for $f''(x)$. A function f satisfies the differential equation
$$f''(x) = 2f(x) - 4x,$$
subject to the conditions $f(0) = 0$, and $f(1) = 1$. Show that
$$f(x+h) - 2(h^2+1)f(x) + f(x-h) \approx -4h^2 x.$$
Use the value 0.25 for h to calculate approximate values for $f(0.25)$, $f(0.5)$ and $f(0.75)$, and for $f'(0.25)$, $f'(0.5)$ and $f'(0.75)$.   [MEI]

55. Integrate with respect to x

(i) $\dfrac{1}{\sqrt{(6x-x^2)}}$;

(ii) $\dfrac{1}{\sqrt{(x^2-6x)}}$;

(iii) $\dfrac{1}{\sqrt{(x^2+4x+3)}}$;

(iv) $\dfrac{1}{\sqrt{(x^2+4x+5)}}$;

(v) $\cos 4x \cos x$;

(vi) $\cos^7 x \sin^9 x$;

(vii) $\dfrac{x-2}{x^2+3}$;

(viii) $\dfrac{1-e^x}{1+e^x}$;

(ix) $\dfrac{\sin x}{1+\cos^2 x}$;

(x) $\dfrac{1}{1+\sin x + 3\cos x}$.

56 to 60 on chapter 8

56. The point A has coordinates $(2, 0, -1)$ and the plane π has equation $x + 2y - 2z = 8$. The line through A parallel to the line $\dfrac{x}{-2} = y = \dfrac{z+1}{2}$ meets π in the point B and the perpendicular from A to π meets π in the point C.
  (i)   Find the coordinates of B and C.
  (ii) Show that the length of AC is $\frac{4}{3}$.
  (iii) Find $\sin A\hat{B}C$.   [MEI]

57. The vector equation of a straight line in 3 dimensions is $\mathbf{r} = \mathbf{a} + t\mathbf{b}$, where $\mathbf{a}$ and $\mathbf{b}$ are fixed vectors, and $\mathbf{b} \cdot \mathbf{b} = 1$. Show that, if $\mathbf{p}$ is a fixed vector and $s = \mathbf{b} \cdot (\mathbf{p} - \mathbf{a})$, then
$$(\mathbf{a} + t\mathbf{b} - \mathbf{p})^2 = (\mathbf{a} + s\mathbf{b} - \mathbf{p})^2 + (s-t)^2$$
and interpret this result geometrically. Find the shortest distance from the point $(0, 1, 1)$ to the straight line joining the points $(1, 0, 2)$ and $(-1, 1, 3)$.   [OS]

58. The position vector of a particle is given by

$$\mathbf{r} = \hat{\mathbf{r}}\, a \cos(3\theta), \quad \theta = \omega t,$$

where a and ω are constants and t is time. Show that when r is a maximum the speed of the particle is a minimum and its acceleration a maximum and that when r is a minimum the speed is a maximum and the acceleration a minimum. Find these maximum and minimum values in each case.   [MEI]

59. Two planes π_1 and π_2 have equations $2x - 3y - 6z = 0$ and $6x + 2y - 9z = 2$ respectively; the line of intersection of π_1 and π_2 is denoted by l; the point P has coordinates $(3, -2, 1)$. Find
(a) the cosine of the acute angle between π_1 and π_2,
(b) a Cartesian equation of the plane through l containing P,
(c) vector and Cartesian equations of the straight line through P parallel to l.   [L]

60. (i) Find $\mathbf{r} \cdot \mathbf{u}$ and $\mathbf{r} \times \mathbf{u}$ where $\mathbf{r} = 4\mathbf{i} - 5\mathbf{k}$ and $\mathbf{u} = 2\mathbf{i} + \mathbf{j} - 2\mathbf{k}$.
Express the vector $\mathbf{r}$ as the sum of two vectors, one parallel to $\mathbf{u}$ and the other perpendicular to $\mathbf{u}$.
(ii) If $\mathbf{r}$ is the position vector of a variable point, give a geometrical interpretation of each of the equations.
(a) $\mathbf{r} \cdot (\mathbf{p} \times \mathbf{q}) = \mathbf{0}$, where $\mathbf{p} \times \mathbf{q} \neq \mathbf{0}$,
(b) $\mathbf{r} \times (\mathbf{u} \times \mathbf{v}) = \mathbf{0}$, where $\mathbf{u} \times \mathbf{v} \neq \mathbf{0}$.
Show that, if $\mathbf{p} = 3\mathbf{i} + \mathbf{k}$, $\mathbf{q} = 4\mathbf{i} - \mathbf{j}$, and $\mathbf{u} = \mathbf{i} + \mathbf{j}$, $\mathbf{v} = \mathbf{j} - \mathbf{k}$, any vector $\mathbf{r}$ that satisfies equation (b) will satisfy equation (a).   [L]

61 to 65 on chapter 9

61. The mapping $T : \mathbb{R}^2 \to \mathbb{R}^2$ is defined by $T\begin{pmatrix} x \\ y \end{pmatrix} = \begin{pmatrix} -2x + y \\ -5x + 2y \end{pmatrix}$. Show that T is a linear transformation.
(a) The vectors $\begin{pmatrix} 1 \\ 0 \end{pmatrix}, \begin{pmatrix} 0 \\ 1 \end{pmatrix}$ form a basis for $\mathbb{R}^2$. Write down the matrix $\mathbf{M}$ which describes the transformation T in terms of this basis.
(b) Evaluate $\mathbf{M}^2$ and $\mathbf{M}^4$ and deduce that T^4 is the identity transformation on $\mathbb{R}^2$.
(c) Show that the transformation T^2 maps the line $x = 0$ in $\mathbb{R}^2$ onto itself and maps the line $y = 0$ in $\mathbb{R}^2$ onto itself but that the transformation T does not. Show that no line in $\mathbb{R}^2$ is mapped onto itself by the transformation T.   [L]

62. Solve the equation $\mathbf{A} \begin{pmatrix} x \\ y \\ z \end{pmatrix} = \begin{pmatrix} a \\ b \\ c \end{pmatrix}$ for the column vector $\begin{pmatrix} x \\ y \\ z \end{pmatrix}$ in terms of

the constants a, b, c where $\mathbf{A}$ is the matrix

$$\begin{pmatrix} 3 & 0 & -1 \\ 2 & -1 & 2 \\ 4 & -1 & 1 \end{pmatrix}.$$

Hence, or otherwise, obtain the inverse $\mathbf{A}^{-1}$ of $\mathbf{A}$. Obtain also the inverse of the matrix $\mathbf{A}^2$.   [O & C]

63. Prove that the equations

$$x - 5y + 2z = 1$$
$$x + ky + 4z = 2$$
$$kx + y + (3k + 1)z = 5$$

have a unique solution except when $k = 1$ or $k = 3$. Show that, when $k = 1$, the equations have no solution. Show also that, when $k = 3$, the equations are not independent, and in this case express each of x and z in terms of y. Interpret these three cases geometrically. [L]

64. (a) Given $\mathbf{M} = \begin{pmatrix} 1 & 2 & -1 \\ 3 & -1 & 0 \\ 0 & 3 & 1 \end{pmatrix}$ and $\mathbf{N} = \begin{pmatrix} 1 & 5 & 1 \\ 3 & -1 & 3 \\ -9 & 3 & 7 \end{pmatrix}$, find $\mathbf{MN}$ and hence write down $\mathbf{M}^{-1}$.

(b) Show that the values of λ for which the equations

$$-(1 + \lambda)x + 3y + z = 0$$
$$x + (1 - \lambda)y + z = 0$$
$$2x - 2y + (1 - \lambda)z = 0$$

have a solution other than $x = y = z = 0$ are the roots of the cubic equation $\lambda^3 - \lambda^2 - 4\lambda + 4 = 0$. Prove that $\lambda = 1$ is a root of this equation and find the other two roots. Find the general solution of the original equations when $\lambda = 1$. [MEI]

65. Consider the simultaneous differential equations $\dot{x} = 2x + 3y$, $\dot{y} = 2x + y$ (where x and y are functions of t and dots denote differentiation).

These are written in the form $\begin{pmatrix} \dot{x} \\ \dot{y} \end{pmatrix} = \mathbf{A} \begin{pmatrix} x \\ y \end{pmatrix}$, where $\mathbf{A} = \begin{pmatrix} 2 & 3 \\ 2 & 1 \end{pmatrix}$.

Find a matrix $\mathbf{P}$ such that $\mathbf{P}^{-1}\mathbf{A}\mathbf{P} = \begin{pmatrix} 4 & 0 \\ 0 & -1 \end{pmatrix}$.

Deduce that $\dot{u} = 4u$, $\dot{v} = -v$, where $\begin{pmatrix} u \\ v \end{pmatrix} = \mathbf{P}^{-1} \begin{pmatrix} x \\ y \end{pmatrix}$.

Given that $x = 2$ and $y = 3$ when $t = 0$, find the values of u and v when $t = 0$. Express u and v as functions of t, and deduce that the solution of the equations for x and y is

$$x = 3e^{4t} - e^{-t}, \quad y = 2e^{4t} + e^{-t}. \qquad \text{[SMP]}$$

66 to 70 on chapter 10

66. If P is the point $(ap^2, 2ap)$ of the parabola $y^2 = 4ax$ and $p^2 > 8$, show that there are two points Q, R of the parabola (apart from P) at which the normals to the parabola pass through P.
Show that the tangents to the parabola at Q and R meet on the line $x = 2a$. [O]

67. The point P is the point of intersection of the tangents to the ellipse with equations $x = a \cos t, y = b \sin t$ at the points $t = \theta + \alpha$ and $t = \theta - \alpha, \alpha \neq \pi/2$.
Show that P has coordinates $(a \cos \theta / \cos \alpha, b \sin \theta / \cos \alpha)$.

Deduce that if α is constant the locus of the point P is an ellipse. Prove also that if QR is the normal at a point Q on an ellipse and R is on the major axis, then the locus of the midpoint of QR is also an ellipse.

[MEI]

68. Prove that the normal to the hyperbola $xy = c^2$ at the point $P(ct, c/t)$

    has equation $y = t^2 x + \dfrac{c}{t} - ct^3$.

    If the normal at P meets the line $y = x$ at N, and O is the origin, show that $OP = PN$ provided that $t \neq 1$.

    The tangent to the hyperbola at P meets the line $y = x$ at T. Prove that $OT \cdot ON = 4c^2$. [O & C]

69. Find the gradient of the tangent to the hyperbola $x^2/a^2 - y^2/b^2 = 1$ at the point $(a \sec\theta, b \tan\theta)$ and deduce that the equation of this tangent is $bx \sec\theta - ay \tan\theta = ab$. If this tangent passes through a focus of the ellipse $x^2/a^2 + y^2/b^2 = 1$, show that it is parallel to one of the lines $y = x, y = -x$ and that its point of contact with the hyperbola lies on a directrix of the ellipse. [L]

70. A chord AB of the conic $ax^2 + by^2 = 1$ passes through the point $P(x', y')$ and makes an angle ϕ with the x-axis. Find the quadratic equation whose roots are the lengths PA, PB.
    (i) Find the slope of the chord whose midpoint is P.
    (ii) Prove that, if CD is the chord through P perpendicular to AB, and if a, b, x', y' remain constant while ϕ varies, then

    $$\frac{1}{PA \cdot PB} + \frac{1}{PC \cdot PD} \text{ is constant.}$$

    [O & C]

71 to 80 on chapters 8, 9, 10

71. The equations
    $$\begin{aligned} x + 5y + 2z &= 6 \\ x + 3y + z &= 4 \\ 2x + 4y + z &= 6 \end{aligned}$$

    represent three planes π_1, π_2 and π_3 respectively. By reducing the system of equations to echelon form, or otherwise, show that the planes belong to a sheaf (i.e. a set of planes with a line in common).
    Explain why, for any value of λ, the equation

    $$x + 3y + z - 4 + \lambda(2x + 4y + z - 6) = 0$$

    represents a plane belonging to the sheaf. By making a suitable choice for λ, find the plane π_4 of the sheaf that is perpendicular to π_1.
    Write down a unit vector normal to π_4, and hence show that the perpendicular distance from $(0, 0, 0)$ onto π_4 is $2/\sqrt{5}$. [SMP]

72. Show that the vector equation of the sphere with centre $\mathbf{c}$ and radius a may be written as $\mathbf{r} \cdot \mathbf{r} - 2\mathbf{r} \cdot \mathbf{c} = a^2 - c^2$, where $c = |\mathbf{c}|$.
    The tangent cone from a point P, with position vector $\mathbf{p}$, to the sphere consists of all tangent lines drawn from P to the sphere.
    Show that $\mathbf{r} \cdot \mathbf{p} - \mathbf{c} \cdot (\mathbf{r} + \mathbf{p}) + c^2 - a^2 = 0$ is the equation of a plane π, and deduce that the points of contact of this cone with the sphere lie in π.

The line $\mathbf{r} = t\mathbf{u} + \mathbf{p}$ through P in the direction of the unit vector $\mathbf{u}$ meets π in a point with parameter t_1 and the sphere in points with parameters t_2, t_3. Show that $\dfrac{1}{t_2} + \dfrac{1}{t_3} = \dfrac{2}{t_1}$, and interpret the case $t_2 = t_3$.

[MEI]

73. If $A = \begin{pmatrix} 1 & 2 & 3 \\ 0 & 1 & 2 \\ -1 & 1 & -1 \end{pmatrix}$ show that $\mathbf{A}^3 = a\,\mathbf{A}^2 + b\,\mathbf{I}$ where a, b are numbers

to be determined. Deduce that $\mathbf{A}$ is non-singular. Hence, or otherwise, find the matrix $\mathbf{X}$ such that

$$\mathbf{AX} = \begin{pmatrix} 1 & 0 \\ 0 & 1 \\ 1 & 1 \end{pmatrix}.$$

[O & C]

74. Prove that, if $\mathbf{A}$ is a non-singular square matrix, then $(\mathbf{A}^T)^{-1} = (\mathbf{A}^{-1})^T$.
A square matrix $\mathbf{A}$ with real elements is said to be orthogonal if $\mathbf{A}^T = \mathbf{A}^{-1}$. Show that the transpose and the inverse of an orthogonal matrix also are orthogonal matrices.
The square matrix $\mathbf{B}$ is skew-symmetric (i.e. $\mathbf{B} = -\mathbf{B}^T$) and the matrix $\mathbf{I} + \mathbf{B}$, where $\mathbf{I}$ is the unit matrix, is non-singular. Using the identity $(\mathbf{I} - \mathbf{B})(\mathbf{I} + \mathbf{B}) \equiv (\mathbf{I} + \mathbf{B})(\mathbf{I} - \mathbf{B})$, or otherwise, show that the matrix $(\mathbf{I} - \mathbf{B})(\mathbf{I} + \mathbf{B})^{-1}$ is orthogonal.

[L]

75. Show that the points of the Argand diagram which represent the complex numbers $z = a(t + j)^2$, where t is a *real* parameter and a is a real constant, lie on a parabola, and that the tangent to the parabola at the point whose parameter is t_1 consists of the points representing the complex numbers $z = a(t_1 + j)(u + j)$ where u is a real parameter. Deduce that the point of intersection of the tangents to the parabola whose parameters are t_1 and t_2 is given by $z = a(t_1 + j)(t_2 + j)$.

Show that the points given by $z = \dfrac{A}{u + j}$, where A is a complex constant

and u is a real parameter, lie on a circle which passes through the origin. Hence show that the circumcircle of the triangle formed by three tangents to the parabola passes through the origin.

[OS]

76. Write down the components of $\mathbf{a} \times \mathbf{r}$, where $\mathbf{a} = \begin{pmatrix} c \\ 1 \\ 0 \end{pmatrix}$, $\mathbf{r} = \begin{pmatrix} x \\ y \\ z \end{pmatrix}$, c is real

and $\times$ denotes a vector product; and hence write down the matrix $\mathbf{A}$ which represents the mapping $\mathbf{r} \mapsto \mathbf{a} \times \mathbf{r}$. Find the eigenvalues of this matrix. What real solutions for $\mathbf{r}$ are there to $\mathbf{Ar} = \lambda \mathbf{r}$?

[SMP]

77. If $\mathbf{X}$ is a 2×2 matrix with real elements, the matrix $\exp(\mathbf{X})$ is defined by

$$\exp(\mathbf{X}) = \mathbf{I} + \mathbf{X} + \tfrac{1}{2}\mathbf{X}^2 + \frac{1}{3!}\mathbf{X}^3 + \dots$$

where $\mathbf{I}$ is the identity matrix. Identify the matrices $\exp(\mathbf{I})$ and $\exp(\mathbf{0})$, where $\mathbf{0}$ is the zero matrix. Show that, if $\mathbf{S}$ is a 2×2 matrix such that $\mathbf{S}^2 = \mathbf{S}$, then
(i) $\exp(\mathbf{S}) = \mathbf{I} + (e - 1)\mathbf{S}$;   (ii) $\exp(-\mathbf{S}) = \mathbf{I} + (e^{-1} - 1)\mathbf{S}$.

Prove, by induction or otherwise, that if $\mathbf{A}$ is a 2×2 matrix satisfying $\mathbf{A}^2 = 2\lambda\mathbf{A} - \lambda^2\mathbf{I}$, then $\mathbf{A}^n = n\lambda^{n-1}\mathbf{A} - (n-1)\lambda^n\mathbf{I}$. Deduce that $\exp(\mathbf{A}) = e^\lambda\mathbf{A} - (\lambda - 1)e^\lambda\mathbf{I}$.

Hence obtain a formula for $\mathbf{A}$, given that $\exp(\mathbf{A}) = \mathbf{I}$.   [MEI]

78.

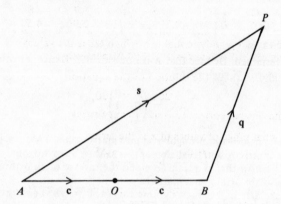

Fixed points A, B lie in a plane and O is their mid-point; the variable point P also lies in the plane. The vectors $\mathbf{q}$, $\mathbf{s}$ are defined in the diagram, and $\mathbf{c} = \mathbf{AO} = \mathbf{OB}$. Curves in the plane are defined by the equation $\alpha(\mathbf{q} \times \mathbf{s})^2 = 4c^2(\mathbf{q}.\mathbf{s} - 3c^2)$ for differing values of the constant α. By taking axes at O, or otherwise, identify the curves for $\alpha > 1$. What is the eccentricity of these curves?   [SMP]

79. (i) If $\mathbf{r}$ is a function of the time t, differentiate with respect to t the the following: (a) $\mathbf{r}.\dfrac{d\mathbf{r}}{dt}$,   (b) $\left(\dfrac{d\mathbf{r}}{dt}\right)^2$,   (c) $\mathbf{r} \times \dfrac{d\mathbf{r}}{dt}$.

(ii) If $\dfrac{d^2\mathbf{r}}{dt^2} = -\mu\mathbf{r}$, show that $\left(\dfrac{d\mathbf{r}}{dt}\right)^2 = c - \mu r^2$, where c and μ are constants.

(iii) If $\dfrac{d^2\mathbf{r}}{dt^2} + 3n\dfrac{d\mathbf{r}}{dt} + 2n^2\mathbf{r} = \mathbf{0}$, find $\mathbf{r}$ in terms of t given that, when $t = 0$, $\mathbf{r} = \mathbf{0}$ and $\dfrac{d\mathbf{r}}{dt} = \mathbf{V}$.   [L]

80. Show that, in matrix form, the coordinates (x', y') representing the image under reflection of a point (x, y) in the line $y = \frac{1}{2}x$ are given by

$$\begin{pmatrix} x' \\ y' \end{pmatrix} = \frac{1}{5}\begin{pmatrix} 3 & 4 \\ 4 & -3 \end{pmatrix}\begin{pmatrix} x \\ y \end{pmatrix}.$$

A curve C is given by the equation

$$21x^2 - 24xy + 14y^2 - 12x - 16y + 15 = 0.$$

Show that the image of C under reflection in $y = \frac{1}{2}x$ is a curve C' having the equation $(x - 2)^2 + 6y^2 = 1$.

For all points on C', show that $|y| \leqslant 1/\sqrt{6}$ and state a corresponding inequality for x. Sketch a graph of C' and use it to sketch a graph of C. Determine the coordinates of points P and Q on C such that the length of OP is greatest and the length of OQ is least, O being the origin.

[JMB]

81 to 85 on chapter 11

81. (a) Assuming that third differences of the polynomial f are constant,
      find $f(2)$ from the data

x:	4	5	6	7	8
$f(x)$:	0.35	0.88	1.71	2.90	4.51

   (b) Find the cubic polynomial in x which takes the values
      $1, -3, -1, 13$ when $x = 1, 2, 3, 4$ respectively.   [MEI]

82. Find, by any method, the solution of the equation

$$x^2 + 20 \ln x = 400,$$

giving your answer correct to 3 significant figures.   [O & C]

83. (i) For what range of values of x is the series

$$1 + \left(\frac{1 - 2x}{1 + x}\right) + \left(\frac{1 - 2x}{1 + x}\right)^2 + \dots \qquad \text{convergent?}$$

   If, for suitable x, the sum to infinity is denoted by $f(x)$, find an
   explicit formula for $f(x)$ in terms of x.
   (ii) Find the sum of the infinite series

$$\sum_{r=0}^{\infty} (r + 1)(2x)^r, \quad |x| < \tfrac{1}{2}. \qquad \text{[O \& C]}$$

84. The function $ae^x + b \ln(1 + x)$ is expressed approximately in the form
   of the periodic function $c \sin x + 5 \cos x + d$, for values of x sufficiently
   small for the expansions of the two functions in ascending powers of x
   to agree up to and including the terms in x^3. Find a, b, c, d.
   By considering the terms in x^4, estimate the difference between these
   two functions when $x = 0.1$.   [C]

85. Show that $\displaystyle\sum_{r=n+1}^{N} \frac{1}{r^2 - \frac{1}{4}} = \frac{1}{n + \frac{1}{2}} - \frac{1}{N + \frac{1}{2}}$, where n and N are positive

integers. Deduce that $\displaystyle\sum_{r=n+1}^{N} \frac{1}{r^2} < \frac{1}{n + \frac{1}{2}}$.

Show that $\dfrac{1}{r^2} > \dfrac{1}{r - 0.48} - \dfrac{1}{r + 0.52}$ if $r \geqslant 7$.

Deduce that $\displaystyle\sum_{r=7}^{\infty} \frac{1}{r^2}$ lies between $\dfrac{1}{6.5}$ and $\dfrac{1}{6.52}$.

Deduce a value, to three decimal places, for $\displaystyle\sum_{r=1}^{\infty} \frac{1}{r^2}$.   [MEI]

86 to 90 on chapter 12

86. Show that if a, b, p, q are all positive numbers then $\dfrac{pa + qb}{p + q}$ lies
   between a and b.
   Hence, or otherwise, prove that for $0 < x < \frac{1}{2}\pi$, $\sin^3 x + \cos^3 x$ lies
   between $\sin x$ and $\cos x$.

Show that $\int_0^{\frac{1}{2}\pi} (\sin^3 x + \cos^3 x)\, dx = 2 \int_0^{\frac{1}{4}\pi} (\sin^3 x + \cos^3 x)\, dx$.

Given that $\int_0^{\frac{1}{2}\pi} \sin^3 x\, dx = \int_0^{\frac{1}{2}\pi} \cos^3 x\, dx = \frac{2}{3}$, verify that

$\int_0^k (\sin^3 x + \cos^3 x)\, dx$ lies between $\int_0^k \sin x\, dx$ and $\int_0^k \cos x\, dx$
if k is $\frac{1}{4}\pi$ but not if $k = \frac{1}{2}\pi$.
Explain, with the aid of a diagram, why this is so.   [MEI]

87. Sketch the '$2m$-rose' defined in polar coordinates by $r = |\sin m\theta|$, for
$m = 1, 2, 3$. Show that for all integers $m > 0$ the total area of the petals
is independent of m, and evaluate this area.   [CS]

88. Defining $\ln x^n$ as $\int_1^{x^n} \frac{1}{t}\, dt$, for $x > 0$, use the substitution $t = u^n$
to prove that $\ln x^n = n \ln x$.
By considering the area under the graph of $y = 1/t$ from $t = 1$ to
$t = 1 + x$, or otherwise, show that, for $x > 0$,

$$\frac{x}{1+x} < \ln(1+x) < x$$

and deduce that, as x decreases to zero, $\dfrac{1}{x} \ln(1+x)$ tends to 1.

A periodic function is defined by

$$\begin{cases} f(x) = \dfrac{1}{x} \ln(1+x) & \text{for } 0 < x \leqslant 1 \\[2mm] f(x+1) = f(x) & \text{for all } x. \end{cases}$$

Sketch the graph of $y = f(x)$ for values of x from -2 to 2.   [L]

89. Prove that

$$\int e^{ax} \sin^n x\, dx =$$
$$\frac{e^{ax} \sin^{n-1} x\, (a \sin x - n \cos x)}{a^2 + n^2} + \frac{n(n-1)}{a^2 + n^2} \int e^{ax} \sin^{n-2} x\, dx,$$

and hence evaluate $\int_0^\pi e^{3x} \sin^4 x\, dx$.

By use of the substitution $x = \frac{1}{2}\pi + y$, or otherwise, find a reduction
formula for $\int e^{ay} \cos^n y\, dy$ of similar form to that given for
$\int e^{ax} \sin^n x\, dx$.   [O & C]

90. (i) Find the value of a for which the integral $\displaystyle\int_1^\infty \frac{a}{x(2x+a)}\, dx$
         converges to the value 1.

    (ii) Write down the derivative, with respect to x, of $x^2 \sin \dfrac{1}{x}$.

         Show that the integral $\displaystyle\int_0^{2/\pi} \left(2x \sin \frac{1}{x} - \cos \frac{1}{x} \right) dx$ converges to the
         value $4/\pi^2$.   [L]

91 to 95 on chapter 13

91. The complex number z satisfies the equation
$$(z - 1)^n = (z + 1)^n \quad (n \neq 0) \tag{1}$$
where n is an integer. Show, geometrically, that the real part of z must be zero. Solve the equation (1) for z.
If $w^n = (w + j)^n$, $(n \neq 0)$ for an integer n, what can you say about the imaginary part of w?   [O & C]

92. Show that, if $x^6 - 2x^3 \cos 3\alpha + 1 = 0$, then $x^3 = \cos 3\alpha \pm j \sin 3\alpha$
and that x has one of the values $\cos\left(\alpha + \dfrac{2k\pi}{3}\right) \pm j \sin\left(\alpha + \dfrac{2k\pi}{3}\right)$,
$k = 0, 1, 2$. Show that
$$x^6 - 2x^3 \cos 3\alpha + 1 =$$
$$\left(x^2 - 2x \cos\alpha + 1\right)\left(x^2 - 2x \cos\left(\alpha + \frac{2\pi}{3}\right) + 1\right)\left(x^2 - 2x \cos\left(\alpha + \frac{4\pi}{3}\right) + 1\right).$$

ABC is an equilateral triangle inscribed in a circle, centre O, radius 1, and P is a point inside the circle, distant x from O and with angle AOP equal to α. Write down the values of the angles BOP and COP and show that
$$PA.PB.PC = (x^6 - 2x^3 \cos 3\alpha + 1)^{\frac{1}{2}}.$$
Show that, if P is the midpoint of AC, then $PA.PB.PC = x^3 + 1$.   [MEI]

93. Simplify the product $(x - e^{\theta j})(x - e^{-\theta j})$.
Write down the eight values of θ, $-\pi < \theta \leqslant \pi$, for which $x = e^{\theta j}$ satisfies the equation $x^8 = 1$. Hence write $x^8 - 1$ as the product of two first-degree and three second-degree factors, all having real coefficients.
                                                                          [SMP]

94. (i) You may assume in this question that multiplication modulo 14 is associative.
Show that the set $\{2, 4, 6, 8, 10, 12\}$ forms a group under the operation of multiplication modulo 14. Name the identity element and the inverse of each of the other elements.
Show that the set $\{1, 3, 5, 7, 9, 11, 13\}$ does *not* form a group under multiplication modulo 14. (Only one reason need be given, but the fact that one of the group postulates is not satisfied must be clearly demonstrated.)

(ii) The tables of two groups G_1, G_2 of order 4 are given.

G_1	a	b	c	d		G_2	p	q	r	s
a	a	b	c	d		p	p	q	r	s
b	b	a	d	c		q	q	r	s	p
c	c	d	a	b		r	r	s	p	q
d	d	c	b	a		s	s	p	q	r

Name a plane figure whose group of symmetries is isomorphic to G_1 and another plane figure whose group of symmetries contains a subgroup isomorphic to G_2.   [L]

95. Show that the set of complex-valued 2 × 2 matrices of the form
$\begin{pmatrix} z & w \\ -w^* & z^* \end{pmatrix}$ satisfying $|z|^2 + |w|^2 = 1$ forms a group G under
matrix multiplication. Determine the subsets G_2 consisting of all
elements of G whose square is the identity matrix, and G_4 consisting of
all elements of G whose fourth power is the identity matrix. Do they
form subgroups of G?   [CS]

96 to 105 on chapters 11, 12, 13

96. Two sequences $(x_0, x_1, x_2, \ldots)$ and $(y_0, y_1, y_2, \ldots)$ of positive integers
are defined inductively by taking $x_0 = 2, y_0 = 1$, and equating rational
and irrational parts in the equations
$$x_n + y_n \sqrt{3} = (x_{n-1} + y_{n-1}\sqrt{3})^2 \quad (n = 1, 2, 3, \ldots).$$
Prove that $x_n^2 - 3y_n^2 = 1$ $(n = 1, 2, 3, \ldots)$, and that when $n \to \infty$, the
sequences x_n/y_n and $3y_n/x_n$ tend to the limit $\sqrt{3}$ from above and
below respectively.

By carrying this process far enough, obtain two rational numbers
enclosing $\sqrt{3}$ and differing from one another by less than 5×10^{-9}.
                                                                      [CS]

97. Denote by S the set of all 2 × 2 matrices and define an operation $\circ$ on S
by $A \circ B = AB - BA$ $A, B \in S$.
In each of the following cases, state whether the proposition is true or
false, giving a proof or counter-example in each case, as appropriate.
You may assume but must state any properties of matrix multiplication
that you require.
   (i) $\circ$ is commutative,   (ii) $\circ$ is distributive over matrix addition,
   (iii) $\circ$ is associative,
   (iv) $(A \circ B) \circ C + (B \circ C) \circ A + (C \circ A) \circ B = 0$, for all $A, B, C \in S$.   [MEI]

98. The function $y = \phi(x)$ is a continuous and strictly increasing function
of x, for $x \geqslant 0$, with $\phi(0) = 0$, and $x = \phi^{-1}(y)$ is the inverse function.
Sketch the graph of ϕ and on it indicate the areas represented by the
integrals
$$\int_0^a \phi(x)\,dx, \quad \int_0^b \phi^{-1}(y)\,dy \quad (a > 0, b > 0),$$
and show that $ab \leqslant \int_0^a \phi(x)\,dx + \int_0^b \phi^{-1}(y)\,dy$.

When does equality hold here? Deduce
   (i) if $p > 1$ and $\dfrac{1}{p} + \dfrac{1}{q} = 1$ then $ab \leqslant \dfrac{a^p}{p} + \dfrac{b^q}{q}$,
   (ii) if $a > 1, b > 0$ then $ab \leqslant a \ln a - a + e^b$.   [OS]

99. If $[r, \theta]$ denotes, in polar form, the complex number with modulus r
and argument θ, prove by induction that, for positive integers n,
$[r, \theta]^n = [r^n, n\theta]$, where $n\theta$ is reduced mod 2π. (You may assume that
$[r, \theta] \times [s, \phi] = [rs, \theta + \phi]$, where the argument $\theta + \phi$ is reduced
mod 2π.)
By writing each side of the equation $z^5 = 1$ in polar form, and using the
result above, find 5 distinct roots of the equation. Deduce the roots of
the equation $z^5 = 32$.

Show that, under multiplication, the set S of distinct roots of the equation $z^5 = 1$ forms an Abelian group. Explain why the set of distinct roots of the equation $z^5 = 32$ is not a group under multiplication. Name a group isomorphic to S. [SMP]

100. The integral $I_{n,m}$ is defined for all non-negative integers m, n by

$$I_{n,m} = \int_0^1 x^n (\ln x)^m \, dx.$$

On the assumption that $x^\alpha \ln x$ can be evaluated as zero when $x = 0$ for any positive constant α, show that

$$I_{n,m} = -\frac{m}{n+1} I_{n,m-1} \text{ for } m > 0.$$

Hence show that, if m is a positive integer,

$$I_{n,m} = \frac{(-1)^m m!}{(n+1)^{m+1}}.$$

Hence, or otherwise, show that if p and q are positive integers

$$\int_0^\infty e^{-pt} t^q \, dt = \frac{q!}{p^{q+1}}. \qquad \text{[OS]}$$

101. (i) The polynomial $f(x)$ of degree 4 is such that $f(1) = 5, f(2) = 0$, $f(3) = 0, f(4) = 8, f(5) = 15$. Form a difference table and use it to find
   (a) the coefficient of x^4 in $f(x)$, (b) the value of $f(6)$.
   (ii) One of the values of y given in the table below is wrong. Given that $y(x)$ is a cubic polynomial in x, find the wrong value and correct it.

x	0	1	2	3	4	5	6	7	
y	33	39	35	27	12	23	39	75	[L]

102. In a river with a flat bottom and vertical sides the speed v of the water at any height h above the bottom is given by $v = ah^2$, a is constant. Under normal conditions the depth of the water is h_0 and the total rate of flow downstream is R_0 (volume per unit time).
Heavy rain occasionally produces flash flooding and during one such period of time T an additional rate of flow

$$R_F = R_0 (1 - \cos(2\pi t/T)), \quad 0 \leqslant t \leqslant T,$$

has to be accommodated.
For the period of the flash flooding find expressions for the depth of the river as a function of time, the maximum and mean depths and the maximum speed of flow at the surface. [MEI]

103. (i) Show that, for $x > 1$, $\ln x < x$. Determine whether the series

$$\sum_{n=1}^\infty \frac{\ln n}{2n^3 - 1} \text{ converges or diverges.}$$

   (ii) Find the set of values of p for which the series $\sum_{n=1}^\infty n^p x^n$

   is convergent (a) when $0 < x < 1$, (b) when $x = 1$, (c) when $x > 1$.

(iii) Show that the series $\sum\limits_{n=1}^{\infty} n \exp(-n^2)$, where $\exp x = e^x$, is convergent. [L]

104. Let $f(x) = 2 \cos x^2 - \dfrac{1}{x^2} \sin x^2$ for $x \geqslant 1$. Find $\displaystyle\int_1^t f(x)\,dx$ for $t \geqslant 1$, and show that it tends to a limit as $t \to \infty$ but that $f(x)$ does not tend to zero as $x \to \infty$.

Give an example of a function $h(x) \geqslant 0$, defined for $x \geqslant 1$ and such that $\displaystyle\int_1^t h(x)\,dx \to \infty$ as $t \to \infty$ but $h(x) \to 0$ as $x \to \infty$.

Give an example of a function $l(x) \geqslant 0$, defined for $x \geqslant 1$, and such that $dl/dx > 0$ for all $x \geqslant 1$ but $l(x) \leqslant 1$ for all $x \geqslant 1$. [CS]

105. By consideration of $\displaystyle\int_a^b (\lambda f(x) + g(x))^2\,dx$, or otherwise, where λ is a constant, prove that $\displaystyle\int_a^b (f(x))^2\,dx \int_a^b (g(x))^2\,dx \geqslant \left(\int_a^b f(x)g(x)\,dx \right)^2$.

By use of this inequality, prove that

$$I = \int_0^{\pi/4} \sqrt{(x \sin x)}\,dx \leqslant \tfrac{1}{8}\pi (2 - \sqrt{2})^{\frac{1}{2}}$$

Prove also, by considering the sign of $x - \sin x$ for $0 \leqslant x \leqslant \tfrac{1}{4}\pi$, or otherwise, that $I > \tfrac{1}{2}(2 - \sqrt{2})$. [O & C]

106 to 115 Problems

106. Three players A, B, C play the following game. On each of three cards an integer is written. These three numbers p, q, r satisfy $0 < p < q < r$. These three cards are shuffled and dealt so that each player has a card. Each then receives the numbers of counters indicated by the card he holds. Then the cards are shuffled again; the counters remain with the players.

This process (shuffling, dealing, giving out counters) takes place for at least two rounds. After the last round A has 20 counters in all, B has 10 and C has 9. At the last round B received r counters. Who received q counters on the first round? [IMO]

107. Given any set of fourteen positive integers each less than 1000, prove that there are two disjoint subsets of the set with the sums of their elements equal.

108. In a finite sequence of real numbers the sum of any seven successive terms is negative, and the sum of any eleven successive terms is positive. Determine the maximum number of terms in the sequence. [IMO]

109. Inside a square of side 7 units there are 101 points. Prove that it is possible to draw a circle of unit radius which contains at least 5 of these points.

110. Six equal rods are joined together to form a regular tetrahedron. Two scorpions are placed at the midpoints of two opposite edges of

this framework, and a beetle is placed at some point of this framework. The scorpions can move along the rods with maximum speed s, and the beetle with maximum speed b. Show that if $b < 2s$ the scorpions can always catch the beetle; and explain in detail how they should manoeuvre to do so.   [CS]

111. When 4444^{4444} is written in decimal notation the sum of its digits is A. Let B be the sum of the digits of A. Find the sum of the digits of B. (A and B are written in decimal notation.)   [IMO]

112. A lattice point in a plane is a point with integer coordinates. Each lattice point (m, n) has four neighbours $(m \pm 1, n \pm 1)$. Let C be a circle of radius $\geqslant 2$ which does not pass through any lattice point. An interior boundary point is a lattice point inside C which has a neighbour outside C; an exterior boundary point is a lattice point outside C which has a neighbour inside C. Prove that there are four more exterior boundary points than interior boundary points.

113. Consider decompositions of an 8×8 chessboard into p non-overlapping rectangles subject to the following conditions:
   (i) each rectangle is to have as many white squares as black;
   (ii) if a_i is the number of white squares in the ith rectangle then
   $$a_1 < a_2 < \ldots < a_p.$$
Find the maximum value of p for which such a decomposition is possible. For this value of p determine all possible sequences $a_1, a_2, \ldots, a_p$.   [IMO]

114. A triangle is called chromatic if all its sides are the same colour. Each pair of n distinct points $P_1, \ldots, P_n$ in space is connected with either a red line or a black line. Prove that if $n = 6$ there must be at least 2 chromatic triangles of the form $P_i P_j P_k$. Deduce that if $n = 7$ there are at least 3 such chromatic triangles.   [CS]

115. Let f be a function defined on the set of all positive integers and taking positive integer values. Prove that if $f(n + 1) > f(f(n))$ for each positive integer n, then $f(n) = n$ for each n.   [IMO]

Answers

§1.1

Qu. 1. (i) $\Rightarrow$; (ii) $\Leftarrow$; (iii) $\Leftarrow$; (iv) $\Rightarrow$.

Qu. 2. (i) false; (ii) true; (iii) false. **Qu. 3.** 2 and 3 are not of this form.

Qu. 5. (ii) $x^2 \leqslant 4 \Rightarrow x \geqslant -2$; (iii) $\tan \theta \neq -1 \Rightarrow \theta \neq \dfrac{7\pi}{4}$;
(iv) $PA \neq PB \Rightarrow P$ is not the midpoint of AB.

Exercise 1A page 5

2. No. **3.** $n^2 - n + 41$ is composite for at least one integer n.

4. (i) if; (ii) only if; (iii) if; (iv) if and only if.

5. (i) necessary; (ii) necessary and sufficient; (iii) sufficient; (iv) neither.

6. $a^2 > b^2 + c^2$. **7.** 5. **8.** 3. **9.** No solution.

10. $b \neq 0, d \neq 0, b \neq d, b \neq -2d$. **11.** (i) true; (ii) false; (iii) true.

15. 'only if', but not 'if'. **16.** $a = 2$, n is prime.

17. (i)

$\times$	A	B
A	A	B
B	B	A

18. (ii) No (in fact there are infinitely many).

§1.2

Qu. 1. $\frac{7}{12}, \frac{4}{7}$. **Qu. 5.** $\frac{1}{2} n(n+1)$.

Exercise 1B page 12

1. n^2. **2.** $\dfrac{n}{n+1}$. **3.** $\dfrac{n}{2n+1}$. **4.** $(n+1)! - 1$. **5.** $\dfrac{1+2n}{1-2n}$.

6. $\dfrac{1}{n}$. **7.** $2^n - 1$. **10.** $(1 - nx^{n-1} + (n-1)x^n)/(1-x)^2$.

11. $(-1)^{n-1} n(n+1)/2$. **16.** $n_0 = 7$. **17.** $\{n : n \neq 3\}$.

19. $11^n + 3.4^{n-1}$ is divisible by 7. **25.** (i) $n+1$; (iii) $t_n = \frac{1}{6}n(n^2+5) + $

26. (iii) $p = 0$, never changes; $p = 1$, alternates; (iv) $\frac{1}{2}$, keep trying.

28. $2^{64} - 1 \approx 1.84 \times 10^{19}$.

§2.1

Qu. 5. No.

Exercise 2A page 20

1. $6x^2 + 14x - 5, 12x + 14, 12$.

2. $\frac{1}{2}x^{-\frac{1}{2}} - 2x^{-3}, -\frac{1}{4}x^{-\frac{3}{2}} + 6x^{-4}, \frac{3}{8}x^{-\frac{5}{2}} - 24x^{-5}$.

3. $\cos x - \sin 2x + \cos 3x, -\sin x - 2\cos 2x - 3\sin 3x,$
   $-\cos x + 4\sin 2x - 9\cos 3x$.

4. $\sec^2 x, 2\sec^2 x \tan x, 2\sec^2 x (\sec^2 x + 2\tan^2 x)$.

5. $2x/(1 + x^2), 2(1 - x^2)/(1 + x^2)^2, 4x(x^2 - 3)/(1 + x^2)^3$.

8. (a) $x < -2, x > 1$; (b) $x > -\frac{1}{2}$.

9. $(-1, 1)$ maximum, $(0, 0)$ minimum; $(-1, -1)$ minimum, $(0, 0)$ inflexion.

10. $(-1, 0)$ minimum, $(0, 7)$ maximum, $(3, -128)$ minimum.

11. (i) -1; (ii) $\frac{1}{2}$; (iii) $-\frac{5}{2}, 0$. 12. $(\pm 2, \frac{1}{16})$. 14. (ii) 0 or 2.

15. $C'(x) > 0, C''(x) < 0$. 16. Point of inflexion. 18. $A = \dfrac{1}{\sqrt{2}}$.

19. (ii) $y = \sqrt{3}\, x$. 20. $\frac{1}{3}\pi a^3 \sin^2 \theta (1 + \cos\theta); \frac{8}{27}$.

21. $|k|$ for $k \leqslant \frac{1}{2}, \sqrt{(k - \frac{1}{4})}$ for $k > \frac{1}{2}$. 22. $(2\pi + \alpha)/3$.

23. (i) $15\sqrt{3} \approx 26.0$ cm; (ii) $20\operatorname{cosec}\theta$. 24. $(a^{\frac{2}{3}} + b^{\frac{2}{3}})^{\frac{3}{2}}$.

§2.2

Qu. 5. Because the graph is discontinuous. Qu. 6. 3, 4.

Qu. 7. 1.42, 1.43. Qu. 8. 3.5, 3.75.

Exercise 2B page 27

1. Between -1 and 0, 1 and 2, 4 and 5. 2. Between 3 and 4.

3. Between -3 and -2, -1 and 0, 0 and 1, 1 and 2.

4. (i) 0, between $\dfrac{\pi}{2}$ and π, between $-\pi$ and $-\dfrac{\pi}{2}$; (ii) between 0 and $\dfrac{\pi}{2}$.

5. One root between $(n - \frac{1}{2})\pi$ and $n\pi$ for each non-positive integer n.

6. 4.0. 7. 0.80, 0.81. 8. Fails if $0 < x < \frac{1}{4}$.

Exercise 2C page 31

1. 57.297. 2. (i) 2.9957; 71.565°.

3. (i) 0.2516, 0.2372; (ii) too small, too large.

5. Between 1.23 and 1.5.

6. -0.83, greater; 1.13, greater; 4.69, less. 7. 1.78.

8. $\ln t = \frac{1}{2}(t - 1)$; 3.5, 3.6; 3.513.

Exercise 2D page 35

1. $\frac{21}{11} \approx 1.91$.   2. -1.54.   3. 25.41.   4. 3.21.

6. (i) (a) $\frac{11}{12}$,   (b) $\frac{7}{6}$;   (a);   (a);   (ii) (a) $\frac{5}{12}$,   (b) $\frac{1}{6}$;   (a);   (b).

7. Because $f'(0) = 0$; $a_2 = -0.8$; between -0.8 and -0.69.

8.

1.1	1.2	1.3	1.4
-0.659	-0.232	0.287	0.904

; 1.24; 1.247.

9. 1.639.   10. 2.03.   11. 1.56.   12. 1.030.

13. $0.89, 0.895$.   14. 1.866.   15. $k = -1, x_3 = \frac{344}{273}$.

§2.5

Qu. 1. $1 + x + \dfrac{x^2}{2} + \dfrac{x^3}{6} + \dfrac{x^4}{24}$; $0.333, 0.375, 1, 2.708, 7$.

Qu. 2. $1 - x + \dfrac{x^2}{2} - \dfrac{x^3}{6}$; $1 - \dfrac{x^4}{12} - \dfrac{x^6}{36}$.   **Qu. 4.** Odd powers only.

Qu. 5. $p_1(x) \equiv 1, p_2(x) \equiv p_3(x) \equiv 1 - \dfrac{x^2}{2}, p_4(x) \equiv p_5(x) \equiv 1 - \dfrac{x^2}{2} + \dfrac{x^4}{24}$,

$\qquad p_6(x) \equiv 1 - \dfrac{x^2}{2} + \dfrac{x^4}{24} - \dfrac{x^6}{720}$.

Exercise 2E page 43

2. (i) $1 - x + \dfrac{x^2}{2} - \dfrac{x^3}{6} + \ldots + (-1)^n \dfrac{x^n}{n!}$;

   (ii) $1 + 2x + 2x^2 + \dfrac{4x^3}{3} + \ldots + \dfrac{2^n x^n}{n!}$;

   (iii) $2 + x^2 + \dfrac{x^4}{12} + \dfrac{x^6}{360} + \ldots + (1 + (-1)^n) \dfrac{x^n}{n!}$;

   (iv) $4x - 4x^2 + \dfrac{14x^3}{3} - \dfrac{10x^4}{3} + \ldots + (1 - (-3)^n) \dfrac{x^n}{n!}$.

3. $0.606\,53$.

4. (i) $1 - \dfrac{x^2}{2} + \dfrac{x^4}{8} - \dfrac{x^6}{48}$;   (ii) $1 - \dfrac{x^2}{2} + \dfrac{x^4}{8} - \dfrac{x^6}{48} + \dfrac{x^8}{384}$; 0.855 ± 0.0002.

5. (i) e^{x+y}.   8. 0.5403.   9. 0.3090.

10. (i) $3x - \dfrac{9x^3}{2} + \dfrac{81x^5}{40} - \dfrac{243x^7}{560}$;   (ii) $1 - \dfrac{x^4}{2} + \dfrac{x^8}{24} - \dfrac{x^{12}}{720}$;

   (iii) $1 - x^2 + \dfrac{x^4}{3} - \dfrac{2x^6}{45}$.

12. 0.24%.   **15.** $0, -3, 15$.   **16.** $1 + x + \dfrac{x^2}{2} - \dfrac{x^4}{8}$.

17. Odd function, $x + \dfrac{x^3}{3}$.   **18.** $1 + \dfrac{x^2}{2} + \dfrac{5x^4}{24}$; $x + \dfrac{x^3}{6} + \dfrac{x^5}{24}$.

20. $-\dfrac{x^2}{2} - \dfrac{x^4}{12}$.

22. $4 + \frac{1}{3}(x - 8), 4.667$; $4 + \frac{1}{3}(x - 8) - \frac{1}{144}(x - 8)^2, 4.639$; $10^{2/3} \approx 4.642$.

23. $3 - 2(x - 2)^3 + (x - 2)^4$; inflexion.   **24.** Too high.

25. (i) $f'(a) > 0, f''(a) > 0, f'''(a) < 0$;   (ii) $f(a + h) - f(a), h f'(a)$.

Exercise 2F page 50

1. 1.10.   **2.** 1.105.   **3.** 0.013.   **4.** 62 m.

5. 7.51, 30.0.   **6.** Both 0.879.   **7.** 0.59.   **8.** 423 cm³.

10. 1.776.   **11.** 1.247; 2.494.   **12.** 0.02%.   **13.** 0.0005.

15. 1.174, 1.187.

16. $\alpha = \sqrt{0.6}, p : q = 5 : 8$;

$$\frac{b - a}{18}\left[5f\left(\frac{a + b}{2} - \sqrt{0.6}\,(b - a)\right) + 8f\left(\frac{a + b}{2}\right) + 5f\left(\frac{a + b}{2} + \sqrt{0.6}(b - a)\right) \right].$$

§3.1

Qu. 2. u.   **Qu. 3.** $SQ = a - b$.

Exercise 3A page 60

1. Vector quantities : velocity, weight, force, acceleration, momentum.
Scalar quantities : mass, area, density, volume, time, electrical resistance, energy, temperature.

2. **AE**.

4. Equality if $\mathbf{b} = \mathbf{0}$ or $\mathbf{a}$ and $\mathbf{b}$ have same direction and sense and $|\mathbf{a}| \geqslant |\mathbf{b}|$.

6. $\mathbf{OC} = \mathbf{AB} = \mathbf{FO} = \mathbf{ED}$.   **7.** Zero force.

8. When $\lambda = \mu = 0$; $x = 4, y = 2$.   **9.** $\mathbf{x} = \frac{3}{8}\mathbf{p} - \frac{1}{8}\mathbf{q}, \mathbf{y} = \frac{1}{8}\mathbf{p} - \frac{3}{8}\mathbf{q}$.

10. $7, 59°$.   **11.** $\frac{1}{2}\mathbf{QS}, \frac{1}{2}\mathbf{QS}$; *KLMN* is a parallelogram.

Exercise 3B page 66

1. (i) $\mathbf{a} - \mathbf{b} + \mathbf{c}$;   (ii) $\mathbf{a} - 2\mathbf{b} + \mathbf{c}$;   (iii) $-\mathbf{a} - \mathbf{c}$;   (iv) $\mathbf{a} - 2\mathbf{b} - \mathbf{c}$.

2. (i) $-4\mathbf{d} - \mathbf{e} - 2\mathbf{f}$;   (ii) $3\mathbf{d} + \mathbf{e} + \mathbf{f}$;   (iii) $6\mathbf{d} + 2\mathbf{e} + 10\mathbf{f}$;
(iv) $4\mathbf{d} + 2\mathbf{e}$.

3. (i), (ii), (iii).

4. (i) $2p - 3q + 6r$;　　(ii) $-2p + 6r$;　　(iii) $-2p + 3q - 3r$;
(iv) $-p + 3q + 3r$.

5. $A(n, 0, 0), B(0, n, 0), C(-n, 0, 0), D(0, -n, 0), E(0, 0, n), F(0, 0, -n)$
where $n = \sqrt{\frac{1}{2}}$.

6. $\sqrt{2}\ \mathbf{OF}$; (i) $-\frac{1}{2}\mathbf{i} + \frac{1}{2}\mathbf{j} + \sqrt{\frac{1}{2}}\mathbf{k}$; (ii) $\mathbf{i} + \mathbf{j}$; (iii) $-\frac{1}{2}\mathbf{i} + \frac{1}{2}\mathbf{j}$; (iv) $\sqrt{2}\ \mathbf{k}$.

7. (a) $\sqrt{2}\ \mathbf{OC}$;　　(b) $\sqrt{2}\ \mathbf{OE}$;　(c) $\mathbf{CB}$.

8. (i) $a + c - 2e$;　　(ii) $-a + c$;　　(iii) $-\frac{1}{2}a + c - \frac{1}{2}e$;
(iv) $2a + 3c - 3e$.

9. (i) $\mathbf{i} + 6\mathbf{j} - 2\mathbf{k}$;　　(ii) $12\mathbf{j} - 5\mathbf{k}$;　　(iii) $(3\mathbf{i} - 7\mathbf{j} + 5\mathbf{k})/\sqrt{83}$;
(iv) $(6\mathbf{i} - 2\mathbf{j} + 5\mathbf{k})/\sqrt{65}$.

10. $\frac{9}{25}\mathbf{i} - \frac{12}{25}\mathbf{j} + \frac{4}{5}\mathbf{k}$; $69°, 119°, 37°$.

11. $\frac{3}{13}, -\frac{4}{13}, \frac{12}{13}$; $-\frac{12}{25}, \frac{16}{25}, \frac{3}{5}$.　　**14.** Yes.　　　**15.** No.

§3.3

Qu. 1. (i) $\dfrac{x - 3}{2} = \dfrac{y - 4}{0} = \dfrac{z - 5}{-1}$;　　(ii) $\dfrac{x - 5}{4} = \dfrac{y - 3}{0} = \dfrac{z + 2}{0}$;

(iii) $\dfrac{x - 1}{0} = \dfrac{y + 4}{2} = \dfrac{z - 5}{-3}$;　　(iv) $\dfrac{x + 6}{1} = \dfrac{y - 7}{0} = \dfrac{z - 10}{0}$;

(v) $\dfrac{x}{0} = \dfrac{y}{0} = \dfrac{z}{1}$.

Qu. 2. (i) $(2\lambda + 1, 4\lambda - 3, 5)$; (ii) $(-4, 8, 9\lambda + 2)$;　　(iii) $(0, \lambda, 0)$.

Qu. 3. The two sets of equations represent the same line.

Exercise 3C page 74

1. (i) $\dfrac{x - 5}{3} = \dfrac{y + 2}{1} = \dfrac{z - 6}{4}$;　　(ii) $\dfrac{x - 2}{7} = \dfrac{y + 3}{3} = \dfrac{z + 7}{-2}$;

(iii) $\dfrac{x + 4}{0} = \dfrac{y - 1}{0} = \dfrac{z}{1}$;　　(iv) $\dfrac{x - 5}{0} = \dfrac{y + 1}{1} = \dfrac{z + 3}{1}$;

(v) $\dfrac{x + 3}{1} = \dfrac{y}{\sqrt{2}} = \dfrac{z}{1}$;　　(vi) $\dfrac{x}{0} = \dfrac{y}{1} = \dfrac{z}{0}$.

2. (i) $\dfrac{x - 3}{2} = \dfrac{y - 2}{-15} = \dfrac{z + 7}{3}$;　　(ii) $\dfrac{x - 12}{2} = \dfrac{y - 7}{2} = \dfrac{z - 1}{1}$;

(iii) $\dfrac{x - 3}{3} = \dfrac{y}{0} = \dfrac{z}{-5}$;　　(iv) $\dfrac{x}{5} = \dfrac{y}{-2} = \dfrac{z}{3}$;

(v) $\dfrac{x - a}{1} = \dfrac{y - 3a}{-2} = \dfrac{z - 2a}{-3}$;　　(vi) $\dfrac{x - b}{1} = \dfrac{y - b^2}{a + b} = \dfrac{z - 2ab}{a - b}$.

3. (i) $\mathbf{r} = 3\mathbf{i} + 2\mathbf{j} - 7\mathbf{k} + \lambda(2\mathbf{i} - 15\mathbf{j} + 3\mathbf{k})$;
    (ii) $\mathbf{r} = 12\mathbf{i} + 7\mathbf{j} + \mathbf{k} + \lambda(2\mathbf{i} + 2\mathbf{j} + \mathbf{k})$;   (iii) $\mathbf{r} = 3\mathbf{i} + \lambda(3\mathbf{i} - 5\mathbf{k})$;
    (iv) $\mathbf{r} = \lambda(5\mathbf{i} - 2\mathbf{j} + 3\mathbf{k})$;
    (v) $\mathbf{r} = a\mathbf{i} + 3a\mathbf{j} + 2a\mathbf{k} + \lambda(\mathbf{i} - 2\mathbf{j} - 3\mathbf{k})$;
    (vi) $\mathbf{r} = b\mathbf{i} + b^2\mathbf{j} + 2ab\mathbf{k} + \lambda[\mathbf{i} + (a + b)\mathbf{j} + (a - b)\mathbf{k}]$.

4. (i) $\mathbf{r} = \mathbf{a} + \lambda(\mathbf{b} - \mathbf{a})$; (ii) $\mathbf{r} = \mathbf{a} + \lambda(\mathbf{c} - \mathbf{b})$;   (iii) $\mathbf{r} = \mathbf{a} + \lambda(\mathbf{c} - \mathbf{a})$;
    (iv) $\mathbf{r} = \mathbf{b} + \lambda(\mathbf{a} - 2\mathbf{b} + \mathbf{c})$.

5. (i) $90°, 0°, 90°$;   (ii) $45°, 90°, 45°$;   (iii) $0°, 90°, 90°$;
    (iv) $54.7°, 125.3°, 54.7°$;   (v) $125.3°, 125.3°, 54.7°$.

6. (i) $(3, -7, 4)$; (ii) skew; (iii) parallel; (iv) coincident; (v) $(0, -5, 0)$.

7. 15 units.   8. $9, 12, 13$ units.   9. (b) $\mathbf{r} = \mathbf{a} + \lambda(\hat{\mathbf{p}} - \hat{\mathbf{q}})$.

10. (i) $\dfrac{x - 2}{1} = \dfrac{y - 3}{-2} = \dfrac{z - 7}{1}$;   (ii) $\dfrac{x - 4}{1} = \dfrac{y + 5}{-1} = \dfrac{z - 3}{1}$;

   (iii) $\dfrac{x - 3}{0} = \dfrac{y + 9}{5} = \dfrac{z - 3}{1}$;   (iv) $\dfrac{x}{3\sqrt{30} + 10} = \dfrac{y - 1}{\sqrt{30} - 10} = \dfrac{z - 6}{-10}$.

11. $(4, 6, -4)$.   12. $a = -1$; $(8, -1, b - 4)$.

13. $\dfrac{x}{1} = \dfrac{y}{2} = \dfrac{z}{3}$; $(1, 2, 3), (3, 6, 9)$.

Exercise 3D page 79

2. $\mathbf{a} - \mathbf{b} + \mathbf{c}$.   5. No.   6. Centroids coincide.   10. No.

17. (b) $\sin 2A$, $\sin 2B$, $\sin 2C$.

§ 4.1

Qu. 1. (i) $\sin x - x \cos x + c$;   (ii) $\frac{1}{2}x^2 \ln x - \frac{1}{4}x^2 + c$;
    (iii) $x \tan x + \ln|\cos x| + c$.

Exercise 4A page 84

1. $\frac{1}{2}x \sin 2x + \frac{1}{4} \cos 2x + c$.   2. $\frac{1}{4} \sin 2x - \frac{1}{2}x \cos 2x + c$.

3. $(1 + x)^{20}(20x - 1)/420 + c$.   4. $x \ln 2x - x + c$.

5. $\frac{1}{3}x^3 \ln x - \frac{1}{9}x^3 + c$.   6. $\frac{1}{4}e^{2x}(2x - 1) + c$.   7. $e^x(x^2 - 2x + 2) + c$.

8. $(2 - x^2)\cos x + 2x \sin x + c$.   9. $x(\ln x)^2 - 2x \ln x + 2x + c$.

10. $\frac{1}{3}x(x^2 + 3)\ln x - \frac{1}{9}x(x^2 + 9) + c$. 11. $-\frac{1}{2}e^{-x}(\sin x + \cos x) + c$.

12. $x \tan x + \ln|\cos x| - \frac{1}{2}x^2 + c$.   13. $e^{2x}(4x^3 - 6x^2 + 6x - 3)/8 + c$.

14. $\frac{1}{2}e^x(\cos x + \sin x) + c$.   15. $\frac{1}{8}(2x \sin 2x + \cos 2x + 2x^2) + c$.

16. (i) π; (ii) $-2\pi - 6$.   17. (i) 1; (ii) $\pi(e - 2)$.

18. Omission of constant of integration.   19. $\frac{1}{2}e^{x^2}(x^2 - 1) + c$.

20. $2e^{\sqrt{x}}(\sqrt{x}-1)+c.$ **21.** (iii) $\frac{1}{2}x+\frac{1}{4}\sin 2x+c.$

22. $\frac{1}{2}\sec x\tan x+\frac{1}{2}\ln|\sec x+\tan x|+c.$

23. $e^{2x}(2\sin 3x-3\cos 3x)/13+c.$ **24.** $e^x(x+1)\ln(x+1)-e^x+c.$

§4.2

Qu. 1. (a) One-to-one;   (b) many-to-one, $\{x:x\geqslant 0\}$;

   (c) many-to-one, $\{x:-\pi/2\leqslant x\leqslant\pi/2\}$;   (d) not a function;

   (e) many-to-one, $\{$ any positive number, 0, any negative number $\}$.

Qu. 2. $x=1/(1-y^2)$; $f^{-1}:x\mapsto 1/(1-x^2), x\geqslant 0, x\neq 1.$

Qu. 3. (i) $f^{-1}:x\mapsto(x-5)/3$;   (ii) $f^{-1}:x\mapsto\ln x$;   (iii) $f^{-1}:x\mapsto\sqrt{x}.$

Exercise 4B page 89

1. (i) $f^{-1}(x)\equiv\sqrt[3]{(x-2)}$; (ii) $f^{-1}(x)\equiv\sqrt{(1-x^2)}, 0\leqslant x\leqslant 1$;
   (iv) $f^{-1}(x)\equiv 1/(x^2-1).$

3. $-2(x+1)^{-3}, -\frac{1}{2}(x+1)^3, 6(x+1)^{-4}, \frac{3}{4}(x+1)^5.$

4. $x=y^2/(1-y^2)$; (i) $\frac{1}{2}x^{-\frac{1}{2}}(1+x)^{-\frac{3}{2}}$; (ii) $(1-y^2)^2/2y.$

5. $(2+x)^{-2}$; $(1-2x)/(x-1).$

6. Range of f = domain and range of f^{-1} = set of real numbers; 1; -1; 2.

§4.3

Qu. 1.

	f	g	h
Domain	$\{x:-1\leqslant x\leqslant 1\}$	$\{x:-1\leqslant x\leqslant 1\}$	all real numbers
Range	$\{x:-\pi/2\leqslant x\leqslant\pi/2\}$	$\{x:0\leqslant x\leqslant\pi\}$	$\{x:-\pi/2<x<\pi/2\}$

Qu. 3. (a) $2n\pi\pm\frac{1}{4}\pi$;   (b) $n\pi-\frac{1}{3}\pi$ where n is any integer.

Exercise 4C page 91

1. (i) $\frac{1}{3}\pi$;   (ii) $\frac{1}{6}\pi$;   (iii) $-\frac{1}{2}\pi$;   (iv) $\frac{1}{2}\pi$;   (v) $\frac{2}{3}\pi$;   (vi) $-\frac{1}{3}\pi.$

2. (i) 0.340; (ii) 0.983; (iii) 2.21; (iv) -0.253; (v) 1.16; (vi) 1.37.

3.

	f	g
Domain	$\{x:x\leqslant -1$ or $x\geqslant 1\}$	$\{x:x\leqslant -1$ or $x\geqslant 1\}$
Range	$\{x:0\leqslant x\leqslant\pi, x\neq\frac{1}{2}\pi\}$	$\{x:-\frac{1}{2}\pi\leqslant x\leqslant\frac{1}{2}\pi, x\neq 0\}$

	h
Domain	all real numbers
Range	$\{x:-\frac{1}{2}\pi<x\leqslant\frac{1}{2}\pi, x\neq 0\}$

4. $\sin^{-1}(x/\sqrt{2}) - \frac{1}{4}\pi, \ 0 \leqslant x \leqslant \sqrt{2}$.   **5.** $\cos^{-1}(-x) = \pi - \cos^{-1}x$.

6. $-\frac{1}{2}\pi \leqslant x \leqslant \frac{1}{2}\pi$.

7. (a) $n\pi + (-1)^n \sin^{-1}k, \ -1 \leqslant k \leqslant 1$;   (b) $2n\pi \pm \sec^{-1}k, \ |k| \geqslant 1$;
(c) $n\pi + \tan^{-1}k$.

8. $n\pi$ or $2n\pi \pm \frac{1}{3}\pi$.   **9.** $2n\pi - \frac{1}{4}\pi$.   **10.** $2n\pi \pm \frac{1}{3}\pi + \sin^{-1}\frac{4}{5}$.

11. $n\pi$ or $n\pi \pm \tan^{-1}(\sqrt{\frac{1}{2}})$. **12.** $4n\pi$ or $4n\pi \pm \frac{4}{3}\pi$.

13. $(2n+1)\pi$ or $2n\pi + 2\sin^{-1}(1/\sqrt{5})$.

§4.4

Qu. 2. (i) $5/\sqrt{(1-25x^2)}$;   (ii) $3/\sqrt{[1-(3x-4)^2]}$; (iii) $6/(4+9x^2)$;
(iv) $-3/(5-12x+9x^2)$.

Qu. 3. (i) $\sin^{-1}\frac{1}{2}x + c$;   (ii) $\frac{1}{3}\sin^{-1}\frac{3}{2}x + c$;   (iii) $\frac{1}{4}\tan^{-1}\frac{1}{4}x + c$.

Exercise 4D page 95

1. (i) $2/\sqrt{(1-4x^2)}$;   (ii) $5/(1+25x^2)$;   (iii) $6x/\sqrt{(1-9x^4)}$;
(iv) $x^{-1}(4x^2-1)^{-\frac{1}{2}}$;   (v) $e^x/(1+e^{2x})$;   (vi) $2(1-x^2)^{-1}(x^2-2)^{-\frac{1}{2}}$.

2. (i) $\sin^{-1}\frac{1}{6}x + c$;   (ii) $\frac{2}{5}\tan^{-1}\frac{2}{5}x + c$;
(iii) $(1/\sqrt{2})\sin^{-1}(\sqrt{2}x/3) + c$;   (iv) $\sin^{-1}(x+4) + c$;
(v) $(1/\sqrt{6})\tan^{-1}(\sqrt{2}x/\sqrt{3}) + c$;   (vi) $\frac{1}{2}\tan^{-1}(2x+4) + c$;
(vii) $\sin^{-1}(\frac{1}{2}x - \frac{1}{2}) + c$;   (viii) $\frac{1}{2}\tan^{-1}(\frac{3}{2}x + \frac{1}{2}) + c$.

3. $c_1 + \frac{1}{2}\pi$.

4. (i) $\pi/12$;   (ii) $\pi/(2\sqrt{6})$;   (iii) $\frac{2}{3}\tan^{-1}3$;   (iv) $\pi/(12\sqrt{10})$.

5. (i) $x\sin^{-1}x + \sqrt{(1-x^2)} + c$;   (ii) $x\cos^{-1}x - \sqrt{(1-x^2)} + c$;
(iii) $x\tan^{-1}x - \frac{1}{2}\ln(1+x^2) + c$;   (iv) $x\cot^{-1}x + \frac{1}{2}\ln(1+x^2) + c$.

6. (i) $x - \frac{1}{3}x^3 + \frac{1}{5}x^5$;   (ii) $x + \frac{1}{6}x^3 + \frac{3}{40}x^5$; $\frac{1}{2}\pi - x - \frac{1}{6}x^3 - \frac{3}{40}x^5$.

7. $\frac{1}{2}b\sqrt{(a^2-b^2)} + \frac{1}{2}a^2\sin^{-1}(b/a)$; area of triangle + area of sector.

8. (i) $\frac{1}{2}\tan^{-1}(\frac{1}{2}x - \frac{3}{2}) + c$;   (ii) $\frac{1}{2}\sin^{-1}(\frac{1}{2}x + \frac{3}{4}) + c$;
(iii) $\frac{1}{4}\tan^{-1}(x + \frac{5}{2}) + c$; (iv) $1/(3-x) + c$;   (v) $\sin^{-1}(\frac{1}{2}x - 1) + c$.

9. (i) $\frac{1}{2}x\sqrt{(4-3x^2)} + (2/\sqrt{3})\sin^{-1}(\sqrt{3}x/2) + c$;
(ii) $\frac{1}{4}(2x+1)\sqrt{(3-4x-4x^2)} + \sin^{-1}(x + \frac{1}{2}) + c$;
(iii) $-\frac{1}{3}(1-x^2)^{\frac{3}{2}} + \frac{1}{2}x(1-x^2) + \frac{1}{2}\sin^{-1}x + c$.

10. (a) (i) $\cos^{-1}(1-2x) + c$;   (ii) $2\sin^{-1}\sqrt{x} + c$.

11. (b) $-1/[|x|\sqrt{(x^2-1)}]$;   (c) $(1/a)\sec^{-1}(x/a) + c$.

12. (b) (i) $x\sec^{-1}x - \ln[x + \sqrt{(x^2-1)}] + c$;
(ii) $x\,\mathrm{cosec}^{-1}x + \ln[x + \sqrt{(x^2-1)}] + c$.

§4.5

Qu. 1. Tangents are parallel to the y-axis.

Qu. 2. (i) $-x/y$; (ii) $(2-y^2)/(2xy-3)$; (iii) $-\cos x/\cos y$.

Exercise 4E page 99

1. (i) $-x^2/y^2$; (ii) $(xy-y)/(x-xy)$; (iii) $(4+y)/(3-x)$;
   (iv) $y \ln k$; (v) $y(1+\ln x)$.

2. $-\frac{7}{8}$. 3. $-r/(2r+h)$. 4. $p(b-cT)/T^2$. 5. $(\gamma+1)V/\gamma^2$.

6. $a/27b^2, 3b, 8a/27bR$. 7. $-\frac{1}{2}$; $(\pm 8/\sqrt 6, \mp 1/\sqrt 6)$.

8. $x/(1-y)$. 9. ± 6. 11. $(2\cos B + \cos C)/(\cos A - \cos C)$.

13. $2(y e^{2x} - e^{-2x})/(3y^2 - e^{2x})$; $h = -\frac{1}{10}\ln 2, k = 2^{\frac{2}{5}}$.

14. $-\pi [(2n-1)!/(n-1)!]^2/[(2n-1). 2^{2n-1}], 2^{2n-1}. n[(n-1)!]^2$.

§4.6

Qu. 2. $3(1-t^2)/16t^7$.

Exercise 4F page 103

1. (i) $(x/a)^2 + (y/b)^2 = 1, -b\cot t/a$; (ii) $y^2 = 4ax, 1/t$;
   (iii) $xy = c^2, -1/t^2$; (iv) $(x/a)^2 - (y/b)^2 = 1, b\,\text{cosec}\,t/a$

2. $2y = 3t_1 x - at_1^3$. 4. π units^3. 5. $x^{2/3} + y^{2/3} = a^{2/3}$.

6. $y = -x\tan p + a\sin p$. 7. $-(1+t^2)^3/4t^3$. 8. $y = \frac{1}{2}x + 2$.

9. $X = (t_1^2 + t_1 t_2 + t_2^2)/3, Y = -t_1 t_2 t_3/2$. 10. 8.1 units^2.

11. $yf'(\theta) = xg'(\theta) + f'(\theta) g(\theta) - f(\theta) g'(\theta)$; $g(\theta) = a(\sin\theta - \theta\cos\theta)$.
    $x\cos\theta + y\sin\theta = a$; $a, (a\cos\theta, a\sin\theta)$.

§4.7

Qu. 1. The set of all real numbers is the domain for both functions, and the range for the sinh function. The range for the cosh function is $\{x : x \geqslant 1\}$.

Qu. 3. e^{-x}.

Exercise 4G page 109

1. (i) $\cosh 2A \equiv 1 + 2\sinh^2 A$;
   (ii) $\cosh A - \cosh B \equiv 2\sinh\dfrac{A+B}{2}\sinh\dfrac{A-B}{2}$;
   (iii) $\tanh(A+B) \equiv (\tanh A + \tanh B)/(1 + \tanh A \tanh B)$;
   (iv) $\sinh 3A \equiv 3\sinh A + 4\sinh^3 A$.

2. (i) domain = { reals }, range = $\{x : 0 < x \leqslant 1\}$;
   domain = range = $\{x : x \neq 0\}$;
   (iii) domain = $\{x : x \neq 0\}$, range = $\{x : |x| > 1\}$.

4. (i) $-\ln 3$;   (ii) $\ln \frac{3}{4}, \ln 2$;   (iii) $\pm \frac{1}{2} \ln 3$;   (iv) $0, \ln 7$.

5. $\ln 3, \ln 2$.

8. (i) $3 \cosh 3x$;   (ii) $5 \sinh(5x-2)$;   (iii) $\sinh 2x$;   (iv) $\frac{1}{2}(1+x^{-2})$;
   (v) $(1-x^2)^{-1}$;   (vi) $\sec x$;   (vii) $\sec x$;   (viii) $\operatorname{sech} x$.

9. (i) $\frac{1}{4}\sinh 2x - \frac{1}{2}x + c$; (ii) $\frac{1}{4}e^{2x} + \frac{1}{2}x + c$; (iii) $x \cosh x - \sinh x + c$;
   (iv) $x \sinh^{-1} x - \sqrt{(1+x^2)} + c$;   (v) $x \cosh^{-1} x - \sqrt{(x^2-1)} + c$;
   (vi) $(\sinh 6x + 3 \sinh 2x)/12 + c$; (vii) $\frac{1}{3}\cosh^3 x - \cosh x + c$;
   (viii) $x - \coth x + c$.

10. (i) $2 \tan^{-1}(e^x) + c$;   (ii) $e^x - 2 \tan^{-1}(e^x) + c$.

11. (i) $\sinh^{-1}(x/2) + c$;   (ii) $\cosh^{-1}(x/3) + c$;
    (iii) $\frac{1}{3}\sinh^{-1}(3x/4) + c$;   (iv) $(1/\sqrt{7})\cosh^{-1}(\sqrt{7}x/\sqrt{5}) + c$;
    (v) $\sinh^{-1}(\frac{1}{2}x - \frac{3}{2}) + c$; (vi) $\frac{1}{3}\cosh^{-1}(x + \frac{1}{3}) + c$;
    (vii) $\frac{1}{5}\ln|5x + 2| + c$.

12. (i) $1 + x^2/2! + x^4/4!$; (ii) $x + x^2/3!$;   (iii) $x + x^3/3 + x^5/5$.

16. (i) $1/(1-x^2)$, $\tanh^{-1} x + c$;   (ii) $\frac{1}{2}\ln[(1+x)/(1-x)] + c$.

17. $2, 1$; $\sqrt{3}$ min at $x = -\ln 3$.   18. $1 - e^{-1}$.

19. $\frac{5}{4} + \frac{3}{4}y + \frac{5}{8}y^2 + \frac{1}{8}y^3 + \frac{5}{96}y^4$ where $y = x - \ln 2$.

§5.1

Qu. 1. Identity for multiplication only. No inverses.

Qu. 2. $+, \times, \div$; q/p.   **Qu. 10.** (i), (iii) true; (ii), (iv) false.

§5.2

Qu. 2. $\dfrac{x}{x^2+y^2} - \dfrac{yj}{x^2+y^2}$.

Exercise 5A page 116

1. $11 + 2j$.   2. $3 + 6j$.   3. $3 - 10j$.   4. $-1 + 17j$.

5. $60 + 15j$.   6. $10 + 16j$.   7. $13 + 84j$.   8. 24.

9. $-37 - 185j$.   10. $-117 - 44j$.   11. $3 + 2j$.   12. $0.7 - 0.4j$.

13. $1.9 - 1.7j$.   14. $4 + 9j$.   15. $\pm 1, \pm j$.

16. (i) $-j$; (ii) 1; (iii) j; (iv) $-j$; (v) -1.   17. (i) 0; (ii) 1.

18. (i) $2 \pm 3j$;   (ii) $(-2 \pm 5j)/3$;   (iii) $(-3 \pm \sqrt{26}j)/5$;   (iv) $\pm 2 - j$.

19. (i) $4z^2 + 9 = 0$;　(ii) $z^2 - 10z + 29 = 0$;　(iii) $z^2 + 14z + 50 = 0$;
(iv) $z^2 - 8z + 22 = 0$;　(v) $z^2 + (2 - 2j)z - 3 - 6j = 0$;
(vi) $z^2 + (-9 + 2j)z + 23 \div 7j = 0$.

20. (i) $a = 1, b = 2$;　(ii) $a = 5, b = 4$;　(iii) $a = b = 1/\sqrt{2}$.

21. $a = -7, b = 11; 5 - 2j$.　　**22.** $3 - j, 2$.

23. $5, -1 \pm \sqrt{3}j$.　　**24.** $-3, (5 \pm \sqrt{3}j)/2$.

26. (ii) $(z + 1 + j)(z + 1 - j)(z - 1 + j)(z - 1 - j)$;
(iii) $(z^2 + 2z + 2)(z^2 - 2z + 2)$.

27. (i) $\pm 1 \pm j$;　(ii) $(\pm 1 \pm j)/\sqrt{2}$.　　**28.** (iv) $\sqrt[3]{2} + \sqrt[3]{4} \approx 2.847$.

29. $b = \pm 1$.

§5.3

Qu. 1. Half turn about O.

Exercise 5B　page 122

3. $0, z, \lambda z$ collinear.　　**4.** (iii) Perpendicular.　　**7.** $6 - 6j, 2, -8 + 4j$.

8. $1 + 5j; 5 - 3j; 4 + 10j, -2$.　　**12.** $13, 7$.　　**16.** $259^2 + 286^2$.

20. (i) Diagonals of rhombus are perpendicular;　(ii) angle in semicircle is $90°$.

Exercise 5C　page 125

1. $\sqrt{74} \approx 8.60, \sqrt{193} \approx 13.89, \sqrt{505} \approx 22.47$.

2. (i) $-3 + 4j$;　(ii) $3 - 4j$.

§5.5

Qu. 1. Points for which $\arg z = \pi$.　　**Qu. 2.** $\arg z + \pi$.

Exercise 5D　page 129

1. $1, 0$.　　**2.** $2, \pi$.　　**3.** $3, \dfrac{\pi}{2}$.　　**4.** $4, -\dfrac{\pi}{2}$.　　**5.** $\sqrt{2}, \dfrac{\pi}{4}$.　　**6.** $4, \dfrac{3\pi}{4}$.

7. $2, -\dfrac{\pi}{3}$.　　**8.** $2, \dfrac{5\pi}{6}$.　　**9.** $7, \dfrac{\pi}{7}$.　　**10.** $\frac{1}{12}, 2$.　　**11.** $6, -2.28$.

12. $5, -0.93$.　　　**13.** $13, -1.97$.　　　**14.** $\sqrt{157} \approx 12.53, 2.07$.

15. $\sqrt{58} \approx 7.62, -2.74$.

16. (i) $\alpha - \pi$;　(ii) $\dfrac{\pi}{2} - \alpha$;　(iii) $\pi - \alpha$;　(iv) $\alpha - \dfrac{\pi}{2}$.

7. (i) $5\sqrt{2}\left(\cos\left(-\frac{\pi}{4}\right)+j\sin\left(-\frac{\pi}{4}\right)\right)$; (ii) $2\sqrt{3}\left(\cos\frac{\pi}{6}+j\sin\frac{\pi}{6}\right)$;

(iii) $\cos(-\theta)+j\sin(-\theta)$; (iv) $7\left(\cos\left(\frac{\pi}{2}-\theta\right)+j\sin\left(\frac{\pi}{2}-\theta\right)\right)$;

(v) $\sec\theta\,(\cos\theta+j\sin\theta)$.

8. $2\cos\frac{\theta}{2},\frac{\theta}{2}$.   21. $\frac{2\pi}{3}$.   22. $k\leqslant|z+2k|\leqslant 3k,-\frac{\pi}{6}\leqslant\arg(z+2k)\leqslant\frac{\pi}{6}$.

§5.6

Qu. 4. $\dfrac{c-a}{b-a}=-\dfrac{c'-a'}{b'-a'}$.

Exercise 5E   page 133

3. In a straight line (i) from $-3j$ to $3j$; (ii) from 2 to -2;
(iii) from $3+j$ to $3+3j$; (iv) from $2-3j$ to $-2+3j$;
(v) from j to $-j$; (vi) from -1 to 0 then back to -1.

4. (i) From 3 once anticlockwise round $|z|=3$;
(ii) from $2j$ once anticlockwise round $|z|=2$;
(iii) from $4+2j$ once anticlockwise round $|z-3-2j|=1$;
(iv) from $3+2j$ once anticlockwise round $|z|=\sqrt{13}$;
(v) from 1 once clockwise round $|z|=1$;
(vi) from 1 twice anticlockwise round $|z|=1$.

5. $2\cos(\theta-\phi)$.

6. $x=\dfrac{\sqrt{3}-1}{4},y=\dfrac{\sqrt{3}+1}{4}$; $\sqrt{2}\left(\cos\frac{3\pi}{4}+j\sin\frac{3\pi}{4}\right),2\left(\cos\frac{\pi}{3}+j\sin\frac{\pi}{3}\right)$;

$\cos\left(\frac{5\pi}{12}\right)=\dfrac{\sqrt{3}-1}{2\sqrt{2}},\sin\left(\frac{5\pi}{12}\right)=\dfrac{\sqrt{3}+1}{2\sqrt{2}}$.

11. (i) Inclined at $\frac{2\pi}{3}$; (ii) $3-\sqrt{3}+(3+2\sqrt{3})j,3+\sqrt{3}+(3-2\sqrt{3})j$.

§5.7

Qu. 2. $16\sin^5\theta-20\sin^3\theta+5\sin\theta$.

Qu. 4. $2^{-6}(4\theta+\sin2\theta-\sin4\theta-\frac{1}{3}\sin6\theta)+k$.

Qu. 5. $\sum_{r=1}^{n}{}^nC_r\sin r\theta=2^n\cos^n\frac{\theta}{2}\sin\frac{n\theta}{2}$.

Exercise 5F   page 138

1. 1.   2. $-j$.   3. $(1-\sqrt{3}j)/2$.   4. $(1+\sqrt{3}j)/162$.

5. $\cos7\alpha-j\sin7\alpha$.   6. $\cos^{10}\theta(\cos10\theta+j\sin10\theta)$.

7. $\cos(5\alpha + 3\beta) + j\sin(5\alpha + 3\beta)$.   8. $\cos 12\phi - j\sin 12\phi$.   9. $-8j$.

10. $(4t - 4t^3)/(1 - 6t^2 + t^4)$, $t \equiv \tan\theta$.

11. $64c^7 - 112c^5 + 56c^3 - 7c$, $64c^6 - 80c^4 + 24c^2 - 1$, $c \equiv \cos\theta$.

12. (i) $\cos\dfrac{(2k+1)\pi}{12}$, $k = 0, 1, \ldots, 5$;   (ii) $\cos\dfrac{(6k\pm 1)\pi}{18}$, $k = 0, 1, 2$.

13. $({}^nC_1 t - {}^nC_3 t^3 + \ldots)/(1 - {}^nC_2 t^2 + {}^nC_4 t^4 - \ldots)$.

14.   (i) $2^{-3}(\cos 4\theta + 4\cos 2\theta + 3)$;
      (ii) $2^{-5}(-\cos 6\theta + 6\cos 4\theta - 15\cos 2\theta + 10)$;
      (iii) $2^{-6}(\cos 7\theta - \cos 5\theta - 3\cos 3\theta + 3\cos\theta)$.

15.   (i) $2^{-2}(3\sin\theta - \sin 3\theta)$;   (ii) $2^{-4}(\sin 5\theta - 5\sin 3\theta + 10\sin\theta)$;
      (iii) $2^{-6}(3\sin\theta + 3\sin 3\theta - \sin 5\theta - \sin 7\theta)$.

17.   (i) $2^{-5}(-\frac{1}{6}\sin 6\theta + \frac{3}{2}\sin 4\theta - \frac{15}{2}\sin 2\theta + 10\theta) + k$;
      (ii) $\frac{2}{35}$;   (iii) $\frac{4}{35}$.

18. $(k + \frac{1}{2})\pi$ or $(k \pm \frac{1}{3})\pi$.   20. $\sin 2n\theta/2\sin\theta$.

21. $\sqrt{3}\left(\cos\dfrac{\pi}{6} + j\sin\dfrac{\pi}{6}\right)$, $3^{n/2}\sin\dfrac{n\pi}{6}$. 22. $2^n\cos^n\dfrac{\beta}{2}\cos\left(\alpha + \dfrac{n\beta}{2}\right)$.

23. $(\sin\theta + \sin(n-1)\theta - \sin n\theta)/(2 - 2\cos\theta)$.

24. $(1 - k\cos\theta - k^n\cos n\theta + k^{n+1}\cos(n-1)\theta)/(1 - 2k\cos\theta + k^2)$,
    $(k\sin\theta - k^n\sin n\theta + k^{n+1}\sin(n-1)\theta)/(1 - 2k\cos\theta + k^2)$;
    $(1 - k\cos\theta)/(1 - 2k\cos\theta + k^2)$, $k\sin\theta/(1 - 2k\cos\theta + k^2)$.

§6.1

Qu. 2. $2x^2 + x - \frac{5}{2}$, $-\frac{21}{2}x + \frac{21}{2}$.

Exercise 6A page 144

1. $p = 4, q = -16$; $x + 3, x - 2$.   2. $2, -3, -7$.

4. $x + a, x + 2a, x^2 + xa + a^2$, etc.   5. $a = -3$; $(x + 3)^2(2x - 1)$.

6. $R = (P(a) - P(b))/(a - b)$, $S = (aP(b) - bP(a))/(a - b)$;
   $P(a) + (x - a)P'(a)$.

8. $y^4 - y + y/(y^3 + 1)$.   9. $p = 1, q = 3, r = 1, s = 0$.

10. $a = b = \pm 2, c = 2$.   11. $\frac{1}{4}n(n + 1)(n + 2)(n + 3)$.

12. $-6, 7, -1, 0$; $\frac{1}{30}n(n + 1)(2n + 1)(3n^2 + 3n - 1)$.

13. $y = \sum \dfrac{A(x - b)(x - c)(x - d)}{(a - b)(a - c)(a - d)}$.

17. (i) $|h| < 2|k^3|$;   (ii) $-2|k^3| < h < 0$;   (iii) $h > 2|k^3|$; $1, \frac{1}{2}(-1 \pm 3\sqrt{5})$.

18. $k = -1$; $-1 \pm \sqrt{6}, -1 \pm \sqrt{2}$.

19. 1; just one real root if the order is odd, no real roots otherwise.

§6.2

Qu. 1. $-\frac{5}{3}, 0, \frac{8}{3}$.

Qu. 2. (i) $z^3 - 5z^2 - 22z + 56 = 0$; (ii) $z^3 - 11z^2 + 44z - 60 = 0$.

Qu. 3. $a\,S_{n+4} + b\,S_{n+3} + c\,S_{n+2} + d\,S_{n+1} + e\,S_n = 0$.

Exercise 6B page 150

1. (i) 2;   (ii) 3;   (iii) -1;   (iv) -2;   (v) -13;
   (vi) -22;   (vii) -3;   (viii) -2;   (ix) 9;   (x) -9.

2. (i) $z^3 - 8z^2 - 4z + 24 = 0$;   (ii) $z^3 - 10z^2 + 27z - 19 = 0$;
   (iii) $z^3 - 8z^2 + 15z + 1 = 0$.

3. (i) $-5, -2, 1$;   (ii) $2, 2(1 \pm \sqrt{3})$;   (iii) $3, 3 \pm 2j$.

4. $k = \frac{47}{2}$; $\frac{3}{2}, 2, \frac{5}{2}$.   5. $\frac{1}{4}, \frac{1}{4}, -\frac{3}{4}$.

6. (i) $a^2 z^3 - (b^2 - 2ac)z^2 + (c^2 - 2bd)z - d^2 = 0$;
   (ii) $d^2 z^3 - (c^2 - 2bd)z^2 + (b^2 - 2ac)z - a^2 = 0$;
   (iii) $da^2 z^3 + a(c^2 - 2bd)z^2 + d(b^2 - 2ac)z + ad^2 = 0$.

7. $b^3 d = ac^3$; $\frac{1}{2}, \frac{3}{2}, \frac{9}{2}$.   8. $0, 0, 0$ or $7, 8, -1$.   9. $-p, \pm\sqrt{(-q)}$.

11. $qz^3 + 9p^2 z^2 - 6pq z + q^2 = 0$; $\frac{3}{2}, \frac{1}{4}(-3 \pm j\sqrt{15})$.

12. (i) $\frac{9}{4}$;   (ii) $\frac{59}{4}$;   (iii) $-\frac{135}{8}$;   (iv) -14.   14. $(3b^2 - 8ac)/a^2$.

15. (a) 14, (b) -22, (c) $z^4 - 5z^2 + 6 = 0$; $\pm\sqrt{2} - 1, \pm\sqrt{3} - 1$.

§6.3

Qu. 2. $(2, 0), (-5, 0), (0, \frac{5}{2})$.

Qu. 3. (i) As $x \to 3$ from the left, $y \to -\infty$; as $x \to 3$ from the right, $y \to +\infty$.
      As $x \to 5$ from the left, $y \to -\infty$; as $x \to 5$ from the right, $y \to +\infty$.
   (ii) As $x \to -1$ (from the left or from the right), $y \to +\infty$.

Qu. 5. (a) $x - a$ is a common factor of numerator and denominator.
   (b) $(2, -\frac{1}{3})$ is not a point on the first graph.

Exercise 6C page 157

1. Minimum.

23. $\alpha > 0 \Rightarrow$ no roots; $\alpha = 0 \Rightarrow 1$ repeated root; $-\frac{1}{16} < \alpha < 0 \Rightarrow 4$ distinct roots;
   $\alpha = -\frac{1}{16} \Rightarrow 1$ repeated root and 2 other distinct roots;
   $\alpha < -\frac{1}{16} \Rightarrow 2$ distinct roots.

24. $u + v = u^2/(u - f), u \neq 0, u \neq f$.

448

25. Hyperfocal distance is
    (i) near limit of depth of field when lens is focussed at infinity,
    (ii) the focussing distance with the greatest depth of field.

26. 88.8% when $i = 19.8$.

§6.4

Qu. 5. $-1 < x < 4$ or $x > 5$.

Exercise 6D page 161

1. $-7 < x < -2$ or $x > 2$.   2. $-5 \leqslant x < -2$ or $-1 < x \leqslant 3$.

3. $x < 3$.   4. $x = 0$ or $1 \leqslant x < 2$.   5. $-\frac{3}{4} \leqslant x < 0$ or $0 < x \leqslant 1$.

6. (i) $\{x : -5 \leqslant x \leqslant -\frac{1}{3}\}$; (ii) $\{x : -5 \leqslant x < 2\}$.   7. 3.

8. $\{x : -1 \leqslant x \leqslant 2 \text{ or } 4 \leqslant x \leqslant 7\}$.   9. $(y-1)x^2 + x + 2 = 0$; $\frac{9}{8}$.

10. $\{y : -1 \leqslant y \leqslant 3\}$.   11. 3, 1.

14. (a) (i) $x < -6$ or $-2 < x < \frac{2}{3}$; (ii) $1 \leqslant x \leqslant 3$. (b) (i) 1; (ii) $-\sqrt{2}$.

15. (i) The sum of a negative number and its reciprocal is at most -2.

17. & 18. When the two numbers are the same.

21. (iii) When $a_1 = a_2 = \ldots = a_n$.

§6.5

Qu. 1. (i) $\dfrac{\frac{2}{3}}{x-1} + \dfrac{\frac{1}{3}}{x+2}$; (ii) $\dfrac{1}{x+1} + \dfrac{1}{x-1} - \dfrac{2}{x+3}$.

Qu. 4. (i) $\dfrac{3}{x-1} - \dfrac{2}{x+2}$; (ii) $\dfrac{2}{x-5} - \dfrac{\frac{3}{2}}{x-2} - \dfrac{\frac{1}{2}}{x+2}$.

Qu. 5. $\dfrac{2x-1}{x^2-x+1} - \dfrac{1}{x-3}$.   Qu. 7. $\dfrac{\frac{1}{4}}{x-1} - \dfrac{\frac{1}{4}}{x+3} - \dfrac{1}{(x+3)^2}$.

Qu. 8. $\dfrac{25}{(x-1)^3} - \dfrac{5}{(x-1)^2} + \dfrac{1}{x-1} - \dfrac{1}{x+4}$.

Qu. 9. $\dfrac{1}{x} - \dfrac{\frac{1}{2}(1-j)}{x+j} - \dfrac{\frac{1}{2}(1+j)}{x-j}$.

Exercise 6E page 167

1. $\dfrac{2}{x-2} + \dfrac{1}{x+1}$.   2. $\dfrac{2}{x+2} - \dfrac{1}{x}$.   3. $\dfrac{0.3}{2x-5} - \dfrac{0.3}{2x+5}$.

4. $\dfrac{\frac{1}{2}}{1-x} + \dfrac{\frac{1}{2}}{1+x}$.   5. $\dfrac{\frac{5}{4}}{x-5} - \dfrac{\frac{1}{4}}{x+3}$.   6. $\dfrac{\frac{7}{10}}{x-2} + \dfrac{\frac{7}{15}}{x+3} - \dfrac{\frac{1}{6}}{x}$.

7. $\dfrac{4}{x+2} + \dfrac{x-2}{x^2+1}.$ 8. $\dfrac{4}{2x-3} - \dfrac{x-3}{x^2+4}.$ 9. $\dfrac{\frac{1}{3}x+\frac{1}{3}}{x^2-x+1} - \dfrac{\frac{1}{3}}{x+1}.$

10. $\dfrac{\frac{1}{4}}{2-x} + \dfrac{\frac{1}{4}x+1}{4+2x+x^2}.$ 11. $\dfrac{4}{x-1} - \dfrac{4}{x} - \dfrac{4}{x^2}.$

12. $\dfrac{\frac{1}{32}}{x-2} - \dfrac{\frac{1}{32}}{x+2} + \dfrac{\frac{1}{8}}{x^2+4}.$ 13. $\dfrac{3}{x+1} - \dfrac{6}{(x+1)^2} + \dfrac{4}{(x+1)^3}.$

14. $\dfrac{-2}{x-2} + \dfrac{3-j}{x+2j} + \dfrac{3+j}{x-2j}.$

15. $\dfrac{4}{x} + \dfrac{9}{1-x}$; $(-2, 1)$ maximum; $(\frac{2}{3}, 25)$ minimum.

16. $\dfrac{\frac{4}{3}}{x-2} - \dfrac{\frac{1}{3}}{x+1}$; $-\frac{4}{3}(x-2)^{-2} + \frac{1}{3}(x+1)^{-2}, \frac{8}{3}(x-2)^{-3} - \frac{2}{3}(x+1)^{-3}$;
$a = 0, b = -4$; $-1, \frac{1}{81}$; max. at $x = 0$, min. at $x = -4$; $-\frac{5}{3}\ln 2$.

§6.6

Qu. 1. (i) $3\ln\left|\dfrac{x+1}{x+2}\right| - \dfrac{1}{x+2} + c$;

       (ii) $\ln|x-2| + \frac{3}{2}\ln|x^2+4| + \tan^{-1}\dfrac{x}{2} + c$.

Qu. 2. (i) $\ln|\tan\frac{1}{2}\theta| + c$;   (ii) $\ln\left|\dfrac{1+\tan\frac{1}{2}\theta}{1-\tan\frac{1}{2}\theta}\right| + c$.

Qu. 3. $\frac{11}{5}\ln|2\sin x + \cos x| + \frac{13}{5}x + c$.

Qu. 4. $\frac{1}{3}[4(1-2x)^{-1} - (1+x)^{-1}]$; $1 + 3x + 5x^2 + 11x^3$.

Exercise 6F   page 171

1. $-x - 2\ln|1-x| + c$.   2. $\frac{1}{2}x^2 - \frac{1}{2}\ln(x^2+1) + c$.

3. $\frac{1}{2}x^2 + \frac{3}{2}\ln|x-1| - \frac{1}{2}\ln|x+1| + c$.

4. $-\frac{1}{4}\ln|x+1| + \frac{5}{4}\ln|x-3| + c$.   5. $\frac{1}{4}x + \frac{5}{48}\ln\left|\dfrac{2x-3}{2x+3}\right| + c$.

6. $\frac{11}{14}\ln|2x-1| - \frac{2}{7}\ln|3x+2| + c$.   7. $\dfrac{1}{2\sqrt{3}}\ln\left|\dfrac{x-\sqrt{3}}{x+\sqrt{3}}\right| + c$.

8. $\dfrac{1}{\sqrt{3}}\tan^{-1}\dfrac{x}{\sqrt{3}} + c$.   9. $-\dfrac{1}{2(x+1)} + \frac{1}{4}\ln\dfrac{(x+1)^2}{x^2+1} + c$.

10. $\sqrt{2}\ln\left|\dfrac{x-1-\sqrt{2}}{x-1+\sqrt{2}}\right| + c$.   11. $\ln|\tan\frac{1}{2}x + 1| + c$.

12. $\dfrac{2}{\sqrt{3}}\tan^{-1}\left(\dfrac{1}{\sqrt{3}}\tan\frac{1}{2}x\right) + c$.   13. $\frac{1}{3}\ln|1 - \cos 3x| + c$.

14. $\ln|\tan x| + c$.   15. $\frac{1}{4}\ln\left|\dfrac{3+\tan x}{1-\tan x}\right| + c$.

16. $\frac{1}{29}[41x - \ln|2\cos x - 5\sin x|] + c.$

17. $\frac{1}{25}[29x - 3\ln|3\cos x + 4\sin x|] + c.$ **18.** $n(n+2)/(n+1)^2.$

19. $\frac{1}{2} - \dfrac{1}{(n+1).2^{n+1}}.$ **20.** (i) $\dfrac{n(5n+7)}{6(2n+1)(2n+3)}$; (ii) $\dfrac{n(n+2)}{3(2n+1)(2n+3)}.$

21. (i) $x + 3x^2 + 7x^3$; (ii) $\frac{1}{6} - \frac{13}{36}x + \frac{19}{216}x^2 - \frac{97}{1296}x^3.$

22. $-\dfrac{1}{x+2} + \dfrac{3}{x+3} - \dfrac{2}{x+4}$;

    (a) $\ln\left|\dfrac{(x+3)^3}{(x+2)(x+4)^2}\right| + c$;   (b) $\dfrac{n(n+1)}{6(n+3)(n+4)}$;

    (c) $(-1)^n n! \left[-(x+2)^{-(n+1)} + 3(x+3)^{-(n+1)} - 2(x+4)^{-(n+1)}\right].$

23. $(-1)^n n! \left[(x+a)^{n+1} - (x-a)^{n+1}\right]/2a(x^2 - a^2)^{n+1}.$

24. $(-1)^{n-1}(n-1)!\, r^{-n}\sin n\theta.$

§7.1

Qu. 1.   (i) $\dfrac{d^2y}{dx^2} = -y$;   (ii) $\dfrac{dy}{dx} = y + \dfrac{y}{x}$;

        (iii) $\dfrac{d^2y}{dx^2} - 3\dfrac{dy}{dx} + 2y = \sin x - 3\cos x$; (iv) $2xy\dfrac{dy}{dx} = y^2 - x^2.$

Exercise 7A page 177

  7. $y = kx + k^2.$

§7.3

Qu. 1. $y = \frac{1}{2}(\sin x - \cos x) + c\,e^{-x}.$ **Qu. 2.** $x^2 + 2y^2 = A/x^2.$

Exercise 7B page 180

  1. $\frac{1}{4}y^4 = \frac{1}{3}x^3 + c.$ **2.** $y = (x+c)e^x.$ **3.** $y = \frac{1}{2}x + \frac{1}{4} + c\,e^{-2x}.$

  4. $y = A\exp(\frac{1}{2}\tan^{-1}\frac{x}{2}).$ **5.** $4(x^2+1)y = x^4 + 2x^2 + c.$

  6. $x^2 - xy + y^2 = A$ **7.** $xy = x\sin x + \cos x + c, x > 0.$

  8. $\sec y = \ln\sec x + c, |x| < \frac{1}{2}\pi.$

  9. $y = (3 + e^{x^2})/(3 - e^{x^2}), x^2 < \ln 3.$

  10. $dy/dx = (x+y)/2$; $x + y + 2 = e^{x/2}.$ **11.** $2y^3 = x + 1/x, x > 0.$

  13. 49 weeks.   **15.** 2.1 metres.   **16.** $21.25°C.$

  17. $1000\,e^{4t^2/(t^2+125)}$; 75 weeks; £54 600.   **18.** $CE(1 - e^{-t/CR}).$

19. $-(E/R) e^{-t/CR}$; 1.38×10^{-5} seconds.   21. $y = \frac{1}{4} x^2, y = Ax - A^2$; yes.

22. $g(1 - e^{-kt})/k$.   23. $[(u + g/k)(1 - e^{-kt}) - gt]/k$ metres.

24. (i) $y = (x - k)^3$.

§7.4

Qu. 4. (i) $x = a \sin \omega t$;   (ii) $x = a \cos \omega t$.

Qu. 5. $A = \dfrac{1}{\omega} \times$ initial velocity; $B =$ initial displacement.

Qu. 6. $x = b + a \sin(\omega t + \epsilon)$.

Exercise 7C   page 185

2. (i) $y = -4 \sin 2x$;   (ii) $y = 4 \cos 2x$;   (iii) $y = 10 \sin(2x + \pi/6)$;
(iv) $y = 2 \sin 2x - 3 \cos 2x$.

3. $y = \frac{1}{3}(2 + 4 \cos 3x)$.   4. $y = \frac{1}{2}(1 - \cos 2x - \sin 2x)$.

5. $\pi/\omega, a/2, x = a/2$.   6. $\pm \omega \sqrt{(x^2 + k)}$.   8. 99.4 cm.

13. $\sqrt{[2a^2 + 2a^2 \cos(\epsilon - \epsilon' - t\delta)]}$; $2\pi/\delta$.

§7.5

Qu. 1. (i) $y = Ae^{-3x} + \frac{1}{5} e^{2x}$; (ii) $y = Ae^{-3x} + \frac{2}{5} \sin x + \frac{1}{5} \cos x$.

Qu. 3. (i) $y = Ae^{2x} + Be^x + x + 2$; (ii) $y = Ae^{-3x} + Bx e^{-3x} + \frac{1}{16} e^x$.

Qu. 5. $y = Ae^{-bx/2a} \sin \left(\dfrac{\omega x}{2a} + \epsilon \right)$ where $\omega = \sqrt{(4ac - b^2)}$.

Exercise 7D   page 191

1. $y = Ae^{2x} + Be^{5x} - 2e^{3x}$.   2. $y = Ae^{3x/2} + Be^{-x} - 2$; $y = \frac{4}{5} e^{3x/2} + \frac{6}{5} e^{-x} - 2$.

3. $y = Ae^{(1 - \sqrt{2})x} + Be^{(1 + \sqrt{2})x} - \frac{1}{2} e^x - x + 2$.

4. $y = e^x (A \sin 2x + B \cos 2x) + x + \frac{2}{3}$;
$y = e^x (-\frac{3}{10} \sin 2x + \frac{3}{5} \cos 2x) + x + \frac{2}{3}$.

5. $y = A + Be^{2x/3} - (9 \sin 3x - 2 \cos 3x)/255$.

6. $y = e^x (A \sin x + B \cos x) + \frac{4}{5} \cos x + \frac{2}{5} \sin x$;
$y = \frac{4}{5}(1 - e^x) \cos x + \frac{2}{5}(1 + e^x) \sin x$.

7. $x = Ae^{-2t} + Be^{-3t} + (\sin t - \cos t)/10$.

8. $y = (Ae^{-x} + 2) \sin x + (Be^{-x} + 1) \cos x$; $y = 2(1 - e^{-x}) \sin x + \cos x$.

9. (a) (ii) $\frac{1}{5}$;   (iii) $y = Ae^{2x} + Be^{-3x} + \frac{1}{5} x e^{2x}$.
(b) $y = Ae^{2x} + Be^{-3x} - \frac{1}{5} x e^{-3x}$.

10. $y = Ae^{3x} + Bx\,e^{3x}$; $\frac{1}{2}$; $y = (A + Bx + \frac{1}{2}x^2)e^{3x}$.

11. $y = Ae^{-3x} + Be^x - \frac{1}{2}x\,e^{-3x}$. **12.** $y = A + Be^x - 3x$; $y = e^x - 3x + 2$.

13. $y = A\sin 2x + B\cos 2x + \frac{1}{4}x\sin 2x$. **14.** $y = (A + Bx + \frac{1}{2}x^2)e^{-2x}$.

15. $y = Ae^x + Be^{-x} + \frac{1}{2}x\sinh x$; $y = \frac{1}{2}(e^x - e^{-x} + x\sinh x)$. **16.** $4c/27$.

17. $e^{-2t}(\ddot{y} - \dot{y})$; $y = Ax + B/x^3$.

18. (i) $x = A\sin(\omega t + \epsilon) + (a\sin kt)/(\omega^2 - k^2)$;
   (ii) $x = A\sin(\omega t + \epsilon) - (\frac{1}{2}at/\omega)\cos\omega t$.

§7.6

Numbers in brackets indicate values obtained from an analytical solution
(only shown when different from that obtained from a step-by-step method).

	x	0	0.1	0.2	0.3
Qu. 1.	y	1	1.1 (1.11)	1.22 (1.24)	1.36 (1.40)
Qu. 2.	y	1	1.1052	1.2210 (1.2214)	1.3494 (1.3499)
Qu. 3.	y	1	1.0050	1.0201	1.0453

	x	0.4	0.5
Qu. 1.	y	1.53 (1.58)	1.72 (1.80)
Qu. 2.	y	1.4909 (1.4918)	1.6476 (1.6487)
Qu. 3.	y	1.0810 (1.0811)	1.1275 (1.1276)

Exercise 7E page 197

1. 0.7431, 0.7907, 0.8362.   **2.** (i) 0.9; (ii) 0.92.   **4.** 0.4073.

5. (i) $y_{n+1} \approx y_n + h\,y_n'$;   (ii) $y_n \approx y_{n+1} - h\,y_{n+1}'$.

6. $y = x + x^3/6 + x^5/120$; $y_1 = 0.100\,167$; $y_2 = 0.2010$;
   check $y_2 = 0.2013$.

7. 0, 0.2, 0.4003, 0.6018, 0.8060, 1.015.   **8.** 0.264 66, 0.529 76, 0.795 45.

9. 4.299 21, 5.912 63, 4.013 48; $y = 3\cosh x$; 4.299 26, 5.912 74, 4.006 94.

10. 0.517, 0.889.

11. $f''(x) \approx [f(x + h) - 2f(x) + f(x - h)]/h^2$; $-0.3547, -0.6690, -0.9001$;
   -1.338; $y = 3e(e^x - e^{-x})/(e^2 - 1) - 4x$; check: $f'(0.25) = -1.367$.

§8.1

Qu. 2. (i) undefined; (ii) vector **c** is multiplied by the scalar **a . b**; no.

Qu. 6. 3; $3\mathbf{i} + 2\mathbf{j}$.   **Qu. 11.** $\hat{\mathbf{a}} . \mathbf{b}$.

Exercise 8A page 203

1. Scalars: (ii), (viii); vectors: (iv), (v), (vi), (vii); undefined: (i), (iii).

2. (ii) 70;   (iv) $10\mathbf{i} + 25\mathbf{j}$;   (v) $19\mathbf{i} + 38\mathbf{k}$;
   (vi) $0.4\mathbf{i} + \mathbf{j}$;   (vii) $-40\mathbf{i} - 40\mathbf{j} + 40\mathbf{k}$; (viii) $\mathbf{0}$.

3. (i) $-22/147$;   (ii) $10/27$;   (iii) $2/29$.

4. (i) $7/27$;   (ii) $2/5\sqrt{66}$;   (iii) $18/\sqrt{462}$;   (iv) $\frac{1}{2}$.

5. (i) $\cos^{-1}\frac{1}{3} \approx 70.5°$;   (ii) $\cos^{-1}\sqrt{\frac{2}{3}} \approx 35.3°$.   12. $\mathbf{b} \cdot \hat{\mathbf{a}}$.

13. $\begin{pmatrix} x \\ y \end{pmatrix} = \begin{pmatrix} 0 \\ -c/b \end{pmatrix} + \lambda \begin{pmatrix} -b \\ a \end{pmatrix}$; $(-b\lambda, a\lambda - c/b)$; $\begin{pmatrix} a \\ b \end{pmatrix}$;
   $|ax_1 + by_1 + c|/\sqrt{(a^2 + b^2)}$.

14. $3\sqrt{10}$ units.   15. (i) 11 units;   (ii) 13 units;   (iii) 7 units.

16. $R\cos\alpha\cos\beta\,\mathbf{i} + R\cos\alpha\sin\beta\,\mathbf{j} + R\sin\alpha\,\mathbf{k}$; 9290 km.

§8.2

Qu. 1. $-5x + 7y + 2z + 9 = 0$.   Qu. 2. $\mathbf{b} = \lambda\mathbf{a} \Rightarrow \mathbf{n} = \mathbf{0}$.

Qu. 5. $7x - 13y - 2z - 39 = 0$.

Exercise 8B page 209

1. (i) $3x - 2y + 7z - 18 = 0$;   (ii) $x + 2y - 5z + 6 = 0$;
  (iii) $3x - 2z + 5 = 0$;   (iv) $y + 3 = 0$;
   (v) $2x - y - z - 9 = 0$;   (vi) $3x - 7y - 6 = 0$;
  (vii) $x + y - 3z = 0$;   (viii) $z = 0$.

2. (i) $x - y - 2z - 5 = 0$;   (ii) $7x + 7y - 3z - 21 = 0$;
  (iii) $x - y + 4z + 30 = 0$;   (iv) $y + 3z + 16 = 0$;
   (v) $15x - 2y - 6z = 0$;   (vi) $x - 5 = 0$;
  (vii) $23x + 10y + 17z - 30 = 0$; (viii) $z = 0$.

3. (i) $x + 3y - 2z + 1 = 0$;   (ii) $x + 4y - z + 20 = 0$;
  (iii) $x - 7y - 4z + 13 = 0$;   (iv) $35x + 21y + 15z - 105 = 0$;
   (v) $5x + 3y = 0$.

4. (i) $7x - 5y - 4z - 59 = 0$;   (ii) $3x + 12y + 17z - 43 = 0$;
  (iii) $x + 10y + 13z = 0$;   (iv) $4x - 2y + 11z - 51 = 0$;
   (v) $y = 0$.

5. (i) $(3, 1, 1)$; (ii) $(2, 0, -5)$; (iii) $(-22, 16, -7)$; (iv) $(3, -2, -7)$;
  (v) $(4, -3, 1)$;   (vi) $(-2, 3, -7)$.

7. $13x - 11y - 7z + 74 = 0$.   8. $13x + 12y - 5z + 57 = 0$.

9. $46x + 55y + 5z - 33 = 0$.   10. $5x - y - 7z + 5 = 0$.

11. (ii) $(2\lambda - 6, -4\lambda + 5, -3\lambda + 7)$; (iii) $(-2, -3, 1)$; (iv) $(2, -11, -5)$;
  (v) $(0, 1, -3)$;   (vi) $x/1 = (y - 1)/(-6) = (z + 3)/(-1)$.

12. (i) $(x-8)/7 = (y-4)/(-17) = (z-3)/14$;
    (ii) $(x-3)/10 = (y-5)/5 = (z-19)/7$;
    (iii) $(x+17)/3 = (y-8)/2 = (z+7)/1$.

13. (i) $2x+y-3z+11 = 0$ or $5x-y+3z-25 = 0$;
    (ii) $4x-5y+z-39 = 0$ or $x+y+z+9 = 0$.

14. (a) $x/a + y/b + z/c = 1$;   (b) (i) $14, 7, 14/3$;   (ii) $686/9$ units^3.

§8.3

Qu. 2. $x - 11y + z = 0$.   **Qu. 3.** A plane parallel to both π_1 and π_2.

Exercise 8C page 216

1. (i) $(x-3)/15 = (y-1)/27 = z/7$;
   (ii) $(x+1)/5 = (y+1)/23 = (z+1)/22$;
   (iii) $(x-8)/9 = (y-11)/24 = z/(-16)$;
   (iv) $(x-1)/1 = (y-1)/(-1) = (z-1)/0$.

2. (i) $(3,-1,-5)$;   (ii) $(-5,2,3)$;   (iii) $(1,3,11)$.

3. Planes form a triangular prism parallel to $2i-j-2k$.

4. (i) $x-y-z = 0$;   (ii) $x+2y+4z+1 = 0$;
   (iii) $6x-3y-z+1 = 0$;   (iv) $3x+2z+1 = 0$;
   (v) $4x-7y-9z-1 = 0$.

5. (i) $x+2y+2z-5 = 0$;
   (ii) $x+2y+2z-5 = 0$ or $4x+4y+7z-15 = 0$.

6. (i) $6x-y-z-16 = 0$;   (ii) $4x+2y+5z-9 = 0$;
   (iii) $2x-4y-z-5 = 0$;   (iv) $5x-21y-3z+5 = 0$.

7. (i) $5/\sqrt{3782}$;   (ii) $2/63$;   (iii) $56/165$.

8. (i) $18i-36j+45k$;   (ii) $-i-k$;   (iii) $(2i+13j+5k)/6$.

9. (i) 1; (ii) $1/7$; (iii) $\sqrt{33}/6$.   12. $x = y-1 = 4-z$.

13. 9 units.   14. $\sqrt{69}$ m; $(9,8,-5), (11,0,-6)$.   15. 818 m; $117°$.

16. (i) $(-24,-18,0)$;   (ii) $\dfrac{x}{4} = \dfrac{y}{3} = \dfrac{z-12}{2}$;   (iii) $4/\sqrt{29}, 3/\sqrt{29}, 2/\sqrt{29}$;
    (iv) $26\frac{1}{2}°, 33\frac{1}{2}°$.

17. $(1,-5,1), (-1,-1,5), (-1,5,-1), (1,1,-5), (-3,-3,-3), (-5,1,1)$.

§8.4

Qu. 1. (i) Straight line through A parallel to d;
        (ii) circle in plane parallel to $y = 0$, centre C, radius a.

Qu. 3. (i) $-3\sin t\,i + 3\cos t\,j + 2k$;   (ii) $ai + (b-gt)\,j$.

Qu. 4. $r = -ai + b(t+\pi)\,j - at\,k$.

Qu. 6. When $\dot\theta$ is constant, the acceleration is directed towards the centre of
        the circle.

Exercise 8D page 222

3. $4t^3$.   4. (i) 1; (ii) 2.   7. Parallel to **k**, constant; directed towards the origin.

8. Circle, centre at origin, radius 5; $10 \begin{pmatrix} -\sin(t^2) \\ \cos(t^2) \end{pmatrix} - 20\,t^2 \begin{pmatrix} \cos(t^2) \\ \sin(t^2) \end{pmatrix}$;

$-20t^2, 10$;   (a) $20t^2$;   (b) $2k$;   (c) 10.

9. $k = 3$, time $= 500$ s; $\sqrt{10}$ km.   10. (ii) $3\mathbf{i} + 4\mathbf{j}$ ms^{-1}.

11. $\frac{1}{5}[(t-6)\mathbf{i} + (16-t)\mathbf{j} + 5\mathbf{k}]$; 16 seconds.

12. $\dot{\theta}\mathbf{v}, -\dot{\theta}\mathbf{u}$; $[b\dot{\phi}(\cos\phi\,\mathbf{w} - \sin\phi\,\mathbf{v}) - b\dot{\theta}\cos\phi\,\mathbf{u}] + [a\dot{\theta}\mathbf{v}]$.

13. $-(\mathbf{v_1} - \mathbf{v_2}).(\mathbf{a_1} - \mathbf{a_2})/|\mathbf{v_1} - \mathbf{v_2}|^2$.

§8.5

Qu. 2. 0.   **Qu. 3.** No.   **Qu. 6.** $\frac{1}{2}|(\mathbf{q} - \mathbf{p}) \times (\mathbf{r} - \mathbf{p})|$.

Qu. 7. $2\mathbf{a} \times \mathbf{b}$.   **Qu. 12.** 0.

Exercise 8E page 227

1. (i) $-11\mathbf{i} + 14\mathbf{j} + \mathbf{k}$; (ii) $2\mathbf{j} - 2$.   2. $\sqrt{41}/21, 20/21$.

3. $\pm \frac{1}{13}(3\mathbf{i} + 4\mathbf{j} + 12\mathbf{k})$.   4. $11\frac{1}{2}$ units^2.   5. 2 units^3.   7. No.

8. $\mathbf{a} \times \mathbf{b} + \delta\mathbf{a} \times \mathbf{b} + \mathbf{a} \times \delta\mathbf{b} + \delta\mathbf{a} \times \delta\mathbf{b}$.   10. $[(\mathbf{a} - \mathbf{b}).(\mathbf{d} \times \mathbf{e})]/|\mathbf{d} \times \mathbf{e}|$.

11. $3/2\pi$ revs. per second; $\sqrt{2}$ ms^{-1}.   12. $2x + y - 3z = 5$; $(5, 3, -2)$.

15. $500/\sqrt{6}$.

§9.1

Qu. 1. (i) $\begin{pmatrix} -1 & 0 & 0 \\ 0 & -1 & 0 \\ 0 & 0 & 1 \end{pmatrix}$;   (ii) $\begin{pmatrix} 1 & 0 & 0 \\ 0 & 0 & 1 \\ 0 & 1 & 0 \end{pmatrix}$

(iii) $\begin{pmatrix} -\frac{1}{3} & 0 & 0 \\ 0 & -\frac{1}{3} & 0 \\ 0 & 0 & -\frac{1}{3} \end{pmatrix}$;   (iv) $\begin{pmatrix} 1 & 0 & 0 \\ 0 & 1 & 0 \\ 0 & 0 & 2 \end{pmatrix}$.

Qu. 2. (i) enlargement, centre O, scale factor 2;
(ii) rotation $180°$ about x-axis;
(iii) reflection in plane $x = y$.
(iv) shear, $z = 0$ invariant, $(0, 0, 1) \mapsto (3, 0, 1)$.

Qu. 3. $\begin{pmatrix} 0.01 & 0 & 0 \\ 0 & 0 & 0.01 \end{pmatrix}, \begin{pmatrix} 0 & 0.01 & 0 \\ 0 & 0 & 0.01 \end{pmatrix}$.

Qu. 4. (i) $\begin{pmatrix} 3x_1 + 2x_2 \\ 3y_1 + 2y_2 \end{pmatrix}$;   (ii) $\begin{pmatrix} 2x_1 + 3y_1 \\ x_1 - y_1 \\ 4x_1 + 2y_1 \end{pmatrix}$;

(iii) $\begin{pmatrix} 2x_2 + 3y_2 \\ x_2 - y_2 \\ 4x_2 + 2y_2 \end{pmatrix}$;   (iv) $\begin{pmatrix} 6x_1 + 9y_1 + 4x_2 + 6y_2 \\ 3x_1 - 3y_1 + 2x_2 - 2y_2 \\ 12x_1 + 6y_1 + 8x_2 + 4y_2 \end{pmatrix}$.

Qu. 5. (i) $\begin{pmatrix} \lambda x_1 + \mu x_2 \\ \lambda y_1 + \mu y_2 \end{pmatrix}$;   (ii) $\begin{pmatrix} ax_1 + dy_1 \\ bx_1 + ey_1 \\ cx_1 + fy_1 \end{pmatrix}$;   (iii) $\begin{pmatrix} ax_2 + dy_2 \\ bx_2 + ey_2 \\ cx_2 + fy_2 \end{pmatrix}$;

(iv) $\begin{pmatrix} \lambda ax_1 + \lambda dy_1 + \mu ax_2 + \mu dy_2 \\ \lambda bx_1 + \lambda ey_1 + \mu bx_2 + \mu ey_2 \\ \lambda cx_1 + \lambda fy_1 + \mu cx_2 + \mu fy_2 \end{pmatrix}$.

Qu. 7. (i), (iv).   **Qu. 10.** $(\lambda T)(\mathbf{p}) = \lambda T(\mathbf{p})$; $(\lambda \mathbf{A})_{ij} = \lambda a_{ij}$.

Exercise 9A page 237

1. (ii) Rotation through ϕ about origin;
   (iii) the product of two reflections is a rotation through twice the angle between the mirror lines about their intersection.

2. $x = y = z$; $\begin{pmatrix} 0 & 1 & 0 \\ 0 & 0 & 1 \\ 1 & 0 & 0 \end{pmatrix}$, $\begin{pmatrix} 1 & 0 & 0 \\ 0 & 1 & 0 \\ 0 & 0 & 1 \end{pmatrix}$; $120°$.

5. $\begin{pmatrix} X \\ Y \end{pmatrix} \mapsto \begin{pmatrix} (X + 2Y)/5 \\ (2X + 4Y)/5 \end{pmatrix}$; yes.

6. (a) (i) $(4, -3, 0)$; (ii) $(10, 1, 1)$;   (iii) $(4, -1, 3)$;   (iv) $(-4, 5, -2)$;
   (v) $(60, -20, 20)$;   (vi) $(X + 3Z, Y - Z, Z)$;

   (b) $\begin{pmatrix} 1 & 0 & 3 \\ 0 & 1 & -1 \\ 0 & 0 & 1 \end{pmatrix}$.

7. $\mathbf{r} = \mathbf{a} + \lambda \mathbf{d}$; $\mathbf{r} = T(\mathbf{a}) + \lambda T(\mathbf{d})$.

10. Point if $T(\mathbf{a}) = T(\mathbf{b}) = \mathbf{0}$.   12. Identity transformation.

13. Two distinct invariant points not collinear with the origin.

14. Rotations $\pm 90°$ or $180°$ about coordinate axes, $\pm 120°$ about $x = \pm y = \pm z$, $180°$ about $x = 0, y = \pm z$ or $y = 0, z = \pm x$ or $z = 0, y = \pm x$; or any of these 24 rotations followed by enlargement, centre O, scale factor -1.

15. $x' = ax + by + c$, $y' = dx + ey + f$;
    (i) $a = e = 1$, $b = d = 0$, $c = 3$, $f = 5$;
    (ii) $a = c = e = f = 0$, $b = -1$, $d = 1$;
    $a = e = 0$, $b = -1$, $c = -5$, $d = 1$, $f = 3$.

16. (i) $\sqrt{(\mathbf{x} \cdot \mathbf{x})}$;   (ii) $\cos^{-1}\left(\dfrac{\mathbf{x} \cdot \mathbf{y}}{|\mathbf{x}| \, |\mathbf{y}|}\right)$; $74.4°$, $\begin{pmatrix} 88 \\ -57 \\ -3 \\ 0 \end{pmatrix}$ etc.

§9.2

Qu. 2. A = 0.   **Qu. 4.** A, B both skew-symmetric of same order
$$\Rightarrow \text{A} + \text{B} \text{ skew-symmetric.}$$

Exercise 9B   page 241

1. C B A^{T}.   2. A^k is symmetric.

3. (b) **BC** symmetric $\Leftrightarrow$ **BC = CB**;
   (c) symmetric if k even, skew-symmetric if k odd.

§9.3

Qu. 1. $-\begin{vmatrix} b_1 & c_1 \\ b_3 & c_3 \end{vmatrix}, \; -\begin{vmatrix} a_1 & c_1 \\ a_2 & c_2 \end{vmatrix}, \; +\begin{vmatrix} a_1 & b_1 \\ a_2 & b_2 \end{vmatrix}.$

Qu. 2. Adjacent edges of 'parallelepiped' coincide.

Qu. 3. Sign of determinant is reversed.

Exercise 9C   page 246

1. (i) -135;   (ii) -28;   (iii) 0.

2. (i) $5(h-2)(4-k)$;   (ii) 0;   (iii) $abc + 2fgh - af^2 - bg^2 - ch^2$.

4. (i) $x = 4, y = -3, z = 2$;   (ii) $x = \frac{1}{4}, y = \frac{3}{4}, z = -\frac{1}{2}$.

7. (i) 630;   (ii) -7560;   (iii) $189x^2y^2/2$.   8. $-2.3^2.5^2.7^2.13$.

9. $|\text{a c e}| + |\text{a c f}| + |\text{a d e}| + |\text{a d f}| + |\text{b c e}| + |\text{b c f}| + |\text{b d e}| + |\text{b d f}|$.

12. (i) 16;   (ii) -146;   (iii) 0.   13. (iii) $\Sigma a_i B_i, \Sigma b_i C_i, \Sigma c_i A_i, \Sigma c_i B_i$.

15. 1.   17. $-2, 3$.

19. (i) $(y-z)(z-x)(x-y)$;   (ii) $(x+y+z)(y-z)(z-x)(x-y)$;
    (iii) $x(x+1)(x-1)^3$.

§9.4

Qu. 1. $\begin{pmatrix} 4 & -12 & 14 \\ 2 & -12 & 16 \\ -4 & 9 & -11 \end{pmatrix}, \begin{pmatrix} -\frac{2}{3} & 2 & -\frac{7}{3} \\ -\frac{1}{3} & 2 & -\frac{8}{3} \\ \frac{2}{3} & -\frac{3}{2} & \frac{11}{6} \end{pmatrix}.$ **Qu. 5.** (i) No; (ii) Yes.

Exercise 9D   page 251

1. $\begin{pmatrix} 0 & 14 & -7 \\ -2 & 3 & 1 \\ 3 & -8 & 2 \end{pmatrix}$; 7I; $\det \text{M} = 7$.   2. 49; 7M.

3. (i) $\frac{1}{5}\begin{pmatrix} -3 & 4 & -7 \\ -2 & 1 & 2 \\ 2 & -1 & 3 \end{pmatrix}$; (ii) $\frac{1}{25}\begin{pmatrix} 86 & 32 & 3 \\ -13 & -6 & 1 \\ 11 & 7 & 3 \end{pmatrix}$; (iii) singular;

(iv) $\frac{1}{30}\begin{pmatrix} 27 & 33 & -63 \\ -4 & -6 & 16 \\ 30 & 30 & -60 \end{pmatrix}$.

4. $49\mathbf{I}$; $x = \frac{2}{7}, y = -\frac{1}{7}, z = \frac{1}{7}$.   **5.** $\frac{1}{4}\begin{pmatrix} 6 & 0 & -2 \\ 2 & -1 & 1 \\ 0 & -1 & 3 \end{pmatrix}$; $x = -2, y = \frac{13}{4}, z = \frac{33}{4}$.

6. $\begin{pmatrix} d & -b \\ -c & a \end{pmatrix}$; $\dfrac{1}{ad - bc}\begin{pmatrix} d & -b \\ -c & a \end{pmatrix}$.

7. (i) -10;   (ii) $-2, 4$;   (iii) $2, 3$;   (iv) $-1, 1, 2$.

8. (i) $\begin{pmatrix} -4 & 1 & 11 \\ -10 & 1 & -1 \\ -5 & -7 & 22 \end{pmatrix}$; (ii) 990; (iii) $\begin{pmatrix} 23 & 9 & -11 \\ 4 & -10 & -14 \\ 2 & 11 & 6 \end{pmatrix}$; (iv) 990;

(v) -33;   (vi) -30.

9. -22; taking conjugates gives the transpose.

12. Yes.   **13.** (i) ± 1;   (ii) rotation θ about O, reflection in $y = x \tan \frac{1}{2}\theta$.

14. $\mathbf{M}^{-1} = \mathbf{M}^{\mathrm{T}}$.

15. $\mathbf{B}^{\mathrm{T}} = \begin{pmatrix} 2 & -2 & 1 \\ 2 & 1 & -2 \\ 1 & 2 & 2 \end{pmatrix}$, $\mathbf{B}\mathbf{B}^{\mathrm{T}} = 9\mathbf{I}$, $\mathbf{B}^{-1} = \frac{1}{9}\mathbf{B}^{\mathrm{T}}$, $\mathbf{C} = \frac{1}{9}\begin{pmatrix} 1 & -4 & 8 \\ 4 & -7 & -4 \\ 8 & 4 & 1 \end{pmatrix}$.

16. $\det \mathbf{A} = 1$, $\mathbf{A} = \frac{1}{15}\begin{pmatrix} 5 & -10 & 10 \\ -14 & -5 & 2 \\ 2 & -10 & -11 \end{pmatrix}$.

§9.5

Qu. 2. a, b, c parallel; $\mathbf{c} = \mathbf{0} \Rightarrow x = y = 0, z = 1$ is a solution;
$\mathbf{c} \neq \mathbf{0} \Rightarrow \mathbf{a} = \alpha\mathbf{c}$ for some $\alpha \Rightarrow x = 1, y = 0, z = -\alpha$ is a solution.

Exercise 9E page 256

1. Triangular prism.

2. (i) Inconsistent, triangular prism;
(ii) $x = 2 + 4\lambda, y = 1 + 13\lambda, z = 1 + 5\lambda$.

3. $k = -2, \emptyset$; $k = 6, \{(\lambda, 4 - \lambda, \lambda - 3)\}$.

4. 0; $x = -1 - 3\lambda, y = \lambda, z = 5 + 2\lambda$.

5. (i) $x = 5, y = 1, z = -15$;   (ii) inconsistent;
(iii) $x = 0, y = \lambda, z = 7 - 2\lambda$;   (iv) $x = \lambda, y = \mu, z = 7 - 4\lambda - 2\mu$.

7. $k = \frac{13}{2}$; $x = -4\lambda, y = 2\lambda, z = \lambda$. **8.** $\lambda = -2$; $x = 1, y = \frac{8}{5}, z = \frac{2}{5}$.

9. $b = c \cos A + a \cos C$, $c = a \cos B + b \cos A$.

11. $k = 13$; $x = 1, y = 2$.   **12.** $a = 4, x = 5, y = -1$; $a = \frac{32}{3}, x = \frac{15}{7}, y = \frac{3}{7}$.

13. (iv) All cofactors zero.

14. (i) $x = -\frac{9}{35}, y = \frac{8}{35}$; inconsistent;   (ii) $x = 2z$;   (iii) $(\frac{2}{3}, 1, \frac{1}{3})$.

§9.6

Qu. 2. $x = -7 + 7\lambda, y = 6 - 5\lambda, z = \lambda$.

Qu. 3. (i) $\begin{pmatrix} 0 & 0 & 1 \\ 0 & 1 & 0 \\ 1 & 0 & 0 \end{pmatrix}$;   (ii) $\begin{pmatrix} 1 & 0 & 0 \\ 0 & -1 & 0 \\ 0 & 0 & 1 \end{pmatrix}$;   (iii) $\begin{pmatrix} 1 & 0 & 0 \\ 0 & 1 & 0 \\ 0 & -4 & 1 \end{pmatrix}$.

Qu. 4. Elementary column operation.

Exercise 9F   page 264

1. $x = \frac{7}{6}, y = -\frac{4}{3}, z = 0$.   2. $p = 1, q = 3, r = -2, s = 5$.

3. $t = 6, u = 1, x = 4, y = 12, z = 10$.   4. $x = 1.97, y = -0.03$.

5. $x_1 = 0.077, x_2 = 0.115, x_3 = 0.577$; residuals $0, -0.001, 0$; solutions correct to 3 D.P.

6. $x_1 = 0.897, x_2 = 0.764, x_3 = 0.615$; residual -0.0013, suggests that solutions are correct to within ± 0.001.

7. 0.3 seconds, 5×10^{146} years.

8. (i) $\begin{pmatrix} -\frac{5}{14} & \frac{4}{7} \\ \frac{3}{14} & -\frac{1}{7} \end{pmatrix}$;   (ii) $\begin{pmatrix} \frac{1}{2} & -\frac{1}{6} & -\frac{1}{6} \\ -\frac{1}{6} & -\frac{1}{18} & \frac{5}{18} \\ -\frac{2}{3} & \frac{7}{9} & \frac{1}{9} \end{pmatrix}$;   (iii) $\begin{pmatrix} -\frac{1}{3} & \frac{11}{57} & \frac{2}{19} \\ 1 & -\frac{14}{19} & \frac{1}{19} \\ -\frac{4}{3} & \frac{65}{57} & -\frac{2}{19} \end{pmatrix}$.

10. (i) $\begin{pmatrix} 2 & 0 & 0 \\ -1 & 1 & 0 \\ 4 & 2 & 3 \end{pmatrix}, \begin{pmatrix} 1 & 5 & -1 \\ 0 & 1 & 2 \\ 0 & 0 & 1 \end{pmatrix}$;   (iii) $x = 12, y = 0, z = -1$.

11. (i) $x = -\frac{258}{7}, y = \frac{16}{7}, z = \frac{88}{7}$;   (ii) $x = 7 - 4\lambda, y = \lambda - 3, z = \lambda$.

12. $\begin{pmatrix} 3 & 0 & 0 \\ 2 & -\frac{14}{3} & 0 \\ 5 & -\frac{17}{3} & \frac{3}{2} \end{pmatrix} \begin{pmatrix} 1 & \frac{4}{3} & \frac{1}{3} \\ 0 & 1 & -\frac{1}{2} \\ 0 & 0 & 1 \end{pmatrix}$; $\frac{1}{21} \begin{pmatrix} 15 & 23 & -14 \\ -3 & -13 & 7 \\ -12 & -17 & 14 \end{pmatrix}$.

13. (i) $x = -198, y = 100$;   (ii) $x = 202, y = -100$.

14. (ii) $x = -19 - 11\lambda, y = 8 + 4\lambda, z = \lambda$;   (a) $x = 10.16, y = z = -2.58$;
                                        (b) $x = 10.52, y = z = -2.76$;
(iii) (a) $x = -178.4, y = z = 80$;   (b) $x = 181.8, y = z = -80$.
(iv) Determinant 'small', equations 'nearly inconsistent'.

§9.7

Qu. 1. (i) $\begin{pmatrix} 4 \\ 1 \end{pmatrix}$;   (ii) $\begin{pmatrix} 10 \\ 5 \end{pmatrix}$;   (iii) $\begin{pmatrix} 2 \\ 3 \end{pmatrix}$;   (iv) $\begin{pmatrix} 0 \\ 5 \end{pmatrix}$;   (v) $\begin{pmatrix} -2 \\ 2 \end{pmatrix}$;   (vi) $\begin{pmatrix} -6 \\ 1 \end{pmatrix}$.

Qu. 2. (i) $y = \frac{1}{4}x$;   (ii) $y = \frac{1}{2}x$;   (iii) $y = \frac{3}{2}x$;   (iv) $x = 0$;
        (v) $y = -x$;   (vi) $y = -\frac{1}{6}x$.

Qu. 3. (i) $3, \begin{pmatrix} k \\ 0 \end{pmatrix}$; $-4, \begin{pmatrix} 0 \\ k \end{pmatrix}$; (ii) $1, \begin{pmatrix} k \\ 0 \end{pmatrix}$; (iii) no real eigenvectors;

(iv) 3, all vectors; (v) $0, \begin{pmatrix} 2k \\ -k \end{pmatrix}$; $8, \begin{pmatrix} 2k \\ 3k \end{pmatrix}$.

Qu. 4. $\begin{pmatrix} \frac{7}{4} & -\frac{3}{4} \\ -\frac{1}{4} & \frac{1}{4} \end{pmatrix} \begin{pmatrix} 6 & -3 \\ 7 & -4 \end{pmatrix} \begin{pmatrix} 1 & 3 \\ 1 & 7 \end{pmatrix} = \begin{pmatrix} 3 & 0 \\ 0 & -1 \end{pmatrix}$.

Qu. 6. (i) Yes; (ii) Yes.

Exercise 9G page 271

1. $4, \pm\begin{pmatrix} 1/\sqrt{2} \\ 1/\sqrt{2} \end{pmatrix}$; $5, \pm\begin{pmatrix} 2/\sqrt{5} \\ 1/\sqrt{5} \end{pmatrix}$; $y = x, y = \frac{1}{2}x$.

2. $y = -x$, repels; $y = 3x$ attracts.

3. $y = \frac{7}{2}x$; ± 2. 4. $\begin{pmatrix} \frac{1}{9} & -\frac{1}{9} \\ \frac{5}{9} & \frac{4}{9} \end{pmatrix} \begin{pmatrix} 3 & 4 \\ 5 & 2 \end{pmatrix} \begin{pmatrix} 4 & 1 \\ -5 & 1 \end{pmatrix} = \begin{pmatrix} -2 & 0 \\ 0 & 7 \end{pmatrix}$.

5. $1, \begin{pmatrix} 3k \\ 4k \end{pmatrix}$, $-0.4, \begin{pmatrix} -k \\ k \end{pmatrix}$; (i) $\begin{pmatrix} \frac{1}{7} & \frac{1}{7} \\ -\frac{4}{7} & \frac{3}{7} \end{pmatrix} \begin{pmatrix} 0.2 & 0.6 \\ 0.8 & 0.4 \end{pmatrix} \begin{pmatrix} 3 & -1 \\ 4 & 1 \end{pmatrix} = \begin{pmatrix} 1 & 0 \\ 0 & -0.4 \end{pmatrix}$;

(ii) $\frac{1}{7}\begin{pmatrix} 3+4(-0.4)^n & 3-3(-0.4)^n \\ 4-4(-0.4)^n & 4+3(-0.4)^n \end{pmatrix} \rightarrow \frac{1}{7}\begin{pmatrix} 3 & 3 \\ 4 & 4 \end{pmatrix}$.

6. $\cos\theta \pm j\sin\theta$.

7. $\begin{pmatrix} k\cos\theta \\ -\sin\theta \end{pmatrix}$, $\begin{pmatrix} k\sin\theta \\ \cos\theta \end{pmatrix}$; $k = \pm 1$, reflection in $y = \mp x\tan\theta$.

8. If $a = c$ and $b = 0$ all vectors are eigenvectors.

9. $1, \begin{pmatrix} 5k \\ -3k \end{pmatrix}$; $5, \begin{pmatrix} -2k \\ k \end{pmatrix}$; $\begin{pmatrix} 3 \\ -2 \end{pmatrix} = \begin{pmatrix} -2 \\ 1 \end{pmatrix} + \begin{pmatrix} 5 \\ -3 \end{pmatrix}$.

10. $\begin{pmatrix} 1 \\ -2 \end{pmatrix}, \begin{pmatrix} 6 \\ 2 \end{pmatrix}$; $\begin{pmatrix} \frac{11}{7} \\ \frac{6}{7} \end{pmatrix}, \begin{pmatrix} \frac{9}{7} \\ -\frac{4}{7} \end{pmatrix}$; $\begin{pmatrix} \frac{11}{7} & \frac{9}{7} \\ \frac{6}{7} & -\frac{4}{7} \end{pmatrix}$.

11. (i) (a) $\frac{1}{1}, \frac{4}{2} = 2, \frac{10}{6} = 1.667, \frac{28}{16} = 1.750, \frac{76}{44} = 1.727, \frac{208}{120} = 1.733$;

(b) $\frac{10}{1}, \frac{13}{11} = 1.182, \frac{46}{24} = 1.917, \frac{118}{70} = 1.686, \frac{328}{188} = 1.745, \frac{892}{516} = 1.729$.

(ii) $1 \pm \sqrt{k}, \begin{pmatrix} \pm\sqrt{k} \\ 1 \end{pmatrix}$.

12. $(9, 18, 18), 1:9$; $(2, -2, 1)$; $k = -3, (2, 1, -2)$.

13. $5; 2, \begin{pmatrix} k \\ 0 \\ -k \end{pmatrix}$; $4, \begin{pmatrix} k \\ k \\ k \end{pmatrix}$.

14. $1, -2, 3$; line $x = y = z$ is enlarged from O with scale factor 3;

$\frac{1}{6}\begin{pmatrix} 4 & -7 & 5 \\ -2 & 5 & -1 \\ -2 & 8 & -4 \end{pmatrix}$; transformation $(\mathbf{A} - 3\mathbf{I})$ is singular, mapping all points of $x = y = z$ to O.

5. (i) $\begin{pmatrix} 4 & 9 & 2 \\ 3 & 5 & 7 \\ 8 & 1 & 6 \end{pmatrix}$;   (ii) 15;   (iii) $\begin{pmatrix} 1 \\ 1 \\ \vdots \\ 1 \end{pmatrix}$ is eigenvector with eigenvalue $\frac{1}{2}n(n^2+1)$.

16. (ii) all zero.

17. $\begin{pmatrix} 1 \\ 1 \\ -1 \end{pmatrix}$, $\begin{pmatrix} 1 \\ 1 \\ -2 \end{pmatrix}$, $\begin{pmatrix} 0 \\ 1 \\ -1 \end{pmatrix}$; $\mathbf{P} = \begin{pmatrix} 1 & 1 & 0 \\ 1 & 1 & 1 \\ -1 & -2 & -1 \end{pmatrix}$, $\mathbf{P}^{-1} = \begin{pmatrix} 1 & 1 & 1 \\ 0 & -1 & -1 \\ -1 & 1 & 0 \end{pmatrix}$;

$\mathbf{C} = \begin{pmatrix} 1 & 0 & 0 \\ 0 & 2 & 0 \\ 0 & 0 & 3 \end{pmatrix}$, etc.; $\mathbf{B} = \begin{pmatrix} 1 & -1 & -1 \\ -2 & 2 & -1 \\ 2 & 0 & 3 \end{pmatrix}$, etc.

20. $2, \begin{pmatrix} 0 \\ 1 \\ 0 \end{pmatrix}$; $2 + \sqrt{2}, \dfrac{1}{\sqrt{(4+2\sqrt{2})}} \begin{pmatrix} 1 \\ 0 \\ 1+\sqrt{2} \end{pmatrix}$; $2 - \sqrt{2}, \dfrac{1}{\sqrt{(4-2\sqrt{2})}} \begin{pmatrix} 1 \\ 0 \\ 1-\sqrt{2} \end{pmatrix}$.

21. $\frac{1}{2} \begin{pmatrix} 1 & \sqrt{2} & 1 \\ \sqrt{2} & 0 & -\sqrt{2} \\ 1 & -\sqrt{2} & 1 \end{pmatrix}$ or $\frac{1}{2} \begin{pmatrix} 1 & 1 \\ \sqrt{2} & -\sqrt{2} \\ 1 & 1 \end{pmatrix}$.

22. (i) $\ddot{x} = -11x + 6y$, $\ddot{y} = 6x - 6y$;   (ii) $\sqrt{2}, \begin{pmatrix} 2k \\ 3k \end{pmatrix}$; $\sqrt{15}, \begin{pmatrix} -3k \\ 2k \end{pmatrix}$.

23. $\begin{pmatrix} 1 & \frac{1}{4} & 0 \\ 0 & \frac{1}{2} & 0 \\ 0 & \frac{1}{4} & 1 \end{pmatrix}$; 6 generations.

24. (a) $\begin{pmatrix} \frac{1}{4} & \frac{1}{3} & 0 \\ \frac{3}{4} & 0 & \frac{1}{2} \\ 0 & \frac{2}{3} & \frac{1}{2} \end{pmatrix}$;   (b) $\frac{3}{8}$;   (c) $A:B:C = 4:9:12$.

§10.1

Qu. 2. (i) $\left(2, -\frac{3\pi}{4}\right)$;   (ii) $\left(3, -\frac{\pi}{6}\right)$;   (iii) $\left(1, \frac{2\pi}{3}\right)$;   (iv) $\left(4, \frac{\pi}{2}\right)$.

Qu. 3. (i) $(-\sqrt{2}, -\sqrt{2})$; (ii) $\left(\frac{3\sqrt{3}}{2}, -\frac{3}{2}\right)$;   (iii) $\left(-\frac{1}{2}, \frac{\sqrt{3}}{2}\right)$;   (iv) $(0, 4)$.

Qu. 4. (i) $\left(4, \frac{3\pi}{4}\right)$;   (ii) $\left(2, -\frac{\pi}{3}\right)$;   (iii) $(5, \pi)$;   (iv) $\left(13, \tan^{-1}\frac{5}{12}\right)$.

Qu. 5. Two circuits anticlockwise.

Exercise 10A page 280

6. $x \cos\alpha + y \sin\alpha = p$.   **8.** $r^2 - 2ar \cos(\theta - \alpha) + a^2 - b^2 = 0$.

9. (i) $\left(2, \frac{\pi}{3}\right)$;   (ii) $r \cos\left(\theta + \frac{\pi}{3}\right) + \frac{3}{2} = 0$.

10. (i) $r(3\cos\theta + 4\sin\theta) = 7$;   (ii) $r = 8\sin\theta$;
  (iii) $r = 6\tan\theta\sec\theta$;   (iv) $r^2 = \cos 2\theta$.

11. (i) $y = \sqrt{3}x$;   (ii) $x^2 + y^2 - 4x - 10 = 0$;   (iii) $xy = 1$;
(iv) $x^2 + y^2 = \sqrt{(2xy)}$.

12. $y^2 = a^2 - 2ax$.   **13.** $\left(a, \pm\dfrac{\pi}{3}\right); r = \frac{1}{2}a \sec\theta$.

15. (i) Enlargement $\times k$, rotation α.   **16.** $(x^2 + y^2)(x - a)^2 = b^2 x^2$.

Exercise 10B   page 288

2. (i) $16, (0, 0), (4, 0), x = -4$;   (ii) $1, (0, 0), (-\frac{1}{4}, 0), x = \frac{1}{4}$;
   (iii) $4, (-2, 3), (-1, 3), x = -3$;   (iv) $6, (0, 0), (0, \frac{3}{2}), y = -\frac{3}{2}$;
    (v) $8, (5, -1), (3, -1), x = 7$;   (vi) $2, (-4, 7), (-4, \frac{13}{2}), y = \frac{15}{2}$.

3. (i) $(y - 4)^2 = 8(x - 1)$;   (ii) $(x + 5)^2 = 16(y - 2)$;
   (iii) $(y + 1)^2 = 12(2 - x)$.

4. 45.2 m.   **5.** $(a/m^2, 2a/m)$.   **6.** $(apq, a(p + q))$.

9. $t_1 + t_2 = Y/a, t_1 t_2 = X/a$;   tangent.   **11.** $(2at^2, 3at)$.

12. $y^2 = 2ax - 8a^2$;   parabola.   **13.** $a(2 - 2^{-n}), 2a$.

17. (ii) $y = kx/t + ht$;   (iii) $(hk, 0)$.

19. (i) $y + rx = a(p + q + pqr)$;   (ii) $(-a, a(p + q + r + pqr))$.

20. $(uv/g, v^2/2g), 2u^2/g$.

§10.3

Qu. 2. From a circle ($e = 0$) to a straight line segment ($e = 1$).

Qu. 4. (i) 0.8;   (ii) 0.28;   (iii) $\sqrt{3}/2 \approx 0.866$.   **Qu. 6.** $2a$.

Qu. 7. At the centre of the circle.   **Qu. 11.** Stretch $\begin{pmatrix} a/b & 0 \\ 0 & 1 \end{pmatrix}$.
Qu. 12. b/a.

Exercise 10C   page 300

1. $e = 0.6$.   **2.** 1.14×10^{10} km.   **5.** $\frac{8}{481} \approx 0.0166$.

7. (i) centre $(3, 1)$, foci $(0, 1), (6, 1)$, directrices $x = -\frac{2}{3}, x = \frac{20}{3}$;
   (ii) centre $(-2, 2)$, foci $(-4, 2), (0, 2)$, directrices $x = -6, x = 2$.

16. $y = -x + 3, (2, 1); y = 11x + 27, (-\frac{22}{9}, \frac{1}{9})$.

25. Similar ellipse with axes parallel to the original and diameter OL.

27. (i) Ellipse, foci S, S'.

§10.4

Qu. 3. (a) branch near S';   (b) branch near S.

Qu. 4. Line SS' excluding line segment SS'.   **Qu. 5.** Becomes less pointed.

Qu. 11. Away from O in quadrant 1, towards O in quadrant 3, away from O in quadrant 2, towards O in quadrant 4.

Qu. 14. (i) line pair $\dfrac{x}{a} = \pm\dfrac{y}{b}$;

(ii) hyperbola in regions above and below asymptotes.

Exercise 10D page 312

1. $e = 1.25$.   **4.** $\dfrac{x^2}{9} - \dfrac{y^2}{27} = 1$.   **5.** $9x^2 - y^2 = 20$.

6. $9x - 20y = 1$, $20x + 9y = 216$.

7. (i) $(b^2 - a^2 m^2)x^2 - 2a^2 mcx - a^2(b^2 + c^2) = 0$;
(ii) $(b^2 - a^2 m^2)x^2 - 2a^2 mcx - a^2 c^2 = 0$.

9. $a^2 y_0(y - y_0) = b^2 x_0(x - x_0)$.

13. $a^2 m^2 = b^2 + c^2 \Rightarrow y = mx + c$ touches H or is an asymptote.

14. The perpendicular tangents are asymptotes, intersecting at the point-circle O.

15. $x = -a \cosh\phi, y = b \sinh\phi$.   **18.** $(at, bt), \left(\dfrac{a}{t}, -\dfrac{b}{t}\right)$.   **19.** $t = e^\phi$.

21. $(\sqrt{2}, 2\sqrt{2}), (-\sqrt{2}, -2\sqrt{2}); y + 2x = \pm 4\sqrt{2}$;
$(9/2\sqrt{2}, -1/\sqrt{2}), (-9/2\sqrt{2}, 1/\sqrt{2})$.

22. $(c\sqrt{2}, c\sqrt{2}), (-c\sqrt{2}, -c\sqrt{2}), x + y = \pm c\sqrt{2}$.   **23.** $\left(-ct, -\dfrac{c}{t}\right)$.

24. $(h, k), x = h, y = k; (-\frac{1}{2}, -1), (\frac{3}{2}, 3)$.

§10.5

Qu. 1. Plane through vertex.

Qu. 3. No; angle between asymptotes cannot exceed vertical angle of cone.

Qu. 4. Circle.

Exercise 10E page 318

1. $2\pi/3$.   **6.** $r = l \sec\theta/e$.

13. Cone, vertex O, axis OC, semi-vertical angle α;
$(ax + by + cz)^2 = (a^2 + b^2 + c^2)(x^2 + y^2 + z^2) \cos^2 \alpha$.

15. (i) No;   (ii) Yes, when $\lambda = -(a^2 + b^2)/2$;   (iii) $\emptyset$.

§11.1

Qu. 1. 2; 3.5, 1.25.   **Qu. 2.** (i) $0, 10^{-n}$; (ii) $\pm 5 \times 10^{-n-1}$.

Qu. 3. 6.   **Qu. 4.** 8%.   **Qu. 6.** B.

Qu. 8. $f(5) = -108, f(6) = -476, f(-1) = -70$.

Exercise 11A page 326

2. $1, -5$.

4. $2.59, 0.016\,125$; $5.90, 0.0487$; $-3.31, 0.067\,575$.

5. $0°, 30°$; $28.99° < x < 30.98°$.   7. $-3kx\,\epsilon\,(a^2 + x^2)^{-5/2}$; (i) 0; (ii) $\pm a/2$

8. $e^\epsilon - 1 \approx \epsilon$.   9. $3.430, -2.455$.

10.

x	f	Δf	$\Delta^2 f$	$\Delta^3 f$	$\Delta^4 f$
0.1	3.0169				
		375			
0.2	3.0544		-10		
		365		-240	
0.3	3.0909		-250		-24.
		115		-264	
0.4	3.1024		-514		
		-399			
0.5	3.0625				

11. $y = 72$ when $x = 6$.   12. 5.

13. (i) $y = 133.40$ when $x = 11$, transposition error;
    (ii) $a = 103.32, b = 7.30, c = 0.06, d = 0.006$; 121.05, 121.07.

14. 5.480.

§11.2

Qu. 3. (a) 13.5;   (b) -2.45;   (c) 0.076.

Exercise 11B page 335

1. 1.96.   2. 0.865.   3. $\sqrt{20}$, no.   5. Converges to 0.

6. 1.1, 4.8; converges to larger root; 4.78.   7. 1.17, 0.70.

8. 0.876 73.   9. (i) $X := X$ NEW;   (ii) 2.10.

10. $a = 0.5, b = 10$; $x_2 = 3.833, x_3 = 3.109, x_4 = 3.717$.

11. (a) is better;   (a) $x_2 = 2.333, x_3 = 2.290$;   (b) $x_2 = 2.5, x_3 = 2.21$.

12. (iii) $0 < x_1 < \dfrac{2}{a}$.   13. $x_n = a + 4^{1-n}\epsilon, y_n = 2a + 2^{1-2n}\epsilon$.

14. $(1 + \sqrt{13})/2$.   **15.** Yes.   **16.** $\frac{199}{89}$.

17. (i) $\alpha_n \to 0.299, \beta_n \to 16.701$;   (ii) no limits;   (iii) sequences oscillate.

§11.3

Qu. 1. (i), (ii) to infinity;   (iii) oscillates finitely;   (iv) oscillates infinitely.

Exercise 11C page 342

1. $\frac{1}{2}$; 250.   **4.** $(1-x)^{-2}$.   **6.** 21.   **7.** 0.965.

8. $\ln[(n+2)!/2(n+3)^n]$; no.   **9.** $(4x^2 + 5x + 6)/(x^3 - 1)$.

10. $x > 1, x < 0$; $x = -1 - \sqrt{2}$.   **14.** 14.4.   **17.** No.

§11.4

Qu. 2. $\ln 2 = 1 - \frac{1}{2} + \frac{1}{3} - \frac{1}{4} + \dots$.

Exercise 11D page 348

1. (i) 0.197;   (ii) -0.465;   (iii) 0.775.   **2.** (i) $\frac{120}{119}$;   (ii) $\frac{1}{239}$.

7. $-1 \leqslant x < 1$.

10.   (i) $3x - \frac{9}{2}x^2 + 9x^3 - \frac{81}{4}x^4$, $-\frac{1}{3} < x \leqslant \frac{1}{3}$;
   (ii) $-2x - 2x^2 - \frac{8}{3}x^3 - 4x^4$, $-\frac{1}{2} \leqslant x < \frac{1}{2}$;
   (iii) $5x - \frac{5}{2}x^2 + \frac{35}{3}x^3 - \frac{65}{4}x^4$, $-\frac{1}{3} < x \leqslant \frac{1}{3}$;
   (iv) $x - \frac{13}{2}x^2 + \frac{19}{3}x^3 - \frac{97}{4}x^4$, $-\frac{1}{3} < x \leqslant \frac{1}{3}$;
   (v) $\frac{3}{2}x - \frac{9}{4}x^2 + \frac{9}{2}x^3 - \frac{81}{8}x^4$, $-\frac{1}{3} < x \leqslant \frac{1}{3}$;
   (vi) $\frac{2}{3}x + \frac{2}{3}x^2 + \frac{8}{9}x^3 + \frac{4}{3}x^4$, $-\frac{1}{2} \leqslant x < \frac{1}{2}$.

11. $y + \frac{1}{2}y^2 - \frac{2}{3}y^3 + \frac{1}{4}y^4$; $1/n$ if $n \neq 3k, -2/n$ if $n = 3k$.

12. (i) 0.5833;   (ii) 0.6823;   (iii) 0.6931.   **13.** 2.3026.

17. 0.092.   **19.** 0.3103.

§12.1

Qu. 4. No.

Exercise 12A page 357

1. 7.   **2.** $26\frac{1}{3}$.   **3.** $4(\sqrt{3} - 1) - \pi/3 \approx 1.88$.

4. $(1 + e^2)/2 \approx 4.19$.   **5.** 0.   **6.** 0.   **7.** $4\pi/3 \approx 4.19$.

8. $9\frac{1}{3}$.   **9.** $0, (15 \pm \sqrt{33})/4$.   **10.** π.   **12.** $\pi^3/96$.   **14.** 0,0.

17. 0.509.   **19.** $P(x) \equiv x^6 - 4x^5 + 5x^4 - 4x^2 + 4, A = -4$.

20. $\frac{2}{\lambda}(1 - e^{-\pi\lambda/2}), \frac{\pi}{\lambda}(1 - e^{-\lambda})$.

Exercise 12B page 361

1. 54.
2. $\frac{1}{2}$.
4. (i) 2.60; (ii) 5.75.

6. (i) $2\pi/3$; (iii) (a) $8/\pi \approx 2.55$, (b) $\sqrt{8} \approx 2.83$.
8. 90.

9. 75 m.
11. (ii) (b); (iii) $4R/\pi$.
13. (i) $5\frac{1}{3}$ cm; (ii) $8\frac{1}{6}$ cm.

14. $3R/4$.
16. $\frac{3}{4}, \frac{6}{5}, \frac{2}{5}$.
17. $\mu = \frac{97}{60}; \frac{77}{160}$.
18. 0.727 units from A.

20. $(2k/(k+1), 0)$; $\rightarrow (2, 0)$.

Exercise 12C page 365

1. $210\frac{4}{7}$.
2. $75\pi/8 \pm 25$.
3. $a^2/2$.
5. 103%.
6. $\frac{5}{4}\pi a^2$.

8. (i) Circle, area πb^2; (ii) square, area $4b^2$.
10. $7 + 3\pi/4$.

13. $2\pi a^2$.
14. $\frac{16}{35}$.

§12.4

Qu. 1.

x	10	10^3	10^6	10^{10}	10^{50}
(i)	2.3×10^{-1}	6.9×10^{-3}	1.4×10^{-5}	2.3×10^{-9}	1.2×10^{-48}
(ii)	1.8	3.5	3.5	2.3	1.2×10^{-3}
(iii)	2.3	6.9	14	23	100

Qu. 2. (i) 230; (ii) 2.3×10^{-6}.
Qu. 4. 96.

Exercise 12D page 368

1. $\ln \ln x$, $(\ln x)^{20}$, $x \ln x$, $\sinh x$, $(\ln x)^x$, e^{x^3}, e^{e^x}.

4. $2x(1 - x^2)e^{-x^2}$;
   increasing for $x < -1$ and $0 < x < 1$, decreasing for $-1 < x < 0$ and $x > 1$.

5. Inflexions when $x = 0$ or $3 \pm \sqrt{3}$; $y = 0$, $y = 4e^{-2}x$.

6. Maximum $(0, c)$ if $c \geqslant 2$;
   minimum $(0, c)$, maxima $(\pm \sqrt{(2 - c)}, e^{(c-2)}(4 - c))$ if $c < 2$.

7. $1/e$; $(e^{3/2}, 3e^{-3/2}/2)$; (a) 2, (b) 0; 1.

10. (ii) $2e^{-1/x^2}/x^3$; $e^{-1/x^2}\left(\dfrac{4}{x^6} - \dfrac{6}{x^4}\right)$.

Exercise 12E page 373

1. (i) $m < -1$; (ii) $m > -1$; (iii) none.
2. $\frac{1}{8}$.
3. $\pi/3\sqrt{3}$.

4. Divergent.
5. $\frac{1}{2}$.
6. $\frac{1}{3}\ln 2$.
7. $\frac{5}{4}$.
8. Divergent.

9. 2.
10. $\pi/2$.
13. (iii) No.

15. (i) $A = \lambda/2$; (ii) $B = 1/\pi$.
16. $2\lambda^{-2}$, infinite.

17. $-\dfrac{\pi}{2}\ln 2.$ 18. (i) $\sinh u$;   (ii) $1/\cosh u$; $\pi/2.$

Exercise 12F   page 377

1. (i) $\tfrac{3}{16}\pi$; (ii) $\tfrac{8}{15}$; (iii) $\tfrac{128}{315}$; (iv) $\tfrac{5}{16}\pi$; (v) $\tfrac{4}{3}$; (vi) 0; (vii) $-\tfrac{16}{35}$.

3. $I_n = x(\ln x)^n - n I_{n-1}$; $9e - 24 \approx 0.465.$

4. $I_n = -n\pi^{n-1} - n(n-1)I_{n-2}.$ 5. $\pi/64.$

9. (i) $\tfrac{2}{35}$; (ii) $\tfrac{1}{120}$; (iii) $\tfrac{3}{512}\pi$; (iv) $\tfrac{3}{512}\pi.$

13. $I_{2n} = \ln|\sec\phi + \tan\phi| - \displaystyle\sum_{r=1}^{n} \frac{\sin^{2r-1}\phi}{2r-1},$

$I_{2n-1} = \ln|\sec\phi| - \displaystyle\sum_{r=1}^{n-1} \frac{\sin^{2r}\phi}{2r}.$

§13.1

Qu. 2. $1.19 + 0.61j.$

Qu. 3. (i) $1, -1$; (ii) $1, (-1 \pm \sqrt{3}j)/2$; (iii) $1, j, -1, -j.$

Qu. 6. $(-1 + j)/\sqrt{2}.$ **Qu. 8.** $-\pi < p\arg z \leqslant \pi.$

Exercise 13A   page 384

1. $\pm(2.74 - 0.73j).$ 2. $-2, 1 \pm \sqrt{3}j$; $(z+2)(z-1-\sqrt{3}j)(z-1+\sqrt{3}j).$

3. (i) $\pm\sqrt{3}, \pm\sqrt{3}j$; (ii) $\pm\sqrt{2} \pm \sqrt{2}j$; (iii) $\pm(\sqrt{3}+j)/2, \pm(-1+\sqrt{3}j)/2.$

4.   (i) $\cos\dfrac{\pi}{15} + j\sin\dfrac{\pi}{15}$;   (ii) $8\left(\cos\dfrac{\pi}{5} + j\sin\dfrac{\pi}{5}\right)$;

(iii) $\tfrac{1}{4}\left(\cos\left(-\dfrac{\pi}{4}\right) + j\sin\left(-\dfrac{\pi}{4}\right)\right)$; (iv) $\dfrac{1}{2187}\left(\cos\dfrac{7\pi}{12} + j\sin\dfrac{7\pi}{12}\right).$

5. $-119 - 120j$; $-3 + 2j, \pm(2 + 3j).$ 7. $(z - 1 - 4j)^5 = 1024.$

10. $\tfrac{1}{2}(a + b + c + d).$ 12. $-\alpha, \pm\alpha\omega, \pm\alpha\omega^2.$

14. (iv) m, n have no common factor.   16. $ma^2/2, 3ma^2/2.$

§13.2

Qu. 3. $\tfrac{1}{2}(n - r).$ **Qu. 4.** $n - 1$; $\tan\dfrac{k\pi}{n}, k = 0, 1, 2, \ldots, n-1, k \neq \dfrac{n}{2}.$

Exercise 13B   page 389

1. $1 - j, -6.$ 2. $2 - \sqrt{3}j, (-3 \pm \sqrt{11}j)/2.$

3. $z^5 - 18z^4 + 135z^3 - 502z^2 + 914z - 680 = 0.$ **4.** $\frac{2}{3}, \pm \sqrt{2}j.$

6. $\cos \dfrac{2k\pi}{5} + j \sin \dfrac{2k\pi}{5}, k = 1, 2, 3, 4;$

$$\left(z^2 - 2\cos \frac{2\pi}{5} z + 1\right)\left(z^2 - 2\cos \frac{4\pi}{5} z + 1\right).$$

7. $\frac{1}{3}\left(\cos \dfrac{k\pi}{3} + j \sin \dfrac{k\pi}{3}\right), k = 1, 2, 3, 4, 5.$

8. $(z^2 + 1)(z^2 - \sqrt{3}z + 1)(z^2 + \sqrt{3}z + 1).$ **9.** $\pm 1, \pm j, \frac{1}{2}(1 \pm \sqrt{3}j).$

11. $\cot \dfrac{(2k + 1)\pi}{2n}, k = 0, 1, 2, \ldots, n - 1.$

12. $\cos(\pm\theta) + j \sin(\pm\theta)$, where $\theta = \alpha + \dfrac{2k\pi}{n}, k = 0, 1, 2, \ldots, n - 1.$

13. $x^3 + x^2 - 14x + 2, 5x^2 + x - 16$, etc.

14. $2 - j, -3 - 2j; (2, -1), (-3, -2).$ **15.** $(-2, 0), (1, \sqrt{3}), (1, -\sqrt{3}).$

§13.3

Qu. 8. $\cosh x \cos y + j \sinh x \sin y.$

Exercise 13C   page 395

1. (i) -1;   (ii) $(1 + j)/\sqrt{2}$;   (iii) $\frac{1}{2}e^{5/6}(\sqrt{3} + j).$ **2.** $-7.32 + 1.04j.$

4. $-3 + (2k - \frac{1}{3})\pi j, k = 0, \pm 1, \pm 2, \ldots .$

6. Velocity $a\dot{\theta}$ in transverse direction;
acceleration has radial component $-a\dot{\theta}^2$, transverse component $a\ddot{\theta}.$

7. $\dot{z} = \dot{r}e^{j\theta} + r\dot{\theta}je^{j\theta}$, velocity has radial component $\dot{r}$, transverse
component $r\dot{\theta}$;
$\ddot{z} = (\ddot{r} - r\dot{\theta}^2)e^{j\theta} + (2\dot{r}\dot{\theta} + r\ddot{\theta})je^{j\theta}$, acceleration has radial component
$\ddot{r} - r\dot{\theta}^2$, transverse component $2\dot{r}\dot{\theta} + r\ddot{\theta}.$

8. $C = \frac{1}{13}e^{3x}(3\cos 2x + 2\sin 2x) + c, S = \frac{1}{13}e^{3x}(3\sin 2x - 2\cos 2x) + c.$

9. $e^{ax}(a\cos bx + b\sin bx)/(a^2 + b^2) + c;$
$e^{ax}(a\sin bx - b\cos bx)/(a^2 + b^2) + c.$

11. $\sinh x \cos y + j \cosh x \sin y.$ **13.** (i) $\sin x \cosh y, \cos x \sinh y.$

14. $(2n + 1)\pi j.$ **15.** $z = \dfrac{4n - 1}{2}\pi \pm j \cosh^{-1}(-k).$

16. $\ln |z| + j(\arg z + 2n\pi).$

17. (i) $j\pi$;   (ii) $j\pi/2$;   (iii) $\ln 4$;   (iv) $\ln 4 + j\pi$;   (v) $\ln 3 - j\pi/2$;
    (vi) $\ln 5 + j \tan^{-1}\frac{3}{4}.$ **19.** $0.208.$

20. (i) 1;   (ii) 0.0432;   (iii) $2.06 + 0.745j$;   (iv) $0.129 + 0.0339j.$

21. $-\pi < \arg z_1 + \arg z_2 \leqslant \pi.$

§13.4

Qu. 2. (i) (A5) fails; (ii) (M5) fails. **Qu. 3.** No; (M3) and (M5) fail.

Qu. 6. Multiplication.

Qu. 7. Enlargement from O, $\times \sqrt{(x^2 + y^2)}$, followed by rotation about O through α, where $\cos\alpha : \sin\alpha : 1 = x : y : \sqrt{(x^2 + y^2)}$.

Qu. 8. Rotation about O through $\pi/2$.

Exercise 13D page 401

4. Identities: 1 for $\oplus$, 0 for $\otimes$; inverse $2 - m$ for $\oplus$; only 2 has inverse for $\otimes$ in Z.

5. $(2 + j)z + 4 - 5j = (1 + 3j)z$; $\begin{pmatrix} -2.8 & 0.6 \\ -0.6 & -2.8 \end{pmatrix}$.

6. $\begin{pmatrix} 5 & -2 \\ 2 & 5 \end{pmatrix}$, $\begin{pmatrix} 5 & 2 \\ -2 & 5 \end{pmatrix}$.

7. $\begin{pmatrix} 1/\sqrt{2} - 1 & 1/\sqrt{2} - 1 \\ -1/\sqrt{2} + 1 & 1/\sqrt{2} - 1 \end{pmatrix}$, $\begin{pmatrix} -1/\sqrt{2} - 1 & -1/\sqrt{2} - 1 \\ 1/\sqrt{2} + 1 & -1/\sqrt{2} - 1 \end{pmatrix}$.

8. $\begin{pmatrix} -1 & 0 \\ 0 & -1 \end{pmatrix}$.   **9.** Taking conjugates.   **11.** $\det(Z_1 Z_2) = \det Z_1 \det Z_2$.

12. (i), (iii).   **13.** $66^2 + 10^2 + 22^2 + 12^2$, $17^2 + 41^2 + 45^2 + 33^2$, and 32 254 other solutions!

§13.5

Qu. 2. $xa = ya \Rightarrow x = y$.   **Qu. 5.** $QH = HQ = H, WH = HU = \{U, V, W\}$.

Qu. 6. $IK = UK = K, PK = WK = \{P, W\}, QK = VK = \{Q, V\}$

Exercise 13E page 408

1. Yes; (ii) Yes; e.g. $S = \{x : x \geqslant 0\}$; No.

2. $(\mathbb{R}, *)$ is closed, associative, has identity, but 1 has no inverse. $(S, *)$ is closed, associative, has identity, and every element has an inverse. $x = \frac{5}{4}$.

3. Not associative; $P \equiv (0, 0)$, $Q \equiv (a_1 - a_2, a_2 - a_1)$; $I = (0, 0)$, no J; argument assumes associativity, conclusion true.

4. $x = a^{-1}b, y = ab$.

5. (i)

	I	*R*	*H*	*S*	*X*	*Y*	*A*	*B*
I	*I*	*R*	*H*	*S*	*X*	*Y*	*A*	*B*
R	*R*	*H*	*S*	*I*	*A*	*B*	*Y*	*X*
H	*H*	*S*	*I*	*R*	*Y*	*X*	*B*	*A*
S	*S*	*I*	*R*	*H*	*B*	*A*	*X*	*Y*
X	*X*	*B*	*Y*	*A*	*I*	*H*	*S*	*R*
Y	*Y*	*A*	*X*	*B*	*H*	*I*	*R*	*S*
A	*A*	*X*	*B*	*Y*	*R*	*S*	*I*	*H*
B	*B*	*Y*	*A*	*X*	*S*	*R*	*H*	*I*

(the rows are labelled "second transformation", the columns "first transformation", with "followed by" at top left)

(ii) $\{I, R, H, S\}$ and either $\{I, H, X, Y\}$ or $\{I, H, A, B\}$.
Different numbers of self-inverse elements.

(iii) $IK = AK = K, RK = YK = \{R, Y\}, HK = BK = \{H, B\}$,
$SK = XK = \{S, X\}$.

7. (i) $j, -j$; (ii) rotational symmetry group of regular heptagon;
(iii) 3.

8.

*	*e*	*a*	*b*
e	*e*	*a*	*b*
a	*a*	*b*	*e*
b	*b*	*e*	*a*

10. $ba = c, b^2 = a$, also generated by c.

11. Yes.

14. Matrices of form $\begin{pmatrix} 1 & 0 & q \\ 0 & 1 & 0 \\ 0 & 0 & 1 \end{pmatrix}$.

15. All elements of G are self-inverse.

Miscellaneous exercise

1. (i) False; (ii) true; (iii) true; (iv) false.

2. $\begin{pmatrix} 1 & 3 \\ 0 & 4 \end{pmatrix}, \begin{pmatrix} 1 & 7 \\ 0 & 8 \end{pmatrix}, \begin{pmatrix} 1 & 2^n - 1 \\ 0 & 2^n \end{pmatrix}$. 3. $\frac{2}{3}n(n+1)(2n+1)$.

6. Minimum at $(-1, -e^{-\frac{1}{2}})$; maximum at $(1, e^{-\frac{1}{2}})$.

7. Maximum at $((a + b)/(a - b), \frac{1}{2}(b - a)^2)$;
minimum at $((a - b)/(a + b), \frac{1}{2}(a + b)^2)$.

8. 1.2, 1.202, less than root.

9. $1 + [x/10 - x^2/200 + \ldots + (-1)^{n+1}x^n/10^n n + \ldots]/\ln 10$.

10. 2.0046, 0.23%. 12. $\mathbf{t} = \frac{1}{3}\mathbf{a} + \frac{2}{3}\mathbf{b}$; $\mathbf{m} = \frac{1}{2}\mathbf{a} + \mathbf{b}$; 1:1.

13. $\mathbf{r} = \lambda(2\mathbf{i} + 3\mathbf{j} + 4\mathbf{k})$; $\mathbf{p} + \lambda\mathbf{a} = \mathbf{q} + \mu\mathbf{b}$ for some λ, μ; $3\mathbf{i} + 3\mathbf{j} + \frac{9}{2}\mathbf{k}$.

14. $(x - 6)/8 = (5 - y)/4 = -(z + 4)/11$. 15. $3\sqrt{3} : 3 : 1$. 16. 0.16.

17. f_1: range $= \{y : 0 < y \leqslant 1\}$; no inverse; restricted domain $= \{x : x \geqslant 0\}$;
f_2: range $= \{y : y \geqslant 1\}$; no inverse; restricted domain $= \{x : 0 \leqslant x < 1\}$;
f_3: range $= \mathbb{R}$; $f^{-1}(x) = e^x - 1$.

18. -2.

19. (i) Circle, centre $(1, 3)$, radius 2; (ii) line segment from $(1, 5)$ to $(3, 3)$.

21. (i) Maximum at $x = \frac{3}{4}\pi$, minimum at $x = \frac{7}{4}\pi$; (ii) $x + x^2 + \frac{1}{3}x^3 + \dots$.

22. If $f'(x_0) = 0$, then $f''(x_0) < 0 \Rightarrow f(x_0)$ is a maximum
and $f''(x_0) > 0 \Rightarrow f(x_0)$ is a minimum.
When $0 < a < 1$, maximum at $x = (-1 - \sqrt{(1 - a^2)})/a$, minimum at
$x = (-1 + \sqrt{(1 - a^2)})/a$; when $a = 1$, stationary point of inflexion at
$x = -1$; when $a > 1$, no stationary values.

23. Uses repeated bisection to locate root of $f(x) = 0$ with an error of less
than 0.05; $0.25, 0.125, 0.1875, 0.21875$.

24. $3py = -2x + 2p^2 + 3p^4$; $A(\frac{1}{3}p^2, 0), B(0, \frac{2}{3}p + p^3)$; $3y^2 = 2x(9x + 1)^2$.

25. (i) $1 + x^2/2! + x^4/4! + \dots$; $\cosh\sqrt{x} - 1$; (ii) $\frac{1}{2}e^2$.   **27.** $-e^{-a\pi}$.

28. $a' = -d, b' = b, c' = c, d' = -a$;
either $a = -d$, or $b = c = 0$ and $a = \pm d \neq 0$.

29. $\frac{1}{2}(e^x - e^{-x})$; $(e^x - e^{-x})/(e^x + e^{-x})$.

30. $u_n(t) = \begin{cases} [(-1 + s)^n - (-1 - s)^n]/2s \text{ where } s = \sqrt{(1 - t)}, -1 \leqslant t < 1. \\ -n.(-1)^n, t = 1. \end{cases}$

31. (b) $|z| = \sqrt{(x^2 + y^2)}$, $\arg z = \theta$ such that $\cos\theta : \sin\theta : 1 = x : y : |z|$
where $z = x + jy$ and x, y are real;
(c) $(\cos\theta + j\sin\theta)^n = \cos n\theta + j\sin n\theta$.

32. (i) $z^* = x - jy$.   **33.** (i) $3 + 4j$; (ii) $83°$.

35. (a) $z_1 = 5 + 4j, z_2 = 2 - 3j$;
(b) $|z_2| = 2\cos\frac{1}{2}\theta$, $\arg z_2 = \frac{1}{2}\theta$, $-\pi < \theta \leqslant \pi$; $\sqrt{3}, \frac{1}{6}\pi$; $27, \pi$.

36. (b) $3(x - y)(y - z)(z - x)$.

38. $y = x + 6, x = -1$; $\{x : x < -3 \text{ or } -1 < x < 1\}$.   **39.** 0.275.

40. $9(x + 1)^{-1} - (x - 3)^{-1}$; $f'(2) = f'(5) = 0$; $f(2) = 4, f(5) = 1$;
$9\ln 1.4 - \ln 3$.

41. $y = -\sqrt{[(x - \ln x)^2 - 1]}$.   **42.** $y = (x^2 + 1)e^{3x}$.

43. (ii) before 5.10 a.m., between 10.50 a.m. and 5.40 p.m., after 11.20 p.m.

44. $p = 2$; $y = e^{3x}(\sin 2x - 2\cos 2x) + 2e^{2x}$.   **45.** (i) 1.181.

46. (iii) Enlargement $\times |w|$ and rotation through $\arg w$, both with centre O.

47. (ii) (a) $-1, -5$; (b) $-1, -5, 3$; $k = 45$.   **48.** $\{w : \operatorname{Re}(w) > \frac{1}{2}\}$.

50.   (i) $x + \ln|(x - 1)/(x + 1)| + c$;   (ii) $\ln(1 + \sin x) + c$;
(iii) $x\cos^{-1}x - \sqrt{(1 - x^2)} + c$;   (iv) $-\sqrt{(1 - x^2)} + 2\sin^{-1}x + c$;
(v) $-\frac{1}{2}e^{-x}(\sin x + \cos x) + c$;   (vi) $\frac{1}{2}\ln|(x + 1)/(x + 3)| + c$;
(vii) $\tan^{-1}(x + 2) + c$;   (viii) $\frac{1}{2}\ln|1 + 2x| - \tan^{-1}x + c$;
(ix) $\frac{1}{3}\ln|x^3/(1 + 3x^2)| + c$;   (x) $\ln|\tan\frac{1}{2}x| + c$.

51. (i) $c < \frac{1}{8}(a - b)^4$; (ii) $c > \frac{1}{8}(a - b)^4$.

52. $\{x : x < -1 \text{ or } x > 2\}$; 2 real roots.

53. Any rational root of a polynomial equation with integer coefficients and leading coefficient 1 is an integer; $4, 1 \pm \sqrt{3}$.

54. $0.3129, 0.6025, 0.8423$; $1.2050, 1.0588, 0.7950$.

55. (i) $\sin^{-1}(\frac{1}{3}x - 1) + c$; (ii) $\cosh^{-1}(\frac{1}{3}x - 1) + c$;
    (iii) $\cosh^{-1}(x + 2) + c$; (iv) $\sinh^{-1}(x + 2) + c$;
    (v) $\frac{1}{6}\sin 3x + \frac{1}{10}\sin 5x + c$;
    (vi) $\frac{1}{10}\sin^{10}x - \frac{1}{4}\sin^{12}x + \frac{3}{14}\sin^{14}x - \frac{1}{16}\sin^{16}x + c$;
    (vii) $\frac{1}{2}\ln(x^2 + 3) + (2/\sqrt{3})\tan^{-1}(x/\sqrt{3}) + c$;
    (viii) $x - 2\ln(e^x + 1) + c$; (ix) $-\tan^{-1}(\cos x) + c$;
    (x) $\frac{1}{3}\ln|(\tan\frac{1}{2}x + 1)/(\tan\frac{1}{2}x - 2)| + c$.

56. (i) $B(4, -1, -3), C(\frac{22}{9}, \frac{8}{9}, -\frac{17}{9})$; (iii) $\frac{4}{9}$.   57. $\sqrt{\frac{7}{3}}$.

58.

	$|\mathbf{r}|$	$|\dot{\mathbf{r}}|$	$|\ddot{\mathbf{r}}|$
maximum	a	$3\omega a$	$10\omega^2 a$
minimum	0	ωa	$6\omega^2 a$

59. (a) $60/77$; (b) $10x + 7y - 12z = 4$;
    (c) $\mathbf{r} = (3\mathbf{i} - 2\mathbf{j} + \mathbf{k}) + \lambda(39\mathbf{i} - 18\mathbf{j} + 22\mathbf{k})$,
    $(x - 3)/39 = -(y + 2)/18 = (z - 1)/22$.

60. (i) $\mathbf{r}.\mathbf{u} = 18, \mathbf{r} \times \mathbf{u} = 5\mathbf{i} - 2\mathbf{j} + 4\mathbf{k}, \mathbf{r} = 2\mathbf{u} - 2\mathbf{j} - \mathbf{k}$;
    (ii) (a) R is in plane OPQ;
    (b) R is on line through O perpendicular to OPQ.

61. (a) $\begin{pmatrix} -2 & 1 \\ -5 & 2 \end{pmatrix}$; (b) $-\mathbf{I}, \mathbf{I}$.

62. $\mathbf{A}^{-1} = \begin{pmatrix} 1 & 1 & -1 \\ 6 & 7 & -8 \\ 2 & 3 & -3 \end{pmatrix}$, $(\mathbf{A}^2)^{-1} = \begin{pmatrix} 5 & 5 & -6 \\ 32 & 31 & -38 \\ 14 & 14 & -17 \end{pmatrix}$.

63. $k \neq 1, k \neq 3$: three distinct, non-parallel planes with a unique common point;
    $k = 1$: two distinct, parallel planes intersected by third plane;
    $k = 3$: three planes with common line: $x = 13y, z = (1 - 8y)/2$.

64. (a) $\mathbf{MN} = 16\mathbf{I}$; $\mathbf{M}^{-1} = \frac{1}{16}\mathbf{N}$; (b) ± 2; $(k, k, -k)$.

65. $\mathbf{P} = \begin{pmatrix} 3 & 1 \\ 2 & -1 \end{pmatrix}$; at $t = 0, u = 1, v = -1$; $u = e^{4t}, v = -e^{-t}$.

69. $b \csc\theta/a$.

70. $r^2(a\cos^2\phi + b\sin^2\phi) + 2r(ax'\cos\phi + by'\sin\phi) + ax'^2 + by'^2 = 1$;
    (i) $-ax'/by'$.

71. $2x - z = 2$; $(2\mathbf{i} - \mathbf{k})/\sqrt{5}$.   73. $a = 1, b = -4$; $\frac{1}{2}\begin{pmatrix} 1 & -3 \\ 2 & 0 \\ -1 & 1 \end{pmatrix}$.

76. $z\mathbf{i} - cz\mathbf{j} + (cy - x)\mathbf{k}$; $0, \pm j\sqrt{(1 + c^2)}$; $\alpha(c\mathbf{i} + \mathbf{j})$.

77. $\exp(\mathbf{I}) = e\mathbf{I}$, $\exp(\mathbf{0}) = \mathbf{I}$; $\mathbf{A} = (e^{\lambda} + \lambda - 1)\mathbf{I}$.

78. Hyperbolas; $e = \sqrt{[\alpha/(\alpha - 1)]}$.

79. (i) (a) $(\dot{\mathbf{r}})^2 + \mathbf{r}.\ddot{\mathbf{r}}$; (b) $2\dot{\mathbf{r}}.\ddot{\mathbf{r}}$; (c) $\mathbf{r} \times \ddot{\mathbf{r}}$; (iii) $\mathbf{r} = n^{-1}(e^{-nt} - e^{-2nt})\,\mathbf{V}$.

80. $1 \leqslant x \leqslant 3$; $P(1.8, 2.4), Q(0.6, 0.8)$.

81. (a) $f(2) = -0.05$; (b) $x^3 - 3x^2 - 2x + 5$.   82. 18.5.

83. (i) $0 < x < 2, f(x) = (1 + x)/3x$; (ii) $(1 - 2x)^{-2}$.

84. $a = -3, b = 2, c = -1, d = -8$; difference $\approx 0.000\,083$.   85. 1.645.

87. $\pi/2$.

89. $\frac{8}{325}(e^{3\pi} - 1) \approx 305$;
$e^{ay}\cos^{n-1}y\,(a\cos y + n\sin y)(a^2 + n^2)^{-1} + n(n-1)(a^2 + n^2)^{-1}\int e^{ay}\cos^{n-2}y\,dy.$

90. (i) $2e - 2$; (ii) $2x\sin(x^{-1}) - \cos(x^{-1})$.

91. $z = -j\cot(k\pi/n), k = 1, 2, \ldots, n - 1$; $\mathrm{Im}(w) = -\frac{1}{2}$.

92. $\alpha + 2\pi/3, \alpha + 4\pi/3$.

93. $x^2 - 2x\cos\theta + 1$; $\theta = k\pi/4, k = 0, 1, \ldots, 7$;
$(x - 1)(x + 1)(x^2 + 1)(x^2 - \sqrt{2}x + 1)(x^2 + \sqrt{2}x + 1).$

94. (i) 8 is identity; $2^{-1} = 4, 4^{-1} = 2, 6^{-1} = 6, 10^{-1} = 12, 12^{-1} = 10$;
    7 does not have an inverse.
(ii) non-square rectangle, square.

95. $G_2 = \{\mathbf{I}, -\mathbf{I}\}, G_4 = G_2 \cup \left\{ \begin{pmatrix} kj & w \\ -w^* & -kj \end{pmatrix} : k \in \mathbb{R}, k^2 + |w|^2 = 1 \right\}$;
G_2 is a subgroup, G_4 is not.

96. 18817/10864, 32592/18817.

97. (i) False; (ii) true; (iii) false; (iv) true.

98. Equality if and only if $b = \phi(a)$.

99. $\cos\theta + j\sin\theta$ where $\theta = 2k\pi/5, k = 0, 1, \ldots, 4$; $2(\cos\theta + j\sin\theta)$;
no unit, not closed, etc.; cyclic group of order 5.

101. (i) (a) $-\frac{1}{2}$; (b) 0; (ii) $y(4) = 21$.

102. depth $= h_0(2 - \cos(2\pi t/T))^{\frac{1}{3}}$; maximum depth $= h_0 \cdot \sqrt[3]{3}$;
    mean depth $= \dfrac{h_0}{T}\displaystyle\int_0^T (2 - \cos(2\pi t/T))^{\frac{1}{3}}\,dt$; maximum speed $= 3^{2/3}ah_0^2$.

103. (i) Converges; (ii) (a) $\mathbb{R}$; (b) $\{p : p < -1\}$, (c) $\emptyset$.

104. $t^{-1}\sin t^2 - \sin 1$; e.g. $h(x) \equiv x^{-1}, l(x) \equiv \tanh x$.

106. *C*.   108. 16.   111. 7.

113. 7; 1, 2, 3, 5, 6, 7, 8; 1, 2, 3, 4, 6, 7, 9; 1, 2, 3, 4, 5, 8, 9;
    1, 2, 3, 4, 5, 7, 10.

Index

Abel, Niels, 403
  -ian group, 403
abstract algebra, 402
adjoint, 249
adjugate, 249
alien cofactors, 247
Ampère, 224
amplitude, 184
angular velocity, 228
anti-commutative, 224
Apollonius of Perga, 315
approximation
  linear, 37
  Maclaurin, 40
  Newton-Raphson, 33
  Taylor, 42
Archimedes, 54
  spiral of −, 280
Argand diagram, 118
  multiplication in −, 130
argument, principal, 127
arithmetic mean, 162
associativity, 403
  of matrix multiplication, 236
  of scalar product, 199
  of vector addition, 58
astroid, 103
asymptote,
  horizontal, 154
  of hyperbola, 306
  oblique, 155
  vertical, 152
auxiliary
  circle, 295, 305
  equation, 188
axiom, 2
  -s of group, 403
axis
  conjugate − of hyperbola, 304
  major − of ellipse, 292
  minor − of ellipse, 293
  of parabola, 283
  transverse − of hyperbola, 304

back-substitution, 259
base vector, 62
beats, 186
Bombelli, Rafael, 117
boundary conditions, 174
bounded above, 339
Brahmagupta, 112

Cardan, Geronimo, 113
  Tartaglia method for cubics, 117
cardioid, 280, 365
Cartesian equations
  of line, 69
  of plane, 207
Cauchy, Augustin, 245, 395
  distribution, 374
Cayley, Arthur, 235, 245
centroid, 78, 361, 363
  of tetrahedron, 80
Ceva, Giovanni, 229
characteristic
  equation, 268
  value, 268
  vector, 267
Chebyshev polynomial, 348
Choleski's method, 265
codomain, 86
coefficient, 140
  equating −s, 142
cofactor, 244
  alien −s, 247
complementary function, 187
complex
  conjugate, 115, 118
  number, 113
  number field, 401
  number, polar form, 127
  plane, 117
  roots, 380
  roots of unity, 382
  variable, functions of, 391
component vectors, 62
  uniqueness of −, 61
concave upwards, downwards, 16
conchoid, 281
confocal system of conics, 320
conic, 315
  central −, 317
  polar equation of −, 318
conjugate,
  axis, 304
  complex number, 115, 118
  diameter, 298
  hyperbola, 313
  roots, 387
contrapositive, 4
convergence
  of improper integrals, 370, 372
  of iteration, 332
  of series, 337

converse, 2
correlation, 239
$\cos^{-1}$, 90
    differentiation of, 92
cosec^{-1}, 91
cosech, 105
coset, 407
cosh, 104
$\cosh^{-1}$, 107
$\cot^{-1}$, 91
Cotes, Roger, 392
coth, 105
counter-example, 4
counting numbers 111
cover-up method, 163
Cramer's rule, 245, 265
cubic equation
    Cardan-Tartaglia method, 113, 117
    properties of roots, 147, 253
cyclic group, 409
cycloid, 103

D'Alembert's ratio test, 341
damping, 190
Dandelin, Germinal, 316
decimal search, 25
deduction, 1, 8
definite integrals, 351
degree of polynomial, 140
deltoid, 367
De Moivre, Abraham, 390
    —'s Theorem, 135, 392
dense, 111
depth of field, 158
derivative
    of implicit function, 97
    of inverse function, 88
    second, 15
    with parameter, 100
Desargues Théorem, 81
Descartes, René, 114
determinant, 243
    of product, 250
    of transpose, 249
diagonal matrix, 241
    reduction to —, 270
diameter
    conjugate —, 298
    of hyperbola, 304
    of parabola, 289
differences, 323
differential equations, 173
    first order, 177
    graphical method, 174
    homogeneous, 179
    linear, 186
    second order, 196

differentiation of a vector, 218
direction
    cosines, 66, 69
    vector, 69
director circle, 302, 314
directrix,
    of ellipse, 295
    of hyperbola, 306
    of parabola, 283
distributivity, 59, 201, 225
divergence
    of improper integrals, 370, 372
    of series, 337
domain, 86
dummy variable, 354

eccentric angle, 298
eccentricity, 318
    of ellipse, 293
    of hyperbola, 304
échelon form, 259
eigen
    value, 268
    vector, 267
Einstein, Albert, 1
elementary matrix, 263
ellipse, 292, 300, 315
enlargement matrix, 398
equating
    coefficients, 142
    real and imaginary parts, 114
equation
    auxiliary —, 188
    Cartesian — of line, 69
    Cartesian — of plane, 207
    differential —, 173
    ill-conditioned —, 266
    parametric — of line, 69
    polynomial —, 147, 387
    simultaneous linear —s, 242, 253
    vector — of line, 69
equiangular spiral, 282, 365
equivalence, 3
error, 321
    absolute —, 321
    in function value, 323
    in sums and products, 322
    location by differencing, 325
    relative —, 321
    truncation —, 342
escape velocity, 178
Euclid, 1
Eudoxus, 112
Euler, Leonhard, 8, 392
    beta function, 378
    constant, 344
    line, 80

method for differential equations, 192
even function, 355
expectation, 361, 363
exponential, complex, 391

factor theorem, 141
Fermat, Pierre de, 8
    numbers, 146
field, 397
flow diagram, 25
focus
    -directrix property, 283, 295, 306, 317
    of ellipse, 292
    of hyperbola, 304
    of parabola, 283
frequency, 184
function
    complementary −, 187
    even −, 355
    implicit −, 97
    inverse −, 87
    of complex variable, 391
    odd −, 355
    polynomial −, 140
    rational −, 151
    self-inverse −, 88
Fundamental Theorem of Algebra, 387

Galois, Evariste, 408
Gauss, Carl Friedrich, 113, 114, 387
    -ian elimination, 261
generator, 409
geometric mean, 162
general solution
    of differential equation, 173
    of trigonometric equation, 91
Girard, Albert, 386
Grassmann, Hermann, 54
Gregory, James, 42, 48, 346
    series, 346
group, 403
    cyclic −, 409
    generator, 409
    order of −, 403
    symmetry −, 405

Hamilton, William, 54
harmonic
    mean, 162
    series, 340
helix, 218
Heun's method, 198
homogeneous
    differential equation, 179
    linear equations, 256
hyperbola, 304, 312, 315

rectangular, 309
hyperbolic functions, 104, 314, 393
    inverse −, 107
hyperfocal distance, 158

if and only if, 5
ill-conditioned equations, 266
imaginary
    axis, 118
    part, 114
implication, 2
implicit function, 97
    differentiation of −, 97
improper integrals, 370
incentre, 75
inconsistent, 255
induction, 8
    proof by −, 10
inequality, 159
inflexion, 17
initial
    conditions, 174
    line, 276
integrating factor, 178
integration
    by inverse hyperbolic functions, 108
    by inverse trigonometric functions, 93
    by parts, 82
    properties, 351
interpolation
    inverse linear, 32
    linear, 28
interval, 173
invariant line, 267
inverse
    curve, 281
    element, 403
    function, 86
    hyperbolic functions, 107
    matrix, 249, 263
    trigonometric functions, 90
irrational number, 112
    proof that e is −, 350
    proof that π is −, 379
irreducible quadratic factors, 388
isocline, 176
isomorphism, 398, 406
iteration, 330
    for square roots, 328

j, 113

Kepler, Johannes, 294, 413
Kirchhoff's loop law, 181

Lagrange, Joseph, 1, 15, 402
    —'s Theorem, 408
Laplace distribution, 374
latent
    value, 268
    vector, 267
latus rectum, 318
    of ellipse, 293
    of parabola, 288
leading coefficient, 140
Leibniz, Gottfried, 15, 346
lemniscate, 365
limaçon, 280
linear
    convergence of iteration, 332
    differential equation, 186
    homogeneous — equations, 256
    interpolation, 28
    simultaneous equations, 242, 253
    transformation, 233
Linton, Hercules, 50
locating roots of equations, 23
logarithm of complex number, 396

Maclaurin, Colin, 42, 245
    Approximation, 40, 345, 347
many-to-one function, 86
marginal cost and revenue, 20
Markov chain, 275
matrix
    addition, 234
    adjugate (adjoint) —, 249
    elementary —, 263
    enlargement —, 398
    inverse —, 249, 263
    lower triangular —, 259
    multiplication, 235
    nilpotent —, 273
    of transformation, 231
    orthogonal —, 252
    similar —, 273
    singular —, 250
    skew-symmetric —, 239
    spiral, —, 400
    symmetric —, 239
    transpose of — 239
    upper triangular —, 259
mean
    absolute value, 361
    arithmetic —, 162
    geometric —, 162
    harmonic —, 162
    root — square, 162, 361
    value of function, 359
    weighted —, 76, 360
Menelaus, 229
Mercator, Nicolaus, 346

    series, 347
Mersenne primes, 146
mid-ordinate rule, 46
mid-point method, 194
minor, 243
modulus, 120
    -argument form of complex
        number, 127
moment of force
    about line, 229
    about point, 228

Napier, John, 45
Napoleon's Theorem, 134
necessary condition, 5
negation, 3
negative
    number, 112
    of vector, 57
nested multiplication, 33
Newton, Isaac, 1, 33, 276, 294, 346,
    392
    -Raphson approximation, 33, 332
Nicomedes, 281
nilpotent matrix, 273
normal, 206
    subgroup, 411
numbers
    complex, 113
    counting, 111
    Fermat, 8
    irrational, 112
    negative, 112
    prime, 1
    rational, 111
    real, 112

odd function, 355
one-to-one function, 86
order
    of convergence of iteration, 332
    of differential equation, 173
    of element of group, 410
    of group, 403
    of matrix, 232
    of subgroup, 408
orthocentre, 80
orthogonal matrix, 252
Osborn's rule, 105, 393

parabola, 283, 288, 315
    semi-cubical —, 103
parameter, 69
    differentiation with —, 100
parametric equations of line, 69
partial
    fractions, 163, 168
    sum, 337

particular integral, 174, 187
period, 184
pitch, 218
pivot, 261
pivotal condensation, 261
plane, 206, 212
    angle between −s, 215
    Cartesian equation of −, 207
    distance of point from −, 214
polar
    coordinates, 126, 276
    equation, 278
    form of complex number, 127
pole, 276
polynomial, 140
    Chebyshev −, 348
    differences of −, 324
    roots of − equation, 147, 387
position vector, 68
power series, 345
prime number, 1
    Mersenne −, 146
primitive, 351
principal
    logarithm, 396
    polar coordinates, 277
product
    of matrices, 235
    scalar −, 199
    vector −, 224
projection, 200
    orthographic −, 232
proof, 1
    by contradiction, 3
    by induction, 10
Pythagoras, 111

quadratic
    convergence of iteration, 332
    irreducible − factors, 388
quartic equation, 113
    properties of roots, 149

radial
    vector, 221
    velocity and acceleration, 395
range, 86
Raphson, Joseph, 33
ratio
    test, 341
    theorem, 76
rational
    function, 151
    numbers, 111
real
    axis, 117
    numbers, 112
    part, 114

reduction formula, 375
reflector property, 315
    of ellipse, 294
    of hyperbola, 305
    of parabola, 284
regression line, 241
relative velocity, 220
remainder theorem, 140
repeated bisection, 26
residual, 262
resolving vectors, 62
resonance, 192
resultant, 56
rhodonea, 279
right-handed triad, 63, 224
root
    locating −, 23
    mean square, 162
    of complex number, 380
    of equation, 23
rounding, 321

Sarrus, P.F., 246
scalar, 54
    product, 199
    triple product, 227
search
    decimal −, 25
    sign-change −, 24
sec⁻¹, 91
    differentiation of −, 96
sech, 105
second derivative, 15
    test, 18
sector, area of, 363
self-inverse function, 88
semi-cubical parabola, 103
separating the variables, 177
sign-change search, 24
similar matrices, 273
simple harmonic motion, 182
Simpson, Thomas, 48
    −'s rule, 48
sin⁻¹, 90
    differentiation of −, 92
singular
    case, linear equations, 253
    matrix, 250
    solution, 174
sinh, 104
sinh⁻¹, 107
skew
    lines, 72
    -symmetric matrix, 239
solution
    complete − of differential
        equation, 174

general – of differential equation, 174
general – of trigonometric equation, 91
of differential equation, 173
of equation, 23
singular – of differential equation, 174
trivial, 256
spiral
    equiangular, 282, 365
    matrix, 400
    of Archimedes, 280
steady state current, 182
step-by-step solution, 192
Stevin, Simon, 54
subfield, 400
subgroup, 406
    normal –, 411
    proper –, 407
sufficient condition, 5
symmetric matrix, 239
systematic elimination, 259

tan $\frac{1}{2}\theta$ substitution, 170
tan^{-1}, 91
    differentiation of –, 92
    power series for –, 346
tangent field, 174
tanh, 105
tanh^{-1}, 107
Tartaglia, Niccolò, 113
Taylor, Brook, 42
    approximation, 42, 193, 345
terminal velocity, 182
Thales of Miletus, 1
Theon of Smyrna, 272
trace, 6, 273
trammel, elliptic, 301
transformation, linear, 233
transpose, 239
transverse
    vector, 221
    velocity and acceleration, 395
trapezium rule, 46
triad, right-handed, 63
triangle
    inequalities, 125

law, 56
trigonometric functions, complex, 394
trivial solution, 256
truncation, 321
    error, 342, 345
twelfths rule, 186

unique components theorem, 61
unit vector, 61
    i, j, k, 63

Vandermonde, Alexandre-Théophile, 245
Van der Waal's equation, 99
variance, 361, 363, 374
vector, 54, 236
    addition, 56
    base –, 62
    component –, 62
    coplanar –s, 61
    differentiation of –, 218
    direction –, 69
    eigen –, 267
    equality, 54
    equation of line, 69
    integration of –, 219
    multiplication by scalar, 58
    position –, 68
    product, 224
    quantity, 57
    radial –, 221
    scalar product of –s, 199
    transverse –, 221
    unit –, 61
    zero –, 54
vertex
    of hyperbola, 304
    of parabola, 283

Wallis's product, 379
Waring, Edward, 341
weight
    -ed mean, 76, 360
    -ing function, 360
Wessel, Caspar, 118
Wronski, Josef, 386

zero, 112
    polynomial, 140
    vector, 54

MATHEMATICS DEPARTMENT
BARTON PEVERIL COLLEGE
CEDAR ROAD
EASTLEIGH